D0202537

Design of Control Systems

Design of Control Systems

A. Frank D'Souza

Professor, Department of Mechanical
and Aerospace Engineering
Illinois Institute of Technology

PRENTICE-HALL, INC., Englewood Cliffs, New Jersey 07632

Library of Congress Cataloging-in-Publication Data

D'Souza, A. Frank.
Design of control systems.

Includes bibliographies and index.
1. Automatic control. I. Title.
TJ213.D766 1988 629.8'312 87-2514
ISBN 0-13-199951-6

Editorial/production supervision and
interior design: Mary Jo Stanley
Cover design: 20/20 Services, Inc.
Manufacturing buyer: Rhett Conklin

Printed in the United States of America

10 9 8 7 6 5 4 3 2 1

ISBN 0-13-199951-6 025

Prentice-Hall International (UK) Limited, *London*
Prentice-Hall of Australia Pty. Limited, *Sydney*
Prentice-Hall Canada Inc., *Toronto*
Prentice-Hall Hispanoamericana, S.A., *Mexico*
Prentice-Hall of India Private Limited, *New Delhi*
Prentice-Hall of Japan, Inc., *Tokyo*
Prentice-Hall of Southeast Asia Pte. Ltd., *Singapore*
Editora Prentice-Hall do Brasil, Ltda., *Rio de Janeiro*

To

Cecilia

Contents

Preface

As the engineering systems become increasingly complex, their control systems tend to become very sophisticated. The area of control engineering has therefore progressed very rapidly from its early developments. This book has been developed as a text for a first course in control systems at the senior undergraduate level, primarily in the departments of mechanical and aerospace engineering. The material is also quite suitable for an interdisciplinary course to engineering students.

Since the book is intended for an introductory course in the area, advanced topics, such as optimal and adaptive control, stochastic design, and nonlinear control systems, have been omitted. An introductory course may be the only exposure of control systems to engineering students, but hopefully it will motivate their further study of this interesting area. The aim of the book is to present the fundamentals pertaining to the design of control systems employing linear, constant-parameter mathematical models, and to lay the foundation for further study of advanced topics at the graduate level.

Both the transfer-function and state-space matrix methods are used and well integrated. The state-space methods are stressed in most of the analytical development as matrix methods using state-variables representation are well suited for the application of computer-aided design tools. Although the emphasis is on single-input, single-output control systems, multivariable systems are included, and the presentation is generalized to multivariable systems as much as possible.

The integration of transfer-function and state-space methods and inclusion of multivariable systems broadens the coverage of control systems in an introductory course and provides a smooth transition to the study of more advanced topics. All sections of the book are illustrated with examples relating the theory to the controller design. The mathematical models of control systems studied in an earlier chapter are employed in later chapters to clarify the origin of the models and to motivate the analysis and controller design.

A brief survey of the contents is as follows. Chapter 1 is the introduction in which classification of feedback control systems, their performance specifications, mathematical models, and design procedure are discussed. Chapter 2 covers the mathematical models of different components, including actuators used in control systems. Chapter 3 presents the techniques of mathematical modeling of feedback control systems. For simplicity, the examples use output feedback with proportional control law. A control systems engineer should be familiar with mechanical, electrical, hydraulic, pneumatic, and thermal systems. The development of mathematical models of control systems from their conceptual design is one of the important skills to be acquired by a control engineer since it serves as a bridge between control theory and controller design.

The transient response of control systems is covered in Chapter 4. The state-transition matrix is used to obtain the response, and time-domain performance specifications are discussed. The frequency response is presented in Chapter 5. Bode diagrams and frequency-domain performance specifications are included. The frequency-domain analysis plays an important role in mathematical modeling of complex engineering systems from experimental frequency response and in experimental validation of the models. Chapter 6 deals with stability analysis and the root-locus method.

The design of controllers with output feedback is presented in Chapter 7. It includes system types, steady-state errors, PID family of control laws, and compensation. Chapter 8 deals with controller design with state feedback. The properties of controllability and observability are discussed and the design of an observer is included. With the availability of inexpensive microprocessors, the use of digital controllers has become commonplace. Chapter 9 is devoted to digital control with both output feedback and state feedback. A digital computer is used to implement the PID family of control laws of Chapter 7 and the state feedback control laws with an observer of Chapter 8. Both the state-difference equations and z-transformation methods are used for the analysis and synthesis.

An adequate mathematical background required of the reader includes familiarity with linear, ordinary differential equations with constant coefficients, and a working knowledge of matrix algebra and Laplace transformation. The matrix notation is used throughout most of the presentation. Appendices A and B present elements of Laplace transformation and matrix algebra, respectively, for the benefit of those who need to review this material.

It is advisable to assign some problems for computer solution to familiarize the students with computer-aided design of control systems. Several computer software packages with graphics are available commercially for control system design with a personal computer. Some of these are listed in the references of Appendix D. The students may also write their own computer programs. Some sample programs in True BASIC for personal computers are listed in Appendix D.

Depending on the prior preparation of the students, several options are available to use this book as a text. Many departments of mechanical engineering require in the junior undergraduate year, a course on introduction to system dynamics. The course covers most of the material presented in Chapter 2 and some of the material of Chapters 4 and 5. In this case, this book can be covered in a one-semester course in the senior year, provided Chapter 2 is used for a brief review and Chapter 4 and 5 are covered at a fast pace.

For those students without any prior exposure to system dynamics, only the first eight chapters may be covered in a one-semester course in the senior year, and Chapter 9 on digital control may be skipped. It is noted that Chapters 8 and 9 are also suitable for a second course at the beginning graduate level with additional topics such as optimal control, with both analog and digital controllers. Alternatively, the book can be completely covered in a two-quarter course at the senior undergraduate level, in case the students have had no prior exposure to system dynamics.

The development of this book has been influenced by several existing books mentioned in the references. Thanks are due to Professor Devendra Garg of Duke University and Professor Richard Klein of the University of Illinois at Urbana-Champaign for their thoughtful review of the manuscript and many helpful suggestions. I am indebted to Professor Francis Raven, who introduced me to control engineering at the University of Notre Dame, and to the late Professor Rufus Oldenburger, who was my Ph.D research advisor at Purdue University. My sincere thanks to Linda Mashek for her expert typing of the manuscript. I acknowledge the assistance of the editorial staff of Prentice-Hall, Inc., especially that of Doug Humphrey, the Engineering Editor, and that of Mary Jo Stanley.

Finally, I would like to acknowledge my wife, Cecilia, and daughters, Geraldine and Raissa, for their continued encouragement.

A. Frank D'Souza

Glossaries

CONTROL GLOSSARY

Plant State Equations

Continuous-time

$$\dot{\mathbf{x}}(t) = \mathbf{A}\mathbf{x}(t) + \mathbf{B}\mathbf{u}(t) + \mathbf{B}_1\mathbf{v}(t)$$

$\mathbf{x}$ = state-variables vector, $n \times 1$
$\mathbf{u}$ = control inputs vector, $q \times 1$
$\mathbf{v}$ = disturbance and/or noise input vector, $p \times 1$
t = time
$\mathbf{A}$ = system matrix, $n \times n$
$\mathbf{B}$ = control input matrix, $n \times q$
$\mathbf{B}_1$ = disturbance and/or noise input matrix, $n \times p$

Discrete-time

$$\mathbf{x}[(k+1)T] = \mathbf{\Phi}\mathbf{x}(kT) + \mathbf{\Psi}\mathbf{u}(kT) + \mathbf{w}(kT)$$

$\mathbf{x}$ = state variables vector, $n \times 1$
$\mathbf{u}$ = control inputs vector, $q \times 1$
$\mathbf{w}$ = disturbance and/or noise input vector, $n \times 1$
T = sampling period
$\mathbf{\Phi}$ = discrete system matrix, $n \times n$
$\mathbf{\Psi}$ = discrete control input matrix, $n \times q$

Output equations

$$\mathbf{y} = \mathbf{Cx}$$

or $\mathbf{y} = \mathbf{Cx} + \mathbf{Eu}$, when the order of the numerator of the transfer function is equal to the order of the denominator

$\mathbf{y}$ = output vector, $m \times 1$
$\mathbf{C}$ = output matrix, $m \times n$
$\mathbf{E}$ = direct transmission matrix, $m \times q$

Plant Transfer Functions

Continuous-time

$$\mathbf{y}(t) = \mathbf{G}_0(D)\mathbf{u}(t)$$

or

$$\mathbf{Y}(s) = \mathbf{G}_0(s)\mathbf{U}(s)$$

Discrete-time

$$\mathbf{Y}(z) = \mathbf{G}_0(z)\mathbf{U}(z)$$

$\mathbf{y}$ = output vector, $m \times 1$
$\mathbf{u}$ = control input vector, $q \times 1$
$\mathbf{G}_0(D), \mathbf{G}_0(s)$ = continuous transfer function matrix, $m \times q$
$\mathbf{G}_0(z)$ = discrete transfer function matrix, $m \times q$
D = differential operator, d/dt
s = Laplace transform variable
z = z-transform variable

Control Law

$$\mathbf{u} = \mathbf{Nr} - \mathbf{Kx}$$

$\mathbf{r}$ = command input vector, $m \times 1$
$\mathbf{N}$ = command input scaling matrix, $q \times m$
$\mathbf{K}$ = feedback gain matrix, $q \times n$

Closed-Loop Transfer Functions

Continuous-time

$$\mathbf{Y}(s) = \mathbf{G}_c(s)\mathbf{R}(s)$$

Discrete-time

$$\mathbf{Y}(z) = \mathbf{G}_c(z)\mathbf{R}(z)$$

$\mathbf{G}_c$ = closed-loop transfer function matrix, $m \times m$

Closed-Loop Control Systems

Continuous-time

$$\dot{\mathbf{x}}(t) = (\mathbf{A} - \mathbf{BK})\mathbf{x}(t) + \mathbf{BNr}(t) + \mathbf{B}_1\mathbf{v}(t)$$

Discrete-time

$$\mathbf{x}[(k+1)T] = \mathbf{P}\mathbf{x}(kT) + \mathbf{Q}\mathbf{r}(kT) + \mathbf{w}(kT)$$

Closed-Loop Characteristic Equations

Continuous-time

$$|s\mathbf{I} - \mathbf{A} + \mathbf{B}\mathbf{K}| = 0$$

Discrete-time

$$|z\mathbf{I} - \mathbf{P}| = 0$$

Controllability Matrices

Continuous-time

$$\mathbf{Q} = [\mathbf{B} \quad \mathbf{A}\mathbf{B} \quad \mathbf{A}^2\mathbf{B} \cdots \mathbf{A}^{n-1}\mathbf{B}]$$

Discrete-time

$$\mathbf{Q} = [\mathbf{\Psi} \quad \mathbf{\Phi}\mathbf{\Psi} \quad \mathbf{\Phi}^2\mathbf{\Psi} \cdots \mathbf{\Phi}^{n-1}\mathbf{\Psi}]$$

Observability Matrices

Continuous-time

$$\mathbf{M} = [\mathbf{C}^T \quad \mathbf{A}^T\mathbf{C}^T \quad (\mathbf{A}^T)^2\mathbf{C}^T \cdots (\mathbf{A}^T)^{n-1}\mathbf{C}^T]$$

Discrete-time

$$\mathbf{M}^T = [\mathbf{C} \quad \mathbf{C}\mathbf{\Phi} \quad \mathbf{C}\mathbf{\Phi}^2 \cdots \mathbf{C}\mathbf{\Phi}^{n-1}]^T$$

Observer/Estimators

Continuous-time

$$\dot{\hat{\mathbf{x}}}(t) = \mathbf{A}\hat{\mathbf{x}}(t) + \mathbf{B}\mathbf{u}(t) + \mathbf{H}[\mathbf{y}(t) - \mathbf{C}\hat{\mathbf{x}}(t)]$$

Discrete-time

$$\hat{\mathbf{x}}[(k+1)T] = \mathbf{\Phi}\hat{\mathbf{x}}(kT) + \mathbf{\Psi}\mathbf{u}(kT) + \mathbf{H}[\mathbf{y}(kT) - \mathbf{C}\hat{\mathbf{x}}(kT)]$$

ALPHABETICAL GLOSSARY

$\mathbf{A}$ = system plant matrix
$\mathbf{B}$ = control input matrix
$\mathbf{B}_1$ = disturbance and/or noise input matrix

$\mathbf{C}$ = output matrix
D = differential operator, d/dt
$\mathbf{E}$ = direct transmission matrix
$\mathbf{G}(D)$ = transfer function
$\mathbf{G}(s)$ = transfer function
$\mathbf{G}(j\omega)$ = frequency-response function
$\mathbf{H}$ = estimator/observer gain matrix
$\mathbf{K}$ = controller gain matrix
$\mathbf{M}$ = observability matrix
$\mathbf{N}$ = command input scaling matrix
$\mathbf{P}$ = discrete closed-loop matrix
$\mathbf{Q}$ = controllability matrix
$\mathbf{r}$ = command input vector
s = Laplace transform variable
T = sampling period
t = time
$\mathbf{u}$ = control input vector
$\mathbf{v}$ = disturbance and/or noise input, continuous systems
$\mathbf{w}$ = disturbance and/or noise input, discrete systems
$\mathbf{x}$ = state vector
$\mathbf{y}$ = output vector
z = z-transform variable
$\mathbf{\Phi}(t)$ = state transition matrix, continuous systems
$\mathbf{\Phi}(T)$ = discrete-time system matrix
$\mathbf{\Psi}(T)$ = discrete-time control input matrix

Design of Control Systems

1

Introduction

1.1 FEEDBACK CONTROL

The successful and efficient operation of a system, which may be an engineering, economic, social, physiological, or any other class of system, requires some form of control. The degree, quality, and means of control may vary widely. Certain performance objectives are established first, and then proper stimuli or command inputs are applied to the system so that its actual performance meets and satisfies the objectives. In this book, we will deal with the control of engineering systems that are governed by the laws of physics and are therefore called physical systems.

In a system, control may be exercised over one or more variables, which are called the *controlled outputs.* A *command* or *reference input* is the desired value of a controlled output. In a *closed-loop* or *feedback* control system, the controlled output is measured and fed back for comparison with the command input and any error is used to make the controlled output correspond to the command input. In an *open-loop* control system, the controlled output is not measured and is not compared with the command input. Hence, in such control systems, the controlled output can deviate from the command input by a large error due to disturbances and parameter variations. An open-loop control system is simple and economical, but because of large errors that are possible, it is not used when precise control is required. Hence, feedback is incorporated in most control systems.

The engineering development of feedback control systems dates back many years. In the third century B.C., Ktesibios of Alexandria is credited with the development of a water clock where the inflow of water in a tank is maintained constant by a float operated valve (Newton, Gould, and Kaiser, 1957). The use of windmills in Europe for motive power led Holland's Andrew Meikle in 1750 to develop a feedback control system for turning a windmill into the wind (Newton,

Gould, and Kaiser, 1957). Then came the advent of the steam age, and in 1788 James Watt invented his flyball governor in Scotland for regulating the speed of steam engines.

The early engineering developments involved trial and error. Many early control systems were successfully developed by the procedures of invention, construction, testing, and modification. Development was more an art than a science. The scientific approach perhaps began with Maxwell in 1868 with his analysis of governors. The theory of control systems has now reached a high level of sophistication and their design has become quite analytical, but the "art of engineering" always plays an important role.

In this introductory chapter, we discuss the reasons why feedback is incorporated in most control systems and then describe the various classifications of closed-loop control systems. We then discuss the performance requirements commonly specified for control systems and the procedural steps involved in their design. The mathematical models used to analyze the dynamic behavior of control systems are described. The mathematical models used in this book are described by linear, ordinary differential equations with constant coefficients.

1.2 CLASSIFICATION OF FEEDBACK CONTROL SYSTEMS

A strong motivation for using feedback in control systems is to correct the error between the controlled output and its desired value corresponding to the command input. The error is caused by disturbance inputs, parameter variations, and imperfect modeling. Feedback also affects other system performance characteristics, such as stability, sensitivity, and overall gain, and these effects will be clarified in later chapters. The block diagram of an open-loop control system is shown in Fig. 1.1, where the actuator is a control element, such as an electric or hydraulic motor, that supplies a control signal to the system to be controlled.

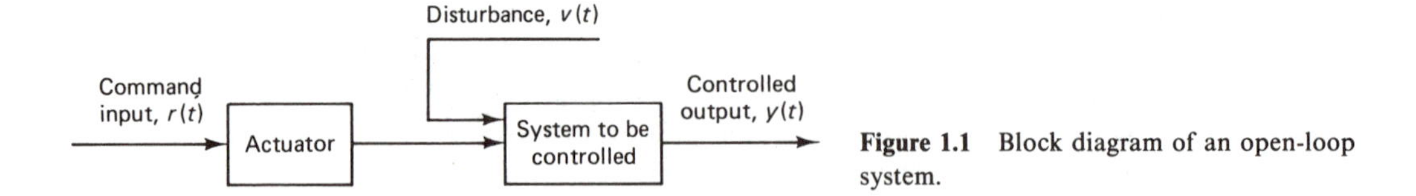

Figure 1.1 Block diagram of an open-loop system.

EXAMPLE 1.1

An open-loop control system for regulating the temperature of a room is shown in Fig. 1.2. The desired temperature, which is the command or reference input temperature T_r, is set on a calibrated dial. This positions the valve that admits hot water for circulation through the radiator. Accordingly, heat flux q_i flows into the room and heat flux q_e flows out from the room to the environment. The outflux q_e depends also on the environmental temperature T_e, which is usually a random variable. The difference between the heat influx and outflux raises the room temperature. Here, the desired room temperature T_r is the

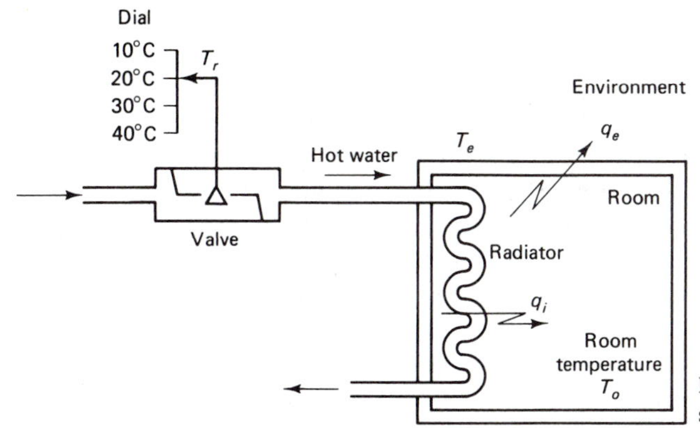

Figure 1.2 Open-loop temperature-control system.

command input, the actual room temperature T_o is the controlled output, and the environmental temperature T_e is the disturbance.

In this example, the actuator for positioning of the valve corresponds to the hands. The valve dial is calibrated when the environmental temperature T_e and system parameters have certain values. When these values change significantly, the controlled temperature T_o will deviate from its command value T_r by a large error and hence precise control will not be realized.

The block diagram of a feedback or closed-loop control system is shown in Fig. 1.3. The controlled output $y(t)$ is measured by a sensor and compared with its desired value, which is the command input $r(t)$, by a comparator or error detector.

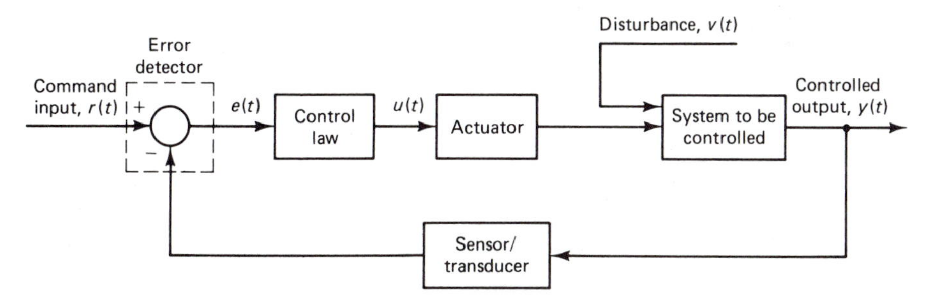

Figure 1.3 Block diagram of a feedback control system.

The error $e(t)$ is used to produce a control input $u(t)$ to the actuator through an appropriate control law with the objective of correcting the error. The functional relationship used to synthesize the control signal $u(t)$ as a function of the error $e(t)$ is called the *control law*. Obviously, a feedback control system has more components than an open-loop system and is therefore more expensive. However, the advantages of precise control outweigh the disadvantages of higher initial cost in most applications and therefore a feedback control system is preferred. The classification of feedback control systems is discussed in the following.

1.2.1 Manual and Automatic Feedback Control System

A *manual control* system results when some of the functions of closed-loop control, such as sensing, error detection, control law synthesis, and/or actuation, are performed by a human operator acting as a controller. When all the functions of the closed-loop are automated and are performed by equipment, the control system is called *automatic.* The early control systems were predominantly manual with a human operator acting as a closed-loop controller. When the human operator is properly trained and experienced, a manual control system can provide a very high level of control and can easily adapt itself to changing situations.

Complete knowledge is still lacking concerning the mathematical modeling and functioning of many sophisticated manual control systems. Considerations of improvements in the quality of control of monotonous, repetitive, or unpleasant tasks and other considerations, such as unsafe jobs or remotely located jobs, have provided economic incentives for the development of automatic control. Recent developments in robotics has led to the replacement of many manual control tasks in manufacturing with automatic control.

EXAMPLE 1.2

A locomotive operator handling a train is an example of a manual speed-control system. The control objective is to maintain the locomotive speed equal to the speed limits that have been set and that may vary with the track and its location. The block diagram of this manual control system is shown in Fig. 1.4.

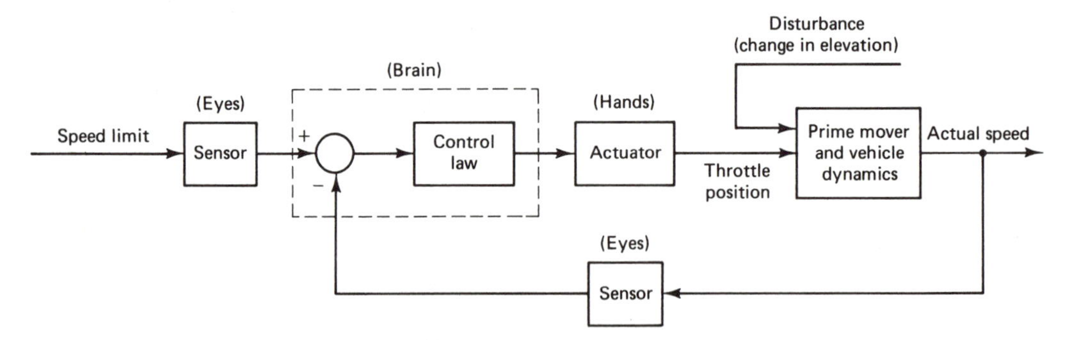

Figure 1.4 Manual speed-control system.

A change in the track elevation is a disturbance that causes the locomotive speed to deviate from the speed limits. An appropriate control law is required for accurate speed control and safe operation. For example, a sudden braking on a curve can cause train derailment.

1.2.2 Active and Passive Feedback Control Systems

When the power necessary to alter the controlled output is supplied primarily from sources other than the command input, the feedback control system is called *active*;

otherwise, the control system is called *passive.* Feedback control systems are predominantly active, but passive control systems are sometimes employed, especially for the isolation and control of vibrations. In the case of vehicles, the main objective of a suspension system, consisting of springs and dampers, is to isolate the sprung mass from the surface irregularities, that is, maintain the sprung mass at a nearly constant level in spite of disturbances.

EXAMPLE 1.3

An example of a passive feedback control system is shown in Fig. 1.5(a). The control objective is to maintain the inverted pendulum at its unstable equilibrium position, $\theta = 0$. A spring is used as sensor and actuator. The angular displacement θ from its equilibrium position is sensed by the spring displacement and a corrective moment is applied at the pendulum pivot to reduce θ

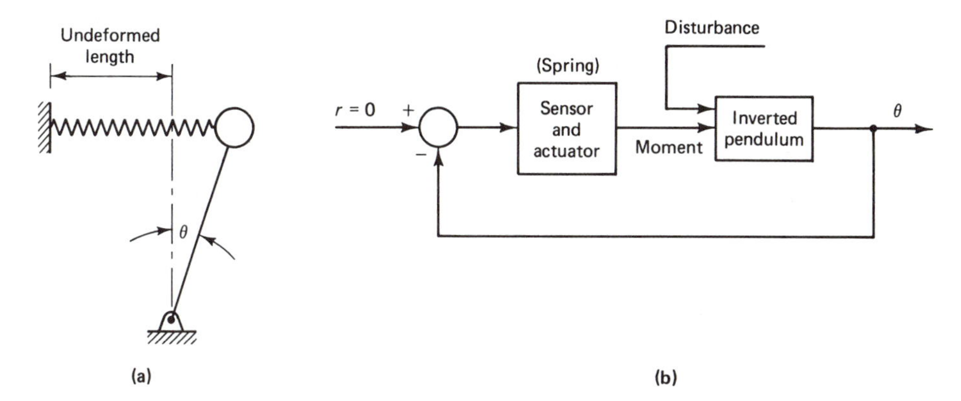

Figure 1.5 Passive control system for inverted pendulum.

to zero. The block diagram of this feedback control system is shown in Fig. 1.5(b). The desired position of θ, i.e., the command input r, is zero. When the spring is sufficiently stiff, the equilibrium position $\theta = 0$ of the closed-loop system becomes stable.

1.2.3 Regulator and Servomechanism (or Tracking System)

When the control objective is to maintain the controlled output at its constant equilibrium position in spite of disturbances, the control system is called a *regulator.* The command input $r(t)$ in Fig. 1.3 for a regulator becomes a constant and is called a *set point.* The set point corresponds to the equilibrium value of the controlled output. The set point may, however, be changed in time from one constant value to another. In a *tracking control* system, the controlled output is required to follow-up or track a time-varying command input. A *servomechanism* is a tracking control system where the controlled output is a mechanical position, velocity, or acceleration. However, from the point of view of classification, whether the controlled output is mechanical or otherwise is immaterial and we shall call any tracking control system a servomechanism.

EXAMPLE 1.4

The autopilot of an aircraft is an example of a regulator. Its objective is to maintain constant direction and altitude in the presence of disturbances such as cross winds and up or down drafts. It is activated after steady flight conditions have been achieved with the desired direction and altitude as set points. The actual direction and altitude are sensed by a compass and altimeter, respectively, and compared with their desired values. Appropriate control laws, which are functions of the errors, supply control signals to hydraulic actuators that move the rudder and elevator, respectively, to correct the errors. But during takeoff and landing, the aircraft is required to follow a time-varying trajectory with variable speed and the pilot is the controller of the feedback system. This manual control system is a servomechanism.

A feedback control system is either a regulator or a servomechanism. A temperature-control system for a house is a regulator since its objective is to maintain the temperature constant, corresponding to the set point of the thermostat in the presence of disturbances. The guidance system of an air-to-air missile is an example of a servomechanism since its objective is to track and home in on an enemy aircraft that may be performing evasive maneuvers. In some missiles, the exhaust of the enemy aircraft is sensed by an infrared sensor and acts as a time-varying command input to the missile.

1.2.4 Single-Input, Single-Output and Multiple-Input, Multiple-Output (Multivariable) Control Systems

In many simple control systems, a single output is controlled by a single input and such systems are called *single-input, single-output* systems. Systems with more than one controlled output and command input are called *multivariable.* Where one control input affects only one controlled output, then the system can be decoupled into n single-input, single-output control systems and there is no longer any need to view the system as multivariable. In multivariable systems, an input that is meant to control a particular output also affects the other controlled outputs. This coupling is called *interaction.* As a result of interaction, the control systems cannot be designed separately and the coupling must be taken into consideration.

EXAMPLE 1.5

The positions y_1 and y_2 of masses m_1 and m_2, respectively, shown in Fig. 1.6 are to be controlled independently by two feedback control systems. In the first feedback control system, the actual position y_1 is sensed and after it is compared with its command input, a force F_1 is generated to correct the error. Similarly, force F_2 is generated to control y_2. However, it is seen that F_1 affects not only y_1 but also y_2. Also, F_2 affects not only y_2 but also y_1. This

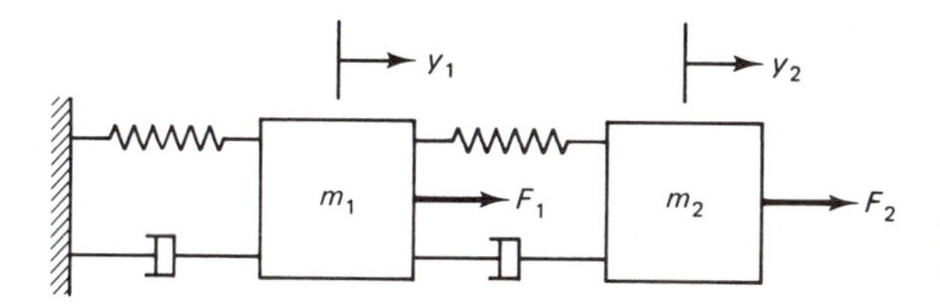

Figure 1.6 Multivariable mechanical system.

system is multivariable and the two control systems cannot be designed independently of each other as the interaction must be considered.

EXAMPLE 1.6

An operator driving an automobile is an example of a multivariable, manual, feedback control system. The system to be controlled has two inputs (steering and acceleration/braking) and two controlled outputs (heading and speed). The two command or reference inputs are the direction of the highway and the speed limits with traffic signals. A block diagram of this two-input, two-output control system is shown in Fig. 1.7.

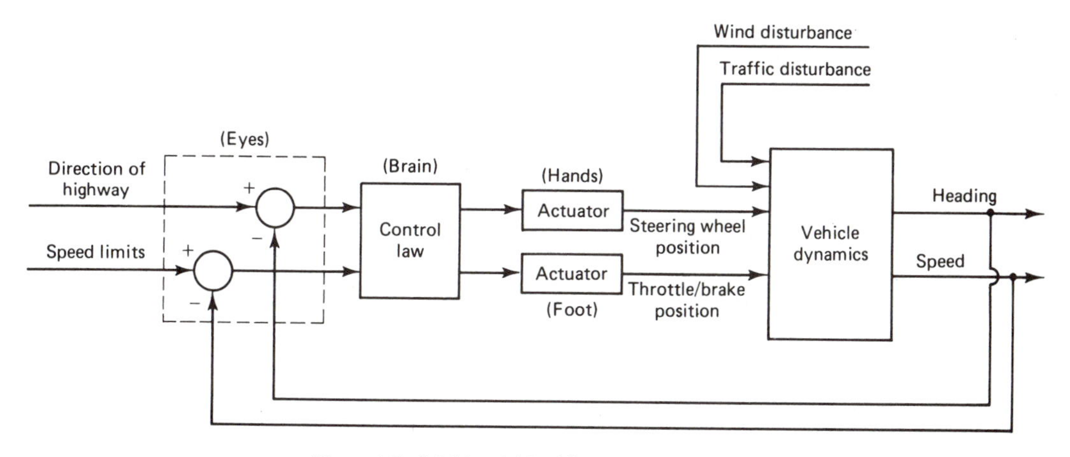

Figure 1.7 Multivariable driver-automobile control system.

The braking of the vehicle for speed control decreases the side forces at the tire–road interface for directional control, and with locked wheels, directional control is completely lost.

1.2.5 Continuous-Data (Analog Control) and Sampled-Data (Digital Control) Systems

In a *continuous-data* system, all the signals in various parts of the system are continuous functions of time. It is also referred to as an *analog control* system. In a *digital control* system, a digital computer is used as the controller. A digital computer operates on digital data and the signals vary only at discrete instants of time; these signals are called discrete-time signals. A system having both discrete-time and continuous signals is called a *sampled-data* system. Hence, when a digital computer is used to control a system whose variables are continuous functions of time, a sampled-data system results. The basic elements of a typical digital control system are shown in the block diagram of Fig. 1.8. In some digital control systems, the command input is an analog signal and the analog-to-digital converter is inserted after the error detector. The control law is implemented by the digital computer.

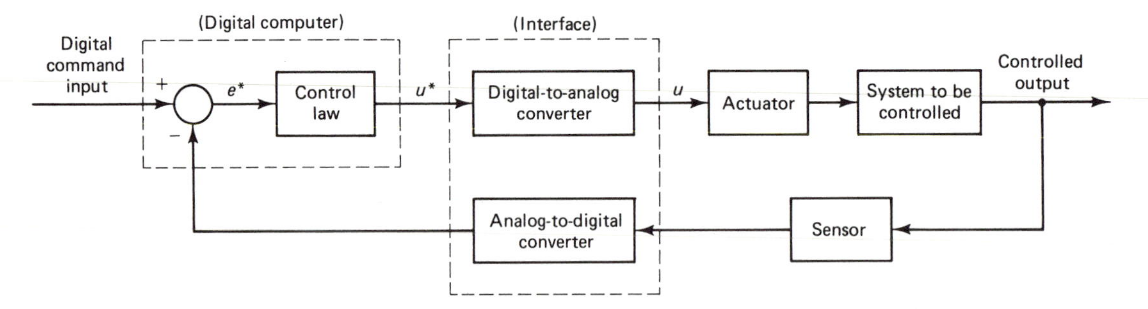

Figure 1.8 Block diagram of a typical digital control system.

Inexpensive microprocessors and microcomputers have reduced the cost of imbedding computers in the control loop for direct computer control. Analog-to-digital and digital-to-analog converters are also becoming inexpensive and easy to interface. The implementation of sophisticated control laws is possible through the microcomputer's capability. Typical plots of the costs of analog and digital control versus the complexity of the control system are shown in Fig. 1.9. This figure is not scaled because the cost of microprocessor-based computers is going down over the years. It is seen that analog control is cost effective for simple control systems.

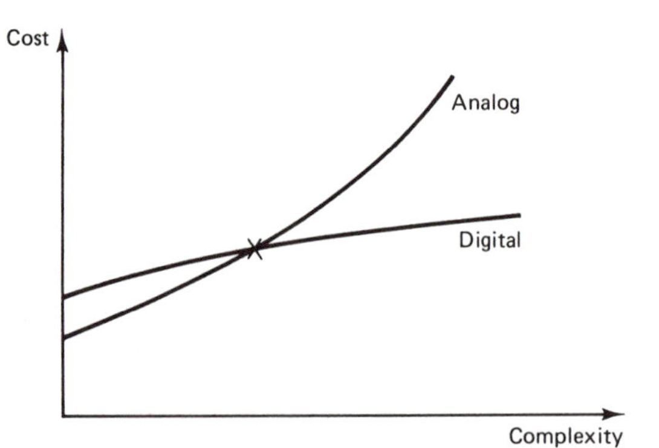

Figure 1.9 Cost versus complexity of analog and digital control.

The advantages of digital control over analog control are the flexibility of changing the control law, time-sharing feature among several control channels, accuracy because many analog components have drift, and decision-making and logic capability. In addition to the control function, the digital computer may be time-shared with other tasks such as optimization, and data reduction and logging. Because of these advantages, digital-computer control is becoming increasingly popular. Computer control is covered in the last chapter of the book. The knowledge of continuous-data systems covered in the earlier chapters is required for the study of digital-computer control systems.

1.3 PERFORMANCE SPECIFICATIONS AND DESIGN PROCEDURE

1.3.1 Performance Specifications

For the successful design of a control system, it is important that the performance objectives be established first. The establishment of performance specifications is not an easy task and usually requires some experience. On the one hand, if the performance specifications are overly stringent, the complexity and cost of the control system will become prohibitive. On the other hand, if the performance specifications are lax, the control system is not likely to function satisfactorily. The factors that require careful consideration include cost, reliability, size and weight, speed of response, stability, accuracy, and ease of operation and maintenance. These factors are briefly discussed in the following.

Hardware Selection. Feedback control systems may use mechanical, electrical, hydraulic, pneumatic, or a combination of components for sensors, actuators, error detectors, and for implementation of the control law. In robotics and automated manufacturing, dc electrical motors are used as actuators when the horsepower requirements are small. But when the power required is large, hydraulic actuators are preferred because of the advantage of the horsepower-to-weight ratio. Electrohydraulic control systems are used in aircraft and missiles because of the same advantage.

Pneumatic systems are commonly used in petrochemical process control because of their relative safety concerning fire hazard and explosion. Compressed air can be supplied from a remote location. Pneumatic systems are slow responding as compared to electrical and hydraulic systems, but this is not a serious disadvantage in process control because most processes themselves are slow responding.

Stable Operation with Sufficient Margin of Stability. Feedback control systems are prone to instability. In active control systems, a supplemental source of power is available at one or more points of the feedback control system so that the possibility exists for instability. In an unstable mode of operation, a rapid and destructive response of the system may render the control useless. In some cases, nonlinearities, such as saturation, may limit the growth of the unstable response and give rise to hunting or self-excited oscillations called limit cycles. Therefore, it is required that not only the control system operate in a stable mode, but also that an adequate margin of stability be specified to accommodate parameter variations. The stability of control systems is studied in Chapter 6.

Acceptable Transient Response. In both regulators and servomechanisms, it is desirable to compare the command input and controlled output response as functions of time to determine the speed of response and the deviation of the response from the input. The specification of an acceptable transient response is often based on a step-command input as a test signal. Typical performance criteria that are used to characterize the transient response to a step input include delay time and rise time for the initial speed of response, maximum overshoot for the deviation, and settling time necessary for the response to settle within certain limits of its final steady-state value. These terms are defined and the transient response of control systems is studied in Chapter 4.

Frequency-Domain Specifications. The specifications in the frequency domain are based on a sinusoidal command input as a test signal. The performance criteria that are used to characterize the frequency response are the peak magnification and bandwidth. The peak magnification is related to the maximum overshoot and settling time of the transient response. The bandwidth is a measure of the system's ability to filter the noise from the signal. These terms are clarified and the frequency response is studied in Chapter 5.

Disturbance Rejection. In a regulator, a good disturbance rejection property refers to the steady-state accuracy where the error between the command input and controlled output is acceptably small in the presence of disturbances. For example, in the case of a speed regulator for an automobile, the error is required to be within acceptable limits when the automobile is traveling uphill or downhill. In a servomechanism, good disturbance rejection refers to the ability of the system to keep the error small as the command input changes with time. For example, in the guidance of a rocket, the effect of wind disturbance must be minimized to keep the rocket close to the desired trajectory. Disturbance rejection is studied in Chapter 7.

Sensitivity to Parameter Changes. The performance of a control system depends on its parameter values, which may change due to wear and aging of components and variations in the environment. A design choice may be made to minimize the effect of parameter variations. A system that has both good disturbance rejection and low sensitivity to parameter variations is called *robust.*

Performance Index. In the design of control systems by the technique of optimal control, the objective is to optimize a figure of merit that is called the performance index. The performance index includes the key variables of the control system that we wish to optimize. The objective may be to minimize the time, fuel, control effort, error, and their combinations. Optimal control theory is, however, beyond the scope of this book.

1.3.2 Design Procedure

The trial-and-error design procedure of construction, testing, and modification has severe drawbacks and an analytical design procedure is definitely preferable. An overview of the analytical procedure for the design of control systems is useful before studying the specific techniques developed and discussed in the following chapters. It is difficult to outline a unique design procedure. In practice, there are many conflicting demands that are not compatible and usually compromises and iteration are required. For example, if high accuracy is desired, then the cost will be high. It will be seen later that a high stability margin reduces the noise filtering capability. A suggested design procedure is given in the following.

1. *Performance Specifications.* The performance objectives are considered first and the specifications to be satisfied are established. The commonly employed specifications have been discussed in the preceding.

2. *Conceptual Design.* The control system components, such as sensor, actuator, and hardware for the implementation of the control law, are selected from electrical, electronic, hydraulic, pneumatic, or their combinations based on some of the considerations discussed earlier. The accuracy, resolution, and sizing of the components require careful consideration. A schematic diagram of the conceptual design is now ready.

3. *Mathematical Modeling.* A mathematical model of the overall control system is obtained so that the schematic diagram of the conceptual design is represented by a set of equations. Different types of mathematical models that may be employed are discussed in the next section. The main objectives of Chapters 2 and 3 are the mathematical modeling of components and that of conceptual design of control systems, respectively.

4. *Model Validation and Parameter Identification.* Usually many simplifications and assumptions are made in obtaining mathematical models. Hence, experimental validation and identification of parameter values are necessary. The frequency-response method discussed in Chapter 5 is very appropriate for experimental model validation and the identification of parameters of models described by ordinary, linear differential equations with constant coefficients. At this stage, the overall control system is not available for testing. But the system to be controlled is usually available. Also, the experimental frequency responses of most components are usually available from their manufacturers.

5. *Analysis of the Mathematical Model.* The performance of the control system is analyzed by employing its mathematical model. Stability and an adequate margin of stability, speed of response, and disturbance rejection are some of the performance criteria that are analyzed by the methods discussed in Chapters 4 to 9.

6. *Modification and Iteration.* The free parameters are optimized, the control law is modified, and iterations are performed among the preceding steps until the specifications are met and the performance is satisfactory.

7. *Construction and Testing.* The last step is the construction and testing of the overall control system, or prototype, to ensure that the actual performance is as predicted by its mathematical model.

1.4 MATHEMATICAL MODELS

The analytical design procedure outlined in the preceding section relies heavily on a mathematical model of the control system to be designed. Modeling constitutes one of the central tasks in the study of control systems. A mathematical model must be detailed enough to include the essential system features to be investigated. Yet, it must not attempt to include insignificant behavior and become more mathematically complicated than needed. In a servomechanism, the command input and disturbances vary with time and hence the system variables are time dependent. In a regulator, the system variables are also time dependent in the presence of disturbances even when the set point has not been changed. Hence, time is an independent variable and the models we consider are dynamic models. Different types of dynamic models used to describe control systems are described in the following.

1.4.1 Distributed-Parameter and Lumped-Parameter Models

The models employed to describe control systems are macroscopic models where the molecular nature of matter considered in microscopic models is ignored. Distributed-parameter models invoke the *continuum hypothesis*, whereby it is assumed that matter is a medium continuously filling space to which properties are assigned to reflect the statistical mean of the molecular effects. The significant variables are distributed in space and they vary with the spatial coordinates and time. The resulting dynamic models, called *distributed-parameter models*, consist of partial differential equations with time and space coordinates as independent variables. The unsteady Navier–Stokes equations of fluid flow, the transient Fourier heat-conduction equation, and the Navier equations of dynamic elasticity for flexible solids are some examples of distributed-parameter models.

In *lumped-parameter models*, the matter is assumed to be lumped at some discrete points of the space or the space is subdivided into cells and matter is assumed to be lumped at these cells. The resulting dynamic models are ordinary differential equations with time as the only independent variable. For example, the assumption that a solid is a rigid body permits lumping all its mass at its mass center and results in lumped-parameter models of dynamics.

Considerable simplification is achieved in mathematical models when lumped-parameter models are used. Since most solids and fluids are in fact distributed-parameter systems, great care must be exercised in obtaining their lumped-parameter representation. It is shown in Chapter 2 that the Biot number plays an important role in lumping thermal systems. The fundamental and higher harmonic modes are to be considered in obtaining lumped-parameter models of flexible solid bodies. In this book, we employ lumped-parameter models of control systems.

1.4.2 Nonlinear and Linear Models

A transformation, operation, function, or in general any mapping, is *linear* if it satisfies the following two properties:

1. Additive property, that is, for any x_1 and x_2 belonging to the domain of the function f, we have

$$f(x_1 + x_2) = f(x_1) + f(x_2)$$

2. Homogeneous property, that is, for any x belonging to the domain of the function f and for any scalar constant α, we have

$$f(\alpha x) = \alpha f(x)$$

Both properties are included in the principle of superposition used in the analysis of linear systems. Any function that does not satisfy both of these properties is *nonlinear.* For example, a differential operator $d/dt(.)$ and an integral operator $\int(.)\,dt$ are linear operators since they satisfy both properties. The Laplace transformation of a function of time $x(t)$ defined by

$$\mathscr{L}\{x(t)\} = \int_0^\infty x(t)\, e^{-st}\, dt$$

is a linear transformation since it satisfies both properties. A function $f(x) = (x)^2$ is nonlinear since $(x_1 + x_2)^2 \neq (x_1)^2 + (x_2)^2$ and $(\alpha x)^2 \neq \alpha(x)^2$.

The differential equations model of a control system may be classified as either linear or nonlinear. The nonlinearities may be introduced by the system to be controlled and/or by the control components, such as sensors, actuators, and control valves. Nonlinearities can cause many adverse effects including self-excited oscillations. For this reason, linear control components are usually preferable to nonlinear ones. However, sometimes nonlinearities may be intentionally introduced as in the case of an on–off control system. In this book, we employ linear lumped-parameter models of control systems. It should be realized that many components behave nonlinearly outside a certain range of operation, and linearization involves an assumption of a small deviation from equilibrium.

1.4.3 Time-Varying and Time-Invariant Models

When the parameters of the differential equation model are varying with time, the model is called *time-varying* or nonstationary. Differential equation models with constant coefficients are called *time-invariant.* For example, in the guidance and control of a rocket, the mass of the rocket changes with time due to the depletion of fuel, and also the aerodynamic damping can change with time as the air density changes with altitude. The complexity of the control system design increases considerably when the parameters are time-varying, even for the case of linear models.

1.4.4 Stochastic and Deterministic Models

In some cases, the system parameters and/or disturbance inputs are random functions of time and are described by probability distribution functions. Also, the sensor measurements may be corrupted by additive random noise. In such cases, the differential equation models are called *stochastic.* When all the inputs are deterministic functions of time and the parameters are either constants or deterministic functions of time, the differential equation model is called *deterministic.*

1.4.5 Continuous-Time and Discrete-Time Models

The signals of the various parts of a continuous-data system vary continuously with time as discussed earlier. The dynamic models are differential equations with time as the independent variable and they are called *continuous-time* models. A sampled-data or digital control system operates on the signal at specific discrete instants of time. The dynamic models are described by difference equations in time and the models are called *discrete-time* models. Hence, discrete-time models are used when a digital-computer control is employed.

1.5 SUMMARY

A major objective of this chapter has been the introduction of the terminology and different classifications of control systems and their mathematical models. It is important to know the commonly employed classifications of control systems so that a proper perspective is obtained and the scope of the book is clarified. This book deals with automatic feedback control systems. The emphasis is on single-input, single-output systems, but multivariable systems are also included. The control systems studied include both regulators and servomechanisms.

Digital-computer control is becoming increasingly popular and it is studied in the last chapter of the book. The introductory material covered earlier is a prerequisite for the study of digital control systems. The mathematical models of control systems that we employ are lumped-parameter, linear, time-invariant, continuous-time, and deterministic. They are described by linear, ordinary differential equations in time with constant coefficients. Linear difference equations with constant coefficients are employed for computer control systems. An overview of the design procedure is given to motivate the study of the specific techniques developed in the following chapters.

REFERENCES

HALE, F. J. (1973). *Introduction to Control System Analysis and Design.* Englewood Cliffs, NJ: Prentice-Hall.

KUO, B. C. (1982). *Automatic Control Systems.* 4th ed. Englewood Cliffs, NJ: Prentice-Hall.

NEWTON, G. C., GOULD, L. A., and KAISER, J. F. (1957). *Analytical Design of Linear Feedback Controls.* New York: John Wiley & Sons.

PALM, W. J. (1983). *Modeling, Analysis, and Control of Dynamic Systems.* New York: John Wiley & Sons.

2

Mathematical Modeling of Control System Components

2.1 INTRODUCTION

One of the important steps in the analysis and design of control systems is the task of mathematical modeling. In this chapter, we develop mathematical models of mechanical, electrical, hydraulic, pneumatic, and thermal elements that are commonly encountered in control systems as components of the system to be controlled and/or as components of actuators, sensors, and controllers. The mathematical models are obtained by the application of the physical laws pertaining to the nature of the components.

Regardless of the nature of the components, it will be seen that passive elements can be classified into three types, two that store energy and one that dissipates it. Each component is said to have terminals corresponding to its interfaces with other components. Elementary components have two terminals and we need two variables to represent them. One variable is called the *through variable*, which flows through the element, and the other is called the *across variable*, which can be measured across the two terminals. This analogy from the energy viewpoint is discussed later in this chapter.

As mentioned in Chapter 1, our mathematical models are restricted to linear, ordinary differential equations with constant coefficients with time as the independent variable. The behavior of many systems is nonlinear, but if the nonlinearities are analytic functions of their arguments, we are able to construct linear models that are valid in a certain operating range about their equilibrium values. Our objective is to represent the system equations in the form of transfer functions and also as state differential equations. Both formulations will be employed in later chapters for analysis and design.

2.2 TRANSFER-FUNCTION AND STATE-VARIABLES REPRESENTATIONS

2.2.1 Transfer Function Representation

For transfer-function representation, we seek the relationship between the input and the output of a component or system. The input is the cause and the output is the effect. Hence, a transfer function relates the effect to the cause. The first step is to obtain the differential equations describing the behavior of a dynamic system by the application of physical laws. In general, we get a set of first- and second-order differential equations. Assuming that the nonlinearities, if any, are analytic functions of their arguments, we linearize the differential equations for small deviations from equilibrium. The second step is to eliminate all the intermediate variables, leaving only the input and the output. Combining the equations in this manner, we obtain the relationship between the output and the input.

Let us suppose that this procedure yields a linear differential equation with constant coefficients relating the input u to the output y by

$$a_3\frac{d^3y}{dt^3} + a_2\frac{d^2y}{dt^2} + a_1\frac{dy}{dt} + a_0y = b_1\frac{du}{dt} + b_0u \tag{2.1}$$

where a's and b's are constant parameters. Letting the ordinary differential operator d/dt be denoted by D, we can express Eq. (2.1) as

$$(a_3D^3 + a_2D^2 + a_1D + a_0)y = (b_1D + b_0)u \tag{2.2}$$

i.e.,

$$y = \left(\frac{b_1D + b_0}{a_3D^3 + a_2D^2 + a_1D + a_0}\right)u \tag{2.3}$$

Defining a transfer operator $G(D)$ as

$$G(D) = \frac{b_1D + b_0}{a_3D^3 + a_2D^2 + a_1D + a_0} \tag{2.4}$$

Eq. (2.3) may be expressed as

$$y = G(D)u$$

The relationship given by Eq. (2.3) is shown in the block diagram of Fig. 2.1. The operator inside the box is the transfer operator $G(D)$ that operates on the input u to yield the output y. The differential equation relating y to u can be obtained from Fig. 2.1 by cross multiplication.

Let $Y(s)$ and $U(s)$ be the Laplace transformations of $y(t)$ and $u(t)$, respectively, where s is a complex variable of the Laplace transformation. Assuming that all initial conditions are zero and Laplace transforming Eq. (2.2), we obtain

$$(a_3s^3 + a_2s^2 + a_1s + a_0)Y(s) = (b_1s + b_0)U(s) \tag{2.5}$$

which is shown in the block diagram of Fig. 2.2. The transfer function relating the

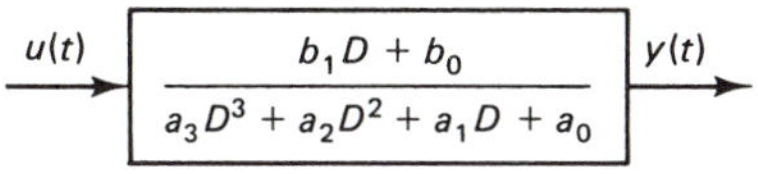

$$u(t) \rightarrow \boxed{\frac{b_1 D + b_0}{a_3 D^3 + a_2 D^2 + a_1 D + a_0}} \rightarrow y(t)$$

Figure 2.1 Block diagram representation.

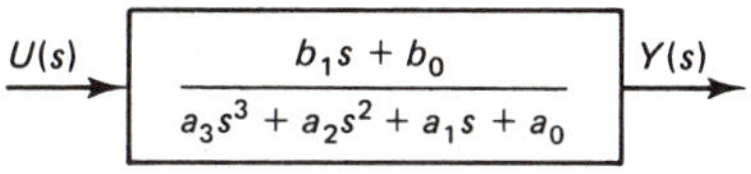

$$U(s) \rightarrow \boxed{\frac{b_1 s + b_0}{a_3 s^3 + a_2 s^2 + a_1 s + a_0}} \rightarrow Y(s)$$

Figure 2.2 Block diagram with transfer function.

output $Y(s)$ to the input $U(s)$ in the Laplace domain is given by

$$Y(s) = G(s)U(s) \tag{2.6}$$

where

$$G(s) = \frac{b_1 s + b_0}{a_3 s^3 + a_2 s^2 + a_1 s + a_0}$$

One of the two formulations that we employ to represent the mathematical models is the transfer-function representation. When some or all of the initial conditions are nonzero, we can replace the variable s in Fig. 2.2 or in Eq. (2.5) by the operator D, and then include the initial conditions. The transfer-function representation is obtained by using the following two steps.

1. Obtain the differential equations of the system by application of physical laws. In general, we get a set of first- and second-order differential equations. Assuming that the nonlinearities are analytic functions of their arguments, the differential equations are linearized for small deviations from equilibrium.
2. Combine the equations by eliminating the intermediate variables so that we obtain a single differential equation relating the output y to the input u.

The preceding discussion is concerned with single-input, single-output systems. In multivariable systems, we get a set of differential equations relating m outputs $\mathbf{Y}(s)$ to q inputs $\mathbf{U}(s)$, where $\mathbf{Y}(s)$ and $\mathbf{U}(s)$ are $m \times 1$ and $q \times 1$ matrices, respectively. The transfer function representation can be extended to this case. The transfer function between the ith output and the jth input is defined by

$$Y_i(s)/U_j(s) = G_{ij}(s) \tag{2.7}$$

and employing the matrix notation, we get

$$\begin{bmatrix} Y_1(s) \\ \\ Y_m(s) \end{bmatrix} = \begin{bmatrix} G_{11}(s) & \cdots & G_{1q}(s) \\ \cdot & & \cdot \\ G_{m1}(s) & \cdots & G_{mq}(s) \end{bmatrix} \begin{bmatrix} U_1(s) \\ \\ U_q(s) \end{bmatrix}$$

i.e.,

$$\mathbf{Y}(s) = \mathbf{G}(s)\mathbf{U}(s) \tag{2.8}$$

where the $m \times q$ matrix $\mathbf{G}(s)$ is called the transfer-function matrix. The procedure for obtaining the transfer-function matrix will be clarified by the illustrative examples.

2.2.2 State-Variables Representation

The second formulation that we employ to represent the mathematical models is the state-variables representation. We consider a multivariable system where there are q inputs $\mathbf{u}(t)$ and m outputs $\mathbf{y}(t)$. In a nth-order system, n variables $x_1(t), x_2(t), \ldots, x_n(t)$ are chosen to represent the dynamic system by a set of first-order coupled equations and the $n \times 1$ matrix $\mathbf{x}(t)$ is called the state vector. For a linear, time-invariant system, the mathematic model is represented by

$$\dot{\mathbf{x}} = \mathbf{A}\mathbf{x} + \mathbf{B}\mathbf{u} \qquad \text{State equations} \tag{2.9}$$

$$\mathbf{y} = \mathbf{C}\mathbf{x} \qquad \text{Output equations} \tag{2.10}$$

where $\mathbf{A}$, $\mathbf{B}$, and $\mathbf{C}$ are $n \times n$, $n \times q$, and $m \times n$ constant matrices, respectively. In a linear, time-varying parameter system, matrices $\mathbf{A}(t)$, $\mathbf{B}(t)$, and $\mathbf{C}(t)$ are time-varying. For a single-input single-output system, u and y are scalars, $\mathbf{B}$ is an $n \times 1$ matrix and $\mathbf{C}$ is a $1 \times n$ matrix.

It will be seen later that in physical systems, the order m of the numerator of the transfer function does not exceed the order n of its denominator, that is, $m \leq n$. These systems are called causal or nonanticipatory, and they yield a response only after the application of an input and not in anticipation. When the order of the numerator of the transfer function is equal to the order of the denominator, the output Eq. (2.10) needs modification to

$$\mathbf{y} = \mathbf{C}\mathbf{x} + \mathbf{E}\mathbf{u}$$

where $\mathbf{E}$ is a constant matrix of appropriate dimension.

We may encounter some components whose transfer functions have the same order numerator as denominator, such as in Example 2.5, Fig. 2.19. However, the systems to be controlled in general have transfer functions where the order of the numerator is less than that of the denominator. Hence, we find that Eq. (2.10) is adequate to represent the output.

The future behavior of a nth-order dynamic system may be specified in terms of n initial conditions at any instant of time and the inputs from that time onward. The knowledge of past inputs is not required to determine future behavior. The n numbers required to specify future behavior represent the initial state of the system, and the variables used to represent these numbers at each instant of time are called state variables. Hence, the state variable vector $\mathbf{x}(t)$ consists of time functions whose values at any specified time represent the state of the dynamic system at that time. The choice of the state variables, however, is not unique. The state-variables formulation is obtained by using the following two steps.

1. Obtain the differential equations of the system by application of physical laws. In general, we get a set of first- and second-order differential equations.
2. Define a sufficient number of state variables so that each variable appears in derivative form of order not higher than the first.

The transfer-function representation has the following two disadvantages. First, it is restricted to mathematical models that are described by linear differential equations with constant coefficients. Second, since the intermediate variables have been eliminated, their dynamic behavior is not directly available. The state-variables representation does not have these disadvantages and can be used for models described by nonlinear differential equations with time-varying parameters in the general form

$$\dot{x}_i = f_i(x_1, x_2, \ldots, x_n, u_1, \ldots, u_q, t) \qquad i = 1, 2, \ldots, n \tag{2.11}$$

$$y_i = g_i(x_1, \ldots, x_n, t) \qquad i = 1, 2, \ldots, m \tag{2.12}$$

where f_i and g_i are nonlinear functions of their arguments and explicit functions of time. Equations (2.9) and (2.10) result from Eqs. (2.11) and (2.12), respectively, by linearization about an equilibrium state when f_i and g_i are not explicit functions of time.

Since we are dealing with linear time-invariant systems, we employ both the transfer-matrix formulation of Eq. (2.8) and the state-variables formulation of Eqs. (2.9) and (2.10), depending upon the method of analysis. It is a simple matter to convert the state equations to the transfer functions and vice versa. In order to convert the state-variables formulation to the transfer-function matrix, we Laplace transform Eq. (2.9) to obtain

$$s\mathbf{X}(s) - \mathbf{x}(0) = \mathbf{AX}(s) + \mathbf{BU}(s)$$

Since the transfer function implies that the initial conditions $\mathbf{x}(0) = \mathbf{0}$, it follows that

$$s\mathbf{X}(s) = \mathbf{AX}(s) + \mathbf{BU}(s)$$

i.e.,

$$\mathbf{X}(s) = (s\mathbf{I} - \mathbf{A})^{-1}\mathbf{BU}(s) \tag{2.13}$$

where $\mathbf{I}$ is an $n \times n$ identity matrix. Using Eq. (2.10), we get

$$\mathbf{Y}(s) = \mathbf{C}(s\mathbf{I} - \mathbf{A})^{-1}\mathbf{BU}(s) \tag{2.14}$$

On comparing Eq. (2.8) with Eq. (2.14), we get the transfer-function matrix

$$\mathbf{G}(s) = \mathbf{C}(s\mathbf{I} - \mathbf{A})^{-1}\mathbf{B} \tag{2.15}$$

We mostly consider single-input, single-output systems, where $G(s)$ is a scalar transfer function of Eq. (2.6), and $\mathbf{B}$ and $\mathbf{C}$ are $n \times 1$ and $1 \times n$ matrices, respectively.

2.2.3 Linearization

In this book, we employ linear time-invariant models expressed either in the form of transfer functions or in the form of state equations, Eqs. (2.9) and (2.10). The behavior of many physical systems is in general nonlinear and nonlinear characteristics may sometimes be employed intentionally to improve system performance. Linearization of nonlinear equations can be justified only for a small departure of the variables from their operating values when the nonlinearities are analytic functions of their arguments.

The two techniques for linearization are the Taylor series expansion method used when the functional equations are available and linearization from the operating curves used when experimental characteristics rather than equations are available.

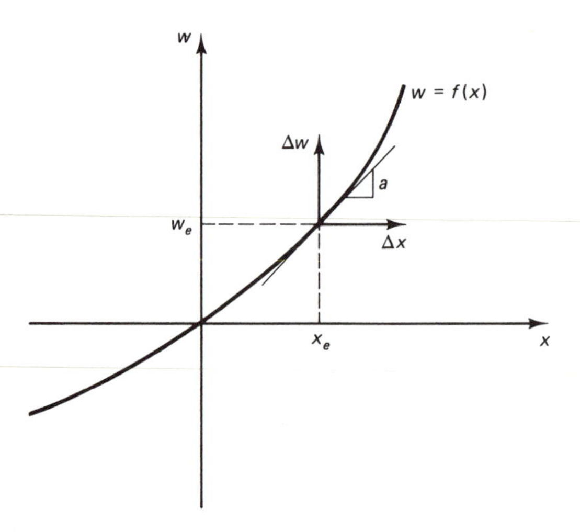

Figure 2.3 Linearization of the function $w = f(x)$.

Consider a scalar analytic function $w = f(x)$ of a single variable x shown in Fig. 2.3. Taylor series expansion about equilibrium value x_e yields

$$w = f(x) = f(x_e) + \left(\frac{df}{dx}\right)_e (x - x_e) + \frac{1}{2!}\left(\frac{d^2f}{dx^2}\right)_e (x - x_e)^2 + \cdots \tag{2.16}$$

where the subscript e denotes that the derivatives are evaluated at the equilibrium value x_e. Letting $w = w_e + \Delta w$ and assuming that $x - x_e = \Delta x$ is small, the terms of second and higher order in the Taylor series are dropped to obtain

$$w_e + \Delta w = f(x_e) + (df/dx)_e \Delta x \tag{2.17}$$

Letting $(df/dx)_e = a$ and noting that $w_e = f(x_e)$, we obtain the linear approximation

$$\Delta w = a\Delta x \tag{2.18}$$

Supposing the analytical form of $f(x)$ is not available, let Fig. 2.3 represent an experimentally obtained steady-state plot of w versus x. Denoting the slope of the curve at the equilibrium point (w_e, x_e) by a, we obtain the linear equation of Eq. (2.18) for a small deviation from equilibrium as shown in Fig. 2.3. This figure also gives a geometric interpretation to the linearization technique by Taylor series expansion.

The technique can be extended to a multivariable function. Let $w = f(x_1, x_2)$ be an analytic function of two variables x_1 and x_2. Taylor series expansion about an equilibrium yields

$$w = f(x_{1e}, x_{2e}) + \left(\frac{\partial f}{\partial x_1}\right)_e (x_1 - x_{1e}) + \left(\frac{\partial f}{\partial x_2}\right)_e (x_2 - x_{2e}) + \cdots \tag{2.19}$$

Letting the two first partial derivatives evaluated at the equilibrium position be denoted by a_1 and a_2, respectively, for small deviations from equilibrium, we obtain

$$\Delta w = a_1 \Delta x_1 + a_2 \Delta x_2 \tag{2.20}$$

Suppose the analytic form of $f(x_1, x_2)$ is not available but the experimentally obtained steady-state operating curves are given in Fig. 2.4. Here, w is plotted

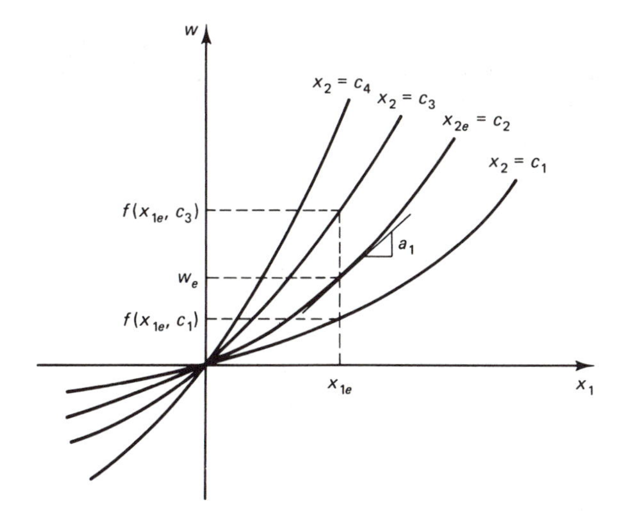

Figure 2.4 Linearization of function $w = f(x_1, x_2)$.

versus x_1 for constant values of $x_2 = c_i$, where c_i is a constant. The partial derivative $(\partial f/\partial x_1)_e$ or a_1 in Eq. (2.20) is the slope of the curve at the equilibrium condition $x_1 = x_{1e}$, $x_2 = x_{2e}$. The partial derivative $(\partial f/\partial x_2)_e = a_2$ can be approximated from Fig. 2.4 with x_1 held constant at x_{1e} and is given by

$$a_2 = \left(\frac{\partial f}{\partial x_2}\right)_e \approx \frac{f(x_{1e}, c_3) - f(x_{1e}, c_1)}{c_3 - c_1} \tag{2.21}$$

When a particular slope has a negative value, we shall denote it by $-a_i$, where $a_i > 0$. This point is not trivial but should be carefully considered in modeling to avoid confusion later during analysis concerning feedback, stability, and dynamic behavior. This technique of linearization fails when the nonlinearities are not analytic functions of their arguments, such as Coulomb friction, hysteresis, and backlash in gears. In those cases, nonlinear analysis cannot be avoided unless these nonlinearities are just ignored in the first approximation.

We can now generalize the procedure to obtain linear approximation to the nonlinear time-invariant state equation

$$\dot{\mathbf{x}} = \mathbf{f}(\mathbf{x}, \mathbf{u}) \tag{2.22}$$

When the inputs $\mathbf{u}$ have constant values $\mathbf{u}_e$, a state $\mathbf{x}_e$ of Eq. (2.22) is called an equilibrium state if, starting at that state, the system will remain in that state in the absence of disturbances or changes in the input. Since for equilibrium $\dot{\mathbf{x}} = \mathbf{0}$, the equilibrium states are obtained from the solution of the nonlinear algebraic equations $\mathbf{f}(\mathbf{x}_e, \mathbf{u}_e) = \mathbf{0}$. We consider an isolated equilibrium state $\mathbf{x}_e$ and let $\Delta\mathbf{x}(t)$ be perturbed state variables defined by

$$\Delta\mathbf{x}(t) = \mathbf{x}(t) - \mathbf{x}_e \tag{2.23a}$$

i.e.,

$$\mathbf{x}(t) = \mathbf{x}_e + \Delta\mathbf{x}(t) \tag{2.23b}$$

Assuming that $\mathbf{f}(\mathbf{x}, \mathbf{u})$ are continuously differentiable with respect to $\mathbf{x}$ and $\mathbf{u}$, we employ Taylor series expansion about the equilibrium state. Let $\mathbf{A}$ and $\mathbf{B}$ denote the $n \times n$ and $n \times m$ Jacobian matrices defined, respectively, by

$$\mathbf{A} = \left(\frac{\partial \mathbf{f}}{\partial \mathbf{x}}\right)_e = \begin{bmatrix} \dfrac{\partial f_1}{\partial x_1} & \cdots & \dfrac{\partial f_1}{\partial x_n} \\ \vdots & \vdots\vdots\vdots & \vdots \\ \dfrac{\partial f_n}{\partial x_1} & \cdots & \dfrac{\partial f_n}{\partial x_n} \end{bmatrix}_e \tag{2.24}$$

$$\mathbf{B} = \left(\frac{\partial \mathbf{f}}{\partial \mathbf{u}}\right)_e = \begin{bmatrix} \dfrac{\partial f_1}{\partial u_1} & \cdots & \dfrac{\partial f_1}{\partial u_m} \\ \vdots & \vdots\vdots\vdots & \vdots \\ \dfrac{\partial f_n}{\partial u_1} & \cdots & \dfrac{\partial f_n}{\partial u_m} \end{bmatrix}_e \tag{2.25}$$

where the subscript e denotes that the partial derivatives are evaluated at the equilibrium values $(\mathbf{x}_e, \mathbf{u}_e)$. Here, an element a_{ij} of matrix $\mathbf{A}$ is given by $(\partial f_i / \partial x_j)_e$. Taylor series expansion then yields

$$\dot{\mathbf{x}}_e + \Delta\dot{\mathbf{x}} = \mathbf{f}(\mathbf{x}_e, \mathbf{u}_e) + \mathbf{A}\Delta\mathbf{x} + \mathbf{B}\Delta\mathbf{u} + \mathbf{h}(\Delta\mathbf{x}, \Delta\mathbf{u}) \tag{2.26}$$

where $\mathbf{h}(\Delta\mathbf{x}, \Delta\mathbf{u})$ contains terms of second or higher order in $\Delta\mathbf{x}$ and $\Delta\mathbf{u}$. Dropping these second- or higher-order terms for small perturbations about an equilibrium and noting that $\dot{\mathbf{x}}_e = \mathbf{f}(\mathbf{x}_e, \mathbf{u}_e) = \mathbf{0}$ for an equilibrium, the linear approximation of Eq. (2.26) is given by

$$\Delta\dot{\mathbf{x}} = \mathbf{A}\Delta\mathbf{x} + \mathbf{B}\Delta\mathbf{u} \tag{2.27}$$

For simplicity of notation, we replace $\Delta\mathbf{x}$ by $\mathbf{x}$ and $\Delta\mathbf{u}$ by $\mathbf{u}$, where $\mathbf{x}$ and $\mathbf{u}$ now denote deviations from an equilibrium and represent Eq. (2.27) by

$$\dot{\mathbf{x}} = \mathbf{A}\mathbf{x} + \mathbf{B}\mathbf{u} \tag{2.28}$$

It should be noted that $\mathbf{A}$ and $\mathbf{B}$ are constant matrices when the nonlinear equation, Eq. (2.22), is time-invariant and the linearization is about an equilibrium. It is obvious that the coefficients of matrices $\mathbf{A}$ and $\mathbf{B}$ depend on the particular equilibrium chosen for linearization.

2.3 MECHANICAL ELEMENTS

The solid bodies that we consider are assumed to be either rigid bodies or particles. This assumption results in lumped-parameter models of dynamics. A body is called rigid if its deformation is negligible when loaded by forces and moments. In general, a rigid body translates and rotates. A particle is defined as a body of any size or shape that only translates without rotation. When a rigid body only translates without rotation, all points of the body have the same velocity and acceleration at any instant of time. Hence, a particle may be considered as a point mass.

The dynamics of solid bodies is important in control engineering because the system to be controlled may be the motion of a missile, vehicle, or robot. In other cases, mechanical elements such as linkages, springs, and dampers are employed as part of the control system. The two methods employed for obtaining the equations of motion are based on the direct application of Newton's second law and Lagrange equations in generalized coordinates, respectively. A detailed coverage of advanced dynamics is beyond our scope and interested readers are referred to the specialized books on the subject such as reference D'Souza and Garg, 1984. Our purpose here is to employ the concepts from an introductory course in engineering mechanics to model simple mechanical systems as a combination of mass or inertia, spring, and damper elements.

An unconstrained rigid body has six degrees of freedom. Three equations may be chosen to represent the translation of the mass center, and three equations for the rotation about the mass center. Let m be the mass of the body, $\sum \mathbf{F}$ the resultant of the external forces acting on the body, and $\sum \mathbf{M}_c$ the resultant moment of external forces and couples about the mass center C. The equations of motion for a rigid body may be written by direct application of Newton's second law as

$$\begin{aligned} \sum \mathbf{F} &= \frac{d}{dt}(m\mathbf{v}_c) \\ &= m\frac{d}{dt}\mathbf{v}_c \\ &= m\mathbf{a}_c \end{aligned} \tag{2.29}$$

$$\sum \mathbf{M}_c = \frac{d}{dt}\mathbf{H}_c \tag{2.30}$$

where $\mathbf{v}_c$ and $\mathbf{a}_c$ are the velocity and acceleration of the mass center C with respect to inertial coordinate system (i.e., the acceleration is absolute), and $\mathbf{H}_c$ is the angular momentum vector. In case the body has a point O fixed in space, then the subscript c in Eq. (2.30) may be replaced by 0 as an alternative. An unconstrained particle has three degrees of freedom and only Eq. (2.29) is sufficient to describe its translation where the subscript c may now be dropped since all points of the body have the same velocity and acceleration. The rotation of a rigid body about an axis fixed in space is a special case. If a coordinate axis is aligned with the axis of rotation, the vector equation, Eq. (2.30), for this special case becomes a scalar

$$\sum M_c = I\frac{d\omega}{dt} \tag{2.31}$$

where I is the mass moment of inertia of the body about the fixed axis of rotation, and ω is the angular velocity.

In passive mechanical networks, the three basic elements are the (1) mass, (2) spring, and (3) damper for translation, and the (1) mass moment of inertia, (2) torsional spring, and (3) rotary damper for rotation. The through variable, which is being transmitted, is the force for translation or moment for rotation. The across variable, which can be measured across the two terminals, is the velocity for translation or angular velocity for rotation. The product of through and across variables is the power. The mass or inertia stores kinetic energy, the spring stores potential energy, whereas the damper dissipates energy.

EXAMPLE 2.1

A single-degree-of-freedom translational system consisting of a mass, spring, and damper (i.e., friction) is shown in Fig. 2.5 where $0x$ is an inertial coordinate with origin at 0. Let $F(t)$ be the applied force and $x(t)$ the position of the mass from its equilibrium value. A freebody diagram of the system is shown in Fig. 2.6, where F_d and F_s are the damper and spring forces, respectively.

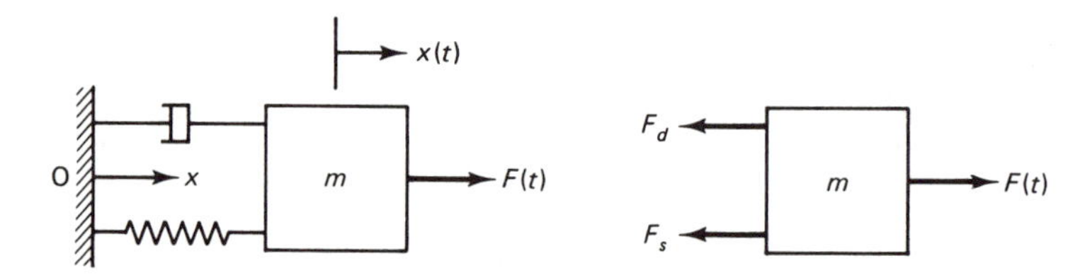

Figure 2.5 Mass, spring, and damper.

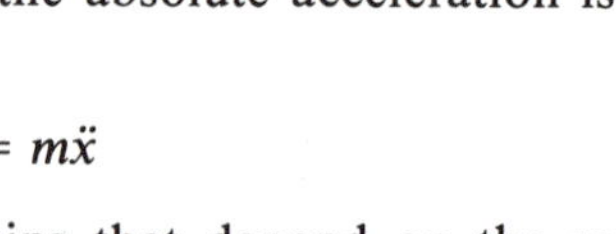

Figure 2.6 Freebody diagram.

Since the coordinate axis is inertial, the absolute acceleration is given by $a = \ddot{x}$ and Eq. (2.29) yields

$$F - F_d - F_s = m\ddot{x} \tag{2.32}$$

We now need constitutive relationships that depend on the material properties to express F_d and F_s as functions of x and its derivative. The friction force is a function of $\dot{x}$. Coulomb friction is shown in Fig. 2.7(a), nonlinear friction force $F_d = c\dot{x}^3$ in Fig. 2.7(b), and linear friction force $F_d = c\dot{x}$ in Fig. 2.7(c). Coulomb friction is not an analytic function of $\dot{x}$ and cannot be linearized even for small deviation from equilibrium. Linear friction of Fig. 2.7(c), which is also called viscous friction, is employed here and is represented by a damper with coefficient c.

The spring force for a "hard" spring is shown in Fig. 2.8(a), for a "soft" spring in Fig. 2.8(b), and for a linear spring in Fig. 2.8(c). Linearizing the hard or soft spring force for small deviations from equilibrium or using the

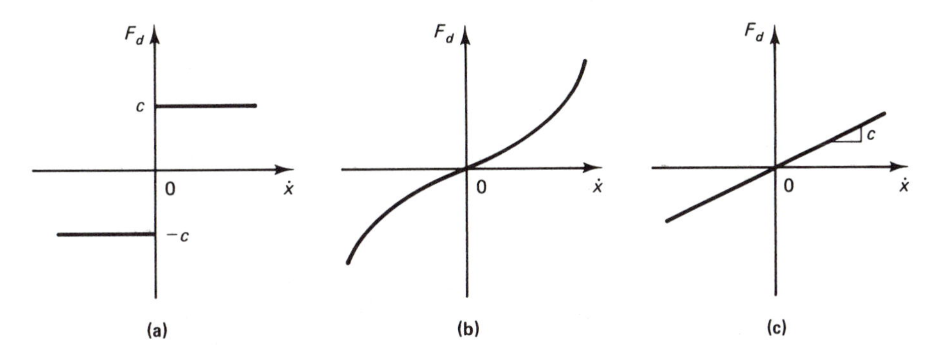

Figure 2.7 (a) Coulomb friction, (b) nonlinear friction, and (c) linear friction.

linear spring, we obtain $F_s = kx$, where k is the spring constant. Hence, Eq. (2.32) now becomes

$$m\ddot{x} + c\dot{x} + kx = F \tag{2.33}$$

which, using the operator notation $D = d/dt$, can be written as

$$(mD^2 + cD + k)x(t) = F(t) \tag{2.34}$$

This equation, which is quadratic in D, can be expressed in the standard form

$$k\left(\frac{1}{\omega_n^2}D^2 + \frac{2\zeta}{\omega_n}D + 1\right)x = F \tag{2.35}$$

where the natural frequency ω_n and damping ratio ζ are defined as

$$\omega_n = \sqrt{\frac{k}{m}} \qquad \text{and} \qquad \zeta = \frac{1}{2}\frac{c}{\sqrt{mk}}$$

where ω_n has the dimensions of rad/s and ζ is dimensionless.

In addition to Eq. (2.35) being in the standard form for a quadratic equation, another reason for employing this form is that it is convenient, as seen in later chapters, to experimentally identify the values of the natural

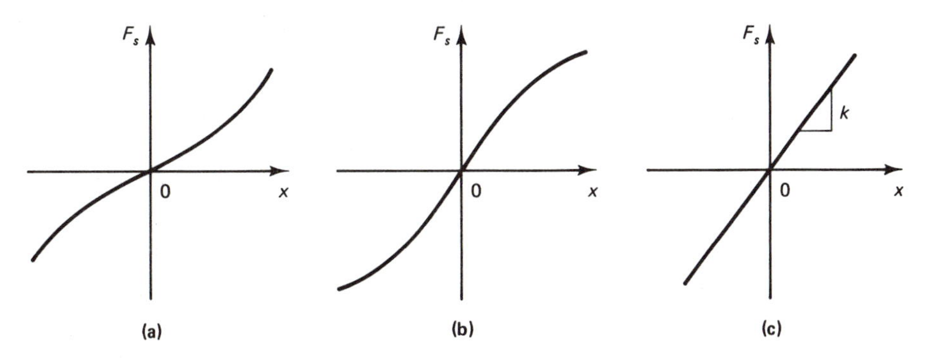

Figure 2.8 (a) Hard spring, (b) soft spring, and (c) linear spring.

frequency and damping ratio rather than the values of the physical parameters. It should be noted that the mass, spring, and damper may represent a conceptual model of a more complicated system, such as a liquid sloshing in a tank. Laplace transformation of Eq. (2.35) with initial conditions $x(0) = 0$ and $\dot{x}(0) = 0$ yields

$$k\left(\frac{1}{\omega_n^2}s^2 + \frac{2\zeta}{\omega_n}s + 1\right)X(s) = F(s) \tag{2.36}$$

which is represented in the block diagram of Fig. 2.9, where $F(s)$ is considered as the input or cause and $X(s)$ as the output or effect. For a steady-state relationship between F and x, we set $D = 0$ in Eq. (2.35) and obtain $x = (1/k)F$. For this reason, the numerator constant $1/k$ is also called the steady-state gain k_g.

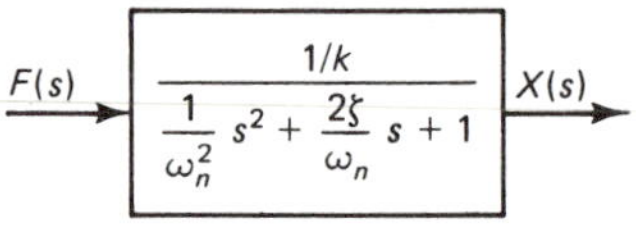

$$F(s) \rightarrow \boxed{\frac{1/k}{\frac{1}{\omega_n^2}s^2 + \frac{2\zeta}{\omega_n}s + 1}} \rightarrow X(s)$$

Figure 2.9 Block diagram representation.

Since Eq. (2.33) is of second order, we need two state variables to represent it in the form of state equations. Choosing x and $\dot{x}$ as state variables, we let $x_1 = x$ and $x_2 = \dot{x}$. Then Eq. (2.35) can be represented as

$$\begin{aligned} \dot{x}_1 &= x_2 \\ \dot{x}_2 &= -\omega_n^2 x_1 - 2\zeta\omega_n x_2 + \frac{1}{m}F \end{aligned} \tag{2.37}$$

where the first equation is obtained from the definition of the state variables and the second from the equation of motion, Eq. (2.35). The **A** and **B** matrices of Eq. (2.9) can be obtained from inspection of Eq. (2.37) as

$$\mathbf{A} = \begin{bmatrix} 0 & 1 \\ -\omega_n^2 & -2\zeta\omega_n \end{bmatrix} \qquad \mathbf{B} = \begin{bmatrix} 0 \\ 1/m \end{bmatrix} \tag{2.38}$$

EXAMPLE 2.2

A single-degree-of-freedom rotational system is shown in Fig. 2.10(a). A disk of mass moment of inertia I about the x-axis is supported by a bearing and a shaft whose one end is fixed. The bearing and other friction are represented by a rotary viscous damper with coefficient c. Within the elastic limit, the shaft behaves as a linear torsional spring. The torsional spring constant for a shaft whose one end is twisted with respect to the other is obtained from strength of materials as

$$k = \frac{\pi d^4 G}{32L}$$

where G is the modulus of elasticity in shear, d the shaft diameter, and L is its length. The conceptual model is shown in Fig. 2.10(b).

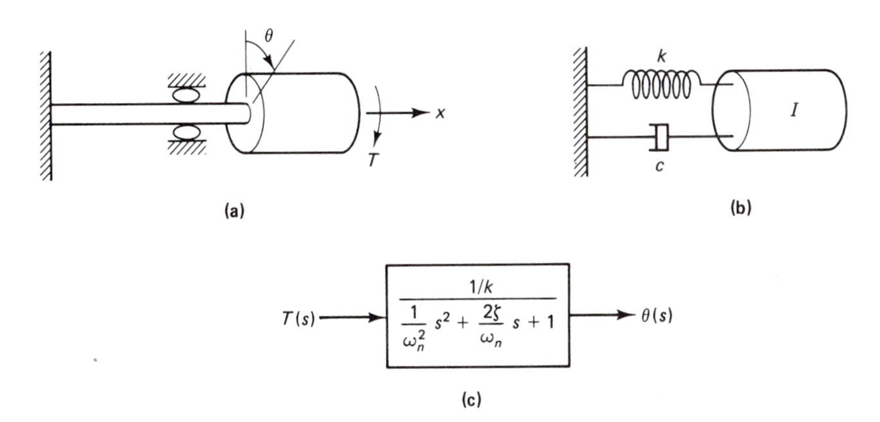

$$T(s) \longrightarrow \boxed{\frac{1/k}{\frac{1}{\omega_n^2}s^2 + \frac{2\zeta}{\omega_n}s + 1}} \longrightarrow \theta(s)$$

(c)

Figure 2.10 Torsional inertia-springer-damper system.

The damper torque T_d and spring torque T_s oppose the applied torque $T(t)$. Letting $\theta(t)$ be the angular displacement of the disk from the equilibrium position about the x-axis, application of Eq. (2.31) yields

$$T - T_d - T_s = I\ddot{\theta} \tag{2.39}$$

Since $T_d = c\dot{\theta}$ and $T_s = k\theta$, we obtain

$$I\ddot{\theta} + c\dot{\theta} + k\theta = T \tag{2.40}$$

Denoting the natural frequency and damping ratio by

$$\omega_n = \sqrt{\frac{k}{I}} \qquad \text{and} \qquad \zeta = \frac{1}{2}\frac{c}{\sqrt{Ik}}$$

we represent Eq. (2.40) as

$$k\left(\frac{1}{\omega_n^2}D^2 + \frac{2\zeta}{\omega_n}D + 1\right)\theta(t) = T(t) \tag{2.41}$$

The transfer function representation is shown in Fig. 2.10(c), whose similarity to Fig. 2.9 is obvious. To represent Eq. (2.40), which is of second order, in the form of state equations, we let $x_1 = \theta$ and $x_2 = \dot{\theta}$ and obtain

$$\begin{aligned} \dot{x}_1 &= x_2 \\ \dot{x}_2 &= -\omega_n^2 x_1 - 2\zeta\omega_n x_2 + \frac{1}{I}T \end{aligned} \tag{2.42}$$

The **A** and **B** matrices are obtained from inspection of Eq. (2.42) as

$$\mathbf{A} = \begin{bmatrix} 0 & 1 \\ -\omega_n^2 & -2\zeta\omega_n \end{bmatrix} \qquad \mathbf{B} = \begin{bmatrix} 0 \\ 1/I \end{bmatrix}$$

Gearing is employed in many rotational systems. In Fig. 2.11(a), a motor which produces a torque T drives a load through gearing with gear ratio $\dot{\theta}_2/\dot{\theta}_1 = n$. Let I_1 be the mass moment of inertia of the motor, c_1 the coefficient

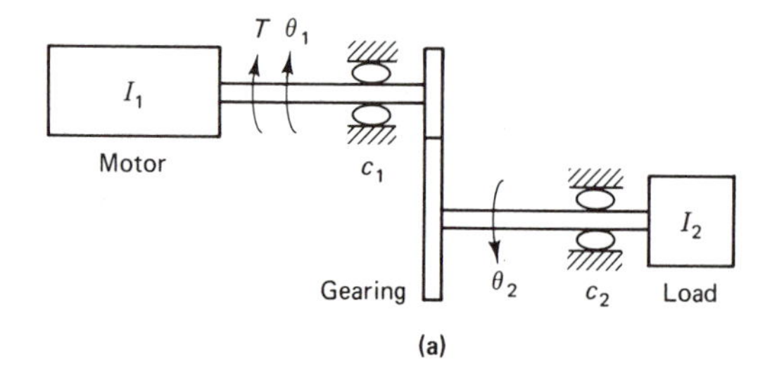

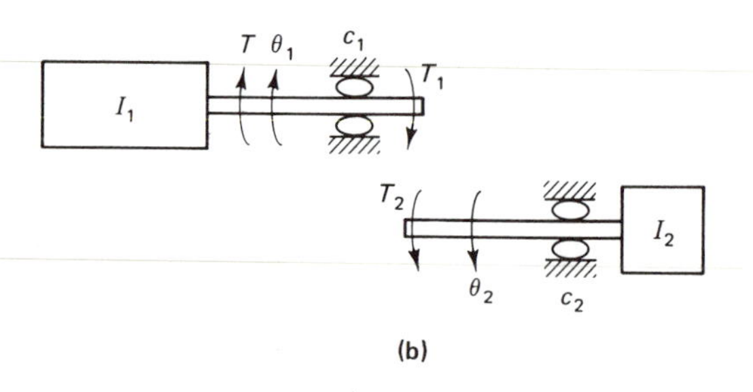

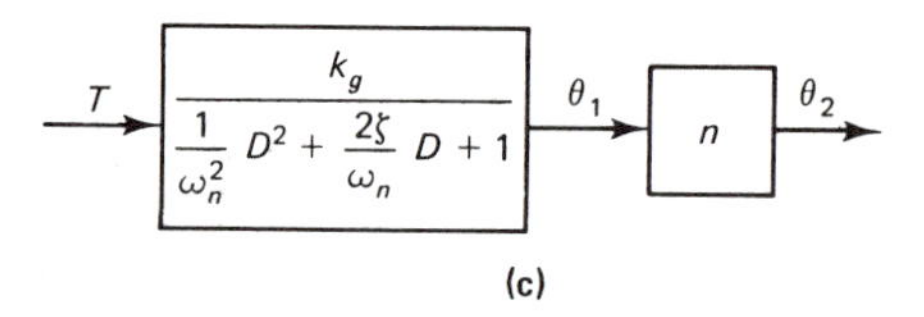

Figure 2.11 (a) Motor driving a load through gearing. (b) Freebody diagram of motor and load. (c) Block diagram.

of damping, and k_1 the torsional stiffness of the motor shaft. Let I_2, c_2, and k_2 be the corresponding parameters of the load shaft. A freebody diagram is shown in Fig. 2.11(b). Balancing the torques on the motor and load shafts, respectively, we get

$$I_1\ddot{\theta}_1 + c_1\dot{\theta}_1 + k_1\theta_1 = T - T_1$$

$$I_2\ddot{\theta}_2 + c_2\dot{\theta}_2 + k_2\theta_2 = T_2$$

Equating the power transmitted by the gearing, we obtain

$$T_1\dot{\theta}_1 = T_2\dot{\theta}_2 \qquad \text{or} \qquad T_1/T_2 = \dot{\theta}_2/\dot{\theta}_1 = n$$

Hence, it follows that

$$\begin{aligned} I_1\ddot{\theta}_1 + c_1\dot{\theta}_1 + k_1\theta_1 &= T - nT_2 \\ &= T - n(I_2\ddot{\theta}_2 + c_2\dot{\theta}_2 + k_2\theta_2) \end{aligned}$$

Substitution of $\theta_2 = n\theta_1$ in this equation yields

$$(I_1 + n^2I_2)\ddot{\theta}_1 + (c_1 + n^2c_2)\dot{\theta}_1 + (k_1 + n^2k_2)\theta_1 = T$$

It is noted that the load inertia I_2 is reflected on the motor shaft as n^2I_2. Similar statements can also be made about the load damping coefficient and the stiffness parameter.

Defining the natural frequency, damping ratio, and gain as

$$\omega_n^2 = \frac{k_1 + n^2 k_2}{I_1 + n^2 I_2} \qquad 2\zeta\omega_n = \frac{c_1 + n^2 c_2}{I_1 + n^2 I_2} \qquad k_g = \frac{1}{k_1 + n^2 k_2}$$

the block diagram representation is shown in Fig. 2.11(c). The state equations can now be obtained as in (2.42).

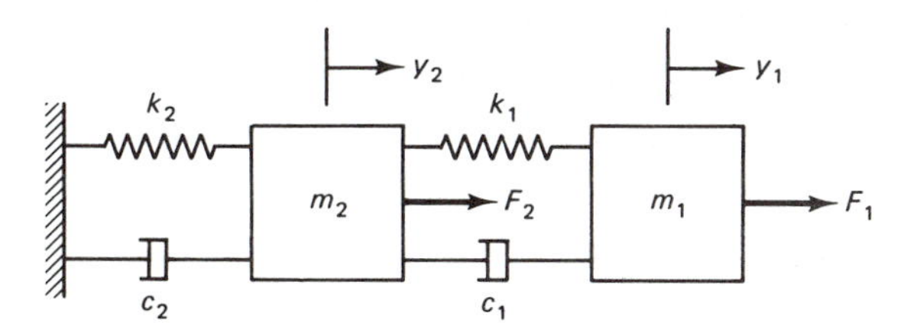

Figure 2.12 A two-degrees-of-freedom system.

EXAMPLE 2.3

A two-degrees-of-freedom translational system is shown in Fig. 2.12. It is an example of a multivariable system where the forces F_1 and F_2 are considered as two inputs and the displacements y_1 and y_2 as two outputs. The freebody diagram of the system is shown in Fig. 2.13.

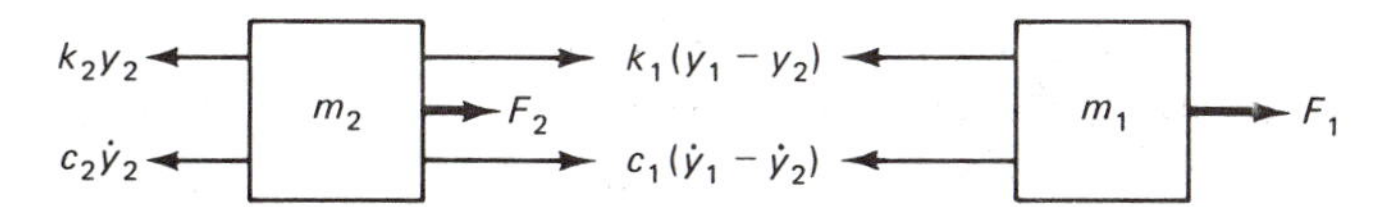

Figure 2.13 Freebody diagram of the system of Fig. 2.12.

Using Eq. (2.29), we obtain the two equations of motion

$$F_1 - c_1(\dot{y}_1 - \dot{y}_2) - k_1(y_1 - y_2) = m_1\ddot{y}_1 \tag{2.43}$$

and

$$F_2 + c_1(\dot{y}_1 - \dot{y}_2) + k_1(y_1 - y_2) - c_2\dot{y}_2 - k_2y_2 = m_2\ddot{y}_2 \tag{2.44}$$

Rearranging the terms, we may express the preceding equations in matrix notation as

$$\begin{bmatrix} m_1D^2 + c_1D + k_1 & -c_1D - k_1 \\ -c_1D - k_1 & m_2D^2 + (c_1 + c_2)D + (k_1 + k_2) \end{bmatrix} \begin{bmatrix} y_1 \\ y_2 \end{bmatrix} = \begin{bmatrix} F_1 \\ F_2 \end{bmatrix} \tag{2.45}$$

To obtain the transfer-function matrix relating the two outputs to the two inputs, we invert the matrix on the left-hand side of Eq. (2.45). Letting $\Delta(D)$ be its determinant, we obtain

$$\begin{bmatrix} y_1 \\ y_2 \end{bmatrix} = \begin{bmatrix} \dfrac{m_2D^2 + (c_1 + c_2)D + (k_1 + k_2)}{\Delta} & \dfrac{c_1D + k_1}{\Delta} \\ \dfrac{c_1D + k_1}{\Delta} & \dfrac{m_1D^2 + c_1D + k_1}{\Delta} \end{bmatrix} \begin{bmatrix} F_1 \\ F_2 \end{bmatrix} \tag{2.46}$$

where

$$\Delta(D) = m_1 m_2 D^4 + (m_1 c_1 + m_1 c_2 + m_2 c_1) D^3 + (m_1 k_1 + m_1 k_2 + m_2 k_1 + c_1 c_2) D^2 + (c_1 k_2 + c_2 k_1) D + k_1 k_2 \tag{2.47}$$

The 2×2 transfer function matrix is shown in block diagram form in Fig. 2.14. Since $\Delta(D)$ of Eq. (2.47) is of fourth order, we can define two natural frequencies and two damping ratios and express them as

$$\Delta(s) = k_1 k_2 \left(\frac{1}{\omega_{n_1}^2} s^2 + \frac{2\zeta_1}{\omega_{n_1}} s + 1 \right) \left(\frac{1}{\omega_{n_2}^2} s^2 + \frac{2\zeta_2}{\omega_{n_2}} s + 1 \right) \tag{2.48}$$

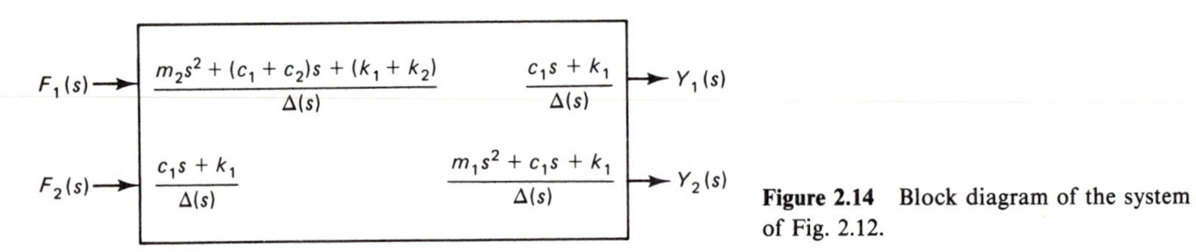

Figure 2.14 Block diagram of the system of Fig. 2.12.

For feedback control of displacement y_1, it is measured by a sensor, compared to its command input, and the error is used by the control law to synthesize the force F_1 by an actuator to correct the error. Similarly, F_2 is synthesized to control y_2. However, as seen in Fig. 2.14, the transfer-function matrix is not diagonal and F_1 affects not only y_1 but also y_2. Similarly, F_2 affects not only y_2 but also y_1. Due to this interaction, we cannot treat the controller design as that of two single-input, single-output systems, but a multivariable design is required to account for the coupling. Multivariable systems are encountered in many applications such as robotics and process control.

For the state-variables representation of the system, we note that it is of fourth order and choose four state variables as $x_1 = y_1$, $x_2 = y_2$, $x_3 = \dot{y}_1$, and $x_4 = \dot{y}_2$. The state equations are then given by

$$\begin{aligned}
\dot{x}_1 &= x_3 \\
\dot{x}_2 &= x_4 \\
\dot{x}_3 &= -\frac{k_1}{m_1} x_1 + \frac{k_1}{m_1} x_2 - \frac{c_1}{m_1} x_3 + \frac{c_1}{m_1} x_4 + \frac{1}{m_1} F_1 \\
\dot{x}_4 &= \left(\frac{k_1}{m_2}\right) x_1 - \left(\frac{k_1 + k_2}{m_2}\right) x_2 + \left(\frac{c_1}{m_2}\right) x_3 - \left(\frac{c_1 + c_2}{m_2}\right) x_4 + \frac{1}{m_2} F_2
\end{aligned} \tag{2.49}$$

The first two equations are obtained from the definition of the state variables, the third is obtained from Eq. (2.43), and the fourth from Eq. (2.44). On comparing Eq. (2.49) to the standard form $\dot{\mathbf{x}} = \mathbf{Ax} + \mathbf{Bu}$, $\mathbf{y} = \mathbf{Cx}$, the matrices **A**, **B**, and **C** are obtained from Eq. (2.49) as

$$\mathbf{A} = \begin{bmatrix} 0 & 0 & 1 & 0 \\ 0 & 0 & 0 & 1 \\ -\dfrac{k_1}{m_1} & \dfrac{k_1}{m_1} & -\dfrac{c_1}{m_1} & \dfrac{c_1}{m_1} \\ \dfrac{k_1}{m_2} & -\dfrac{k_1 + k_2}{m_2} & \dfrac{c_1}{m_2} & -\dfrac{c_1 + c_2}{m_2} \end{bmatrix}$$

$$\mathbf{B} = \begin{bmatrix} 0 & 0 \\ 0 & 0 \\ 1/m_1 & 0 \\ 0 & 1/m_2 \end{bmatrix} \qquad \mathbf{C} = \begin{bmatrix} 1 & 0 & 0 & 0 \\ 0 & 1 & 0 & 0 \end{bmatrix}$$

where $u_1 = F_1$ and $u_2 = F_2$. If we obtain the characteristic equation of matrix **A**, it will be seen that $\det|s\mathbf{I} - \mathbf{A}| = \Delta(s)$, where $\Delta(s)$ is given by Eq. (2.47) with D replaced by s.

2.4 ELECTRICAL ELEMENTS

We are concerned here with those electrical elements used primarily for measurement and control. The inductance, resistance, and capacitance are the three basic elements of passive electrical networks. The across variable, which is measured across the terminals of an element, is the voltage $E(t)$ and the through variable, which flows through the element, is the current $i(t)$. The product of the across and through variables is the power. The inductance and capacitance store energy, whereas the resistance dissipates energy. The mathematical models for electrical circuits are obtained by the application of Kirchhoff's laws.

Kirchhoff's current law states that the sum of the currents at a node in a circuit is zero and is a statement of conservation of charge. Kirchhoff's voltage law states that the instantaneous sum of the voltages around any loop in a circuit is zero and is an expression of conservation of energy. Hence, we have

$$\sum i_i(t) = 0 \qquad \text{at a node}$$

$$\sum E_n(t) = 0 \qquad \text{around a loop}$$

In addition to Kirchhoff's laws, we need constitutive relationships for the across and through variables of an element. An inductance L, resistance R, and capacitance C are shown in Fig. 2.15(a), (b), and (c), respectively, where i is the

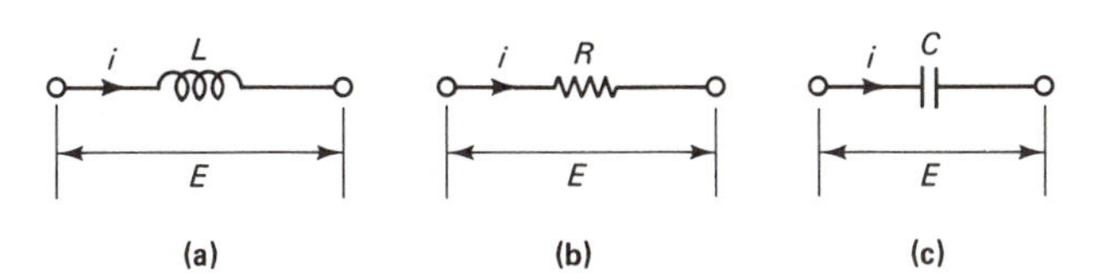

Figure 2.15 (a) Inductance, (b) resistance, and (c) capacitance.

current through the element and E the voltage across it. The relationships between i and E are given in the following.

Inductance: Henry's law $$E = L\frac{di}{dt} \tag{2.50}$$

Resistance: Ohm's law $$E = Ri \tag{2.51}$$

Capacitance: Faraday's law $$E = \frac{1}{C}\int i\,dt \tag{2.52}$$

Hence, the relationships that we employ are linear and L, R, and C are assumed to be constants. The ratio E/i is called the impedance and is denoted by Z. Hence, the impedances* Z_L, Z_R, and Z_C of inductance, resistance, and capacitance, respectively, are given by

$$Z_L = LD \qquad Z_R = R \qquad Z_C = 1/CD$$

There are several methods of employing Kirchhoff's laws to obtain transfer functions or state equations for circuits. These include the impedance method, the branch current method, the loop current method, and the node-pair voltage method. Here, we employ only the impedance method. Electrical elements are said to be in series if the same current flows through each element and the total voltage is the sum of the individual voltages. Electrical elements are said to be in parallel if the same voltage acts across each element and the total current is the sum of the individual currents. Using Kirchhoff's laws, we can verify that the total impedance Z_s of n elements in series is given by

$$Z_s = Z_1 + Z_2 + \cdots + Z_n \tag{2.53}$$

and the total impedance Z_p of n elements in parallel by

$$Z_p = \frac{1}{\dfrac{1}{Z_1} + \dfrac{1}{Z_2} + \cdots \dfrac{1}{Z_n}} \tag{2.54}$$

EXAMPLE 2.4

An electrical circuit is shown in Fig. 2.16. Obtain the transfer function and state equations relating the output voltage $E_o(t)$ to the input voltage $E_1(t)$.

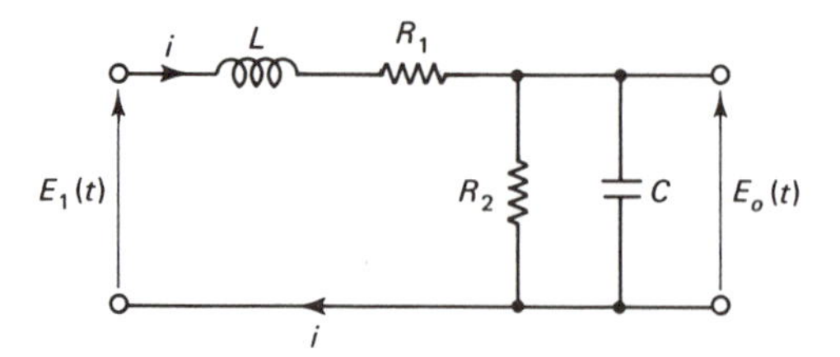

Figure 2.16 Electrical network.

It is assumed that if the terminals across which $E_o(t)$ is measured are connected to any device, then the device draws negligible current. Noting that

*In the frequency domain, covered in Chapter 5, operator D is replaced by $j\omega$.

R_2 and C are in parallel, their combined impedance is given by

$$Z_p = \frac{1}{1/R_2 + CD} = \frac{R_2}{R_2CD + 1}$$

This combination is in series with L and R_1. Hence,

$$Z_s = LD + R_1 + \frac{R_2}{R_2CD + 1}$$

Now,

$$\begin{aligned} E_o &= Z_p i \\ &= \left(\frac{R_2}{R_2CD + 1}\right) i \end{aligned} \tag{2.55}$$

and

$$\begin{aligned} E_1 &= Z_s i \\ &= \left(LD + R_1 + \frac{R_2}{R_2CD + 1}\right) i \end{aligned} \tag{2.56}$$

From Eqs. (2.55) and (2.56), we obtain

$$\frac{E_o}{E_1} = \frac{R_2}{LR_2CD^2 + (R_1R_2C + L)D + (R_1 + R_2)} \tag{2.57}$$

We define the natural frequency, damping ratio, and gain as

$$\omega_n = \left(\frac{R_1 + R_2}{LR_2C}\right)^{1/2} \qquad 2\zeta\omega_n = \frac{R_1R_2C + L}{LR_2C}$$

i.e.,

$$\zeta = \frac{1}{2}\left(\frac{R_1R_2C + L}{[LR_2C(R_1 + R_2)]^{1/2}}\right)$$

and

$$k_g = \frac{R_2}{R_1 + R_2}$$

Equation (2.57) may now be expressed as

$$\frac{E_o}{E_1} = \frac{k_g}{\dfrac{1}{\omega_n^2} D^2 + \dfrac{2\zeta}{\omega_n} D + 1} \tag{2.58}$$

and its block diagram is shown in Fig. 2.17.

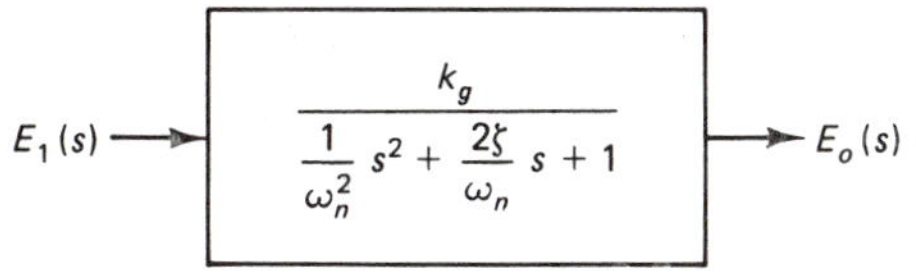

$$E_1(s) \longrightarrow \boxed{\frac{k_g}{\dfrac{1}{\omega_n^2} s^2 + \dfrac{2\zeta}{\omega_n} s + 1}} \longrightarrow E_o(s)$$

Figure 2.17 Block diagram of the circuit of Fig. 2.16.

Since Eq. (2.58) is of second order, we need two state variables. We choose $x_1 = E_o$ and $x_2 = i$ as the state variables. From Eq. (2.55), we obtain

$$(R_2CD + 1)E_o = R_2 i \tag{2.59}$$

and from Eqs. (2.55) and (2.56),

$$E_1 = (LD + R_1)i + E_o \tag{2.60}$$

Now, Eqs. (2.59) and (2.60) are expressed as

$$\dot{x}_1 = -\left(\frac{1}{R_2C}\right)x_1 + \left(\frac{1}{C}\right)x_2$$

$$\dot{x}_2 = -\left(\frac{1}{L}\right)x_1 - \left(\frac{R_1}{L}\right)x_2 + \left(\frac{1}{L}\right)E_1$$

$$y = E_0$$

The **A**, **B**, and **C** matrices of the state-variables formulation can be obtained from the preceding equations as

$$\mathbf{A} = \begin{bmatrix} -1/R_2C & 1/C \\ -1/L & -R_1/L \end{bmatrix} \qquad \mathbf{B} = \begin{bmatrix} 0 \\ 1/L \end{bmatrix} \qquad \mathbf{C} = \lfloor 1 \quad 0 \rfloor$$

It can be verified that the transfer function of Fig. 2.17 can be obtained by employing these matrices in Eq. (2.14).

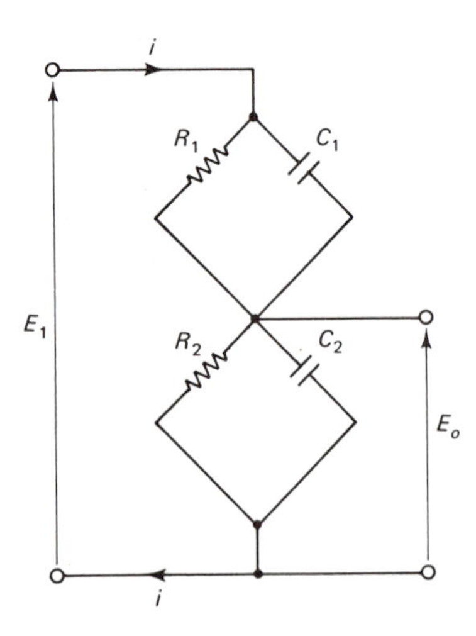

Figure 2.18 Electrical network.

EXAMPLE 2.5

An electrical circuit is shown in Fig. 2.18. Obtain the transfer function relating the output voltage $E_o(t)$ to the input voltage $E_1(t)$.

Since R_1 and C_1 are in parallel, their combined impedance is given by

$$Z_1 = \frac{1}{1/R_1 + C_1D} = \frac{R_1}{R_1C_1D + 1} \tag{2.61}$$

Similarly, since R_2 and C_2 are in parallel, we obtain

$$Z_2 = \frac{1}{1/R_2 + C_2D} = \frac{R_2}{R_2C_2D + 1} \tag{2.62}$$

Now Z_1 and Z_2 are in series. Hence, we get

$$E_1 = (Z_1 + Z_2)i$$

and

$$E_o = Z_2 i$$

It follows that

$$\frac{E_o}{E_1} = \frac{Z_2}{Z_1 + Z_2} \tag{2.63}$$

Substituting for Z_1 and Z_2 in Eq. (2.63) from Eqs. (2.61) and (2.62), respectively, and simplifying the resulting expression, we obtain

$$\frac{E_o}{E_1} = \frac{R_2(R_1C_1D + 1)}{(R_1R_2C_1 + R_1R_2C_2)D + R_1 + R_2} \tag{2.64}$$

We define two time constants and gain as

$$\tau_1 = R_1C_1 \qquad \tau_2 = \frac{R_1R_2C_1 + R_1R_2C_2}{R_1 + R_2} \qquad k_g = \frac{R_2}{R_1 + R_2}$$

and express Eq. (2.64) as

$$\frac{E_o}{E_1} = \frac{k_g(\tau_1 D + 1)}{(\tau_2 D + 1)} \tag{2.65}$$

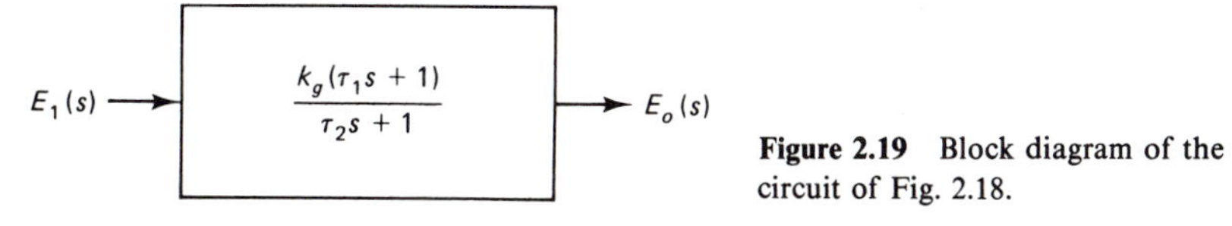

Figure 2.19 Block diagram of the circuit of Fig. 2.18.

The block diagram of the transfer function is shown in Fig. 2.19. It should be noted that a time constant has the dimension of time.

2.5 HYDRAULIC ELEMENTS

The three basic elements of a hydraulic network are the hydraulic resistance, inertance, and capacitance. The inertance (Shearer, Murphy, and Richardson, 1967) represents fluid inertia and is derived from the inertia forces required to accelerate a fluid in a pipe or passage. The across variable, which is measured across an element, is the pressure $p(t)$ and the through variable, which flows through the element, is the volumetric flow rate $q(t)$. The product of the across and through variables is the hydraulic power. The hydraulic inertance and capacitance store energy, whereas resistance dissipates energy.

The equations for a hydraulic network are obtained by using the principles of conservation of mass and momentum, i.e., continuity and force balance. The relationships relating the across and through variables are nonlinear. In the following, we linearize the equations for deviations from the steady-state operating conditions to obtain the hydraulic resistance, inertance, and capacitance.

2.5.1 Hydraulic Resistance

Hydraulic energy is dissipated by a restriction, such as a valve, orifice, or obstruction, by wall friction in the internal flow in a pipe or passage, and by drag in the external flow over a surface. Figure 2.20 shows an obstruction where $(p_1 - p_2)$ is the pressure

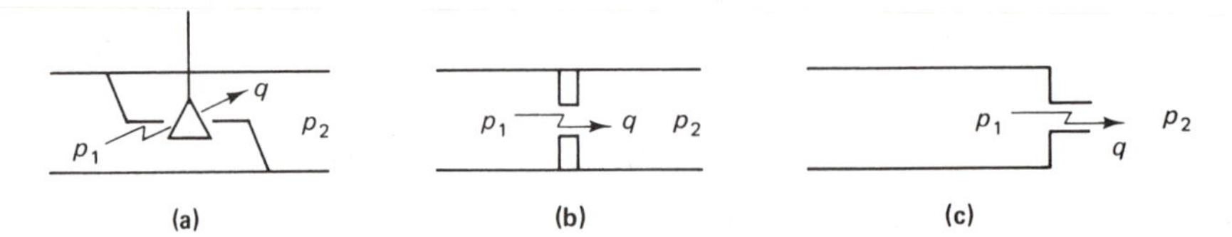

Figure 2.20 Flow obstruction.

drop caused by resistance. We can evaluate the resistance either experimentally or analytically. Suppose that an experimental curve relating the steady-state flow q to the pressure drop is available for the obstruction as shown in Fig. 2.21.

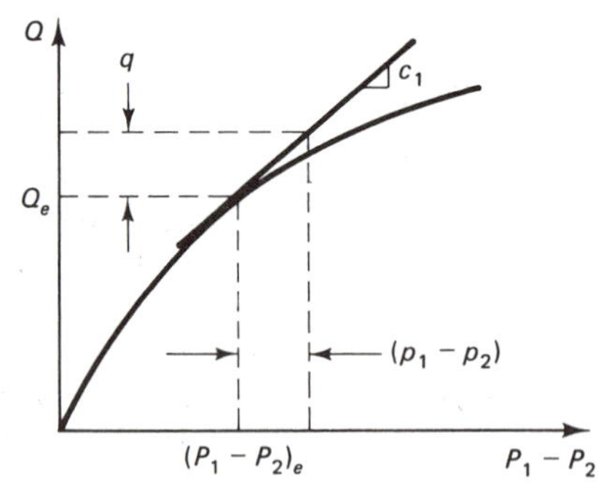

Figure 2.21 Steady-state operating curve.

Following the approach of section 2.2.3, let the slope of the curve at the equilibrium point be denoted by c_1 as shown in Fig. 2.21. Then, for a small deviation from equilibrium, we obtain

$$q = c_1(p_1 - p_2)$$

which may be expressed as

$$p_1 - p_2 = \frac{1}{c_1} q \tag{2.66}$$

On comparing this relationship between the through and across variables to Ohm's law for an electrical resistance, we define the hydraulic resistance as $R = 1/c_1$. Its value depends on the steady-state condition chosen for linearization. Where an equation relating the flow to the pressure drop is available, then the resistance can be obtained analytically. For many valves and orifices, we have

$$Q = c\sqrt{P_1 - P_2}$$

where c is the product of the coefficient of discharge and the area. Expanding this expression in a Taylor series about an equilibrium, we get

$$Q_e + q = c\sqrt{(P_1 - P_2)_e} + \left.\frac{\partial Q}{\partial(P_1 - P_2)}\right|_e (p_1 - p_2) + \left.\frac{\partial^2 Q}{\partial^2(P_1 - P_2)}\right|_e \frac{(p_1 - p_2)^2}{2!} + \cdots$$

Dropping terms of second and higher order for small deviations and noting that the first terms on both sides of the preceding equation are equal, we obtain

$$q = \left(\frac{c}{2(P_1 - P_2)_e^{1/2}}\right)(p_1 - p_2)$$

Hence, the hydraulic resistance is given by

$$R = \frac{2}{c}(P_1 - P_2)_e^{1/2} \tag{2.67}$$

It is difficult to obtain an analytic expression for the resistance due to wall friction in internal or external flow. Consider the steady-state laminar flow of an incompressible fluid in a rigid pipe of radius a and length L as shown in Fig. 2.22.

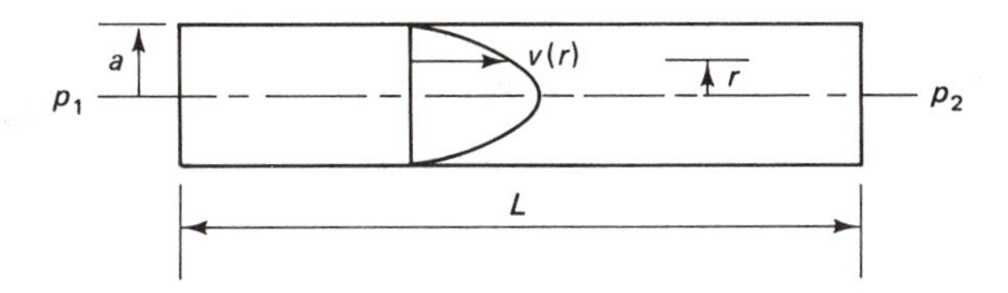

Figure 2.22 Steady-state laminar flow in a rigid pipe.

It is well known (Fox and McDonald, 1978) that after the entrance length, the velocity is independent of the axial direction and is a parabolic function of the radial coordinate r. It can be shown (Fox and McDonald, 1978) that

$$v(r) = \frac{1}{4\mu}(a^2 - r^2)\left(\frac{p_1 - p_2}{L}\right) \tag{2.68}$$

where μ is the fluid viscosity. The average velocity is obtained as

$$\begin{aligned}\bar{v} &= \frac{1}{\pi a^2}\int_0^a v(r)2\pi r\,dr \\ &= \frac{a^2}{8\mu}\left(\frac{p_1 - p_2}{L}\right)\end{aligned} \tag{2.69}$$

Since the volumetric flow rate $q = \pi a^2 \bar{v}$, we obtain

$$p_1 - p_2 = \left(\frac{8\mu L}{\pi a^4}\right) q \tag{2.70}$$

Hence, the hydraulic resistance is $R = 8\mu L/\pi a^4$. However, this expression is not useful to us for several reasons. In transient flow, the velocity profile is not parabolic and the total pressure drop is not caused by resistance alone. In our approach, we employ the hydraulic resistance R and combine it with the other physical parameters to define the standard parameters, which are natural frequency, damping ratio, time constant, and gain. The values of the standard parameters are then identified as discussed in Chapters 4 and 5.

2.5.2 Hydraulic Inertance

The hydraulic inertance is derived from the inertia force required to accelerate a column of fluid in a pipe or passage. Let $(p_1 - p_2)_I$ be the pressure drop required to accelerate a column of fluid, A the cross-sectional area of the pipe, m the fluid mass, and v the fluid velocity. Application of Newton's second law yields

$$A(p_1 - p_2)_I = m\frac{dv}{dt} \tag{2.71}$$

Now, mass $m = \rho LA$, where ρ is the density, L is the pipe length, and $A\dot{v} = \dot{q}$. Hence, from Eq. (2.71), we obtain

$$(p_1 - p_2)_I = \frac{\rho L}{A}\frac{dq}{dt} \tag{2.72}$$

This equation is analogous to Henry's law that relates the across and through variables for an inductance. The hydraulic inertance I is defined as $I = \rho L/A$.

2.5.3 Hydraulic Capacitance

In unsteady flow, the difference between the inflow and outflow is stored. In liquids, there exist the following three storage effects: storage in a gravity field, storage due to compressibility of the liquid, and storage due to the change in volume of the container. First, we consider storage in a gravity field. In Fig. 2.23, let q_1 and q_2 denote the inflow and outflow, respectively, in a tank of cross-sectional area A. Let h be the head of the liquid. It is assumed that the liquid is incompressible and the tank walls are rigid. From the continuity equation, we obtain $q_1 - q_2 = A(dh/dt)$.

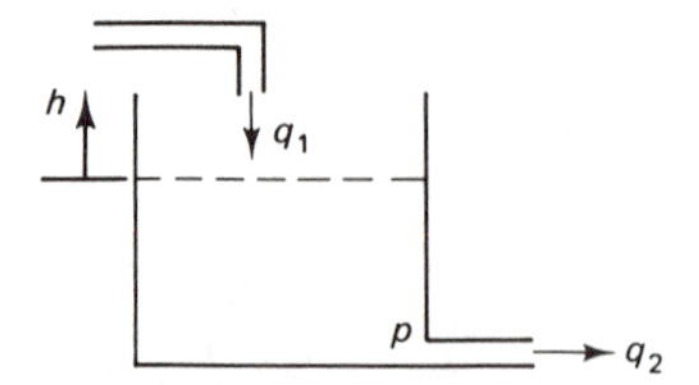

Figure 2.23 Liquid-storage tank.

The head h is related to the pressure p by the relationship $p = \rho g h$, where ρ is the liquid density, and g is the constant of gravity. Hence, it follows that

$$q_1 - q_2 = \frac{A}{\rho g}\frac{dp}{dt}$$

i.e.,

$$p = \frac{1}{A/\rho g}\int (q_1 - q_2)\, dt \tag{2.73}$$

This equation is analogous to Faraday's law relating the across and through variables for a capacitance. The hydraulic capacitance due to storage in a gravity field is defined by $C_1 = A/\rho g$.

Now, we consider jointly the other two storage effects. In Fig. 2.24, let $\dot{m}_1$ and $\dot{m}_2$ be the mass flow rate in and out of a container, respectively, of volume V,

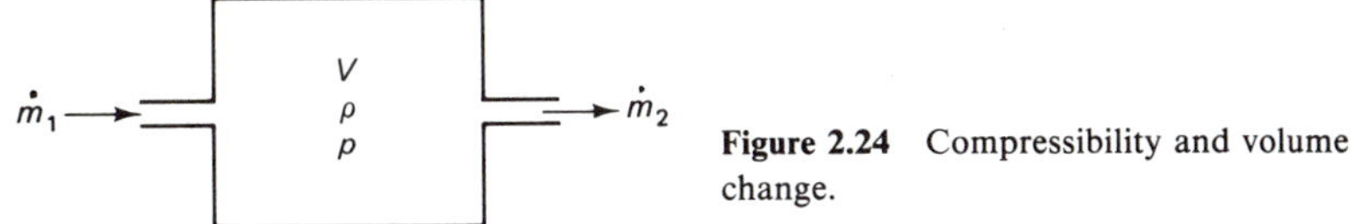

Figure 2.24 Compressibility and volume change.

and ρ be the density of the liquid inside. From the principle of conservation of mass, we get

$$\begin{aligned} \dot{m}_1 - \dot{m}_2 &= \frac{d}{dt}(\rho V) \\ &= \rho\frac{dV}{dt} + V\frac{d\rho}{dt} \end{aligned} \tag{2.74}$$

This equation is linearized by using the mean values ρ_0 and V_0 of the density and volume, respectively. Then, Eq. (2.74) is expressed as

$$\rho_0(q_1 - q_2) = \rho_0\frac{dV}{dp}\frac{dp}{dt} + V_0\frac{d\rho}{dt} \tag{2.75}$$

From the equation of state for a liquid, we have $\Delta\rho/\rho_0 = \Delta p/K$, where K is the liquid bulk modulus. Hence, Eq. (2.75) becomes

$$q_1 - q_2 = \frac{dV}{dp}\frac{dp}{dt} + \frac{V_0}{K}\frac{dp}{dt} \tag{2.76}$$

The hydraulic capacitance due to the change in volume of the container is defined as $C_2 = dV/dp$, and the capacitance due to the compressibility of the liquid by $C_3 = V_0/K$. For example, pressure in a pipe causes hoop stress and strain from which we can evaluate C_2. The liquid compressibility effect for pressures (20.7×10^6 Pa or 3000 psi) used in hydraulic control systems is usually negligible.

Including the hydraulic inertance, resistance, and capacitance effects and letting $(p_1 - p_2)$ be the total pressure drop, we obtain

$$I\frac{dq}{dt} + Rq + \frac{1}{C}\int q\,dt = p_1 - p_2 \tag{2.77}$$

EXAMPLE 2.6

For the U-tube shown in Fig. 2.25, obtain the transfer function and state equations relating the gage pressure $p(t)$ to the displacement $h(t)$. Let ρ be the liquid density, L the length of the liquid column, and A the cross-sectional area of the tube.

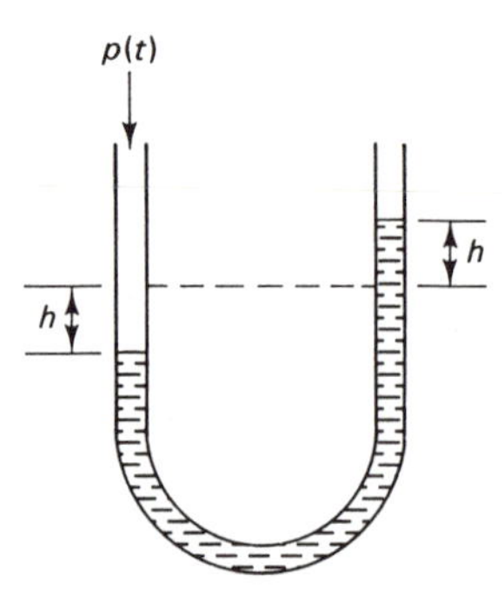

Figure 2.25 U-tube.

It is assumed that the liquid is incompressible and that the tube is rigid. Hence, the only hydraulic capacitance is that due to storage in the gravity field. Using Eq. (2.77), we get

$$\frac{\rho L}{A}\frac{dq}{dt} + Rq + \frac{2}{A/\rho g}\int q\,dt = p \tag{2.78}$$

The coefficient 2 in Eq. (2.78) is clarified, if we express the equation as

$$\frac{\rho L}{A}\frac{dq}{dt} + Rq = p - 2\rho g h$$

$$= p - \frac{2\rho g}{A}\int q\,dt$$

since $q = A(dh/dt)$. Expressing q in terms of h in Eq. (2.78), we get

$$(\rho L D^2 + ARD + 2\rho g)h = p \tag{2.79}$$

Defining the natural frequency and damping ratio as

$$\omega_n = \sqrt{2g/L} \qquad 2\zeta\omega_n = AR/\rho L$$

Eq. (2.79) is expressed as

$$2\rho g\left(\frac{1}{\omega_n^2}D^2 + \frac{2\zeta}{\omega_n}D + 1\right)h = p \tag{2.80}$$

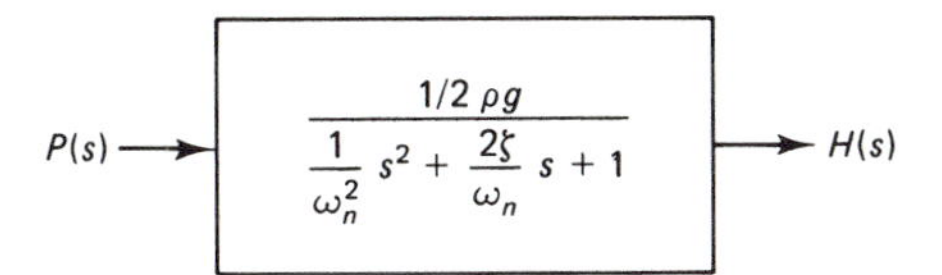

Figure 2.26 Transfer function.

The transfer function is shown in Fig. 2.26. The gain $k_g = 1/(2\rho g)$, the natural frequency, and the damping ratio can now be identified as discussed in Chapters 4 and 5. Choosing $x_1 = h$ and $x_2 = \dot{h}$ as the state variables, the state-variables representation of Eq. (2.80) is obtained as

$$\begin{aligned} \dot{x}_1 &= x_2 \\ \dot{x}_2 &= -\omega_n^2 x_1 - 2\zeta\omega_n x_2 + (1/\rho L)p \\ y &= x_1 \end{aligned} \tag{2.81}$$

EXAMPLE 2.7

For the tank shown in Fig. 2.27, obtain the transfer function relating the deviation in head $h(t)$ as output to the deviation in inflow q_1 as input.

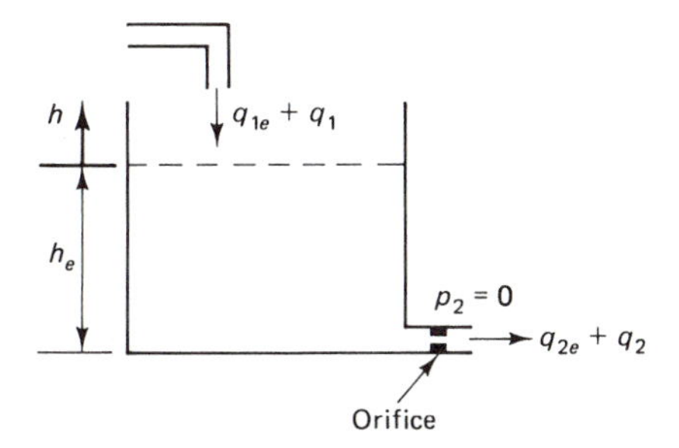

Figure 2.27 Tank and orifice.

For steady-state equilibrium flow, we have $q_{1e} = q_{2e}$ = a constant and also h_e = a constant. Now, q_1, q_2, and h are deviations about this equilibrium. It is assumed that the liquid is incompressible and the tank walls are rigid. Hence, the only capacitance effect is due to storage in the gravity field. From Eq. (2.73), we obtain

$$q_1 - q_2 = \frac{A}{\rho g}\frac{dp}{dt} \tag{2.82}$$

where $p = \rho g h$ is the change in the gage pressure at the tank bottom. Neglecting the fluid inertance, the orifice equation is described by

$$p = Rq_2 \tag{2.83}$$

Eliminating q_2 from Eqs. (2.82) and (2.83), we obtain

$$\frac{A}{\rho g}\frac{dp}{dt} + \frac{1}{R}p = q_1$$

Defining a time constant $\tau = AR/\rho g$, we get

$$\frac{1}{R}(\tau D + 1)p = q_1$$

Since $p = \rho gh$, it follows that

$$\frac{\rho g}{R}(\tau D + 1)h = q_1 \tag{2.84}$$

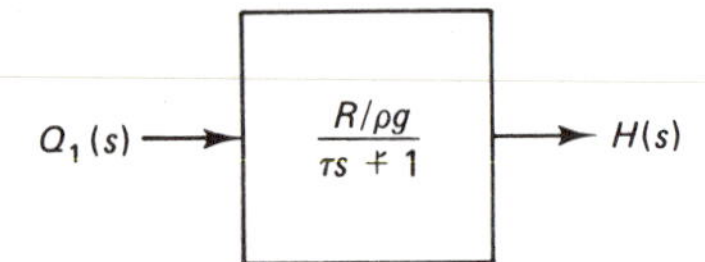

Figure 2.28 Block diagram.

This transfer function is shown in the block diagram of Fig. 2.28. The state-variable representation of Eq. (2.84) is obtained by choosing $x = h$ as the state variable. Hence, we get

$$\dot{x} = -\frac{1}{\tau}x + \frac{R}{\rho g}q_1 \tag{2.85}$$

2.6 PNEUMATIC ELEMENTS

Pneumatic resistance, inertance, and capacitance are the three basic elements of pneumatic networks. But compressible gas flow requires careful consideration as the flow may be subsonic, sonic, or supersonic. Pressure is used as the across variable but mass flow rate should be used as the through variable as the mass and volume flow rates are not readily interchangeable as in liquids, where the compressibility effects are small. However, the product of pressure and mass flow rate is not power, but it is the product of density and power. In the following, we develop expressions for pneumatic resistance, inertance, and capacitance.

2.6.1 Pneumatic Resistance

The mass flow rate through a valve, orifice, or restriction is a complicated function of the pressure drop. Consider the flow through a restriction shown in Fig. 2.29, where $p_1 - p_2$ is the pressure drop. Figure 2.30 is drawn for a constant value of p_1. When $p_2 = p_1$, there is no flow through the orifice. For a constant value of p_1, the

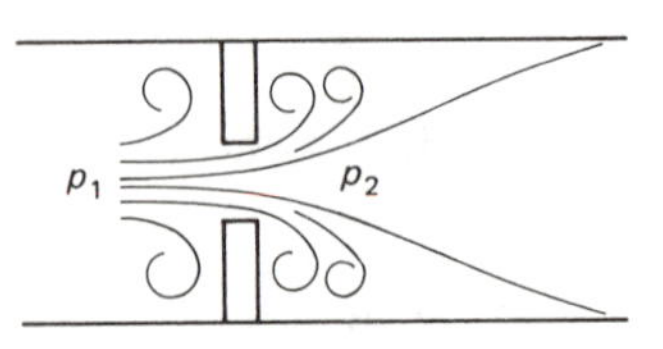

Figure 2.29 Flow through an orifice.

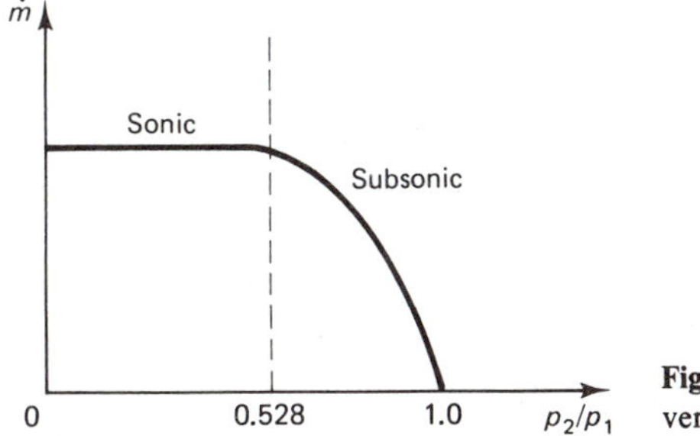

Figure 2.30 Mass flow rate versus p_2.

flow increases as p_2 decreases until it reaches a critical value p_c at which sonic velocity occurs at the throat. For air, where the specific heat ratio $\gamma = 1.4$, it can be shown (John, 1969) that $p_c = 0.528p_1$. When $p_2 < p_c$, there is no further increase in the mass flow rate (Fig. 2.30). This condition is called choking.

It can be shown (John, 1969) that for subsonic flow ($p_2 > p_c$), the mass flow rate is given by

$$\dot{m} = C_d A \left[\frac{2}{R_g T_1} p_2(p_1 - p_2) \right]^{1/2} \tag{2.86}$$

where C_d is the coefficient of discharge, A is the throat area, R_g is the gas constant, and T_1 is the absolute temperature upstream of the orifice. For the case where $p_2 < p_c$, the mass flow rate is independent of p_2 and is given by

$$\dot{m} = C_d A p_1 \left[\frac{\gamma}{R_g T_1} \left(\frac{2}{\gamma + 1} \right)^{(\gamma+1)/(\gamma-1)} \right]^{1/2} \tag{2.87}$$

In typical pneumatic control systems, low-pressure air is employed, the pressure drops are small, and the flow is usually subsonic. Under these conditions, we let $\dot{m}$, p_1, and p_2 be the deviations from their steady-state equilibrium values of $\dot{m}_e$, p_{1e}, and p_{2e}, respectively. Expanding Eq. (2.86) in a Taylor series about the equilibrium and retaining only the first-order term, we define pneumatic resistance R as

$$\dot{m} = \frac{1}{R}(p_1 - p_2) \tag{2.88}$$

where R has the units of $1/\text{m}\cdot\text{s}$ in SI units.

We also use Eq. (2.88) to represent the pressure drop caused by wall friction in internal flow. There is no need to evaluate R since the values of the standard parameters are identified. Use of Eq. (2.88) requires caution since it implies that there is no choking.

2.6.2 Pneumatic Inertance

Let $(p_1 - p_2)_I$ be the pressure drop required to accelerate a column of gas of mass m in a pipe of cross-sectional area A. Application of Newton's second law yields

$$A(p_1 - p_2)_I = \frac{d}{dt}(mv)$$

Since $m = \rho LA$ and $v = q/A$, we obtain

$$\begin{aligned}(p_1 - p_2)_I &= \frac{L}{A}\frac{d}{dt}(\rho q) \\ &= \frac{L}{A}\frac{d}{dt}\dot{m}\end{aligned} \tag{2.89}$$

The pneumatic inertance is defined as $I = L/A$ and has the units of 1/m in SI units. The pneumatic inertance is usually negligible for short lengths of tubing.

2.6.3 Pneumatic Capacitance

The difference between the mass flow rate flowing in and that flowing out is stored. The storage is due to the compressibility of the gas and due to the change in volume of the container. In pneumatics, storage in a gravity field is absent. In Fig. 2.24, let $\dot{m}_1$ and $\dot{m}_2$ be the mass flow rate in and out of a container of volume V and ρ be the density of the gas. From Eq. (2.74), we have

$$\begin{aligned}\dot{m}_1 - \dot{m}_2 &= \frac{d}{dt}(\rho V) \\ &= \rho\frac{dV}{dt} + V\frac{d\rho}{dt}\end{aligned} \tag{2.90}$$

Now, $dV/dt = (dV/dp)(dp/dt)$ and assuming air to be a perfect gas, $p = \rho R_g T$, where T is the absolute temperature. For an isothermal process, T is a constant and from Eq. (2.90), we obtain

$$\dot{m}_1 - \dot{m}_2 = \left(\rho\frac{dV}{dp}\right)\frac{dp}{dt} + \frac{V}{R_g T}\frac{dp}{dt} \tag{2.91}$$

The pneumatic capacitance due to a change in volume of the container is defined as $C_1 = \rho(dV/dp)$ and the capacitance due to the compressibility of the gas as $C_2 = V/R_gT$. For small deviations from equilibrium, C_1 and C_2 may be treated as constants by using average values of ρ and V. Pneumatic capacitance has units of m·s^2 in SI units.

EXAMPLE 2.8

A pneumatic bellows is shown in Fig. 2.31. It is an expandable chamber, where the elasticity of the walls is represented by a spring with spring constant

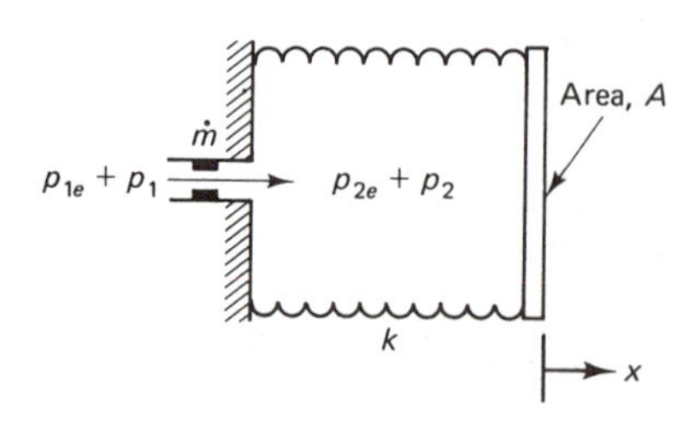

Figure 2.31 Pneumatic bellows.

k. Obtain the transfer function relating the change in pressure p_1 as input to the displacement from equilibrium x as output.

Let p_2 be the pressure inside the bellows. For equilibrium, $p_{1e} = p_{2e}$. Let $p_1(t)$, $p_2(t)$, and $x(t)$ denote small changes from equilibrium. Neglecting the pneumatic inertance and assuming no choking, the mass flow rate flowing into the bellows is given by

$$\dot{m}_1 = \frac{p_1 - p_2}{R} \tag{2.92}$$

where R is the pneumatic resistance. Since the mass flow rate flowing out $\dot{m}_2 = 0$, $\dot{m}_1$ is stored and from Eq. (2.91), we obtain

$$\dot{m}_1 = C\frac{dp_2}{dt} \tag{2.93}$$

where the pneumatic capacitance $C = C_1 + C_2$ and $C_1 = \rho(dV/dp_2)$ and $C_2 = V/R_gT$. Eliminating $\dot{m}_1$ from Eqs. (2.92) and (2.93), we get

$$(RCD + 1)p_2 = p_1 \tag{2.94}$$

Assuming that the bellows expands or contracts slowly, we neglect the product of mass and acceleration and employ the force balance to obtain

$$Ap_2 = kx \tag{2.95}$$

From Eqs. (2.94) and (2.95), it follows that

$$\frac{k}{A}(RCD + 1)x = p_1 \tag{2.96}$$

For this example, we can obtain an analytic expression for the capacitance C_1 due to the change in volume of the bellows as

$$C_1 = \rho\frac{dV}{dp_2} = \rho\frac{d(Ax)}{dp_2}$$

$$= \rho A^2/k$$

where we have used Eq. (2.95). Defining a time constant $\tau = RC$ and gain $k_g = A/k$, Eq. (2.96) is expressed as

$$(\tau D + 1)x = k_g p_1 \tag{2.97}$$

The block diagram is shown in Fig. 2.32. For this first-order system, we choose x as the state variable and obtain the state equation from Eq. (2.97) as

$$\dot{x} = -\frac{1}{\tau}x + \frac{k_g}{\tau}p_1 \tag{2.98}$$

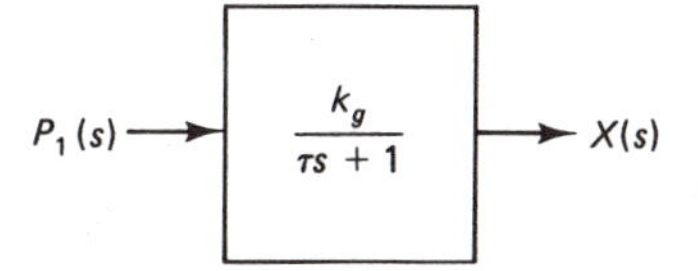

Figure 2.32 Block diagram.

The numerical values of parameters τ and k_g can be obtained as discussed in Chapters 4 and 5.

EXAMPLE 2.9

Various pneumatic devices are commonly used to perform dynamic operations on a pressure signal. A pneumatic system, where the two bellows are identical but the orifice resistances R_1 and R_2 are unequal, is shown in Fig. 2.33. The variables p_i, p_1, p_2, and y are deviations from their equilibrium values. In equilibrium, $p_{ie} = p_{1e} = p_{2e}$ and $y = 0$. The expansion or contraction of each bellow is resisted only by a spring with spring constant k. Obtain a linear model relating output y to the input p_i.

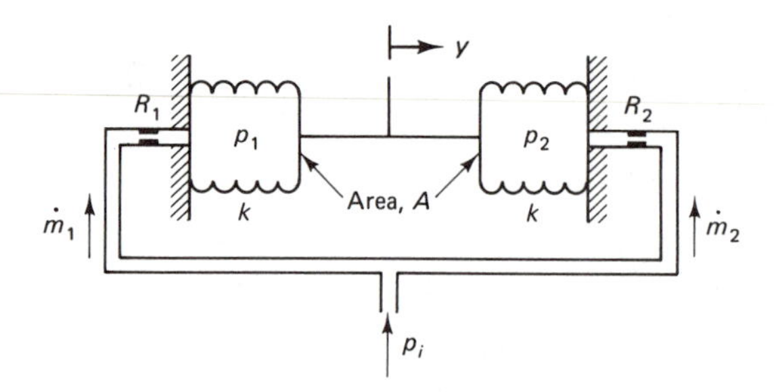

Figure 2.33 Pneumatic device.

For the first bellows, the mass flow rate as in Eqs. (2.92) and (2.93) is given by

$$\dot{m}_1 = \frac{p_i - p_1}{R_1} \tag{2.99}$$

$$= C\frac{dp_1}{dt} \tag{2.100}$$

where the pneumatic capacitance includes the effects of the compressibility of air and the change in volume of the bellows. From Example 2.8, it is seen that $C = \rho A^2/k + V/R_gT$. Eliminating $\dot{m}_1$ from Eqs. (2.99) and (2.100), we obtain

$$(R_1CD + 1)p_1 = p_i \tag{2.101}$$

Similarly, for the second bellows, we get

$$(R_2CD + 1)p_2 = p_i \tag{2.102}$$

Force balance yields

$$(p_1 - p_2)A = 2ky \tag{2.103}$$

Substituting for p_1 and p_2 in Eq. (2.103) from Eqs. (2.101) and (2.102), respectively,

$$A\left(\frac{1}{R_1CD + 1} - \frac{1}{R_2CD + 1}\right)p_i = 2ky$$

Simplifying this expression, we obtain the transfer function

$$y = \left(\frac{\frac{A}{2k}(R_2 - R_1)CD}{(R_1CD + 1)(R_2CD + 1)} \right) p_i \tag{2.104}$$

Defining two time constants and a gain as $\tau_1 = R_1C$, $\tau_2 = R_2C$, and $k_g = (A/2k)(R_2 - R_1)C$, Eq. (2.104) can be expressed as

$$Y(s) = \left(\frac{k_g s}{(\tau_1 s + 1)(\tau_2 s + 1)} \right) P_i(s) \tag{2.105}$$

We note that if $R_1 = R_2$, or p_i is a constant (i.e., $Dp_i = 0$), then $y = 0$. For this second-order system, we choose $x_1 = p_1$ and $x_2 = p_2$ as state variables. Then, from Eqs. (2.101) to (2.103), the state-variables representation is given by

$$\begin{aligned} \dot{x}_1 &= -\frac{1}{\tau_1} x_1 + \frac{1}{\tau_1} p_i \\ \dot{x}_2 &= -\frac{1}{\tau_2} x_2 + \frac{1}{\tau_2} p_i \\ y &= \frac{A}{2k}(x_1 - x_2) \end{aligned} \tag{2.106}$$

The **A**, **B**, and **C** matrices are obtained from inspection of the preceding equations as

$$\mathbf{A} = \begin{bmatrix} -1/\tau_1 & 0 \\ 0 & -1/\tau_2 \end{bmatrix} \qquad \mathbf{B} = \begin{bmatrix} 1/\tau_1 \\ 1/\tau_2 \end{bmatrix} \qquad \mathbf{C} = \lfloor A/2k \quad -A/2k \rfloor$$

2.7 THERMAL ELEMENTS

We encounter thermal elements in connection with the system to be controlled, such as those found in chemical processes, power plants, and heating/air conditioning of buildings. Thermal resistance and capacitance are the two basic elements of a thermal network. The element corresponding to mass, inductance, or inertance is not meaningful. The across variable, which is measured across an element, is the temperature T, and the through variable, which flows through the element, is the heat flux (i.e., heat flow rate) q. However, the product Tq of across and through variables is not power. In the following, we develop expressions for thermal resistance and capacitance.

2.7.1 Thermal Resistance

The resistance that we have encountered so far has been an element that dissipates energy and converts it into heat. Thermal resistance is not an element that dissipates energy but it is a consequence of the fact that a temperature difference is required to cause heat to flow. There are three different thermal resistance effects corresponding to the three modes of heat transfer, namely, conduction, convection, and radiation. Thermal resistance is obtained from the constitutive relationship of the heat transfer mode.

Fourier's law for one-dimensional heat conduction is given by

$$q = Ak\left(\frac{T_1 - T_2}{L}\right) = \frac{T_1 - T_2}{L/Ak} \tag{2.107}$$

where A is the surface area, k is the thermal conductivity of the material, and L is the thickness. This relation is analogous to Ohm's law relating the across and through variables for a resistance. Hence, thermal resistance due to conduction is defined by $R = L/Ak$. Heat transfer by convection is described by Newton's law of cooling as

$$q = Ah(T_1 - T_2) = \frac{T_1 - T_2}{1/Ah} \tag{2.108}$$

where A is the surface area and h is the coefficient of heat transfer. The thermal resistance due to convection is defined from Eq. (2.108) as $R = 1/Ah$. The net heat flux by radiation between two bodies at absolute temperatures T_1 and T_2, respectively, is given by the Stefan-Boltzmann law as

$$q = \sigma F_e F_{12} A_1 (T_1^4 - T_2^4) \tag{2.109}$$

where σ is the Stefan-Boltzmann constant, F_e is the emissivity factor, F_{12} is the geometric view factor, and A_1 is the surface area of the first body. (Note that $F_{12}A_1 = F_{21}A_2$.) A book on heat transfer such as Kreith (1973) should be consulted for determination of F_e and F_{12}. To linearize Eq. (2.109), we express it as

$$q = \frac{1}{\left(\dfrac{T_1 - T_2}{\sigma F_e F_{12} A_1 (T_1^4 - T_2^4)}\right)} (T_1 - T_2)$$

The thermal resistance due to radiation is then defined by

$$R = \frac{T_1 - T_2}{\sigma F_e A_1 F_{12} (T_1^4 - T_2^4)} \tag{2.110}$$

where the right-hand side is evaluated at equilibrium conditions.

2.7.2 Thermal Capacitance

Let q_1 be the heat flux flowing into an element of a body and q_2 be the heat flux flowing out. The difference $(q_1 - q_2)$ is stored by the element in the form of internal energy and we have

$$q_1 - q_2 = cm\frac{dT}{dt} \tag{2.111}$$

where m is the mass of the element and c is the specific heat. This equation is analogous to Faraday's law, relating the through and across variables for a capacitance. Hence, the thermal capacitance is defined as $C = cm$.

2.7.3 Lumped-Parameter Model

To obtain a lumped-parameter model for transient heat transfer, we divide a body into compartments or lumps. The internal temperature gradient for each lump should be small so that we can represent its temperature by a spatial average, time-varying value. When both conduction and convection heat-transfer modes are present, a useful criterion for dividing a body into lumps is that the resistance due to conduction is much less than the resistance due to convection. That is,

$$L/Ak \ll 1/Ah \qquad \text{or} \qquad hL/k \ll 1$$

The dimensionless quantity hL/k is called the Biot number, and L is the characteristic length, such as the ratio of volume to the surface area. The usual criterion employed in dividing a body into lumps is that the characteristic length L must be such that the Biot number is less than 0.1 for each lump. The equation for each lump is obtained from the principle of conservation of energy.

EXAMPLE 2.10

Figure 2.34 shows a slab perfectly insulated on all faces except one which is exposed to a fluid at temperature $T_f(t)$. Obtain its linear mathematical model, considering $T_f(t)$ as input and the slab temperature as output.

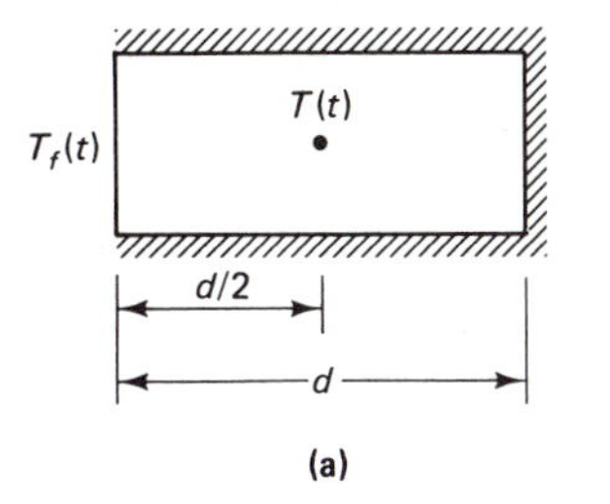

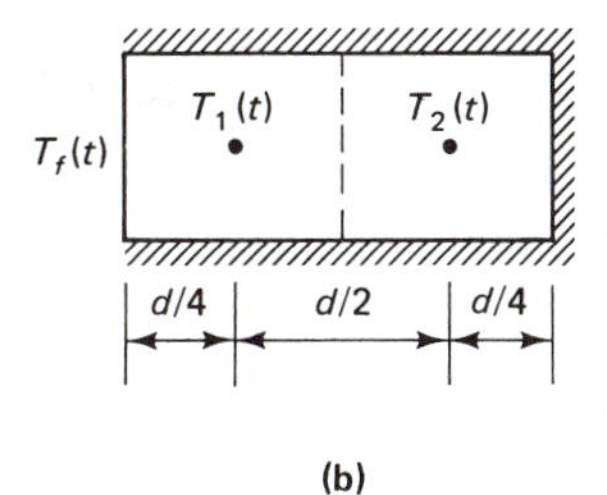

Figure 2.34 (a) One-lump model of slab, and (b) a two-lump model.

Considering the slab as one lump as in Fig. 2.34(a), we choose its center temperature $T(t)$ as the representative temperature. Letting d be the slab thickness, $L = d/2$ and the Biot number is $hd/2k$. In case it is less than 0.1, then one-lump representation is adequate. The heat flux flowing to the slab is given by

$$q = hA(T_f - T) = \frac{T_f - T}{R} \tag{2.112}$$

Since there is no heat flux flowing out, we get

$$q = cm\frac{dT}{dt} = C\frac{dT}{dt} \tag{2.113}$$

Eliminating q from Eqs. (2.112) and (2.113), we obtain

$$(RCD + 1)T = T_f$$

Defining a time constant as $\tau = RC$, the transfer function becomes

$$T(s) = \left(\frac{1}{\tau s + 1}\right) T_f \tag{2.114}$$

For this first-order system, we choose $x = T$ as the state variable and obtain the state equation as

$$\dot{x} = -\frac{1}{\tau} x + \frac{1}{\tau} T_f \tag{2.115}$$

In case $hd/2k$ is not less than 0.1, we divide the slab into two lumps as shown in Fig. 2.34(b). Now, if the new Biot number $hd/4k < 0.1$, then the two-lump representation is adequate. For the first lump, the heat flux flowing in is given by

$$q = hA(T_f - T_1) = \frac{T_f - T_1}{R_1}$$

and the heat flux flowing out by

$$q_0 = \frac{kA}{d/2}(T_1 - T_2) = \frac{T_1 - T_2}{R_2}$$

The heat balance for each lump then yields

$$C_1 \frac{dT_1}{dt} = \left(\frac{T_f - T_1}{R_1}\right) - \left(\frac{T_1 - T_2}{R_2}\right)$$

$$C_2 \frac{dT_2}{dt} = \frac{T_1 - T_2}{R_2} \tag{2.116}$$

Defining time constants $\tau_1 = R_1C_1$, $\tau_2 = R_2C_2$, and $\tau_3 = R_2C_1$, and choosing state variables as $x_1 = T_1$ and $x_2 = T_2$, the state equations are given by

$$\dot{x}_1 = -\left(\frac{1}{\tau_1} + \frac{1}{\tau_3}\right)x_1 + \frac{1}{\tau_3}x_2 + \frac{1}{\tau_1}T_f$$

$$\dot{x}_2 = \frac{1}{\tau_2}x_1 - \frac{1}{\tau_2}x_2 \tag{2.117}$$

EXAMPLE 2.11

The heating of a room is shown in Fig. 2.35. The heater supplies a heat flux q. It is assumed that the inside of the room can be lumped into one lump at temperature T_1 and the walls into another lump at temperature T_2. The outside temperature, which is a disturbance input, is the ambient temperature T_a. Develop a linear model, considering q as input and T_1 as output.

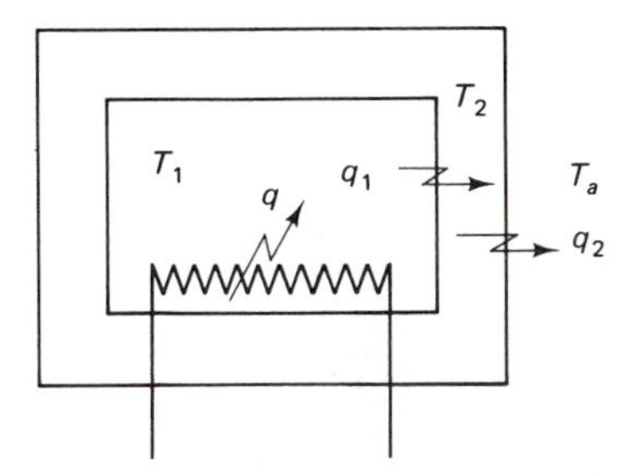

Figure 2.35 Heating a room.

Let h_1 and h_2 be the coefficients of convective heat transfer for the inside and the outside of the wall, respectively, and A_1 and A_2 be the corresponding areas. The heat flux from the room to the inside of the walls is given by

$$q_1 = h_1 A_1 (T_1 - T_2) = \frac{T_1 - T_2}{R_1}$$

and that to the outside by

$$q_2 = h_2 A_2 (T_2 - T_a) = \frac{T_2 - T_a}{R_2}$$

Conservation of energy for each of the two lumps yields

$$C_1 \frac{dT_1}{dt} = q - \left(\frac{T_1 - T_2}{R_1}\right) \tag{2.118}$$

$$C_2 \frac{dT_2}{dt} = \left(\frac{T_1 - T_2}{R_1}\right) - \left(\frac{T_2 - T_a}{R_2}\right) \tag{2.119}$$

Choosing $x_1 = T_1$ and $x_2 = T_2$ as the state variables, the state equations are obtained as

$$\dot{x}_1 = -\left(\frac{1}{R_1 C_1}\right) x_1 + \left(\frac{1}{R_1 C_1}\right) x_2 + \frac{1}{C_1} q$$

$$\dot{x}_2 = \left(\frac{1}{R_1 C_2}\right) x_1 - \left(\frac{1}{R_1} + \frac{1}{R_2}\right) \frac{1}{C_2} x_2 + \left(\frac{1}{R_2 C_2}\right) T_a \tag{2.120}$$

$$y = T_1$$

The **A**, **B**, and **C** matrices are given by

$$\mathbf{A} = \begin{bmatrix} -\dfrac{1}{R_1 C_1} & \dfrac{1}{R_1 C_1} \\ \dfrac{1}{R_1 C_2} & -\left(\dfrac{1}{R_1} + \dfrac{1}{R_2}\right)\dfrac{1}{C_2} \end{bmatrix} \qquad \mathbf{B} = \begin{bmatrix} \dfrac{1}{C_1} & 0 \\ 0 & \dfrac{1}{R_2 C_2} \end{bmatrix} \qquad \mathbf{C} = \lfloor 1 \quad 0 \rfloor$$

To obtain the transfer function, we eliminate T_2 from Eqs. (2.118) and (2.119). Solving for T_2 from Eq. (2.119), we get

$$T_2 = \left(\frac{1}{R_1 C_2 D + 1 + R_1/R_2}\right) T_1 + \left(\frac{R_1/R_2}{R_1 C_2 D + 1 + R_1/R_2}\right) T_a \tag{2.121}$$

Substituting for T_2 from Eq. (2.121) in Eq. (2.118) and simplifying the expression, we obtain

$$T_1 = \left(\frac{R_2(R_1C_2D + 1 + R_1/R_2)}{R_1C_1R_2C_2D^2 + (R_1C_1 + R_2C_2 + R_2C_1)D + 1}\right)q + \left(\frac{1}{R_1C_1R_2C_2D^2 + (R_1C_1 + R_2C_2 + R_2C_1)D + 1}\right)T_a$$

We now define

$$\omega_n = \frac{1}{\sqrt{R_1C_1R_2C_2}} \qquad 2\zeta\omega_n = \frac{R_1C_1 + R_2C_2 + R_2C_1}{R_1C_1R_2C_2}$$

$$\tau_1 = \left(\frac{R_2}{R_1 + R_2}\right)R_1C_2 \qquad \text{and} \qquad \text{gain } k_g = R_1 + R_2$$

Hence,

$$T_1 = \left(\frac{k_g(\tau_1 D + 1)}{\frac{1}{\omega_n^2}D^2 + \frac{2\zeta}{\omega_n}D + 1}\right)q + \left(\frac{1}{\frac{1}{\omega_n^2}D^2 + \frac{2\zeta}{\omega_n}D + 1}\right)T_a \qquad (2.122)$$

where q and T_a are the control and disturbance inputs, respectively.

2.8 ANALOGIES

We now summarize the analogies from the energy point of view among mechanical, electrical, hydraulic, pneumatic, and thermal components. Two variables are involved in the flow of any kind of energy. The across variable $\alpha(t)$ is measured across an element, and the through variable $\theta(t)$ flows through the element. The product $\alpha\theta$ of the across and through variables is the instantaneous power except in pneumatic and thermal systems. It would be power in pneumatic systems if volumetric flow rate instead of mass flow rate were used as the through variable, and in thermal systems if enthropy instead of heat flux were used as the through variable.

We have encountered three types of elements, two of which store energy and one dissipates energy. The two energy-storage elements can be further classified as θ-type and α-type, depending on whether the energy storage is mainly due to the through or across variable (Shearer, Murphy, and Richardson, 1967). The element type can be recognized from the constitutive equation relating the across and through variables as shown in Table 2.1. Thermal resistance however does not dissipate energy.

Another approach to modeling from energy viewpoint is to use bond graphs (Karnopp and Rosenberg, 1975) where the analogies and duals are readily apparent. But bond graphs have not been employed here because to be useful, they must be studied in depth.

TABLE 2.1 ELEMENTARY TWO-TERMINAL COMPONENTS

	Variables	Dissipator	θ-type storage	α-type storage
Generic form	$\alpha(t)$: across $\theta(t)$: through	$\alpha(t) = R\theta(t)$	$\alpha(t) = I\,d\theta/dt$	$C\,d\alpha/dt = \theta(t)$
Mechanical translation	α: velocity θ: force	Damper	Spring	Mass
Mechanical rotation	α: angular velocity θ: torque	Angular damper	Torsional spring	Mass moment of inertia
Electrical	α: voltage θ: current	Resistance	Inductance	Capacitance
Hydraulic	α: pressure θ: volume flow rate	Resistance	Inertance	Capacitance
Pneumatic	α: pressure θ: mass flow rate	Resistance	Inertance	Capacitance
Thermal	α: temperature θ: heat flux	Resistance (not dissipator)	(Not meaningful)	Capacitance

2.9 ACTUATORS

An actuator is a control element that uses power to drive the system to be controlled. The power requirement may be small, as in the case of positioning a control valve, or large, as in the case where a large load is to be moved. Electrical motors, hydraulic servomotors, and pneumatic diaphragm or bellows type actuators are the common examples of actuators used in electrical, hydraulic, and pneumatic control systems, respectively. In the following, we develop linear mathematical models for these actuators.

2.9.1 Electrical DC Motor

A dc motor is an energy converter that converts electrical energy into mechanical energy. There are several types of dc motors, depending on the way the magnetic field is produced and the winding. The magnetic field may be produced by permanent magnets or by a field current. When the field winding is connected in series with the armature winding, the motor has nonlinear characteristics as seen in the following. For this reason, dc motors used in control systems have separately excited field windings.

The basic equations for a dc motor are obtained from Maxwell's electromagnetic theory. The torque produced by a motor is proportional to the product of the magnetic flux ϕ and the armature current i_a, i.e.,

$$T = k_1 \phi i_a \tag{2.123}$$

where k_1 is a constant for a motor. The voltage produced by a motor is proportional to the product of the magnetic flux ϕ and shaft angular velocity ω, i.e.,

$$E_b = k_2 \phi \omega \tag{2.124}$$

This voltage E_b is 180° out of phase with the applied armature voltage E_a and is therefore called back emf. A dc motor with separate field winding is shown in Fig. 2.36. The motor is controlled by varying the armature voltage $E_a(t)$ with the field current i_f kept constant. The combined load and armature mass moment of inertia is I and the viscous friction coefficient is c. The inductance and resistance of the armature are denoted by L_a and R_a, respectively. Let T_d be a disturbance load torque acting on the motor.

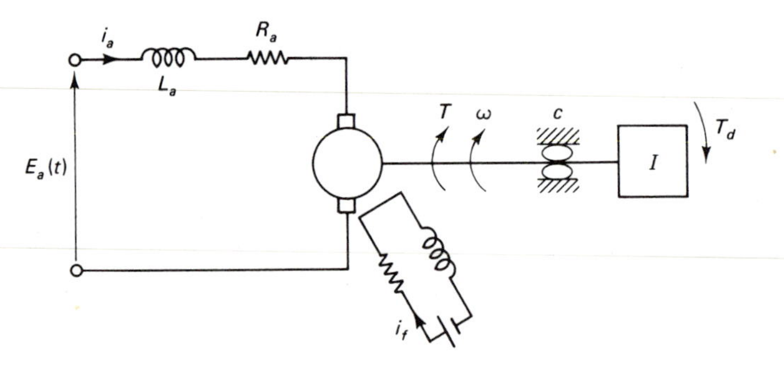

Figure 2.36 Armature-controlled dc motor with fixed field current.

Balancing the voltages and torques, we obtain

$$E_a - E_b = L_a \frac{di_a}{dt} + R_a i_a \tag{2.125}$$

$$T = I\dot{\omega} + c\omega + T_d \tag{2.126}$$

The magnetic flux ϕ is a function of the field current i_f as shown in Fig. 2.37. If saturation is avoided, we obtain $\phi = k_3 i_f$. Then from Eqs. (2.123) and (2.124), it follows that

$$T = (k_1 k_3 i_f) i_a = k_t i_a \tag{2.127}$$

$$E_b = (k_2 k_3 i_f)\omega = k_b \omega \tag{2.128}$$

Defining two time constants as $\tau_1 = L_a/R_a$ and $\tau_2 = I/c$, and substituting for T and E_b from Eqs. (2.127) and (2.128), we can express Eqs. (2.125) and (2.126) as

$$E_a - k_b\omega = R_a(\tau_1 D + 1)i_a \tag{2.129}$$

$$k_t i_a - T_d = c(\tau_2 D + 1)\omega \tag{2.130}$$

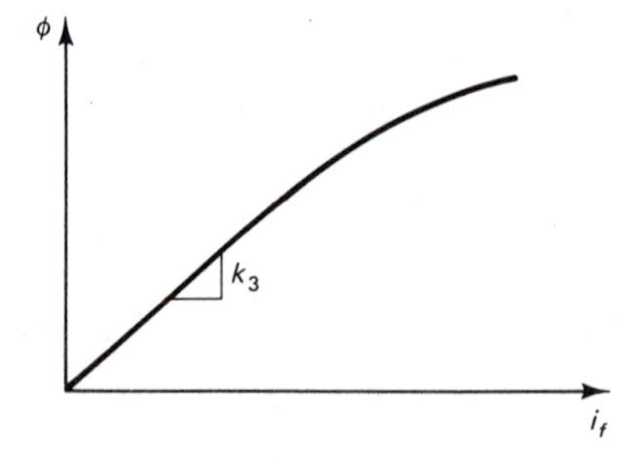

Figure 2.37 Magnetic flux versus field current.

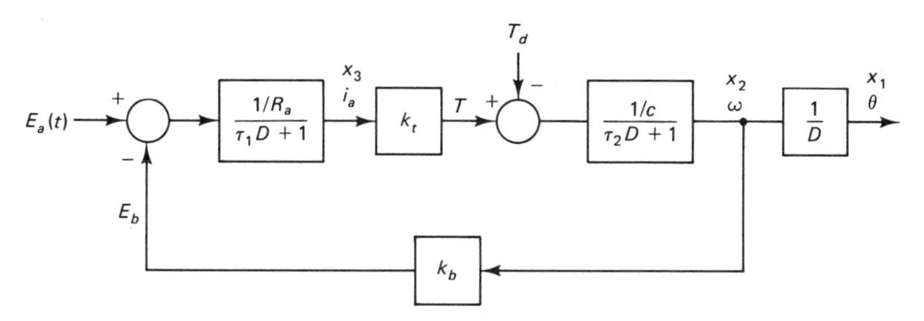

Figure 2.38 Block diagram of a dc motor driving a load.

The block diagram of Eqs. (2.129) and (2.130) is shown in Fig. 2.38, where $\dot{\theta} = \omega$. The dc motor driving a load is an open-loop system with an inherent feedback loop caused by the back emf. In a permanent-magnet dc motor, the field winding is absent and the magnetic flux ϕ is provided by the permanent magnet and is constant. Since in the preceding development, the magnetic flux is a constant, Eqs. (2.129) and (2.130) are also valid for a permanent-magnet dc motor.

To obtain state variables representation, we choose $x_1 = \theta$, $x_2 = \omega$, and $x_3 = i_a$ as state variables. The state equations are obtained either from Eqs. (2.129) and (2.130) or directly from the block diagram of Fig. 2.38 as

$$\begin{aligned} \dot{x}_1 &= x_2 \\ \dot{x}_2 &= -\left(\frac{1}{\tau_2}\right)x_2 + \left(\frac{k_t}{c\tau_2}\right)x_3 - \left(\frac{1}{c\tau_2}\right)T_d \\ \dot{x}_3 &= -\left(\frac{k_b}{R_a\tau_1}\right)x_2 - \left(\frac{1}{\tau_1}\right)x_3 + \left(\frac{1}{R_a\tau_1}\right)E_a \end{aligned} \tag{2.131}$$

If the block diagram is used to obtain the state equations, from Fig. 2.38, we get

$$Dx_1 = x_2$$

$$(\tau_2 D + 1)x_2 = \frac{1}{c}(k_t x_3 - T_d)$$

$$(\tau_1 D + 1)x_3 = \frac{1}{R_a}(E_a - k_b x_2)$$

The state equations, Eq. (2.131), are then obtained by rearranging the preceding equations. In case the angular velocity ω is to be controlled and the angular displacement θ is of no consequence, then in Fig. 2.38 we terminate the block diagram at ω and in the state equations, Eq. (2.131), we disregard the first equation. The order of the system is thus reduced from three to two.

It is possible in theory to control a dc motor by changing the field voltage with fixed armature current. But this control scheme poses great practical difficulties. The armature current cannot be kept constant by supplying a constant voltage to the armature because of the back emf that is a function of ω and i_f. For this reason, field control of a dc motor with fixed armature current is not used in practice.

2.9.2 Electrical AC Motor

Electrical ac motors are less expensive than dc motors and are sometimes employed in control systems for low-power applications.* However, ac motors are nonlinear and hence are difficult to control. The ac motors used in control systems are of the two-phase induction type shown in Fig. 2.39. The stator of the motor has two windings that are displaced by 90 electrical degrees. One winding, the fixed or reference phase, is supplied by a constant ac voltage E_o. The other winding, the control phase, is supplied by a variable ac voltage $E(t)$ that is 90 degrees out of phase with respect to the voltage of the fixed phase.

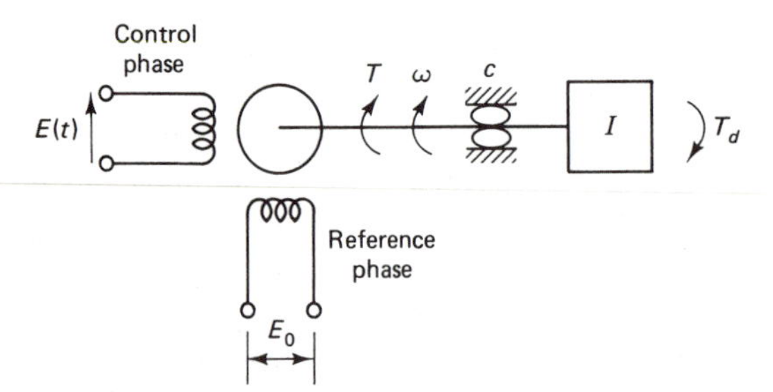

Figure 2.39 Two-phase induction motor.

The typical torque-speed curves of a two-phase induction motor, plotted for different values of the control voltage E, are nonlinear as shown in Fig. 2.40. They are to be linearized for small deviations from a steady-state equilibrium. Let T, ω, and E represent deviations from their steady-state equilibrium values T_e, ω_e, and

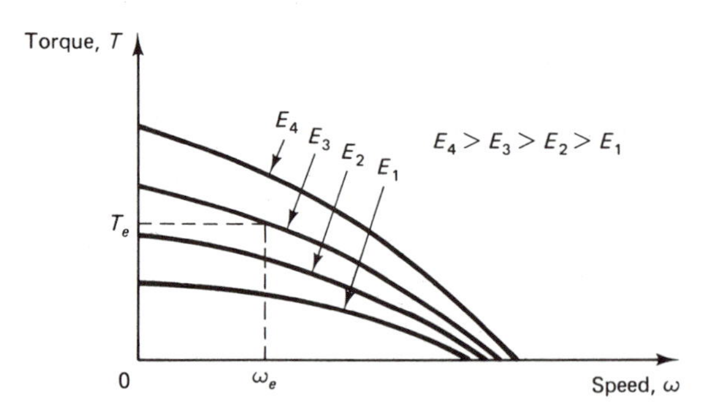

Figure 2.40 Torque-speed curves.

E_e, respectively. Using the linearization techniques discussed in section 2.2.3, we obtain

$$T = -c_1\omega + c_2E \tag{2.132}$$

where $c_1 > 0$ and $c_2 > 0$, since the torque decreases as ω increases but increases as E increases. The combined load and armature mass moment of inertia is I and the viscous friction coefficient is c. The load disturbance torque is denoted by T_d.

*Due to developments in electronics, three-phase ac induction motors with pulse width modulated power amplifiers are recently finding increasing use in control applications.

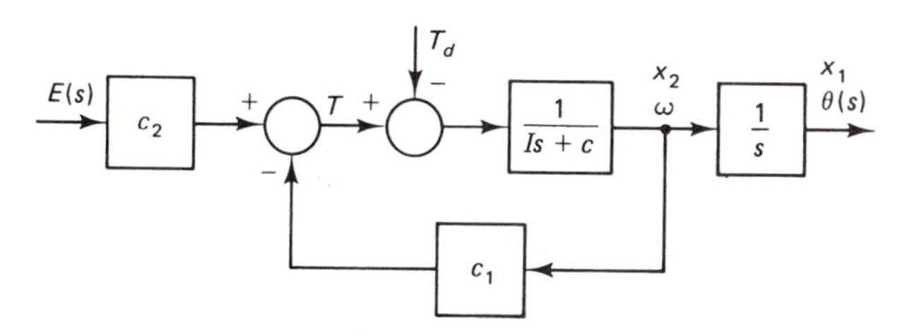

Figure 2.41 Block diagram of an ac motor.

Torque balance then yields

$$T = I\dot{\omega} + c\omega + T_d \tag{2.133}$$

The block diagram of Eqs. (2.132) and (2.133) is shown in Fig. 2.41, where $\dot{\theta} = \omega$. The transfer function is represented by eliminating T from Eqs. (2.132) and (2.133) as

$$\theta(s) = \left(\frac{c_2}{s(Is + c + c_1)}\right)E(s) - \left(\frac{1}{s(Is + c + c_1)}\right)T_d(s) \tag{2.134}$$

or, if ω is considered as the output, as

$$\omega(s) = \left(\frac{c_2}{(Is + c + c_1)}\right)E(s) - \left(\frac{1}{(Is + c + c_1)}\right)T_d(s) \tag{2.135}$$

Choosing $x_1 = \theta$ and $x_2 = \omega$ as the state variables, the state representation becomes

$$\begin{aligned} \dot{x}_1 &= x_2 \\ \dot{x}_2 &= -\left(\frac{c + c_1}{I}\right)x_2 + \frac{c_2}{I}E - \frac{1}{I}T_d \end{aligned} \tag{2.136}$$

where E and T_d are the control and disturbance inputs, respectively. In case ω is to be controlled, then we disregard the first equation in Eq. (2.136).

2.9.3 Hydraulic Servomotor

The hydraulic servomotor shown in Fig. 2.42 consists of a spool valve and a rectilinear actuator (cylinder and piston) that positions a load. The spool valve is a four-port valve. One port is connected to the hydraulic fluid supply at pressure p_s, two control ports are connected one to each side of the cylinder, and the drain port (the two drain ports are joined) is connected to the sump. When the valve is centered, there is no motion. When the spool valve is moved to the left as shown in Fig. 2.42, high-pressure fluid enters the left-hand chamber of the cylinder through the control port and pushes the piston to the right. The fluid from the right-hand side of the cylinder chamber flows to the sump through the drain port. When the spool valve is moved to the right, the action is reversed.

In the following, we obtain a linear model relating the spool displacement x_s to the load displacement y from the centered position. The flow rates q_1 and q_2 of the cyclinder are functions of their respective pressure drops across the valve and

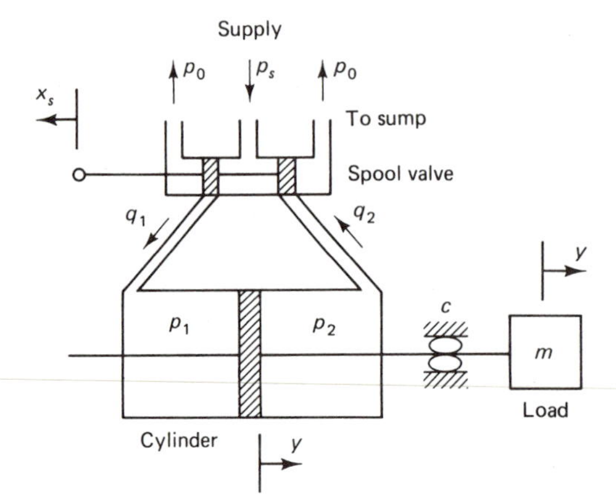

Figure 2.42 Hydraulic servomotor and load.

of x_s. For a symmetric valve, we obtain

$$q_1 = f(x_s, p_s - p_1) \tag{2.137}$$

$$q_2 = f(x_s, p_2 - p_0) \tag{2.138}$$

For pressures found in hydraulic control systems (around 20.7×10^6 Pa or 3000 psi), the compressibility of the liquid is negligible. Hence, $q_1 = q_2 = q$, and from Eqs. (2.137) and (2.138), we see that $p_s - p_1 = p_2 - p_0$. For steady-state equilibrium, $p_{1e} = p_{2e}$ and $x_s = 0$. Linearizing Eqs. (2.137) and (2.138) about this equilibrium for constant p_s and p_0, we obtain

$$q = c_1 x_s - c_2 p_1 \tag{2.139}$$

$$q = c_1 x_s + c_2 p_2 \tag{2.140}$$

where $c_1 > 0$ and $c_2 > 0$. The sign of the coefficients in the preceding equations reflects the fact that the flow increases when x_s increases, decreases when p_1 increases, and increases when p_2 is increased. Adding Eqs. (2.139) and (2.140), we get

$$q = c_1 x_s - c_3(p_1 - p_2) \tag{2.141}$$

The spool valve is usually slightly underlapped (i.e., the land width is less than the port width). The effect of underlapping is reflected in Eq. (2.141), which gives nonzero flow when $x_s = 0$; otherwise, we set $c_3 = 0$. Let A be the piston area, $A\dot{y}$ the volume rate displaced by the piston, and q_L the leakage flow. Neglecting the compressibility of the liquid, the continuity equation yields

$$\begin{aligned} q &= A\dot{y} + q_L \\ &= A\dot{y} + c_4(p_1 - p_2) \end{aligned} \tag{2.142}$$

where the leakage flow is assumed to be proportional to the pressure difference $(p_1 - p_2)$, and c_4 is the leakage coefficient. Eliminating q from Eqs. (2.141) and (2.142), we obtain

$$c_1 x_s = A\dot{y} + (c_3 + c_4)(p_1 - p_2) \tag{2.143}$$

The load on the actuator is assumed to consist of a mass m and a viscous friction with coefficient c. Balancing forces,

$$(p_1 - p_2)A = m\ddot{y} + c\dot{y} \tag{2.144}$$

Substituting for the pressure difference $(p_1 - p_2)$ from Eq. (2.144) in Eq. (2.143), we get

$$c_1 x_s = \left(\frac{c_3 + c_4}{A}\right) m\ddot{y} + \left[A + \left(\frac{c_3 + c_4}{A}\right) c\right]\dot{y} \tag{2.145}$$

Defining a time constant and gain, respectively, as

$$\tau = \frac{(c_3 + c_4)m}{A^2 + (c_2 + c_4)c} \qquad k_g = \frac{Ac_1}{A^2 + (c_3 + c_4)c} \tag{2.146}$$

we express Eq. (2.145) as

$$x_s = \left(\frac{1}{k_g}\right) D(\tau D + 1)y$$

and the transfer function is given by

$$Y(s) = \left(\frac{k_g}{s(\tau s + 1)}\right) X_s(s) \tag{2.147}$$

which is shown in the block diagram of Fig. 2.43(a). When the load on the actuator is negligible, as for example when the actuator positions a control valve, we let both m and c tend to zero and it is seen from Eq. (2.146) that $\tau \to 0$ and $k_g = c_1/A$. The transfer function for this case simplifies to

$$Y(s) = \frac{k_g}{s} X_s(s) \tag{2.148}$$

which is shown in the block diagram of Fig. 2.43(b). Hence, the linear model that we employ is given either by Eq. (2.147) or Eq. (2.148), depending on whether the load on the actuator is significant or negligible.

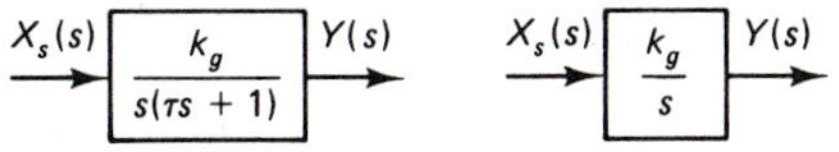

$X_s(s)$ → $\frac{k_g}{s(\tau s + 1)}$ → $Y(s)$

(a)

$X_s(s)$ → $\frac{k_g}{s}$ → $Y(s)$

(b)

Figure 2.43 Block diagram.

2.9.4 Electro Hydraulic Servovalve and Actuator

Electro hydraulic servovalves are frequently used in hydraulic control systems. In a servovalve, the displacement x_s of the spool valve is caused by an electrical, dc torque motor. The servovalve construction depends on the size and the manufacturer, but the basic principles of operation are illustrated by the schematic diagram in Fig. 2.44. It consists of two stages, one of which is an electrical torque motor and the other is the spool valve. The armature consists of a flapper with a winding

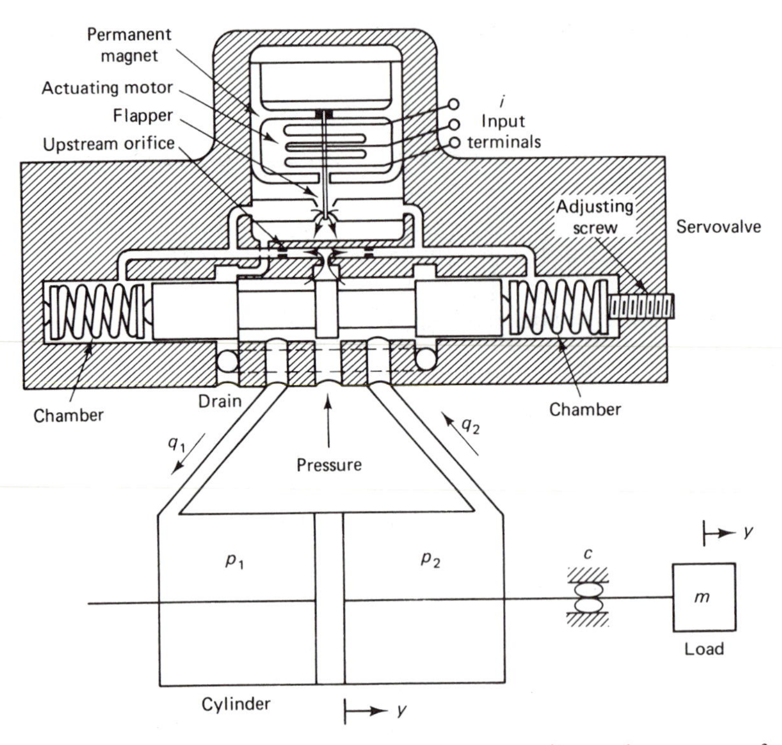

Figure 2.44 Electrohydraulic servovalve and actuator. (Servovalve courtesy of Moog Inc.)

carrying a current. The armature extends into the air gaps of magnetic flux provided by a permanent magnet. The flapper is pivoted and its one end is positioned between two nozzles.

The current i flowing through the armature coil in the presence of magnetic flux ϕ produces a torque T, which from Eq. (2.123) is given by

$$\begin{aligned} T &= k_1\phi i \\ &= k_2 i \end{aligned} \tag{2.149}$$

where constant $k_2 = k_1\phi$. Let θ be the angular rotation of the armature/flapper, I_f its mass moment of inertia, c_f the damping coefficient, and k_f the effective torsional spring constant. Balancing torques, we obtain

$$I_f\ddot{\theta} + c_f\dot{\theta} + k_f\theta = T \tag{2.150}$$

With L as the distance from the position of the nozzles to the pivot point of the flapper, $x_f = L\theta$, where x_f is the displacement of the flapper from its centered position between the two nozzles. Defining the natural frequency and the damping ratio as $\omega_f = (k_f/I_f)^{1/2}$ and $2\zeta_f\omega_f = (c_f/I_f)$, respectively, we express Eq. (2.150) as

$$x_f = \left(\frac{L/k_f}{\dfrac{1}{\omega_f^2}D^2 + \dfrac{2\zeta_f}{\omega_f}D + 1}\right)T \tag{2.151}$$

Now, I_f is very small and consequently ω_f is very large. In some servovalves, ω_f is about 4000 rad/s and $\zeta_f = 0.4$. It is shown in Chapter 5 that time constants and natural frequencies that lie beyond the system bandwidth may be neglected. In other words, because our frequency range of interest is much less than ω_f, it can be assumed that the torque is balanced only by the torsional spring and Eq. (2.151) is closely approximated by

$$x_f = k_3 T \tag{2.152}$$

where $k_3 = L/k_f$ is the gain. This displacement of the flapper partially blocks the flow from one of the nozzles and unblocks it by an equal amount from the other. When a nozzle is completely blocked so that there is no flow, the pressure behind the nozzle is the supply pressure p_s. If the nozzle is completely unblocked, the pressure behind it approaches the ambient pressure as shown in Fig. 2.45. When the flapper is centered, the leakage flow through the nozzles is about 2% of the full-rated flow of the valve and the equilibrium is about the middle of the nearly linear range in Fig. 2.45.

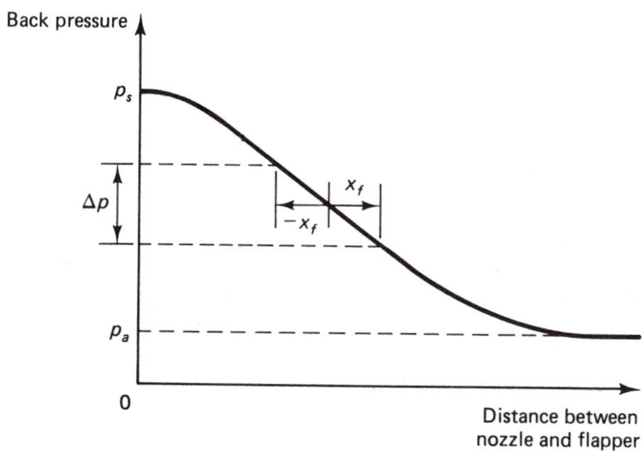

Figure 2.45 Nozzle-flapper characteristics.

For a small deviation x_f from the equilibrium position, the differential pressure Δp between the two nozzles may be linearized as

$$\Delta p = -2c_1(-x_f) = k_4 x_f \tag{2.153}$$

where $c_1 > 0$, and $-c_1$ is the slope of the curve at the equilibrium position. The pressures behind the nozzles are fed one to each side of the spool valve. The differential pressure produces a force on the spool valve given by

$$F = \Delta p A_s \tag{2.154}$$

where A_s is the effective cross-sectional area of the spool. Let m_s be the mass of the spool, c_s the damping coefficient, k_s the effective spring constant, and x_s the displacement of the spool valve from the neutral position. The force balance yields

$$(m_s D^2 + c_s D + k_s)x_s = F$$

Defining the natural frequency ω_n and the damping ratio ζ for the spool valve, the preceding equation yields the transfer function

$$x_s = \left(\frac{1/k_s}{\frac{1}{\omega_n^2} D^2 + \frac{2\zeta}{\omega_n} D + 1} \right) F \tag{2.155}$$

The block diagram for the servovalve relating the input current i to the spool displacement x_s is obtained by combining Eqs. (2.149), (2.152), (2.153), (2.154), and (2.155) and is shown in Fig. 2.46(a). It may be simplified as shown in Fig. 2.46(b), where k is the gain. Choosing $x_1 = x_s$ and $x_2 = \dot{x}_s$ as the state variables,

I(s)
k2
T
k3
xf
k4
Δp
As
F
1/ks
1/ωn² s² + 2ζ/ωn s + 1
Xs(s)
(a)
I(s)
k
1/ωn² s² + 2ζ/ωn s + 1
Xs(s)
(b)

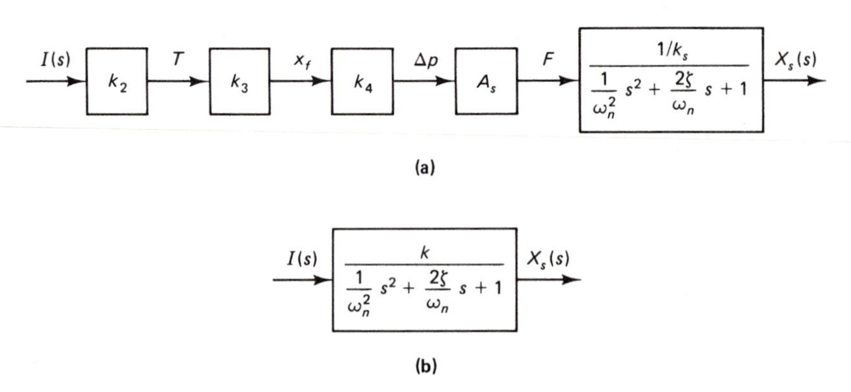

Figure 2.46 Block diagram of a servovalve.

the transfer function can be represented by the state equations. For a typical servovalve, ω_n is around 350 rad/s and ζ is between 0.7 and 0.9. If the frequency range of interest is much less than ω_n, then the model of the servovalve may be approximated by

$$x_s = ki \tag{2.156}$$

where k is the gain. The model for the combined servovalve and actuator is obtained by combining the block diagrams of Figs. 2.46(b) and 2.43, and is shown in Fig. 2.47.

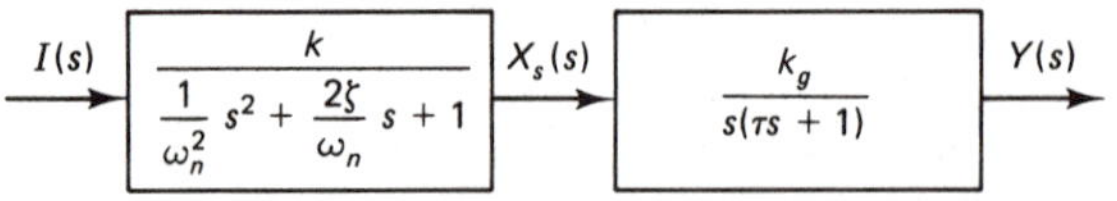

I(s)
k
1/ωn² s² + 2ζ/ωn s + 1
Xs(s)
kg
s(τs + 1)
Y(s)

Figure 2.47 Block diagram of a servovalve and actuator.

For rotary motion, the rectilinear actuator of Fig. 2.44 is replaced by a hydraulic motor. Different types of hydraulic motors, such as radial-piston type and vane type, are available. We consider a servovalve supplying fluid to a hydraulic motor driving a load as shown in Fig. 2.48. The transfer function of the servovalve relating the current i to the spool valve displacement x_s is shown in Fig. 2.46. Hence, a mathematical model of only the hydraulic motor is obtained as follows.

Referring to Fig. 2.48 and neglecting the compressibility of the liquid, we get $q_1 = q_2 = q$. The linearized equations for the flow rate in and out of the motor are

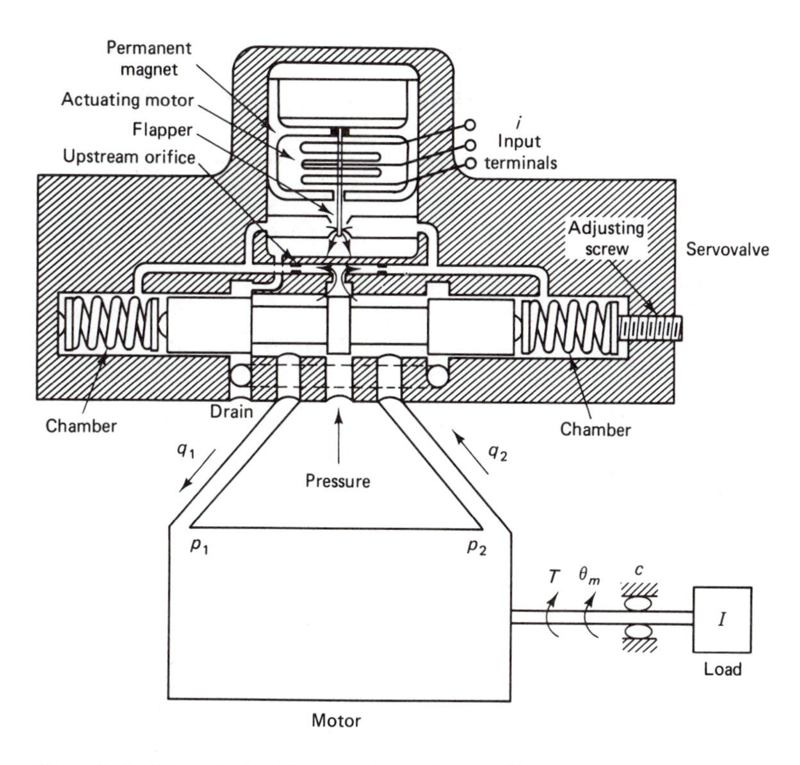

Figure 2.48 Electrohydraulic servovalve and motor. (Servovalve courtesy of Moog Inc.)

given by Eqs. (2.139) and (2.140), respectively, and represented by Eq. (2.141), namely,

$$q = c_1 x_s - c_3(p_1 - p_2)$$

Since the compressibility of the liquid is neglected, the continuity equation yields

$$q = V\dot{\theta}_m + q_L$$

where V is the volume displaced by the motor per radian of rotation, $\dot{\theta}_m$ the angular velocity of the motor, and q_L the leakage flow rate. As in Eq. (2.142), the leakage flow is assumed to be proportional to the pressure difference and is expressed by

$$q_L = c_4(p_1 - p_2)$$

Eliminating q from the preceding equations, we obtain

$$c_1 x_s - c_3(p_1 - p_2) = V\dot{\theta}_m + c_4(p_1 - p_2)$$

or

$$c_1 x_s = V\dot{\theta}_m + (c_3 + c_4)(p_1 - p_2)$$

Letting T be the torque developed by the motor, we have

$$\text{hydraulic power} = \text{mechanical power}$$

$$(p_1 - p_2)V\dot{\theta}_m = T\dot{\theta}_m$$

or

$$(p_1 - p_2) = T/V$$

Thus,

$$c_1 x_s = V\dot{\theta}_m + \left(\frac{c_3 + c_4}{V}\right)T$$

For a load consisting of inertia and linear rotary friction, the torque balance yields

$$T = (ID + c)\dot{\theta}_m$$

Hence, we obtain

$$\begin{aligned} c_1 x_s &= V\dot{\theta}_m + \left(\frac{c_3 + c_4}{V}\right)(ID + c)\dot{\theta}_m \\ &= \left[\left(\frac{I(c_3 + c_4)}{V}\right)D + \left(\frac{c(c_3 + c_4)}{V}\right) + V\right]\dot{\theta}_m \end{aligned}$$

Defining a time constant and gain, respectively, as

$$\tau = \frac{I(c_3 + c_4)/V}{c(c_3 + c_4)/V + V}$$

$$k_g = \frac{c_1}{c(c_3 + c_4)/V + V}$$

we get

$$k_g x_s = (\tau D + 1)\dot{\theta}_m = D(\tau D + 1)\theta_m$$

The combined transfer function of the servovalve and motor is shown in Fig. 2.49, which is seen to be similar to Fig. 2.47.

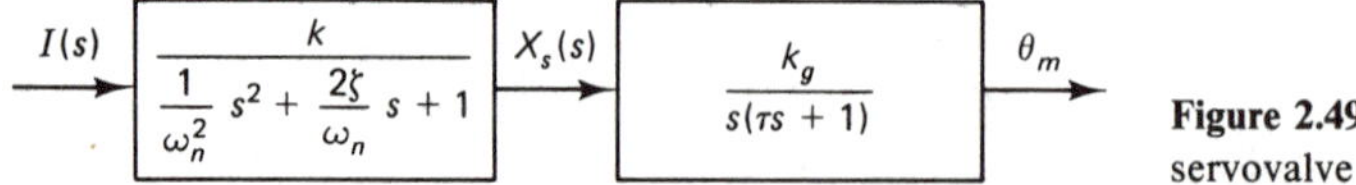

Figure 2.49 Block diagram of a servovalve and motor.

2.9.5 Pneumatic Diaphragm-Type Actuator

The pneumatic actuator with a spring-supported diaphragm shown in Fig. 2.50 is used in pneumatic control systems usually to position a control valve. It is actuated by a pressure signal p_1. A change in p_1 causes a change in p_2, the pressure inside the chamber. The pressure p_2 acting on the diaphragm generates a change in the force that moves the control valve against a spring.

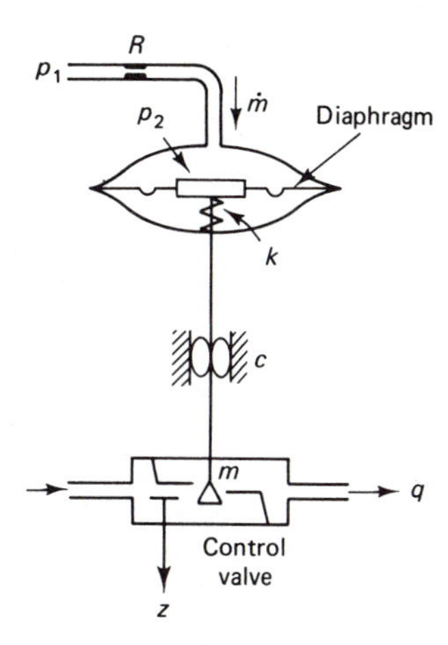

Figure 2.50 Pneumatic actuator and valve.

When the system is in steady-state equilibrium, $p_{1e} = p_{2e} =$ constant, the mass flow rate to or from the chamber is zero, and the valve position z_e and flow q_e through the valve are constants. Let p_1, p_2, and q denote deviations from their steady-state values. There will be no choking for small deviations, and the mass flow rate $\dot{m}$ of air flowing into the chamber is given by

$$\dot{m} = \frac{p_1 - p_2}{R} \tag{2.157}$$

where R is the pneumatic resistance. This flow is stored in the chamber and we get

$$\dot{m} = C\frac{dp_2}{dt} \tag{2.158}$$

where the pneumatic capacitance includes the effects of compressibility and the change in volume of the chamber. Eliminating $\dot{m}$ from Eqs. (2.157) and (2.158) and defining a time constant $\tau = RC$, we obtain

$$(\tau D + 1)p_2 = p_1 \tag{2.159}$$

Let m, c, and k be the mass, coefficient of viscous friction, and spring constant, respectively, and A be the area of the diaphragm. Force balance yields

$$(mD^2 + cD + k)z = p_2 A$$

i.e.,

$$k\left(\frac{1}{\omega_n^2}D^2 + \frac{2\zeta}{\omega_n}D + 1\right)z = p_2 A \tag{2.160}$$

The valve motion is relatively slow and the valve mass is small. Hence, ω_n is quite large and usually beyond the frequency range of interest. Hence, we assume that the force is balanced only by the spring and approximate Eq. (2.160) by

$$kz = p_2 A \tag{2.161}$$

The linear equation for the deviation of the flow rate through the control valve is given by

$$q = c_1 z \tag{2.162}$$

The block diagram of the pneumatic actuator and valve is obtained by combining Eqs. (2.159), (2.161), and (2.162) as shown in Fig. 2.51, where the gain is defined

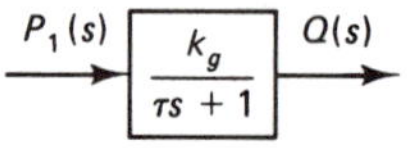

$$P_1(s) \longrightarrow \boxed{\frac{k_g}{\tau s + 1}} \longrightarrow Q(s)$$

Figure 2.51 Block diagram.

by $k_g = Ac_1/k$. Choosing $x = z$ as the state variable, the state equation is obtained from Eqs. (2.159) and (2.161) and the output from Eq. (2.162) as

$$\begin{aligned} \dot{x} &= -\frac{1}{\tau}x + \frac{A}{\tau k}p_1 \\ q &= c_1 x \end{aligned} \tag{2.163}$$

2.10 SUMMARY

In this chapter, we have reviewed and discussed the conservation equations, basic laws and constitutive equations governing the dynamic behavior of mechanical, electrical, hydraulic, pneumatic, and thermal elements of control systems. The use of through and across variables facilitates the classification of physical elements and the functional analogy and duality become apparent from the point of view of energy.

The behavior of many systems is in general nonlinear. We have ignored nonanalytic nonlinearities, such as Coulomb friction and hysteresis, and have linearized the analytic nonlinearities for small deviations from equilibrium. The physical parameters of the systems have been combined into four types of standard parameters, which are natural frequency, damping ratio, time constant, and gain. Techniques of identifying the values of the standard parameters are discussed in later chapters. The linear mathematical models developed in this chapter, in the form of transfer functions or state equations, will be used extensively in the following chapters.

It should be noted that in all the transfer functions that we have developed, the order of the numerator polynomial does not exceed the order of the denominator polynomial. This is a consequence of causality as all physical systems are causal and do not exhibit anticipatory response.

REFERENCES

D'Souza, A. F., and Garg, V. K. (1984). *Advanced Dynamics: Modeling and Analysis.* Englewood Cliffs, NJ: Prentice-Hall.

FOX, R. W., and MCDONALD, A. T. (1978). *Introduction to Fluid Mechanics.* New York: John Wiley & Sons.

JOHN, J. E. A. (1969). *Gas Dynamics.* Boston: Allyn and Bacon.

KARNOPP, D., and ROSENBERG, R. (1975). *System Dynamics: A Unified Approach.* New York: John Wiley & Sons.

KREITH, F. (1973). *Principles of Heat Transfer.* 3d ed. New York: Intext.

OGATA, K. (1978). *System Dynamics.* Englewood Cliffs, NJ: Prentice-Hall.

PALM, W. J. (1983). *Modeling, Analysis, and Control of Dynamic Systems.* New York: John Wiley & Sons.

SHEARER, J. L., MURPHY, A. T., and RICHARDSON, H. H. (1967). *Introduction to System Dynamics.* Reading, MA: Addison-Wesley.

PROBLEMS

2.1. A machine of mass m is mounted on a foundation with a stiffness k and a damping coefficient c as shown in Fig. P2.1. The foundation vibrates with displacement u, and y is the displacement of the mass, where both u and y are measured with respect to the fixed ground level. Obtain the transfer function and state equations relating the output y to the input u.

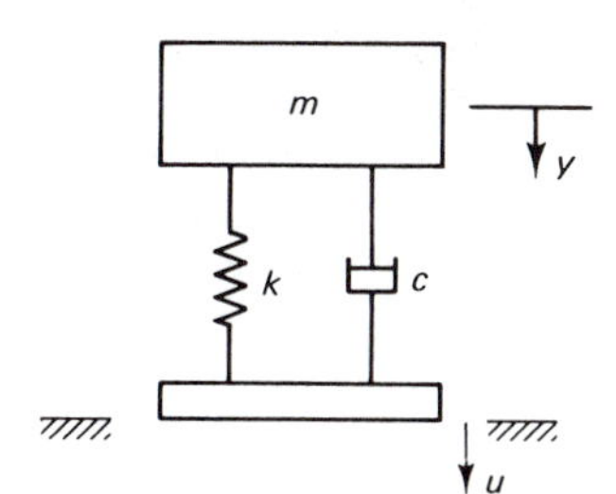

Figure P2.1 Machine mounted on a foundation.

2.2. A slender uniform rod of mass m and length L is hinged at one end and connected as shown in Fig. P2.2 in a vertical plane. The friction moment at the hinge is $c_1\dot{\theta}$. For small θ, determine the transfer function relating the output angular displacement θ to the force input F. Give the state equations.

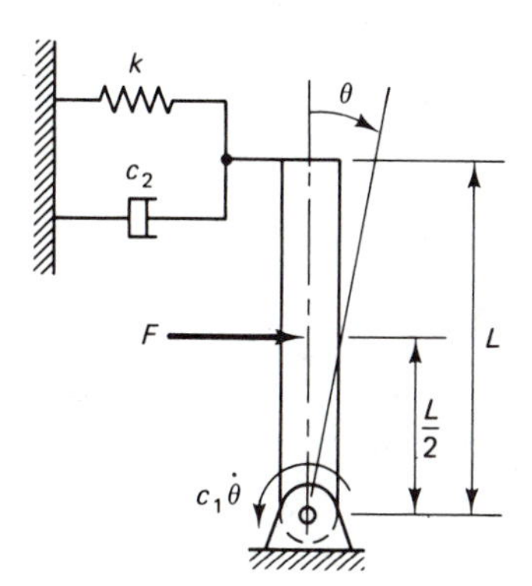

Figure P2.2 Rod hinged in a vertical plane.

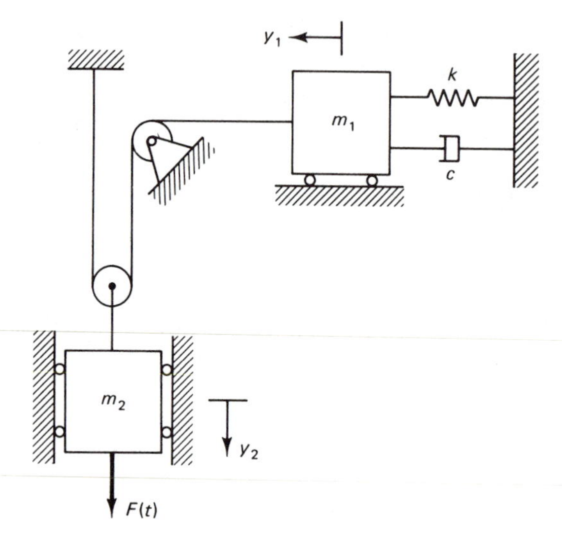

Figure P2.3 System of two connected masses.

2.3. For the system shown in Fig. P2.3, the cable is inextensible and the pulleys are massless and frictionless. Let y_1 and y_2 denote displacements from equilibrium.

(**a**) Give the number of degrees of freedom.

(**b**) Obtain the transfer function relating output y_2 to input force $F(t)$.

(**c**) Determine the natural frequency and damping ratio in terms of system parameters.

2.4. A cylinder of radius R and mass m is free to rotate about its longitudinal axis, which is connected to a support by springs and a damper as shown in Fig. P2.4. The cylinder rolls without slipping on the horizontal surface. Obtain the transfer function relating the output y to the input force F. The mass moment of inertia $I = (1/2)mR^2$. If $k = 100\ N/m$ and $m = 24$ kg, obtain the natural frequency ω_n.

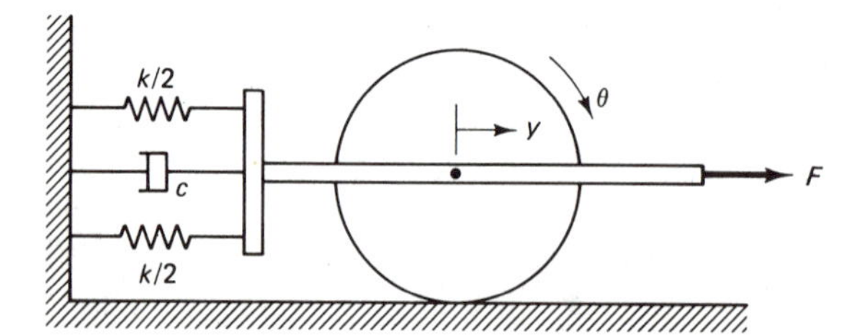

Figure P2.4 Cylinder rolling on a horizontal surface.

2.5. As shown in Fig. P2.5, the torque T produced by a motor is transmitted through two gear trains. The gear ratios are $n_1 = \omega_2/\omega_1$ and $n_2 = \omega_3/\omega_2$. Let I_1, I_2, and I_3 be the

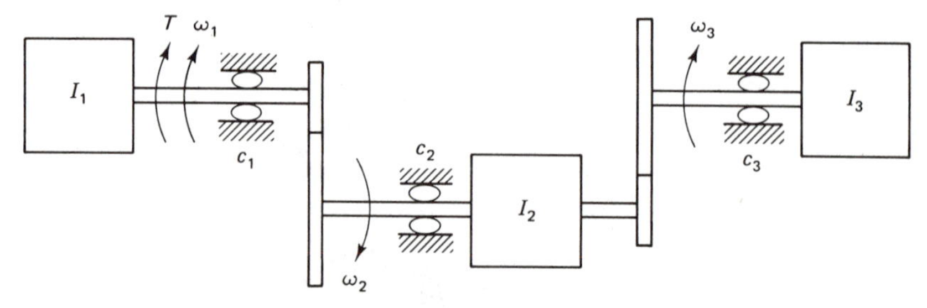

Figure P2.5 Rotors connected by two gear trains.

mass moments of inertia of the motor, intermediate rotor, and load, respectively, and c_1, c_2, and c_3 be the corresponding torsional damping constants. Assume that the shafts are rigid (the angle of twist of each shaft is negligible). Obtain the transfer function relating the output angular velocity ω_3 to the input torque T.

2.6. A rigid body is hinged at its mass center 0 and supported as shown in Fig. P2.6. The mass moment of inertia about the hinge point is I. For small displacements, obtain the transfer function relating the output angular displacement θ to the input displacement u. Express the natural frequency and damping ratio in terms of the system parameters.

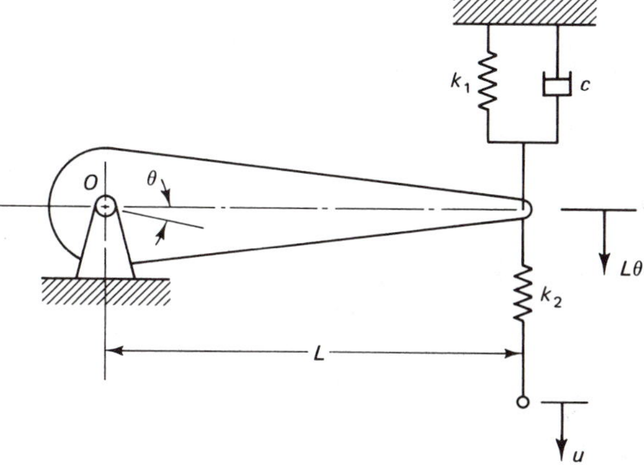

Figure P2.6 Hinged and supported rigid body.

2.7. A disk of mass moment of inertia I is hinged at 0 and connected as shown in Fig. P2.7. Mass m is connected to the disk with a spring. Torque T and force F are applied to the disk and mass, respectively. Choosing y, θ, $\dot{y}$, and $\dot{\theta}$ as the state variables, obtain the state equations for small displacements. Give the transfer-function matrix relating y and θ to F and T.

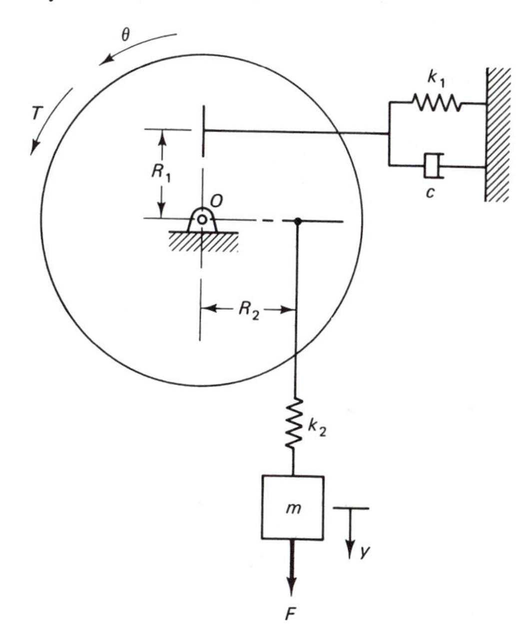

Figure P2.7 System of a connected cylinder and mass.

2.8. A body of mass m is suspended as shown in Fig. P2.8. For this plane motion, let y and θ denote small displacements from equilibrium, and I be the mass moment of inertia about the centroidal axis. Considering y and θ as outputs and force F and moment LF as inputs, obtain the transfer function matrix. Letting y, θ, $\dot{y}$, and $\dot{\theta}$ as state variables, give the state equations.

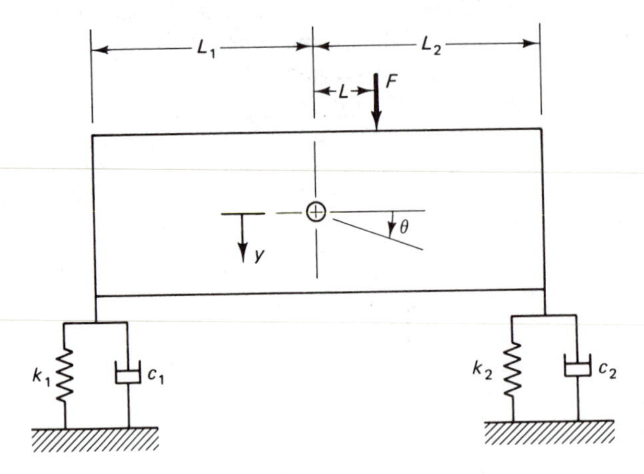

Figure P2.8 Rigid body with two-degrees-of-freedom.

2.9. Obtain the transfer function relating the output voltage E_0 to the input voltage E_1 and give the state equations for the circuit shown in Fig. P2.9.

2.10. Solve Problem 2.9 for the circuit shown in Fig. P2.10.

2.11. Solve Problem 2.9 for the circuit shown in Fig. P2.11.

2.12. Solve Problem 2.9 for the circuit shown in Fig. P2.12.

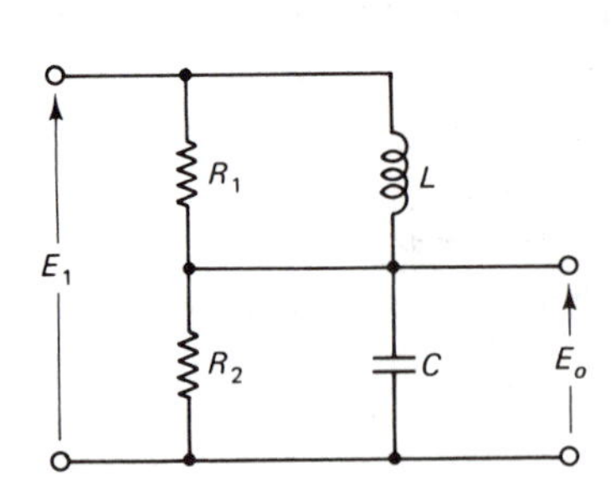

Figure P2.9 Electrical network.

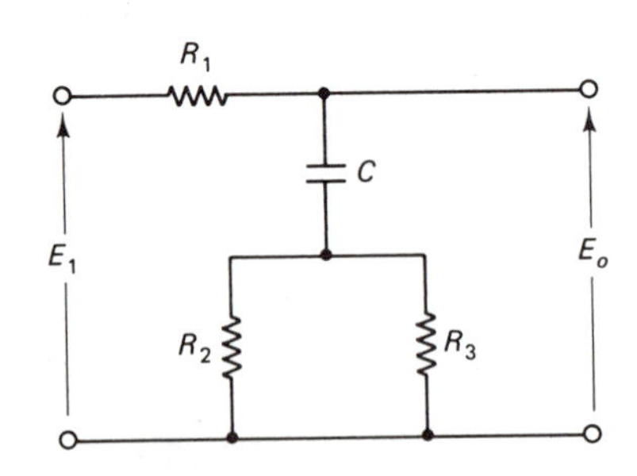

Figure P2.10 Electrical network.

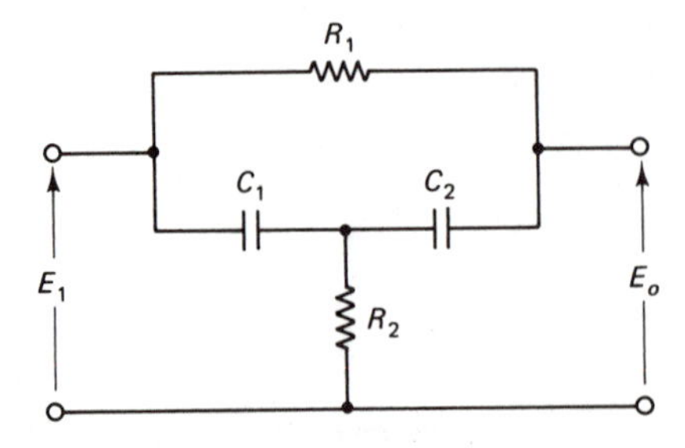

Figure P2.11 Electrical network.

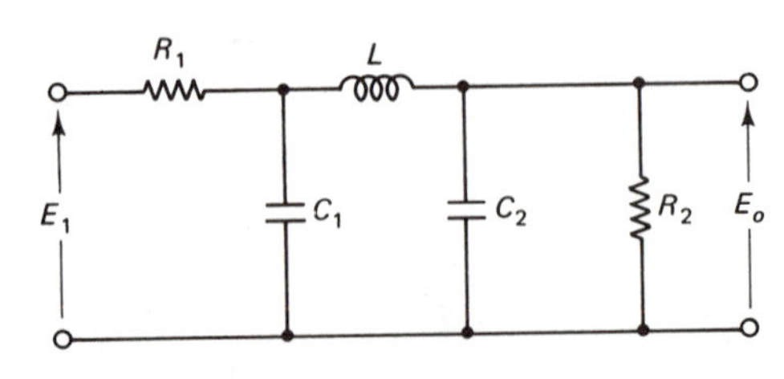

Figure P2.12 Electrical network.

2.13. A two-tank hydraulic system is shown in Fig. P2.13, where q_1, q_2, h_1, and h_2 are deviations from their equilibrium values. Let ρ be the fluid density, A_1 and A_2 the tank cross-sectional areas, and R_1 and R_2 the hydraulic resistances. Obtain the transfer function relating output h_2 to input q_1. Choosing h_1 and h_2 as the state variables, give the state equations.

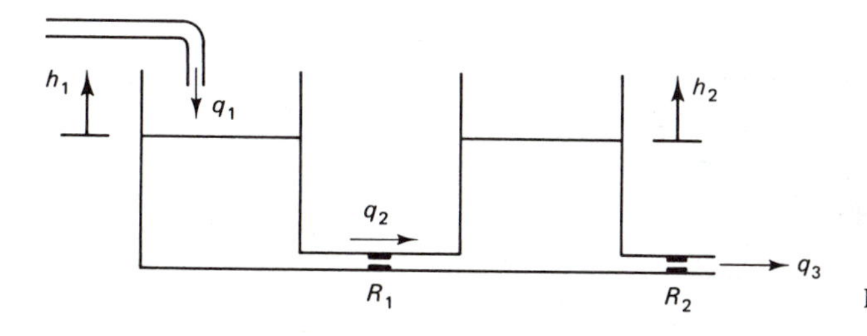

Figure P2.13 Two-tank hydraulic system.

2.14. An instrument for measuring the transient pressure $p_1(t)$ at the discharge of a pump is shown in Fig. P2.14. Neglect the compressibility of the liquid. Obtain the transfer function relating the output displacement y to input p_1.

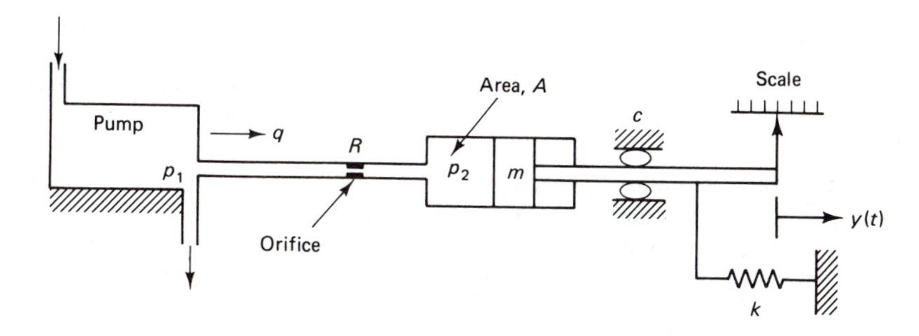

Figure P2.14 Instrument for measuring transient pressure.

2.15. A hydraulic system is shown in Fig. P2.15. The piston displacement u is the input and the cylinder displacement y is the output. Obtain the transfer function relating the output to the input. Neglect the fluid compressibility.

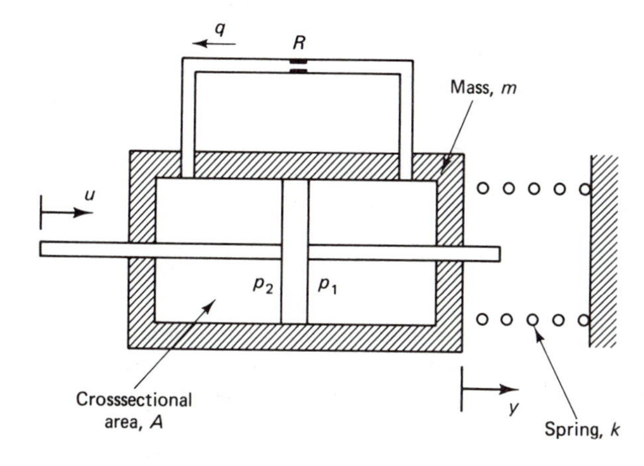

Figure P2.15 Hydraulic system with a piston and cylinder.

2.16. A pneumatic system consists of a nozzle-flapper combination and a bellows actuating a control valve as shown in Fig. P2.16. Obtain a transfer function relating the output flow q to the input displacement z. The only load on the bellows is the spring. All variables represent deviations from equilibrium.

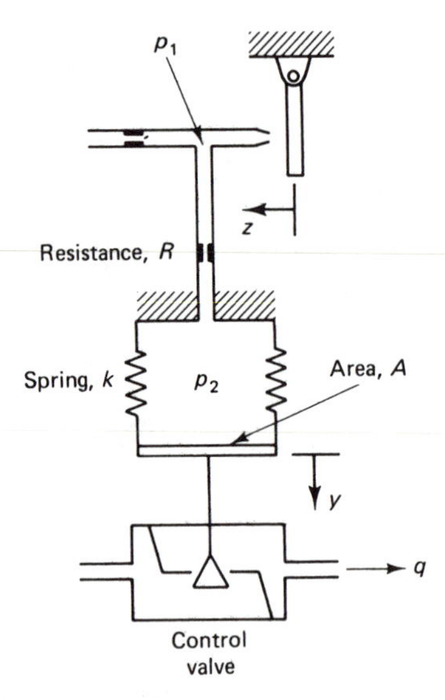

Figure P2.16 Pneumatic system with a nozzle-flapper.

2.17. The pneumatic system shown in Fig. P2.17 consists of two bellows and a nozzle-flapper. The bellows have the same cross-sectional area A but different spring constants k_1 and k_2, and different resistances R_1 and R_2. Obtain the transfer function relating the output pressure p_0 to the input pressure p_i. Give the state equations.

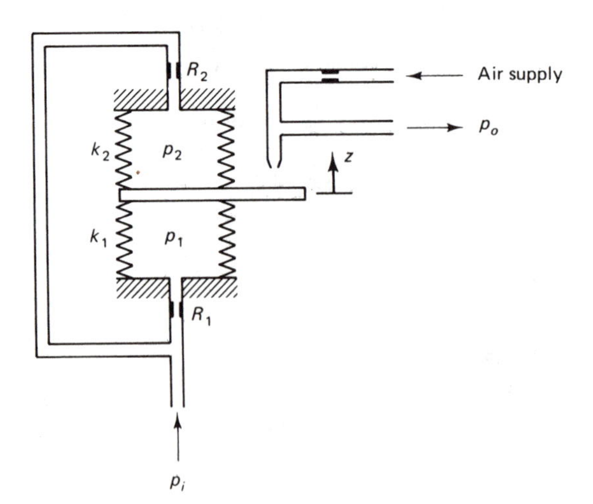

Figure P2.17 Pneumatic system.

2.18. A liquid at temperature $T_1(t)$ is pumped into and flows out of a tank at a constant volume flow rate q as shown in Fig. P2.18. The walls, which are perfectly insulated, have temperature $T_w(t)$, thermal resistance R, and capacitance C. The liquid, which is stirred and considered as one lump, has a temperature $T_2(t)$, volume V, specific heat

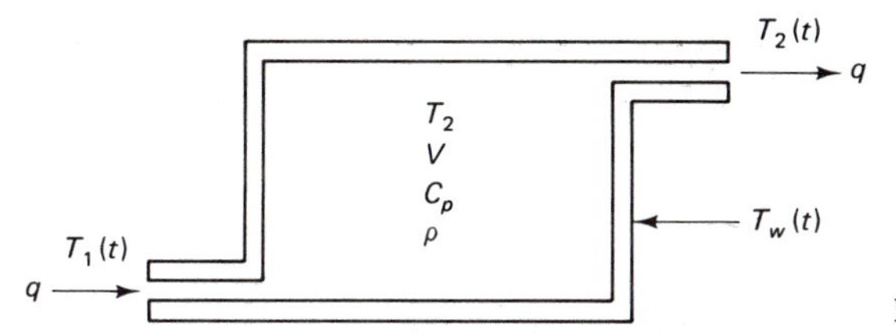

Figure P2.18 Stirred-tank thermal system.

C_p, and density ρ. Obtain the transfer function relating the output temperature $T_2(t)$ to the input $T_1(t)$. Also, give the state equations.

2.19. For the two compartments shown in Fig. P2.19, neglect the thermal capacitances of the walls and treat each compartment as one lump at temperatures $T_1(t)$ and $T_2(t)$ and thermal capacitances C_1 and C_2, respectively. The resistance of the outside walls of each compartment is R_1, that of the partition is R_2, and the ambient temperature is $T_a(t)$. The heat flux input q is supplied only to the first compartment. Obtain the transfer functions relating the output temperature T_1 to the heat flux input q and disturbance input $T_a(t)$, and give the state equations.

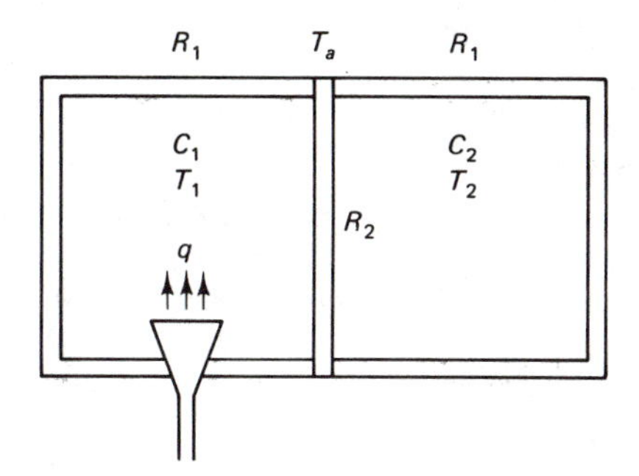

Figure P2.19 Two-compartment thermal system.

2.20. The voltage E_2 produced by a dc generator is controlled by varying the input voltage E_1 to its field. See Fig. P2.20. A prime mover drives the generator at a constant speed ω. Obtain the transfer function and state equations relating the output armature current i_2 to the input voltage E_1.

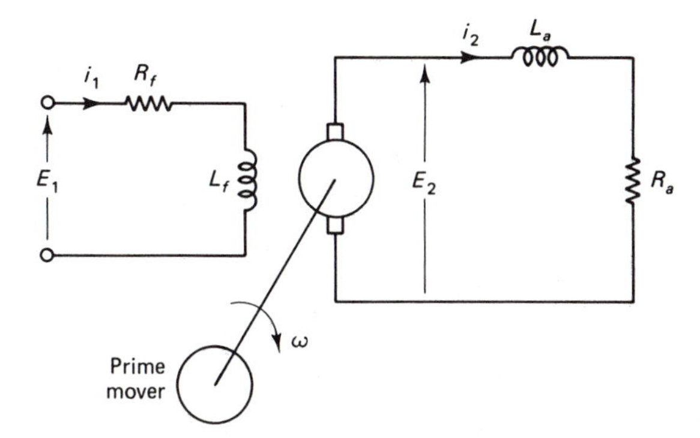

Figure P2.20 Dc generator.

2.21. Two hydraulic spool valve and actuator units shown in Fig. P2.21 are coupled together to form a simple harmonic oscillator. Obtain the differentrial equation describing the

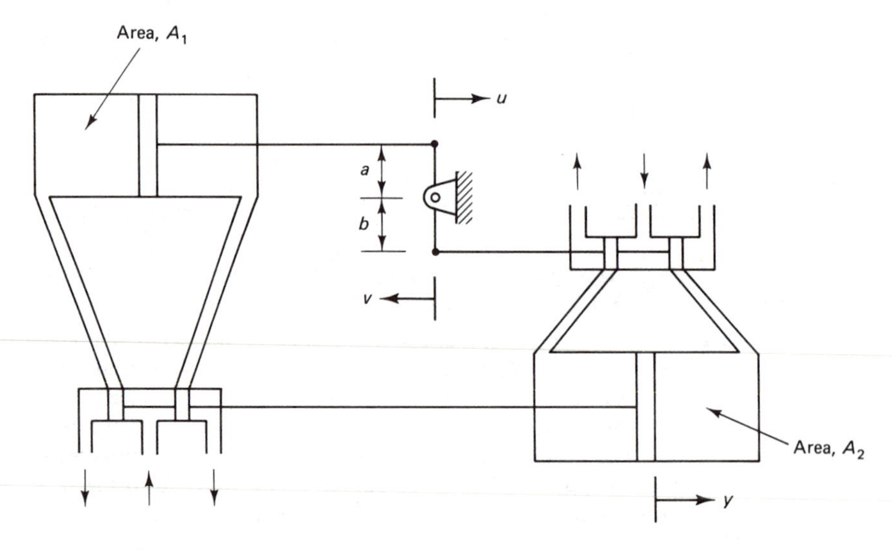

Figure P2.21 Two connected spool valve and actuator units.

motion y and show the dependence of the natural frequency on the system parameters. Neglect load on both actuators.

2.22. A jet-pipe-type hydraulic actuator is shown in Fig. P2.22. When the angle of the jet is displaced by an input displacement u to the left, the oil from the nozzle enters the left port of the actuator and displaces the piston to the right. The oil from the right-hand side of the piston flows out of the right-hand port. The action is reversed when the direction of u is to the right. Obtain a transfer function relating the output displacement y of the load to the input u.

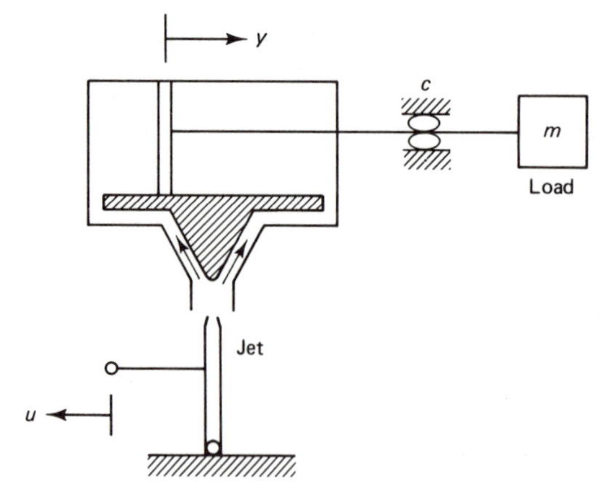

Figure P2.22 Jet-pipe-type hydraulic actuator.

2.23. A hydraulic power-steering unit is shown in Fig. P2.23. The angular position $\theta(t)$ of the steering wheel causes the movement x of the spool valve that supplies the fluid to one side of the piston in the cylinder. The sleeve, which is directly connected to the piston, tends to close the valve opening. Obtain the transfer function relating the output position y to the input θ.

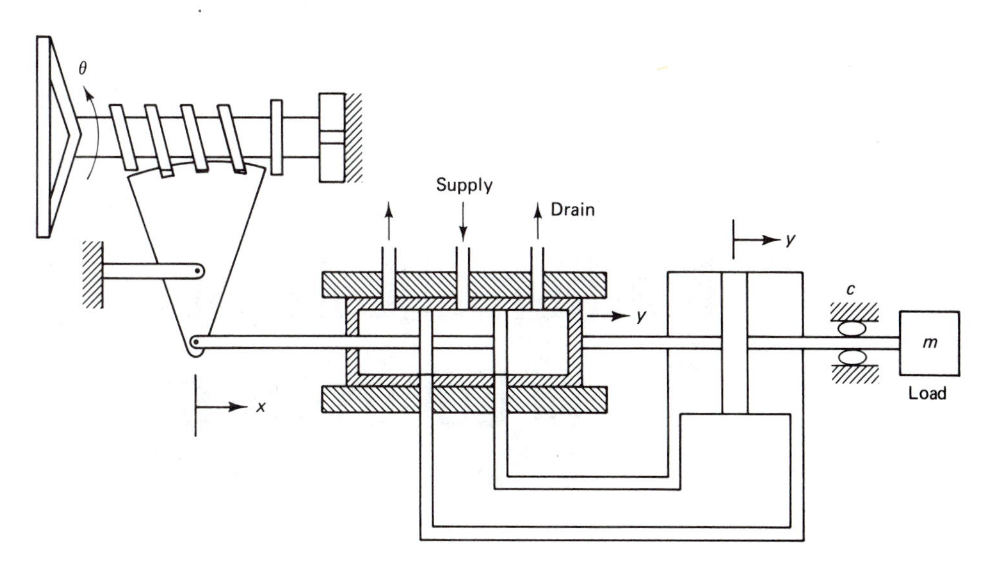

Figure P2.23 Hydraulic power steering unit.

3

Mathematical Modeling of Control Systems

3.1 INTRODUCTION

Control systems may be classified as electrical, hydraulic, or pneumatic, depending on the type of actuator that is employed. However, many control systems are of a hybrid type, such as electrohydraulic and electropneumatic. A control system that employs a hydraulic actuator may also have an electrical sensor/transducer for feedback and a servovalve. In addition, passive control systems are also used in some simple applications where no external source of power is used for actuation.

In some applications, the preferred choice among electrical, hydraulic, and pneumatic systems is very obvious. In other applications, the preference is not very clear and either an electrical, hydraulic, or pneumatic control system may be equally suitable. Thus, the relative importance of accuracy, cost, reliability, size, weight, safety, speed of response, and other factors must be carefully considered in the selection.

In this chapter, given the conceptual design of a control system, our objective is to obtain its linear mathematical model in the form of transfer-function and state-equations representations. The conceptual design of a control system is an early step in its design and is usually undertaken by someone with considerable experience. It requires familiarity with the merits and disadvantages of commonly available control components.

It should be understood that the conceptual designs considered in the examples may not be the final designs to be adopted. It may be necessary to modify a conceptual design after completing a detailed analysis of its mathematical model, and in some cases the entire conceptual design may have to be discarded as unsuitable in favor of another design. The mathematical models of several examples given in this chapter are analyzed later and their dynamic performance studied in the subsequent chapters.

The examples we consider in this chapter employ output feedback to synthesize a control law. Only the controlled output is measured, fed back, and compared to the command input to obtain the error. The controller is simply an electronic, pneumatic, or other amplifier with a gain constant k_p and the control law is obtained by multiplying the error by this gain. This type of control law is known as proportional control and has some disadvantages. A better contol law called proportional-plus-integral-plus-derivative (PID) is studied in Chapter 7 for analog controllers and in Chapter 9 for digital computer control. A further refinement can be obtained by synthesizing a control law produced by feeding back all the state variables. This type of control law is studied in Chapter 8 and its digital computer implementation in Chapter 9.

3.2 TRANSFER-FUNCTION AND STATE-VARIABLES REPRESENTATIONS

In this chapter, starting with the conceptual design of a feedback control system and its schematic diagram, we obtain its linear mathematical model and present it in the form of a block-diagram and state-equations representations. In the block-diagram representation, the objective is to replace the components shown in Fig. 3.1 by their respective transfer functions. The state-equations representation can be obtained either directly from the equations or from the block diagram.

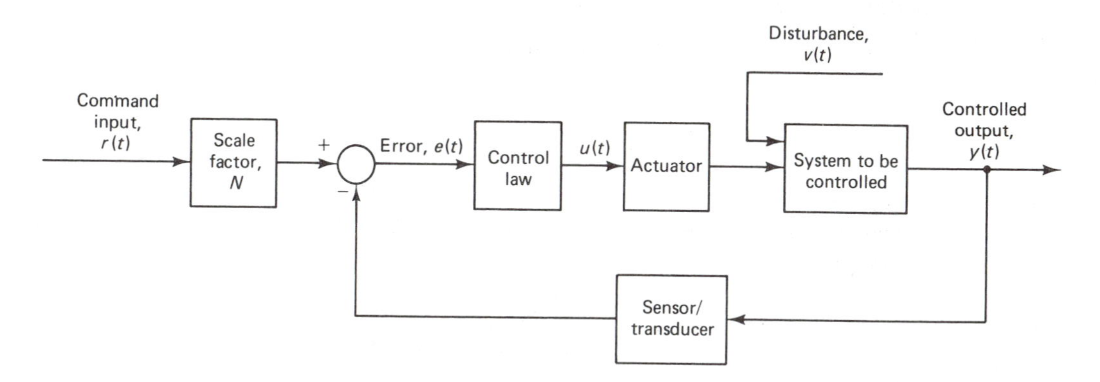

Figure 3.1 Block diagram of a feedback control system.

The standard form of the state representation that we employ is described by the following equations.

$$\text{Plant:} \qquad \dot{\mathbf{x}}(t) = \mathbf{A}\mathbf{x}(t) + \mathbf{B}\mathbf{u}(t) + \mathbf{B}_1\mathbf{v}(t) \tag{3.1}$$

$$\text{Control Law:} \qquad \mathbf{u}(t) = \mathbf{N}\mathbf{r}(t) - \mathbf{K}\mathbf{x}(t) \tag{3.2}$$

$$\text{Output:} \qquad \mathbf{y}(t) = \mathbf{C}\mathbf{x}(t) \tag{3.3}$$

The system to be controlled and the actuator are commonly called the *plant* in control terminology. In Eqs. (3.1) through (3.3), $\mathbf{x}(t)$ is an n-dimensional state vector, $\mathbf{u}(t)$ is the m-dimensional control input, $\mathbf{v}(t)$ represents the disturbance, $\mathbf{r}(t)$ is an m-dimensional command input vector, and $\mathbf{y}(t)$ is an m-dimensional control output vector. It is assumed that the dimensions of $\mathbf{u}$, $\mathbf{r}$, and $\mathbf{y}$ are the same. Here, $\mathbf{A}, \mathbf{B}, \mathbf{B}_1, \mathbf{N}, \mathbf{K}$, and $\mathbf{C}$ are constant matrices of appropriate dimensions. Our emphasis is on single-input, single-output control systems, where u, r, and y are scalars.

The state equations of the closed-loop control system are obtained by substituting for $\mathbf{u}$ from Eq. (3.2) in Eq. (3.1) as

$$\dot{\mathbf{x}}(t) = (\mathbf{A} - \mathbf{BK})\mathbf{x}(t) + \mathbf{BNr}(t) + \mathbf{B}_1\mathbf{v}(t) \tag{3.4}$$

The transfer function relating the output $\mathbf{y}$ to the command input $\mathbf{r}$ of the closed-loop system can be obtained directly from the block diagram or from the state equations, Eqs. (3.3) and (3.4), as follows. Laplace transforming Eq. (3.4) with initial conditions $\mathbf{x}(0) = \mathbf{0}$, we obtain

$$s\mathbf{X}(s) = (\mathbf{A} - \mathbf{BK})\mathbf{X}(s) + \mathbf{BNR}(s) + \mathbf{B}_1\mathbf{V}(s)$$

and it follows that

$$(s\mathbf{I} - \mathbf{A} + \mathbf{BK})\mathbf{X}(s) = \mathbf{BNR}(s) + \mathbf{B}_1\mathbf{V}(s)$$

where $\mathbf{I}$ is a $n \times n$ identity matrix. Hence,

$$\mathbf{X}(s) = (s\mathbf{I} - \mathbf{A} + \mathbf{BK})^{-1}\mathbf{BNR}(s) + (s\mathbf{I} - \mathbf{A} + \mathbf{BK})^{-1}\mathbf{B}_1\mathbf{V}(s)$$

Laplace transforming Eq. (3.3) and substituting for $\mathbf{X}(s)$ from the preceding equation, we obtain

$$\mathbf{Y}(s) = \mathbf{C}(s\mathbf{I} - \mathbf{A} + \mathbf{BK})^{-1}\mathbf{BNR}(s) + \mathbf{C}(s\mathbf{I} - \mathbf{A} + \mathbf{BK})^{-1}\mathbf{B}_1\mathbf{V}(s) \tag{3.5}$$

Since

$$(s\mathbf{I} - \mathbf{A} + \mathbf{BK})^{-1} = \frac{\text{Adj}\,(s\mathbf{I} - \mathbf{A} + \mathbf{BK})}{\det|s\mathbf{I} - \mathbf{A} + \mathbf{BK}|}$$

the closed-loop transfer function relating the output $\mathbf{Y}(s)$ to the command input $\mathbf{R}(s)$ is given by

$$\mathbf{G}_c(s) = \frac{\mathbf{C}[\text{Adj}\,(s\mathbf{I} - \mathbf{A} + \mathbf{BK})]\mathbf{BN}}{\det|s\mathbf{I} - \mathbf{A} + \mathbf{BK}|} \tag{3.6}$$

and that relating the output $\mathbf{Y}(s)$ to the disburbance $\mathbf{V}(s)$ by

$$\mathbf{G}_d(s) = \frac{\mathbf{C}[\text{Adj}\,(s\mathbf{I} - \mathbf{A} + \mathbf{BK})]\mathbf{B}_1}{\det|s\mathbf{I} - \mathbf{A} + \mathbf{BK}|} \tag{3.7}$$

When r, y, and v are scalars, Eqs. (3.6) and (3.7) represent scalar transfer functions. The closed-loop transfer functions can also be obtained from the block diagram as follows. After obtaining the transfer functions of the components, let the block diagram be represented as shown in Fig. 3.2. Since superposition is valid for a linear system, we consider $\mathbf{r}$ and $\mathbf{v}$ separately, one at a time.

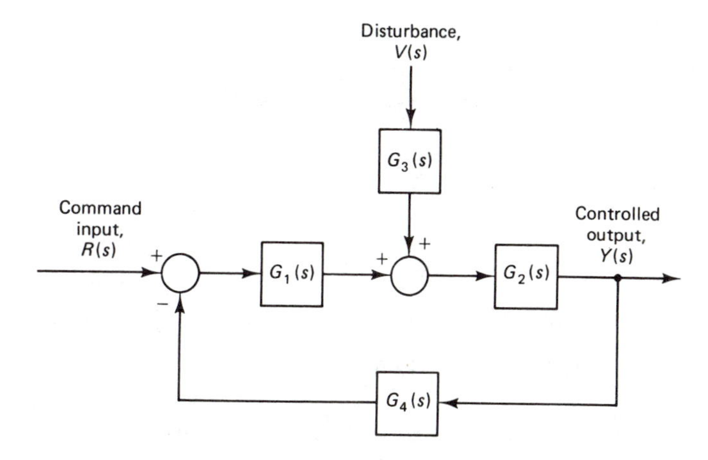

Figure 3.2 Block diagram representation.

First, setting $\mathbf{v} = \mathbf{0}$, from Fig. 3.2, we obtain

$$\mathbf{Y}(s) = \mathbf{G}_2(s)\mathbf{G}_1(s)[\mathbf{R}(s) - \mathbf{G}_4(s)\mathbf{Y}(s)]$$

Hence,

$$[\mathbf{I} + \mathbf{G}_2(s)\mathbf{G}_1(s)\mathbf{G}_4(s)]\mathbf{Y}(s) = \mathbf{G}_2(s)\mathbf{G}_1(s)\mathbf{R}(s)$$

or

$$\mathbf{Y}(s) = [\mathbf{I} + \mathbf{G}_2\mathbf{G}_1\mathbf{G}_4]^{-1}\mathbf{G}_2\mathbf{G}_1\mathbf{R}(s)$$

and the closed-loop transfer function relating output $\mathbf{Y}(s)$ to the command $\mathbf{R}(s)$ is given by

$$\mathbf{G}_c(s) = [\mathbf{I} + \mathbf{G}_2\mathbf{G}_1\mathbf{G}_4]^{-1}\mathbf{G}_2\mathbf{G}_1 \tag{3.8}$$

For a single-input, single-output system, it can be represented by

$$G_c(s) = \frac{G_2G_1}{1 + G_2G_1G_4} \tag{3.9}$$

Henceforth, we can obtain Eq. (3.9) by inspection of Fig. 3.2 by noting that G_2G_1 is the transfer function of the forward path and G_4 is the transfer function of the feedback path. Now setting $\mathbf{r} = 0$, we consider only the disturbance $\mathbf{V}$ and from Fig. 3.2 we get

$$\mathbf{Y}(s) = \mathbf{G}_2(s)[\mathbf{G}_3(s)\mathbf{V}(s) - \mathbf{G}_1(s)\mathbf{G}_4(s)\mathbf{Y}(s)]$$

Hence,

$$(\mathbf{I} + \mathbf{G}_2\mathbf{G}_1\mathbf{G}_4)\mathbf{Y}(s) = \mathbf{G}_2\mathbf{G}_3\mathbf{V}(s)$$

or

$$\mathbf{Y}(s) = (\mathbf{I} + \mathbf{G}_2\mathbf{G}_1\mathbf{G}_4)^{-1}\mathbf{G}_2\mathbf{G}_3\mathbf{V}(s)$$

The transfer function relating the output $\mathbf{Y}(s)$ to the disturbance $\mathbf{V}(s)$ is given by

$$\mathbf{G}_d(s) = (\mathbf{I} + \mathbf{G}_2\mathbf{G}_1\mathbf{G}_4)^{-1}\mathbf{G}_2\mathbf{G}_3 \tag{3.10}$$

and for a scaler Y and V, it can be represented by

$$G_d(s) = \frac{G_2G_3}{1 + G_2G_1G_4} \tag{3.11}$$

We note that Eq. (3.8) is identical to Eq. (3.6) and (3.10) is identical to (3.7).

3.3 PASSIVE CONTROL SYSTEMS

A passive feedback control system does not use any external power source for sensing, error detection, amplification, or actuation. Only the energy available in the input to each and every component of the system is used to produce its output.* For example, the hydraulic servomotor discussed in Chapter 2, Fig. 2.42, is an active component. It uses an external source of hydraulic power to convert its input, which is the spool-valve displacement, to its output, which is the load position. The common passive control systems are regulators and are used in very simple applications. They are usually mechanical and employ components such as springs, dashpots, levers, linkages,and gears to perform the control functions.

EXAMPLE 3.1

A conceptual design of a passive mechanical-feedback system for regulating the liquid level in a tank is shown in Fig. 3.3. Here, q_{1e}, q_{2e}, and h_e are the constant equilibrium values. Let q_1, q_2, and h denote the deviations from their equilibrium values and q_d be a disturbance flow. In steady-state equilibrium, $q_{1e} = q_{2e}$, $q_d = 0$, and h_e is a constant. The control system is a regulator whose purpose is to maintain the head of the liquid equal to its desired or reference value h_e when there is a disturbance flow q_d. Obtain its linear mathematical model.

A change h in the liquid level is sensed by a float that is connected by a mechanical lever to a control valve. A turn screw in the float-lever mechanism is used to change the length L when a change h_r in the set point corresponding to the desired level h_e is required. Here, we assume that $h_r = 0$, that is, there is no change in the desired value of h_e. If the head increases by h, the valve moves an amount z and reduces the flow to the tank and vice versa. For small displacements, the valve displacement z is related to the float displacement h by

$$z = \frac{a}{b}h \tag{3.12}$$

where a and b are the lever lengths shown in Fig. 3.3 For a small deviation,

*There are also semiactive control systems that extract most of the energy requirements from the process to be controlled with minimal energy supplied externally.

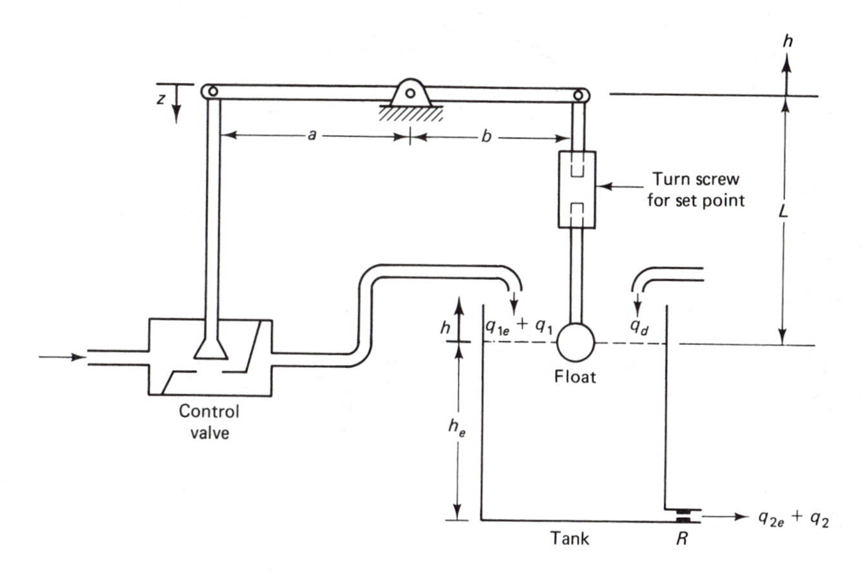

Figure 3.3 Liquid-level regulator system.

the linearized equation for the flow control valve is given by

$$q_1 = -c_1 z \tag{3.13}$$

where $c_1 > 0$. The negative sign in Eq. (3.13) indicates that when z increases, the flow q_1 decreases and vice versa. The continuity equation for the tank yields

$$q_1 + q_d - q_2 = A\frac{dh}{dt} \tag{3.14}$$

where A is the tank cross-sectional area and the outflow q_2 is obtained from Example 2.7 as $q_2 = (\rho g/R)h$. Here, R is the hydraulic resistance of the outlet orifice. Defining a time constant $\tau_1 = AR/\rho g$ as in Example 2.7, Eq. (3.14) becomes

$$q_1 + q_d = \frac{\rho g}{R}(\tau_1 D + 1)h \tag{3.15}$$

The block diagram is obtained from Eqs. (3.12), (3.13), and (3.15) and shown in Fig. 3.4(a). Letting $k_1 = ac_1/b$, the block diagram of Fig. 3.4(a) can be expressed as shown in Fig. 3.4(b), where h_r, which is the reference or desired change in the liquid level, has been set to zero and $-h$ becomes the error. When modeling a regulator, we can represent all variables as deviations from the equilibrium state that is required to be maintained in the presence of disturbances. When the set point is not changed, the reference input is then set to zero.

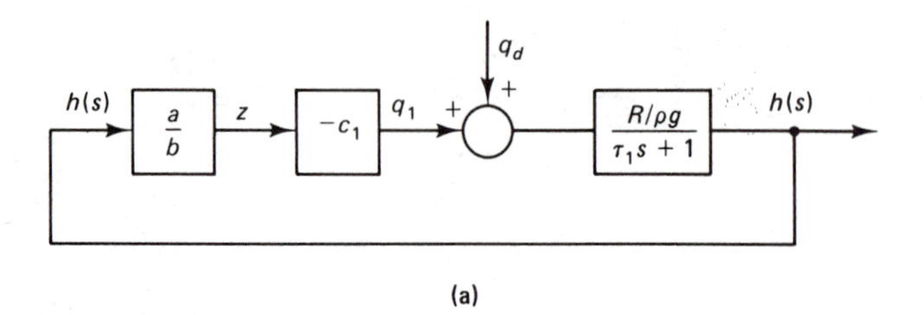

(a)

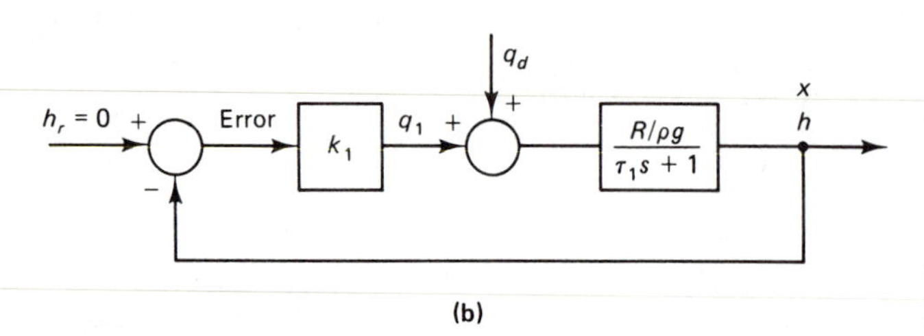

(b)

Figure 3.4 (a) Block diagram, and (b) standard block diagram.

For this system, which is of first order, we choose one state variable $x = h$. From Fig. 3.4(b), we obtain

$$(\tau_1 D + 1)x = \frac{R}{\rho g}(q_1 + q_d)$$

i.e.,

$$\dot{x} = -\left(\frac{1}{\tau_1}\right)x + \left(\frac{R}{\rho g \tau_1}\right)q_1 + \left(\frac{R}{\rho g \tau_1}\right)q_d \tag{3.16}$$

and

$$q_1 = -k_1 x \tag{3.17}$$

We note that Eq. (3.16) can also be obtained directly from Eq. (3.15) and Eq. (3.17) from Eqs. (3.12) and (3.13) with k_1 as defined in the preceding. On comparing Eqs. (3.16) and (3.17) to the generic equations, Eqs. (3.1) to (3.3), we note that $x = h$, $u = q_1$, $v = q_d$, $r = h_r = 0$, and $y = h$. Also, A, B, B_1, K, and C are scalars and are given by

$$A = -1/\tau_1 \qquad B = R/\rho g \tau_1 \qquad B_1 = R/\rho g \tau_1 \qquad K = k_1, \qquad C = 1$$

The state equation for the closed-loop system corresponding to the generic form Eq. (3.4), is obtained by substituting for the control law from Eq. (3.17) in Eq. (3.16) as

$$\dot{x} = -\frac{1}{\tau_1}\left(1 + \frac{Rk_1}{\rho g}\right)x + \left(\frac{R}{\rho g \tau_1}\right)q_d \tag{3.18}$$

where the scalar $A - BK$ of Eq. (3.4) becomes

$$A - BK = -\frac{1}{\tau_1}\left(1 + \frac{Rk_1}{\rho g}\right) \tag{3.19}$$

In case the set point is changed, that is, the desired change in the liquid level h_r is not zero, the control law becomes $q_1 = k_1(h_r - h)$ and Eq. (3.18) is modified to

$$\dot{x} = -\frac{1}{\tau_1}\left(1 + \frac{Rk}{\rho g}\right)x + \left(\frac{Rk_1}{\rho g \tau_1}\right)h_r + \left(\frac{R}{\rho g \tau_1}\right)q_d \tag{3.20}$$

and in Eqs. (3.2) and (3.4), $N = k_1$. It will be seen later in Chapter 7 that this control law, where q_1 is proportional to the error, does not possess a good disturbance-rejection property. The closed-loop transfer functions relating the output $h(s)$ to the command input $h_r(s)$ and the disturbance q_d can be obtained either from Eq. (3.5) or from Eqs. (3.9) and (3.11). It can be verified by employing both methods that

$$h(s) = \left(\frac{k_1 R/\rho g}{\tau_1 s + 1 + k_1 R/\rho g}\right)h_r(s) + \left(\frac{R/\rho g}{\tau_1 s + 1 + k_1 R/\rho g}\right)q_d(s) \tag{3.21}$$

EXAMPLE 3.2

A mechanical, passive, feedback control system was discussed in Example 1.3. A boom, which is modeled as a uniform beam of length L, is held in a bearing at its lower end. A passive regulator is to be designed to maintain the boom in its vertical, unstable equilibrium position. The conceptual design uses a spring as a sensor and actuator as shown in Fig. 3.5(a). Obtain its linear mathematical model.

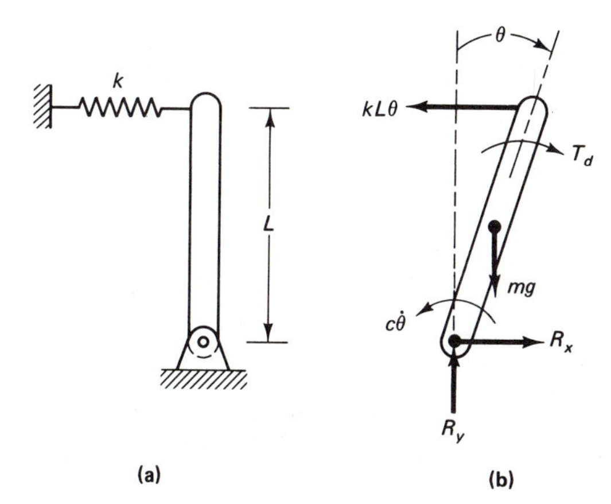

Figure 3.5 (a) A passive regulator, and (b) its freebody diagram.

A freebody diagram of the system is shown in Fig. 3.5(b). It is assumed that for small θ, the spring displacement is $L\theta$, the spring constant is k, that there is a viscous friction torque at the bearing with coefficient c, and T_d is the disturbance torque. The mass moment of inertia of the beam about the bearing is $(1/3)mL^2$. Taking moments about the bearing, we obtain

$$T_d + mg\frac{L}{2}\sin\theta - c\dot{\theta} - kL^2\theta = \frac{1}{3}mL^2\ddot{\theta} \tag{3.22}$$

For small θ, $\sin\theta$ is replaced by θ and it follows from Eq. (3.22) that

$$(\tfrac{1}{3}mL^2D^2 + cD - \tfrac{1}{2}mgL)\theta = -kL^2\theta + T_d \tag{3.23}$$

The left-hand side of this equation represents the system to be controlled and on the right-hand side, $u = -kL^2\theta$ is the control law produced by the spring. The block diagram of Eq. (3.23) is shown in Fig. 3.6, where θ_r, which is the command or desired change in the angular position, has been set to zero and $-\theta$ represents the error.

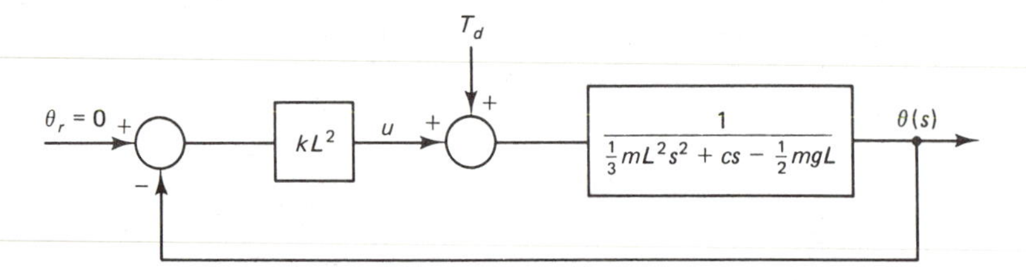

Figure 3.6 Block diagram of the control system.

For this second-order system, we choose $x_1 = \theta$ and $x_2 = \dot{\theta}$ as the state variables to represent Eq. (3.23) as

$$\dot{x}_1 = x_2$$

$$\dot{x}_2 = \left(\frac{3g}{2L}\right)x_1 - \left(\frac{3c}{mL^2}\right)x_2 + \left(\frac{3}{mL^2}\right)u + \left(\frac{3}{mL^2}\right)T_d \tag{3.24}$$

$$\begin{aligned} u &= -kL^2x_1 \\ y &= x_1 \end{aligned} \tag{3.25}$$

The matrices of Eqs. (3.1) to (3.3) are obtained from inspection of the preceding equations as

$$\mathbf{A} = \begin{bmatrix} 0 & 1 \\ 3g/2L & -3c/mL^2 \end{bmatrix} \qquad \mathbf{B} = \begin{bmatrix} 0 \\ 3/mL^2 \end{bmatrix} \qquad \mathbf{B}_1 = \begin{bmatrix} 0 \\ 3/mL^2 \end{bmatrix} \tag{3.26}$$

$$\mathbf{K} = \lfloor kL^2 \quad 0 \rfloor \qquad \mathbf{C} = \lfloor 1 \quad 0 \rfloor$$

The state equations for the closed-loop system corresponding to the form of Eq. (3.4) are obtained by substituting the control law of Eq. (3.25) in Eq. (3.24) to yield

$$\dot{x}_1 = x_2$$

$$\dot{x}_2 = \left(\frac{3g}{2L} - \frac{3k}{m}\right)x_1 - \left(\frac{3c}{mL^2}\right)x_2 + \left(\frac{3}{mL^2}\right)T_d$$

and the closed-loop matrix $\mathbf{A} - \mathbf{BK}$ of Eq. (3.4) is given by

$$\mathbf{A} - \mathbf{BK} = \begin{bmatrix} 0 & 1 \\ 3g/2L - 3k/m & -3c/mL^2 \end{bmatrix} \tag{3.27}$$

We now obtain the transfer function relating the output θ to the disturbance torque T_d by using Eq. (3.7). We have

$$s\mathbf{I} - \mathbf{A} + \mathbf{BK} = \begin{bmatrix} s & -1 \\ -3g/2L + 3k/m & s + 3c/mL^2 \end{bmatrix}$$

$$\text{Adj}\,(s\mathbf{I} - \mathbf{A} + \mathbf{BK}) = \begin{bmatrix} s + 3c/mL^2 & 1 \\ 3g/2L - 3k/m & s \end{bmatrix}$$

and

$$\det|s\mathbf{I} - \mathbf{A} + \mathbf{BK}| = s^2 + \left(\frac{3c}{mL^2}\right)s - \frac{3g}{2L} + \frac{3k}{m} \tag{3.28}$$

Substituting these results in Eq. (3.7), we obtain

$$G_d(s) = \frac{3/mL^2}{s^2 + \left(\dfrac{3c}{mL^2}\right)s - \dfrac{3g}{2L} + \dfrac{3k}{m}} \tag{3.29}$$

We can also obtain this transfer function from Eq. (3.11). On comparing Figs. 3.2 and 3.6, we get

$$G_1(s) = kL^2 \qquad G_2(s) = \frac{1}{\frac{1}{3}mL^2s^2 + cs - \frac{1}{2}mgL}$$

$$G_3(s) = 1 \qquad G_4(s) = 1$$

Substituting these results in Eq. (3.11), we can verify that the transfer function is given by Eq. (3.29).

3.4 ELECTRICAL CONTROL SYSTEMS

The characteristics of electrical control systems can be altered and the control law modified by simple and inexpensive resistance-capacitance networks. Hence, electrical control systems are very versatile. Electrical components have an advantage in applications where control signals are transmitted over long distances as in remote control. Electrical wiring or microwaves can be used for transmitting signals. In applications such as machine tools and robotics, dc electric motors are commonly used as actuators for relatively low power requirements. For large loads, hydraulic actuators are preferred because of size and weight considerations.

The electrical control systems most widely used in industry employ dc motors. The torque-speed characteristics of ac motors are highly nonlinear unlike those of separately excited dc motors. However, ac motors are relatively inexpensive and two-phase induction motors are sometimes used for low-power (fractional horsepower) applications.* But advances in rare-earth permanent magnets for providing

*Three-phase ac induction motors with pulse-width modulated power amplifiers are currently gaining popularity in high-power control applications.

the magnetic flux and improvements in brush and commutator design have made it possible to use dc motors in many control applications that formerly used ac motors. A special dc motor used in digital control systems is called a stepper or step motor. Its input consists of electrical pulses and the motor converts each pulse into a fixed angular displacement. Stepper motors, however, produce a low torque. In the following, we give two examples of dc electrical control systems.

EXAMPLE 3.3

We consider the speed control of a diesel-electric locomotive. The efficiency of a diesel engine is very sensitive to speed. Hence, it is run at a constant speed, where its efficiency is maximum. The locomotive is powered by dc motors, located on each of the axles, because dc motors run efficiently over a wide speed range. The principle of operation of this control system is discussed with reference to the schematic diagram shown in Fig. 3.7, where only one motor is shown for convenience.

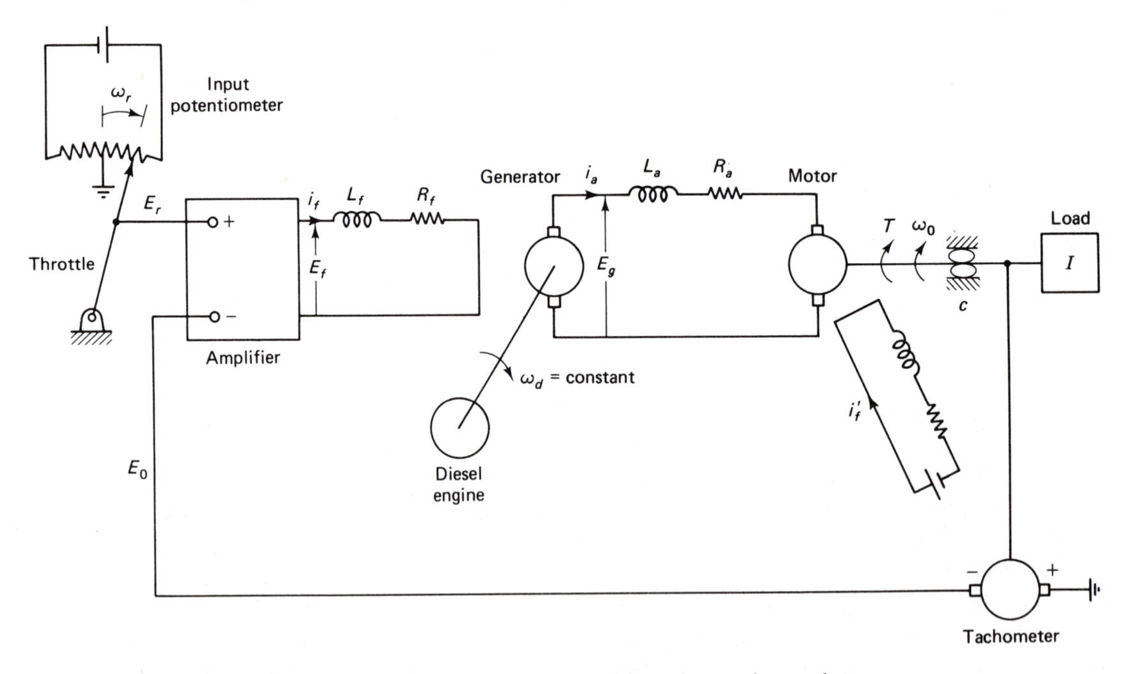

Figure 3.7 Schematic diagram of a speed-control system.

The throttle position corresponding to the desired change in reference speed ω_r is set on the input potentiometer, which provides a reference command voltage E_r. The controlled speed ω_0 is sensed by a tachometer, which supplies a feedback voltage E_0. The tachometer may be belt driven from the motor shaft. An electronic amplifier amplifies the error $(E_r - E_0)$ between the reference and feedback voltage signals and provides a voltage E_f that is supplied to the field winding of a dc generator.

The generator is run at a constant speed ω_d by the diesel engine and generates a voltage E_g that is supplied to the armature of a dc motor. The motor is armature controlled with a fixed current i supplied to its field. As a result, the motor produces a torque T and drives the load connected to its shaft so that the controlled speed ω_0 tends to equal the command speed ω_r.

A linear mathematical model of this control system is now obtained as follows. The command voltage E_r from the potentiometer is proportional to the desired speed ω_r, that is,

$$E_r = c_1\omega_r \tag{3.30}$$

where c_1 is a constant. The actual speed ω_0 is measured by a tachometer whose flux ϕ is provided by a permanent magnet. Referring to Eq. (2.124), the voltage generated by the tachometer is given by

$$E_0 = k_2\phi\omega_0 = c_2\omega_0 \tag{3.31}$$

where the constant $c_2 = k_2\phi$. The amplifier amplifies the error voltage $(E_r - E_0)$, and its output voltage E_f is given by

$$E_f = k_a(E_r - E_0) \tag{3.32}$$

where k_a is the gain of the amplifier. The voltage E_f is supplied to the field of the generator and causes a field current i_f. The voltage balance yields

$$\begin{aligned} E_f &= (L_fD + R_f)i_f \\ &= R_f(\tau_3D + 1)i_f \end{aligned} \tag{3.33}$$

where time constant $\tau_3 = L_f/R_f$. The voltage E_g produced by the generator is obtained by referring to Eq. (2.124) as

$$\begin{aligned} E_g &= k_3\Phi\omega_d \\ &= (k_3k_4\omega_d)i_f \\ &= k_gi_f \end{aligned} \tag{3.34}$$

where $\Phi = k_4i_f$, and $k_g = k_3k_4\omega_d$. The block diagram of the preceding equations, Eqs. (3.30) through (3.34), is shown in Fig. 3.8.

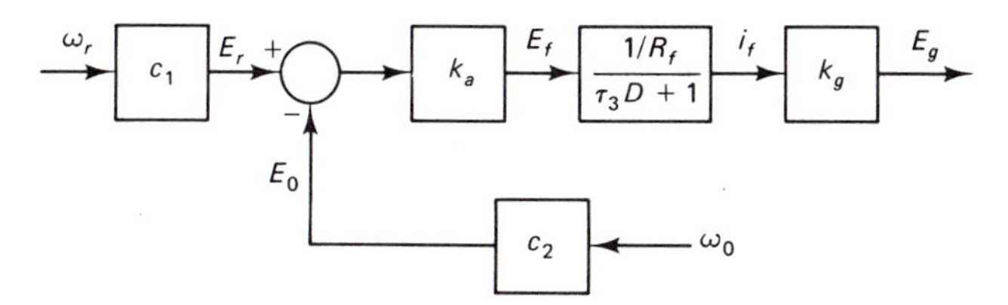

Figure 3.8 Partial block diagram of a speed-control system.

The remaining part of the block diagram may now be completed by using Fig. 2.38 and in that figure replacing E_a by E_g, ω by ω_0, and terminating the diagram at the controlled output speed ω_0. However, the development of the equations is repeated in the following for convenience. Let L_a and R_a be the inductance and resistance of the combined generator and motor armature windings. Balancing the voltages, we obtain

$$\begin{aligned} E_g - E_b &= (L_a D + R_a) i_a \\ &= R_a(\tau_1 D + 1) i_a \end{aligned} \tag{3.35}$$

where $\tau_1 = L_a/R_a$ and the back emf of the motor is obtained from Eq. (2.128) as

$$E_b = k_b \omega_0 \tag{3.36}$$

The torque T produced by the motor and the equation for the torque balance are obtained from Eqs. (2.127) and (2.126), respectively, and are given by

$$T = k_t i_a \tag{3.37}$$

$$\begin{aligned} T - T_d &= (ID + c)\omega_0 \\ &= c(\tau_2 D + 1)\omega_0 \end{aligned} \tag{3.38}$$

The complete block diagram of the control system is obtained from Eqs. (3.35) through (3.38) and Fig. 3.8 and is shown in Fig. 3.9. The inner feedback loop is inherent and is caused by the motor back emf, whereas the outer loop is the feedback control loop. The input scale factor c_1 must be chosen such that $c_1 = c_2$.

It is seen that the control law $u = E_f = k_a(E_r - E_0)$, that is, it is proportional to the error. It will be seen in Chapter 7 that this proportional control law has a poor disturbance-rejection property. To obtain the state representation of the control system, we note from Fig. 3.9 that the system is of third order. Hence, we need three state variables, but their choice is not unique. We select $x_1 = \omega_0$, $x_2 = i_a$, and $x_3 = i_f$ as the state variables. From Fig. 3.9, we obtain

$$(\tau_2 D + 1)\omega_0 = \frac{1}{c}(k_t i_a - T_d)$$

Also,

$$(\tau_1 D + 1) i_a = \frac{1}{R_a}(k_g i_f - k_b \omega_0)$$

and

$$(\tau_3 D + 1) i_f = \frac{1}{R_f} u$$

$$u = k_a(c_1 \omega_r - c_2 \omega_0)$$

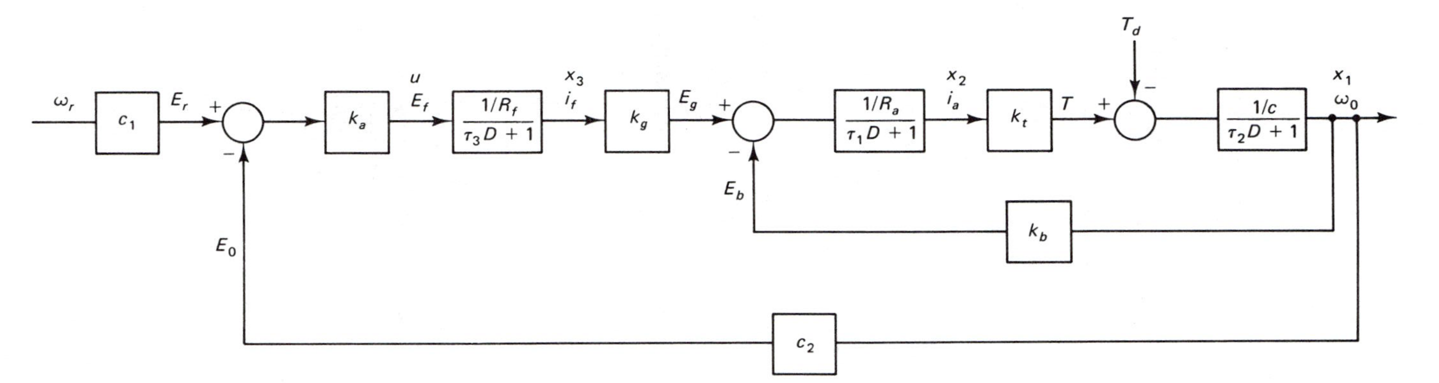

Figure 3.9 Block diagram of a speed-control system.

With our choice of the state variables, the state representation is obtained from the preceding equations as

$$\dot{x}_1 = -\left(\frac{1}{\tau_2}\right)x_1 + \left(\frac{k_t}{c\tau_2}\right)x_2 - \left(\frac{1}{c\tau_2}\right)T_d$$

$$\dot{x}_2 = -\left(\frac{k_b}{R_a\tau_1}\right)x_1 - \left(\frac{1}{\tau_1}\right)x_2 + \left(\frac{k_g}{R_a\tau_1}\right)x_3$$

$$\dot{x}_3 = -\left(\frac{1}{\tau_3}\right)x_3 + \left(\frac{1}{R_f\tau_3}\right)u \tag{3.39}$$

$$u = k_a(c_1\omega_r - c_2x_1)$$

Obviously, the scale factor c_1 must be chosen such that $c_1 = c_2$ and the control law becomes

$$u = k_ac_1(\omega_r - x_1) \tag{3.40}$$

The matrices of Eq. (3.4) are obtained from inspection of Eqs. (3.39) and (3.40) as

$$\mathbf{A} = \begin{bmatrix} -1/\tau_2 & k_t/c\tau_2 & 0 \\ -k_b/R_a\tau_1 & -1/\tau_1 & k_g/R_a\tau_1 \\ 0 & 0 & -1/\tau_3 \end{bmatrix} \qquad \mathbf{B} = \begin{bmatrix} 0 \\ 0 \\ 1/R_f\tau_3 \end{bmatrix} \qquad \mathbf{B}_1 = \begin{bmatrix} -1/c\tau_2 \\ 0 \\ 0 \end{bmatrix}$$

$$\mathbf{K} = \lfloor k_ac_1 \quad 0 \quad 0 \rfloor \qquad N = k_ac_1 \tag{3.41}$$

For the closed-loop system, Eq. (3.4), we get

$$\mathbf{A} - \mathbf{BK} = \begin{bmatrix} -1/\tau_2 & k_t/c\tau_2 & 0 \\ -k_b/R_a\tau_1 & -1/\tau_1 & k_g/R_a\tau_1 \\ -k_ac_1/R_f\tau_3 & 0 & -1/\tau_3 \end{bmatrix} \tag{3.42}$$

EXAMPLE 3.4

The conceptual design of a dc electrical position-control system is shown in the schematic diagram of Fig. 3.10. An application of such a system is in the area of robotics for the position control of arms. Usually, each degree of freedom employs its own actuator and this example has relevance to the control of a single axis as shown in Fig. 3.11. Such systems are also employed for position control in machine tools, radar antenna, and various other applications where the power requirements are not high. Hydraulic actuators are preferred for high-power applications.

The operation of the system is as follows. A load with an inertia I and viscous friction with coefficient c is to be positioned at some desired angle θ_r. The desired angle θ_r may be set as shown in Fig. 3.10 on an input potentiometer, which provides the reference voltage E_r. Alternately, the reference voltage E_r corresponding to θ_r may be provided by a tape reader or computer storage through a digital-to-analog converter. The controlled position θ_0 is measured

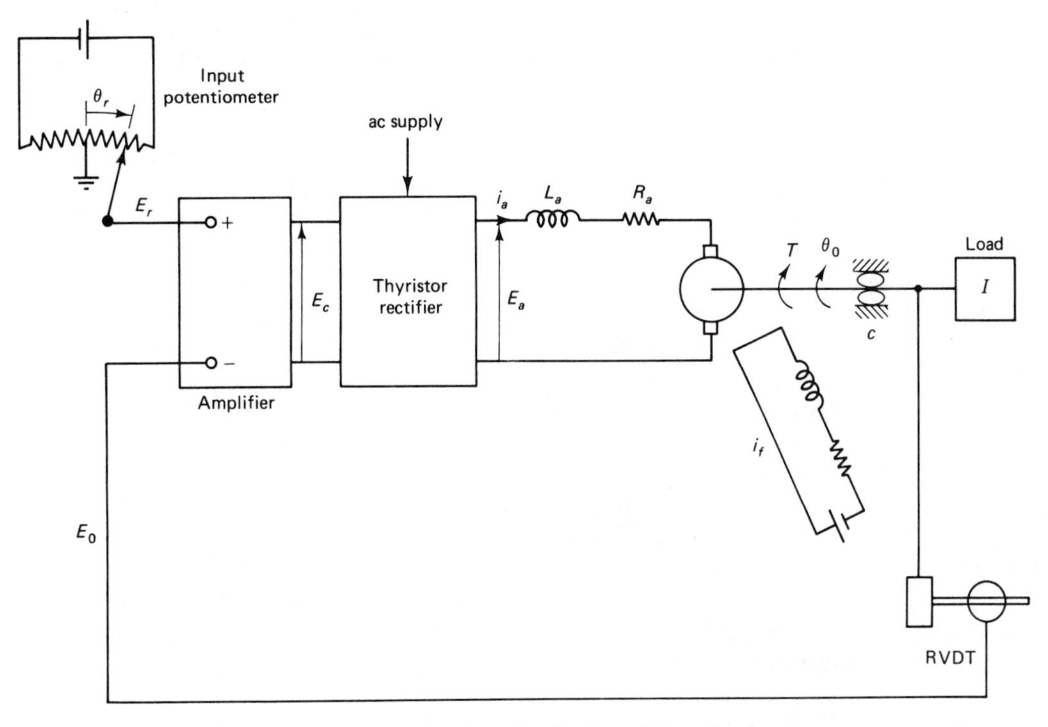

Figure 3.10 Schematic of a dc position-control system.

by a rotary variable-differential transformer (RVDT), which supplies a feedback voltage E_0.

A rotary potentiometer could be employed to sense θ_0, but it is usually avoided for continuous use because the contact of the sliding wiper reduces its life. In RVDT and its rectilinear counterpart, the linear variable-differential

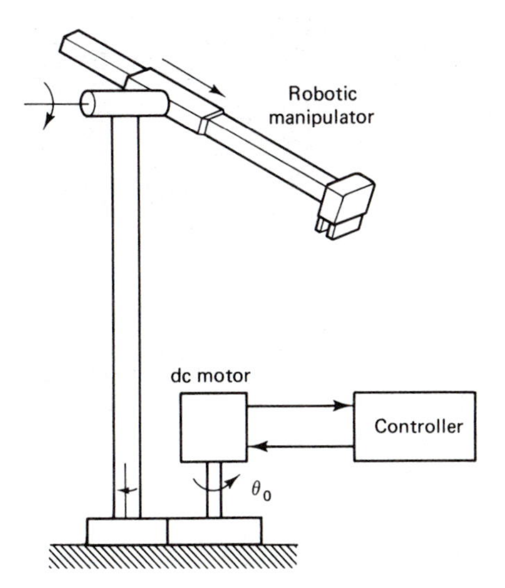

Figure 3.11 Position control of a robot arm.

transformer (LVDT), there is no physical contact between the core and the coil and hence the mechanical components do not wear out or deteriorate. The corresponding absence of friction leads to high resolution, no hysteresis, and high reliability.

The error voltage $(E_r - E_0)$ is amplified by an electronic amplifier that supplies a voltage E_c to a thyristor rectifier, which is a power amplifier. Electronic power amplifiers have drift and do not produce sufficient power to drive a motor when the load is high. The old approach is to use a motor-generator set of Example 3.3. The generator, which is essentially a power amplifier, is run at a constant speed by a prime mover such as a synchronous motor. But in modern drives, the classical motor-generator set is replaced by a thyristor rectifier, which has the advantages of high efficiency, small size, fast response, and lower cost. Its disadvantage is that it produces a high ripple at its output.

The thyristor is supplied by an external single-phase or three-phase ac power and it amplifies its input voltage E_c to produce an output voltage E_a, which is supplied to the armature of a dc motor. A current-limiting protective feature to prevent damage to the thyristors and regenerative braking capabilities can be incorporated if desired (Sen and MacDonald, 1978). The motor is armature controlled with fixed field current. It produces a torque T to position a load connected to its shaft so that the controlled position θ_0 tends to equal the command θ_r.

A linear mathematical model of this control system is now obtained in the following. The reference voltage E_r is proportional to the command position θ_r. Hence, we have

$$E_r = c_1 \theta_r \tag{3.43}$$

The controlled position θ_0 is measured by the RVDT. A typical RVDT (Herceg, 1972) is illustrated in Fig. 3.12. The primary and secondary windings

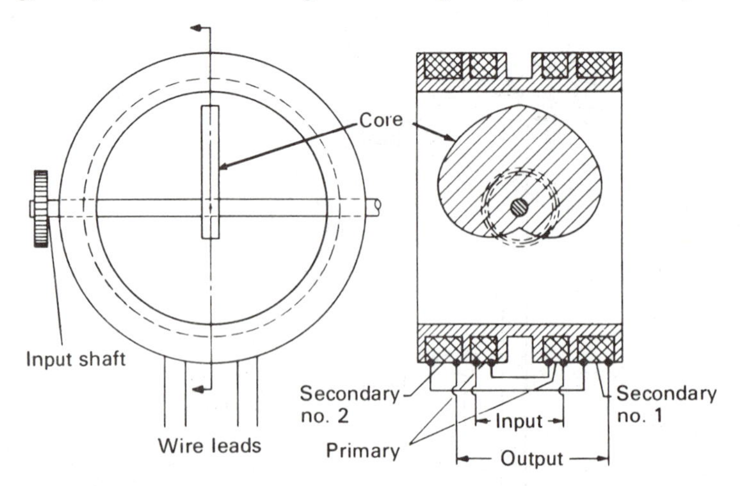

Figure 3.12 Schematic diagram of a RVDT. (Courtesy of Schaevitz Engineering, Pennsauken, N.J.)

are wound symmetrically on a stator. A cardioid-shaped cam of magnetic material is used as a rotor. An input shaft, which is driven by the motor shaft, passes across the middle of the stator at the plane of winding symmetry and is fastened to the cardioidal core. The shape of the rotor is chosen to produce a linear output voltage over a specified range of rotation.

The primary winding is supplied by an ac excitation voltage and the RVDT produces an ac output voltage at the secondary winding, depending on the rotation of the core. Signal conditioning is used to convert the ac output to a dc voltage. The output voltage of a typical RVDT is shown in Fig. 3.13. When the core is centered so that the angular displacement is zero,

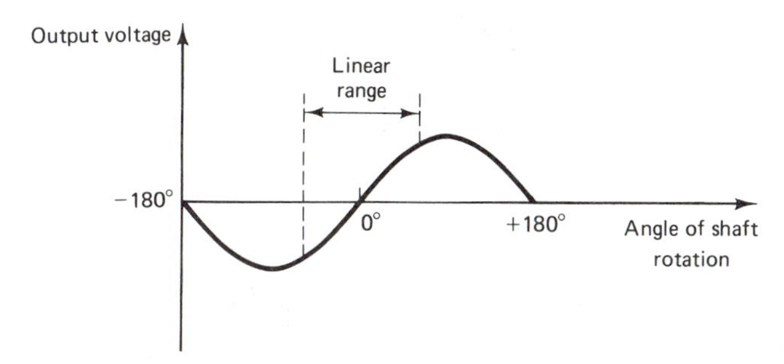

Figure 3.13 Output versus shaft rotation for a RVDT.

there is no output voltage. There are two linear operating ranges 180° apart. The practical upper limit for linearity is about ±60°. Within this linear range, the feedback voltage E_0 provided by the RVDT is given by

$$E_0 = c_2\theta_0 \tag{3.44}$$

The electronic amplifier amplifies the error voltage and produces a voltage E_c so that

$$E_c = k_a(E_r - E_0) \tag{3.45}$$

where k_a is the gain of the amplifier. The voltage E_c is fed to the driver of the thyristor rectifier. The driver produces time gate pulses that control the condution of the thyristors in the rectifier module. The rectified output voltage E_a depends on the firing angle of the pulses relative to the ac supply waveform. A linear relationship between the input voltage E_c and output voltage E_a can be obtained when a proper firing control scheme is used (Pelly, 1971). The time constants associated with the rectifier are negligibly small. Neglecting the dynamics of the rectifier, we get

$$E_a = k_r E_c \tag{3.46}$$

where k_r is the gain of the rectifier. It can be shown (Pelly, 1971) that this gain is given by

$$k_r = \frac{3\sqrt{2}\,V_{LL}}{\pi V_0} \tag{3.47}$$

where V_0 corresponds to the zero firing angle, and V_{LL} is the ac line-to-line

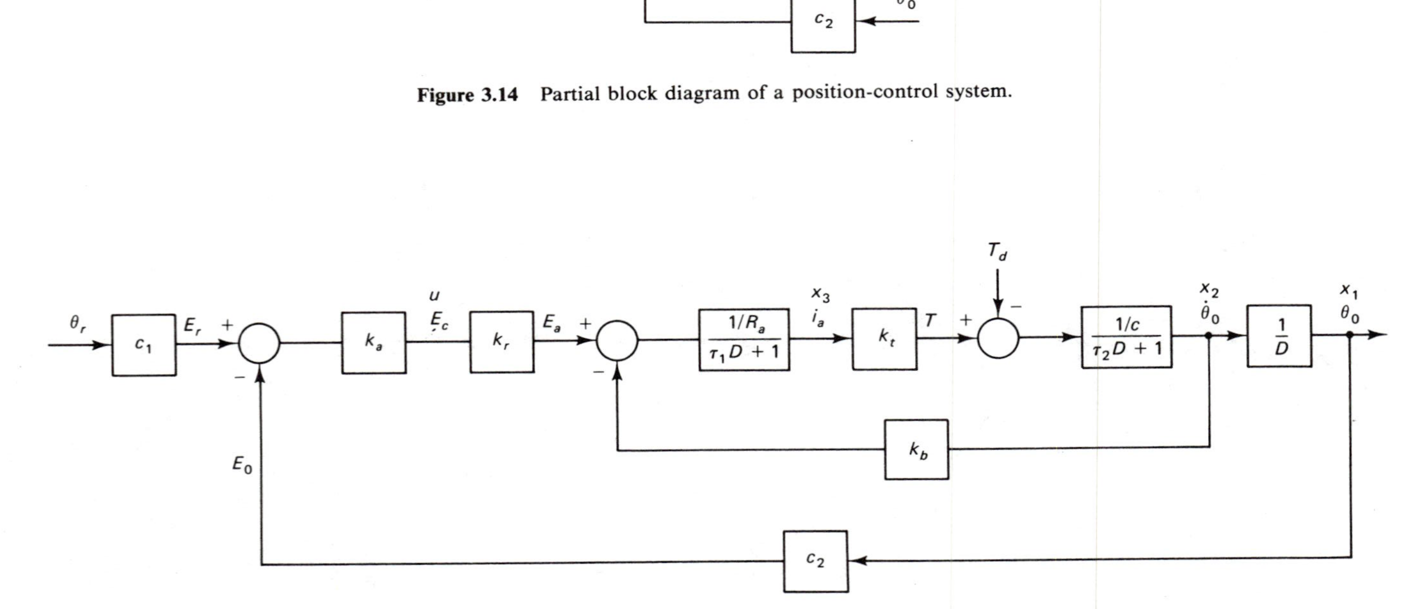

Figure 3.14 Partial block diagram of a position-control system.

Figure 3.15 Block diagram of a position-control system.

rms voltage. The block diagram of the preceding Eqs. (3.43) to (3.46) is shown in Fig. 3.14. The remaining part of the block diagram may now be completed by employing Fig. 2.38 and the complete block diagram is shown in Fig. 3.15.

This system is of third order and we select $x_1 = \theta$, $x_2 = \dot{\theta}$, and $x = i_a$ as the state variables. The control input $u = E_c$. From Fig. 3.15, we obtain

$$\dot{x}_1 = x_2$$

$$(\tau_2 D + 1)x_2 = \left(\frac{k_t}{c}\right)x_3 - \left(\frac{1}{c}\right)T_d$$

$$(\tau_1 D + 1)x_3 = -\left(\frac{k_b}{R_a}\right)x_2 + \left(\frac{k_r}{R_a}\right)u$$

and

$$u = k_a(c_1\theta_r - c_2\theta_0)$$

The scale factor c_1 must be chosen such that $c_1 = c_2$ and the control law can be expressed as

$$u = k_a c_1(\theta_r - x_1)$$

The state equations are obtained from the preceding equations and are given by

$$\begin{aligned}
\dot{x}_1 &= x_2 \\
\dot{x}_2 &= -\left(\frac{1}{\tau_2}\right)x_2 + \left(\frac{k_t}{\tau_2 c}\right)x_3 - \left(\frac{1}{\tau_2 c}\right)T_d \\
\dot{x}_3 &= -\left(\frac{k_b}{R_a\tau_1}\right)x_2 - \left(\frac{1}{\tau_1}\right)x_3 + \left(\frac{k_r}{R_a\tau_1}\right)u \\
u &= k_a c_1(\theta_r - x_1)
\end{aligned} \tag{3.48}$$

and output $y = x_1$.

The matrices of Eq. (3.4) are now obtained from inspection of these equations and we get

$$\mathbf{A} = \begin{bmatrix} 0 & 1 & 0 \\ 0 & -1/\tau_2 & k_t/\tau_2 c \\ 0 & -k_b/R_a\tau_1 & -1/\tau_1 \end{bmatrix} \qquad \mathbf{B} = \begin{bmatrix} 0 \\ 0 \\ k_r/R_a\tau_1 \end{bmatrix} \qquad \mathbf{B}_1 = \begin{bmatrix} 0 \\ -1/\tau_2 c \\ 0 \end{bmatrix} \tag{3.49}$$

$$\mathbf{K} = \lfloor k_a c_1 \quad 0 \quad 0 \rfloor \qquad N = k_a c_1 \qquad \mathbf{C} = \lfloor 1 \quad 0 \quad 0 \rfloor$$

The matrix of the closed-loop system is given by

$$\mathbf{A} - \mathbf{BK} = \begin{bmatrix} 0 & 1 & 0 \\ 0 & -1/\tau_2 & k_t/\tau_2 c \\ -k_r k_a c_1/R_a\tau_1 & -k_b/R_a\tau_1 & -1/\tau_1 \end{bmatrix} \tag{3.50}$$

3.5. HYDRAULIC CONTROL SYSTEMS

Hydraulic control systems have several advantages over other types. A comparatively small-size hydraulic actuator using high-pressure fluid can develop very large forces or torques to provide rapid acceleration or deceleration of heavy loads. Hence, for the same horsepower, hydraulic actuators are lighter than electrical motors and considerable reduction in size and weight can be achieved. Hydraulic components are more rugged and more resistant to vibrations and shocks than electrical components. Availability of both rectilinear actuators and rotary hydraulic motors makes the design versatile. Hydraulic fluid can be used to carry away the heat generated in the system and it also acts as a lubricant.

However, hydraulic control systems also have some disadvantages. A source of pressurized hydraulic fluid with supply and return lines is required. Hydraulic power is not as readily available as electrical power. The initial cost of hydraulic systems is usually higher than that of electrical systems. Leaks can be a problem and cause fire hazards unless fire resistant hydraulic fluids are used. The opening and closing of valves can cause oil hammer and the resulting pressure surges can damage the equipment. Care is required in preventing dirt from contaminating the hydraulic fluid as otherwise the failure of components can result.

EXAMPLE 3.5

The conceptual design of a hydraulic system for the speed control of a prime mover is shown by the schematic diagram of Fig. 3.16. The prime mover may be an engine or a turbine. The operation of this control system is as follows.

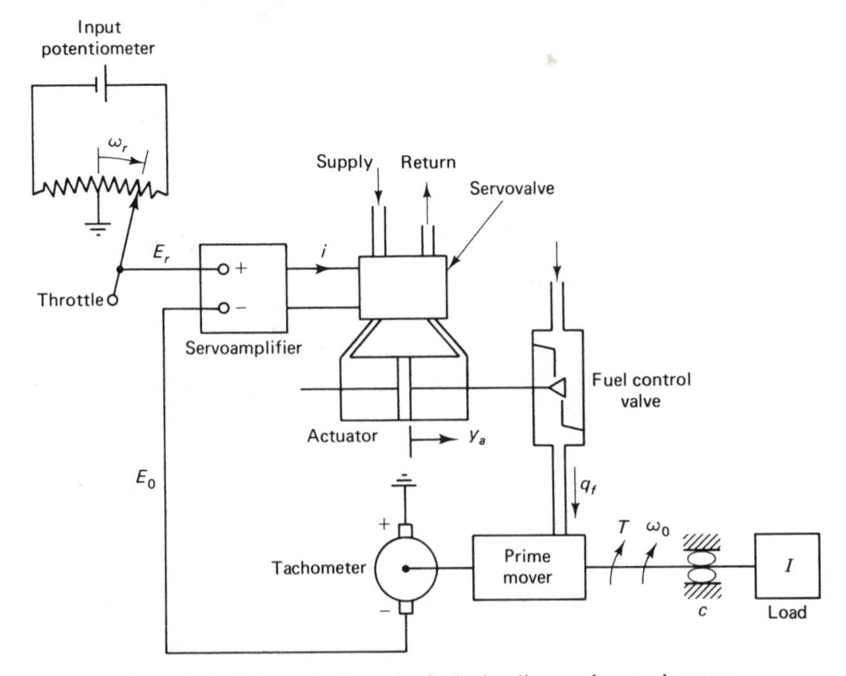

Figure 3.16 Schematic diagram of a hydraulic speed-control system.

The desired change in reference speed ω_r is set by the throttle position on an input potentiometer, which provides a reference command voltage E_r. The controlled speed ω_0 is sensed by a tachometer, which supplies the feedback voltage E_0. The tachometer is driven from the prime mover shaft. The servoamplifier determines the error voltage $(E_r - E_0)$ and supplies a current i to the servovalve torque motor. The servovalve admits hydraulic fluid to one side of the actuator, depending on the direction of the current i. The actuator positions a fuel control valve. The change in the fuel flow q_f causes a change in the speed of the prime mover. A load consisting of inertia I and viscous friction with coefficient c is connected to the prime mover shaft.

A linear mathematical model of this control system is obtained in the following. The command voltage E_r from the input potentiometer is proportional to the change in the desired speed ω_r. Hence, it follows that

$$E_r = c_1 \omega_r \tag{3.51}$$

The controlled speed ω_0 is sensed by a tachometer whose flux is provided by a permanent magnet. From Eq. (2.124), the voltage generated by the tachometer is given by

$$E_0 = c_2 \omega_0 \tag{3.52}$$

The electronic amplifiers may be classified as either voltage, current, or power amplifiers. The servoamplifier is a current amplifier and its output is given by

$$i = k_a(E_r - E_0) \tag{3.53}$$

where k_a is the amplifier gain. The transfer function of the servovalve relating the current i to the spool-valve displacement x_s is shown in the block diagram of Fig. 2.46 and is given by

$$x_s = \left(\frac{k_1}{\dfrac{1}{\omega_n^2} D^2 + \dfrac{2\zeta}{\omega_n} D + 1} \right) i \tag{3.54}$$

The transfer function relating the spool-valve displacement x_s to the actuator displacement y_a is shown in the block diagram of Fig. 2.43. Since the actuator positions a valve, its load is negligible and we get

$$y_a = \frac{k_g}{D} x_s \tag{3.55}$$

For a small displacement y_a of the fuel-control valve from its equilibrium position, a linear relationship for the change q_f in fuel flow is given by

$$q_f = c_3 y_a \tag{3.56}$$

where it is seen from Fig. 3.16 that $c_3 > 0$. A partial block diagram, using Eqs. (3.51) to (3.56), is shown in Fig. 3.17.

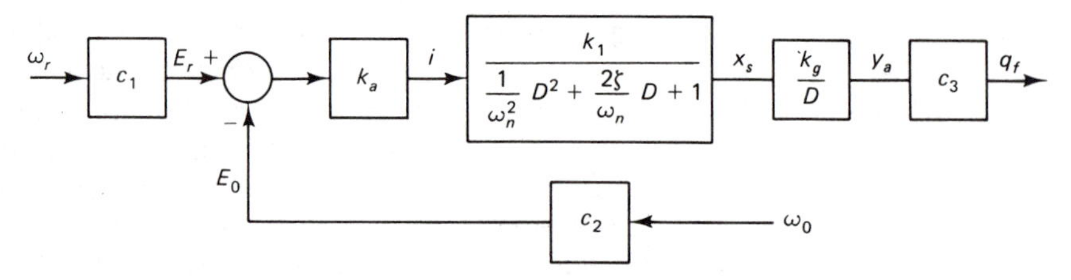

Figure 3.17 Partial block diagram of a speed-control system.

The typical, experimentally obtained characteristic curves of a prime mover are shown in Fig. 3.18, where fuel flow is plotted versus speed for different values of constant torque. We now employ the linearization technique discussed in section 2.2.3 and illustrated in Fig. 2.4. For small deviations from equilibrium, we obtain

$$\omega_0 = c_4 q_f - c_5 \Delta T \tag{3.57}$$

where ω_0, q_f, and ΔT represent deviations in output speed, fuel flow, and torque, respectively, and $c_4 > 0$, $c_5 > 0$. The signs in Eq. (3.57) reflect the fact that the speed increases when the fuel flow is increased and decreases when the load torque is increased. The equation for the torque balance yields

$$\Delta T = I\dot{\omega}_0 + c\omega_0 + T_d \tag{3.58}$$

where I is the load inertia, c is the coefficient of viscous friction, and T_d is the disturbance torque. Substituting for ΔT from Eq. (3.58) in Eq. (3.57), we get

$$\omega_0 = c_4 q_f - c_5(ID + c)\omega_0 - c_5 T_d$$

i.e.,

$$c_4 q_f - c_5 T_d = (c_5 ID + 1 + c_5 c)\omega_0$$

$$= \frac{1}{k_2}(\tau D + 1)\omega_0 \tag{3.59}$$

where the time constant $\tau = c_5 I/(1 + c_5 c)$, and $1/k_2 = 1 + c_5 c$.

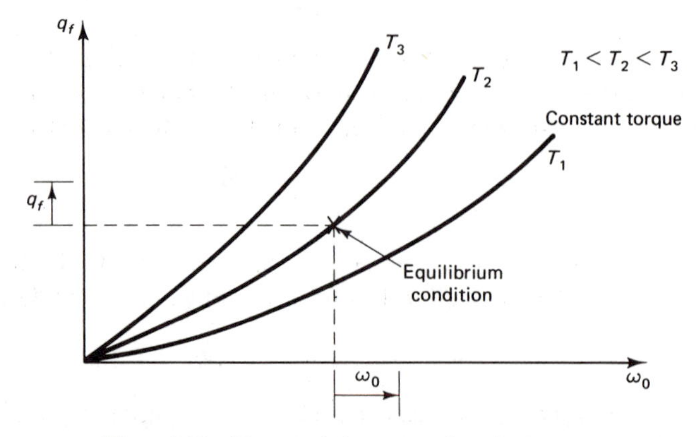

Figure 3.18 Characteristic curves of a prime mover.

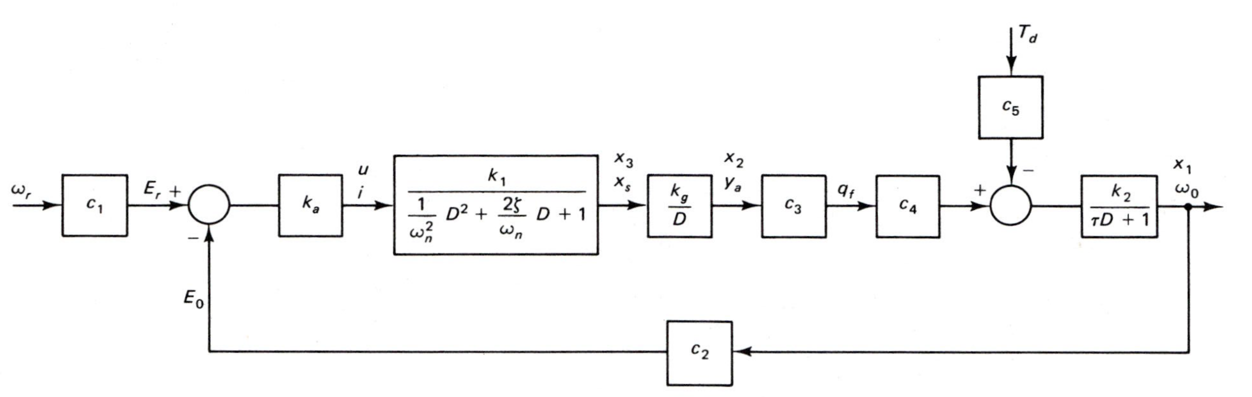

Figure 3.19 Block diagram of a hydraulic speed-control system.

The complete block diagram of the control system is obtained by using the partial block diagram of Fig. 3.17 and Eq. (3.59) and is shown in Fig. 3.19. It is seen that the system is of fourth order. It is expected that the spool-valve natural frequency ω_n would be high, that is, $1/\omega$ is very small compared to the prime mover time constant τ. It may be possible to accurately represent the transfer function of the servovalve relating current i to the spool-valve displacement x_s by a simple gain, that is $x_s = k_1 i$, from the considerations of bandwidth discussed in Chapter 5. The overall control system would thus be reduced to second order. The identification of parameter values is discussed in Chapter 5. Hence, lacking a priori knowledge of parameter values, we retain the servovalve dynamics in this chapter.

The four state variables required to represent this system are chosen as $x_1 = \omega_0$, $x_2 = y_a$, $x_3 = x_s$, and $x_4 = \dot{x}_s$. The control input is given by $u = i$. From Fig. 3.19, we obtain the equations

$$(\tau D + 1)\omega_0 = k_2 c_4 c_3 y_a - k_2 c_5 T_d$$

$$D y_a = k_g x_s$$

$$\left(\frac{1}{\omega_n^2} D^2 + \frac{2\zeta}{\omega_n} D + 1\right) x_s = k_1 i$$

$$i = k_a(c_1\omega_r - c_2\omega_0) = k_a c_1(\omega_r - \omega_0)$$

where in the last equation, the scale factor is chosen such that $c_1 = c_2$. With our choice of the state variables, the preceding equations are represented by

$$\begin{aligned}
\dot{x}_1 &= -\frac{1}{\tau} x_1 + \frac{1}{\tau} k_2 c_4 c_3 x_2 - \frac{1}{\tau} k_2 c_5 T_d \\
\dot{x}_2 &= k_g x_3 \\
\dot{x}_3 &= x_4 \\
\dot{x}_4 &= -\omega_n^2 x_3 - 2\zeta\omega_n x_4 + \omega_n^2 k_1 u
\end{aligned} \tag{3.60}$$

$$\text{Control law:} \qquad u = k_a c_1(\omega_r - x_1)$$

$$\text{Output:} \qquad y = x_1$$

The matrices of Eq. (3.4) are now obtained from inspection of Eq. (3.60) and we get

$$\mathbf{A} = \begin{bmatrix} -1/\tau & (1/\tau)k_2 c_4 c_3 & 0 & 0 \\ 0 & 0 & k_g & 0 \\ 0 & 0 & 0 & 1 \\ 0 & 0 & -\omega_n^2 & -2\zeta\omega_n \end{bmatrix} \qquad \mathbf{B} = \begin{bmatrix} 0 \\ 0 \\ 0 \\ \omega_n^2 k_1 \end{bmatrix}$$

$$\mathbf{B}_1 = \begin{bmatrix} -k_2 c_5/\tau \\ 0 \\ 0 \\ 0 \end{bmatrix}$$

$$\mathbf{K} = \lfloor k_a c_1 \quad 0 \quad 0 \quad 0 \rfloor \qquad N = k_a c_1 \qquad \mathbf{C} = \lfloor 1 \quad 0 \quad 0 \quad 0 \rfloor \tag{3.61}$$

The matrix $\mathbf{A} - \mathbf{BK}$ of the closed-loop system can be obtained from the preceding equations.

EXAMPLE 3.6

Many machine tools, such as lathes, milling machines and grinders, are controlled automatically. The conceptual design of a hydraulic control system for maching a workpiece on a lathe is shown in the schematic diagram of Fig. 3.20. The control system is a servomechanism, where the actual tool position $y_a(t)$ is required to follow a time-varying-command input position $y_r(t)$. The carriage is moved at a constant speed and the control of the tool position is in the direction of the cross slide.

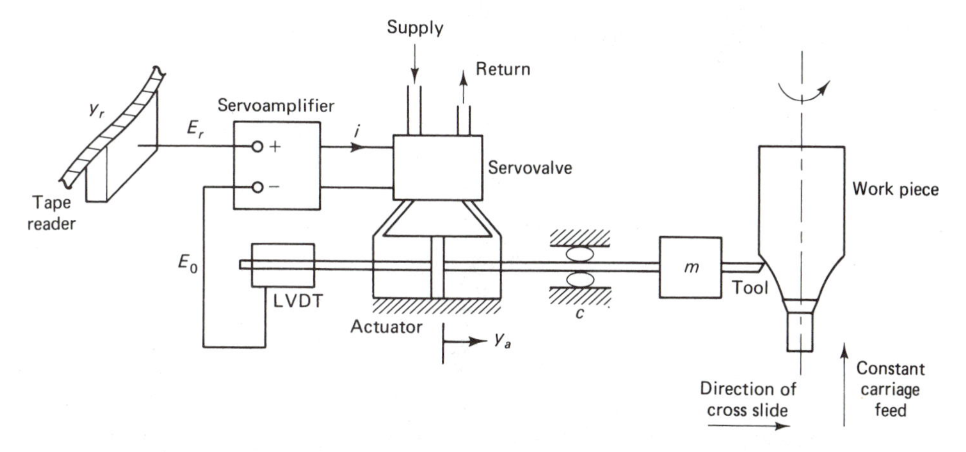

Figure 3.20 Schematic diagram of a hydraulic machining-control system.

The operation of the system is as follows. The command input $y_r(t)$ may be generated by a template, cam drive, a punched-tape reader, or a computer program through a digital-to-analog converter. When the command input is stored digitally, the automatic system is often referred to as a numerical control system. The input device produces a reference voltage E_r proportional to the command input y_r.

The actual position y_a of the tool is sensed by a linear variable-differential transformer (LVDT), which produces a feedback voltage E_0. The servo-amplifier determines the error voltage and supplies a current i to the servovalve. The servovalve admits high-pressure fluid to one side of the actuator such that the actuator piston displacement y_a follows the command input y_r. The tool post is connected directly to the actuator piston and hence its displacement is also given by y_a.

A linear mathematical model of this control system is now obtained in the following. The reference voltage E_r is proportional to the command input displacement y_r so that

$$E_r = c_1 y_r \tag{3.62}$$

The actual displacement y_a is measured by a LVDT, which is a mutual inductance sensor and is the rectilinear counterpart of the RVDT discussed in Example 3.4. The schematic diagram of a LVDT is shown in Fig. 3.21 (Herceg, 1972). It produces an electrical output proportional to the displacement of a moveable contactless core through a primary and two identical secondary windings. An ac voltage is applied to the primary winding. The displacement of the noncontacting magnetic core varies the mutual inductance of each secondary to the primary, which determines the voltage induced from the primary to each secondary.

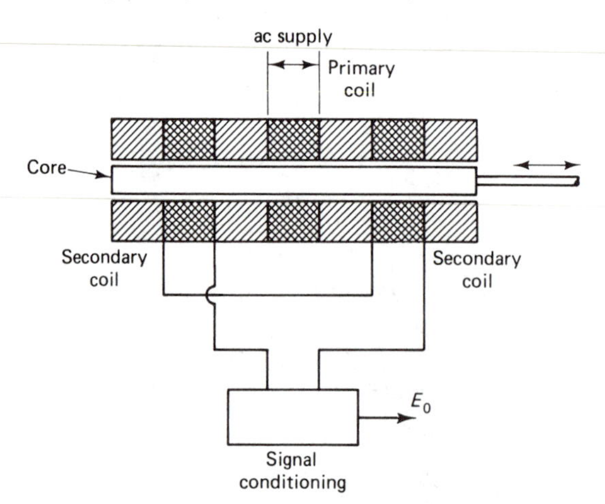

Figure 3.21 Linear variable-differential transformer (LVDT).

If the core is centered between the secondary windings, the voltages induced in each secondary are identical and 180° out of phase, so that there is no net output. If the core is moved off center, the mutual inductance of the primary with one secondary is greater than that with the other, and a differential voltage appears across the secondaries in series. For off-center displacements within the range of operation, this voltage is a linear function of the displacement. Signal conditioning is used to convert the ac output to a dc voltage. Hence, we have

$$E_0 = c_2 y_a \tag{3.63}$$

An LVDT has the same advantages over a linear potentiometer that an RVDT has over a rotary potentiometer. The servoamplifier, which is a current amplifier, yields the current output given by

$$i = k_a(E_r - E_0) \tag{3.64}$$

where k_a is the amplifier gain. The transfer function of the servovalve relating the current i to the spool-valve displacement x_s is shown in the block diagram of Fig. 2.46 and is given by

$$x_s = \left(\frac{k_1}{\frac{1}{\omega_n^2} D^2 + \frac{2\zeta}{\omega_n} D + 1} \right) i \tag{3.65}$$

The transfer function relating the spool-valve displacement x_s to the actuator displacement y_a is shown in the block diagram of Fig. 2.43. The cutting and thrust forces required to machine the workpice can be quite high and hence we do not neglect the load on the actuator. It follows from Eq. (2.147) that

$$y_a = \left(\frac{k_g}{D(\tau D + 1)}\right)x_s \tag{3.66}$$

The disturbance acting on the actuator is the thrust force on the tool and to show it explicitly, we modify the force balance equation expressed by Eq. (2.144) and get

$$(p_1 - p_2)A = m\ddot{y}_a + c\dot{y}_a + F_d$$

where F_d is the thrust force. With the time constant and gain defined by Eq. (2.146), we then obtain

$$y_a = \left(\frac{k_g}{D(\tau D + 1)}\right)(x_s - k_2 F_d) \tag{3.67}$$

where $k_2 = (c_3 + c_4)/c_1 A$ and these constants are defined in section 2.9.3. The block diagram is obtained from Eqs. (3.62) to (3.67) and is shown in Fig. 3.22. The scale factor must be chosen such that $c_1 = c_2$.

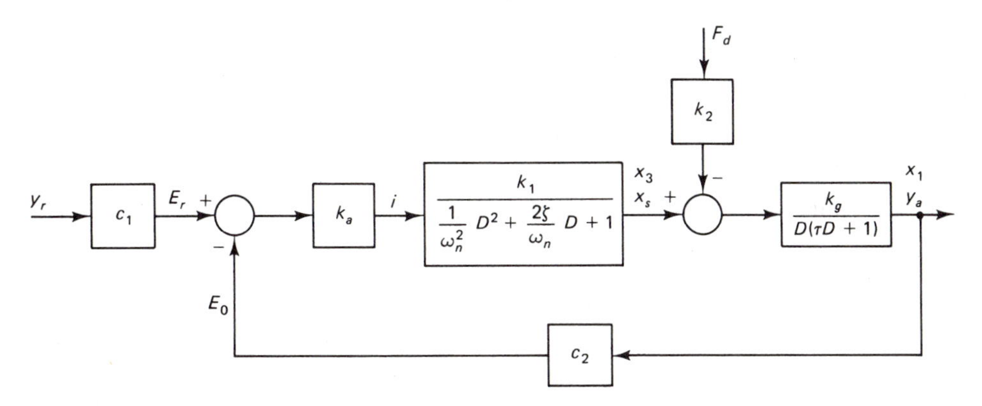

Figure 3.22 Block diagram of a hydraulic machining-control system.

This system is of fourth order and we choose four state variables as $x_1 = y_a$, $x_2 = \dot{y}_a$, $x_3 = x_s$, and $x_4 = \dot{x}_s$. The control input, disturbance, and output are defined as $u = i$, $v = F_d$, and $y = y_a = x_1$, respectively. From Fig. 3.22, we obtain the equations

$$D(\tau D + 1)y_a = k_g x_s - k_g k_2 F_d \tag{3.68}$$

$$\left(\frac{1}{\omega_n^2}D^2 + \frac{2\zeta}{\omega_n}D + 1\right)x_s = k_1 i \tag{3.69}$$

$$i = k_a c_1(y_r - y_a) \tag{3.70}$$

where in the last equation, $c_1 = c_2$. The state equations are given by

$$\begin{aligned}
\dot{x}_1 &= x_2 \\
\dot{x}_2 &= -\left(\frac{1}{\tau}\right)x_2 + \left(\frac{k_g}{\tau}\right)x_3 - \left(\frac{k_g k_2}{\tau}\right)F_d \\
\dot{x}_3 &= x_4 \\
\dot{x}_4 &= -\omega_n^2 x_3 - 2\zeta\omega_n x_4 + \omega_n^2 k_1 u
\end{aligned} \tag{3.71}$$

Control law: $u = k_a c_1(y_r - x_1)$

Output: $y = x_1$

The first and third equations in Eq. (3.71) are obtained from the definition of the state variables and the second and fourth from Eqs. (3.68) and (3.69), respectively. The matrices of Eq. (3.4) are obtained from inspection of the preceding equations and we get

$$\mathbf{A} = \begin{bmatrix} 0 & 1 & 0 & 0 \\ 0 & -1/\tau & k_g/\tau & 0 \\ 0 & 0 & 0 & 1 \\ 0 & 0 & -\omega_n^2 & -2\zeta\omega_n \end{bmatrix} \quad \mathbf{B} = \begin{bmatrix} 0 \\ 0 \\ 0 \\ \omega_n^2 k_1 \end{bmatrix} \quad \mathbf{B}_1 = \begin{bmatrix} 0 \\ -k_g k_2/\tau \\ 0 \\ 0 \end{bmatrix}$$

$$\mathbf{K} = \lfloor k_a c_1 \quad 0 \quad 0 \quad 0 \rfloor, \qquad N = k_a c_1 \qquad \mathbf{C} = \lfloor 1 \quad 0 \quad 0 \quad 0 \rfloor \tag{3.72}$$

The matrix $\mathbf{A} - \mathbf{BK}$ of the closed-loop system can be obtained from Eq. (3.72).

3.6 PNEUMATIC CONTROL SYSTEMS

Pneumatic control systems have some advantages over other types. The working fluid commonly used in pneumatic systems is air. It is readily available and return lines are not required. A leak in the system does not pose any fire hazard as flammable fluids and electrical sparks are absent. Hence, pneumatic control systems are preferred in chemical and petrochemical process-control applications. Pneumatic components are easy to maintain and are quite rugged. Also, the initial cost of pneumatic commonents is less than that of electrical and hydraulic components.

However, pneumatic control systems have some disadvantages. Air is a compressible fluid when compared to hydraulic fluids and hence pneumatic systems are slow acting with large time constants. Also, air lacks any lubricating properties. The normal operating pressure of pneumatic systems is much less than that of hydraulic systems and consequently the output power is also less for the same size. A source of compressed air is required and it is not as readily available as electrical power.

EXAMPLE 3.7

The conceptual design of a pneumatic system for temperature control of a room is shown in the schematic diagram of Fig. 3.23. The control system is

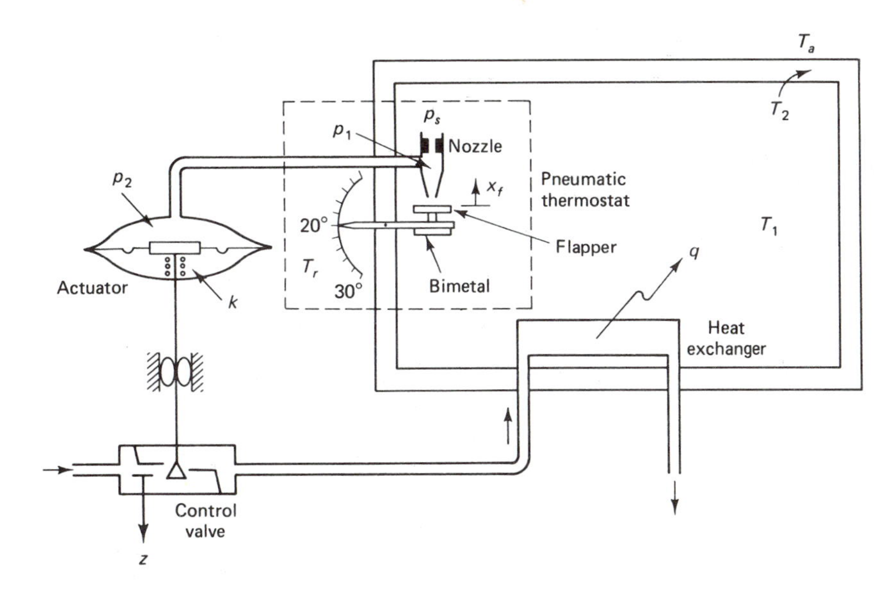

Figure 3.23 Schematic diagram of a pneumatic temperature-control system.

a regulator whose objective is to maintain a constant room temperature T_1 corresponding to the set point, in the presence of disturbance caused by variation of the outside ambient temperature T_a. Such pneumatic systems are used for the temperature control of large office buildings, laboratories, factories, and other facilities where compressed air is available, because of cost effectiveness.

The system employs a pneumatic thermostat for sensing and error detection. It is shown schematically in the diagram of Fig. 3.23. The desired room temperature T_r, which is the command input, is set on a calibrated dial and it positions the initial distance between a nozzle and a flapper. The actual room temperature T_1 is sensed by a bimetal strip that is mounted such that the distance between the nozzle and the flapper increases when the temperature T_1 increases from its equilibrium value and vice versa. Thus, the distance between the nozzle and the flapper corresponds to error $(T_r - T_1)$ between the reference and actual temperatures.

A change in the distance between the nozzle and the flapper causes a change in the pressure p_1 behind the nozzle and consequently a change in the pressure p_2 inside the chamber of the diaphragm-type actuator. The force thus generated changes the position z of the flow-control valve, which admits hot or chilled water to the heat exchanger in the room such that the controlled temperature T_1 corresponds to the set-point temperature T_r.

A linear mathematical model of the system is obtained as follows. Let all variables denote deviations from their constant equilibrium values. The displacement x_f of the flapper as shown in Fig. 2.23 increases when the set point is increased by T_r. Also, the bimetal is mounted such that x_f decreases when the room temperature increases by T_1. Hence, for small deviations from equilibrium, a linear relationship is given by

$$x_f = c_i T_r - c_2 T_1 \tag{3.73}$$

where $c_1 > 0$, and $c_2 > 0$. The pressure behind the nozzle versus the distance between the nozzle and the flapper was shown in Fig. 2.45 and is repeated here in Fig. 3.24 for convenience.

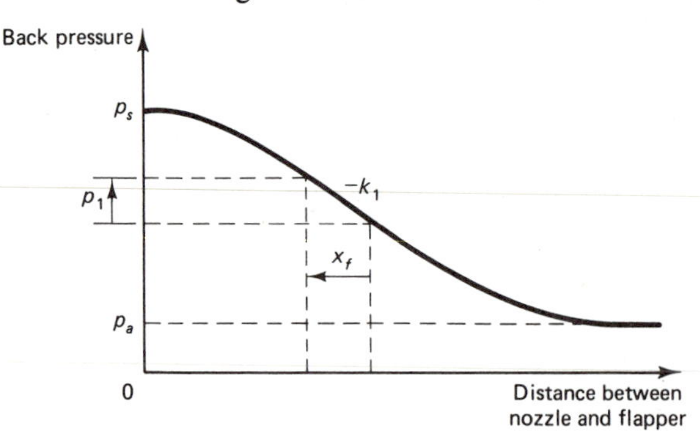

Figure 3.24 Nozzle-flapper characteristics.

When the nozzle is completely blocked so that there is no flow through it, the pressure behind the nozzle is the supply pressure p_s. When the nozzle is completely unblocked, the pressure behind it approaches the atmospheric pressure p_a. When the flapper displacement is x_f, the distance between the nozzle and flapper decreases by x_f. Within the linear range, the increase in the pressure behind the nozzle is given by

$$p_1 = -k_1(-x_f) = k_1 x_f \tag{3.74}$$

where $-k_1$, with $k_1 > 0$, is the slope of the curve at the equilibrium position. The transfer function between p_1 and the pressure p_2 in the actuator chamber is given by Eq. (2.159), that is,

$$(\tau D + 1)p_2 = p_1 \tag{3.75}$$

The control-valve displacement z and the deviation of the fluid flow q_f through it have been described by Eqs. (2.161) and (2.162), respectively. These equations are

$$kz = p_2 A \tag{3.76}$$

and

$$q_f = c_4 z \tag{3.77}$$

Combining Eqs. (3.75) to (3.77), we obtain

$$(\tau D + 1)q_f = k_2 p_1 \tag{3.78}$$

where $k_2 = Ac_4/k$. This transfer function of the pneumatic actuator was also shown in Fig. 2.48. The coefficient of heat transfer in the heat exchanger increases with the fluid flow rate q_f. For small deviations from equilibrium, it is assumed that the heat flux q is proportional to the fluid flow rate q_f. Hence, it follows that

$$q = c_5 q_f \tag{3.79}$$

The model of the system to be controlled was obtained in Example 2.11 by assuming that the room can be lumped into one lump at temperature T_1 and the walls into another at temperature T_2. The state equations are represented by Eq. (2.120) and the transfer function by Eq. (2.122). The block diagram of the control system is obtained from Eqs. (3.73), (3.74), (3.78), (3.79), and (2.122) and is shown in Fig. 3.25.

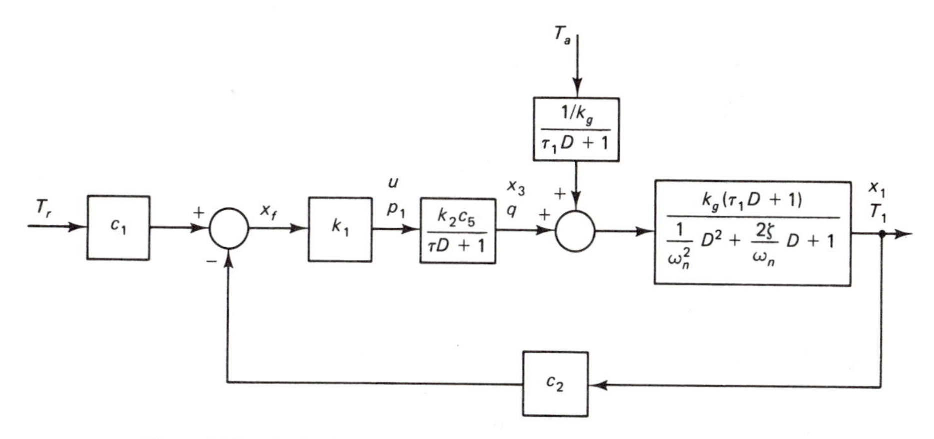

Figure 3.25 Block diagram of a pneumatic temperature-control system.

This system is of third order and we choose the three state variables as $x_1 = T_1$, $x_2 = T_2$, and $x_3 = q$, that is, the controlled temperature, wall temperature, and heat flux, respectively, are selected for the state representation. The control input, disturbance, and output are identified as $u = p_1$, $v = T_a$, and $y = T_1 = x_1$, respectively. From Fig. 3.25, we obtain the equation

$$(\tau D + 1)q = k_2 c_5 p_1$$

i.e.

$$\dot{x}_3 = -\left(\frac{1}{\tau}\right)x_3 + \left(\frac{k_2 c_5}{\tau}\right)u \tag{3.80}$$

The first two state equations are given by Eq. (2.120), and are combined with Eq. (3.80) to obtain

$$\begin{aligned}
\dot{x}_1 &= -\left(\frac{1}{R_1 C_1}\right)x_1 + \left(\frac{1}{R_1 C_1}\right)x_2 + \left(\frac{1}{C_1}\right)x_3 \\
\dot{x}_2 &= \left(\frac{1}{R_1 C_2}\right)x_1 - \left[\left(\frac{1}{R_1} + \frac{1}{R_2}\right)\frac{1}{C_2}\right]x_2 + \left(\frac{1}{R_2 C_2}\right)T_a \qquad (3.81) \\
\dot{x}_3 &= -\left(\frac{1}{\tau}\right)x_3 + \left(\frac{k_2 c_5}{\tau}\right)u
\end{aligned}$$

Control law: $u = k_1 c_1 (T_r - x_1)$ with $c_2 = c_1$

Output: $y = x_1$

The matrices of Eq. (3.4) are obtained from inspection of the preceding equations as

$$A = \begin{bmatrix} -\dfrac{1}{R_1 C_1} & \dfrac{1}{R_1 C_1} & \dfrac{1}{C_1} \\ \dfrac{1}{R_1 C_2} & -\left(\dfrac{1}{R_1} + \dfrac{1}{R_2}\right)\dfrac{1}{C_2} & 0 \\ 0 & 0 & -\dfrac{1}{\tau} \end{bmatrix} \qquad B = \begin{bmatrix} 0 \\ 0 \\ \dfrac{k_2 c_5}{\tau} \end{bmatrix} \qquad B_1 = \begin{bmatrix} 0 \\ \dfrac{1}{R_2 C_2} \\ 0 \end{bmatrix}$$

$$\mathbf{K} = \lfloor k_1 c_1 \quad 0 \quad 0 \rfloor \qquad N = k_1 c_1 \qquad \mathbf{C} = \lfloor 1 \quad 0 \quad 0 \rfloor \tag{3.82}$$

The matrix $\mathbf{A} - \mathbf{BK}$ of the closed-loop system can now be obtained from the preceding equations.

3.7 MULTIVARIABLE CONTROL SYSTEMS

In power plants, process industries, robotics, and other applications, we encounter systems where several variables are to be controlled. In Example 2.3, if the positions of both bodies are to be controlled and a control system is designed separately to control the position of one body, it is seen that it will also affect the position of the other and vice versa. Such interactions must be considered in the controller design of multivariable systems.

In some multivariable systems, where the controlled variables are weakly coupled, it may be possible to ignore the interactions and design separate controllers by considering the overall system as consisting of an appropriate number of separate single-input, single-output systems. In such cases, the coupling effects are considered as disturbances to the separate control systems and may not cause significant degradation in their performance when the coupling is weak. But if the coupling is strong and/or the performance needs improvement, then the interactions must be directly taken into account and the control system design approached from the multivariable viewpoint.

In multivariable systems, the objective is to synthesize a control law vector $\mathbf{u}$ such that each member of the controlled output vector $\mathbf{y}$ follows its corresponding member of the command vector $\mathbf{r}$. Electrical, hydraulic, or pneumatic actuators may be employed for the individual control loops of a multivariable system. Because there is interaction between the inputs and outputs, the synthesis of a control law vector $\mathbf{u}$ is much more complicated than in the case of a scalar control law for a single-input, single-output system.

Because of this complexity, it becomes cost effective to employ a microprocessor-based computer control for implementation of the control law. The synthesis of appropriate control laws for multivariable systems will be studied in Chapters 7, 8, and 9. In this chapter, we give a simple example to illustrate the mathematical modeling of multivariable control systems.

EXAMPLE 3.8

The schematic diagram of Fig. 3.26 shows a two-input, two-output regulator system for controlling the temperatures T_1 and T_2 in two interconnected tanks of a chemical process where a liquid is flowing through the tanks. The desired temperatures T_{r1} and T_{r2} are set on input potentiometers that supply the command voltages E_{r1} and E_{r2}.

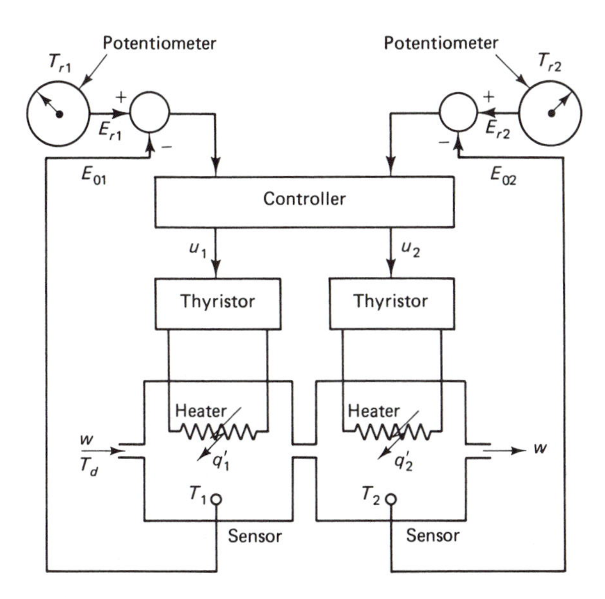

Figure 3.26 Schematic diagram of a multivariable system.

The controlled temperatures are sensed by thermistors whose outputs E_{01} and E_{02} are compared with the command voltages to obtain the errors. The errors are used by the controller to synthesize a control law $\mathbf{u}$. The controller output signals u_1 and u_2 are amplified by thyristor amplifiers, which supply power to the resistance heaters, as shown in Fig. 3.26.

In the linear mathematical model developed in the following, all variables denote deviations from their equilibrium values. We make simplifying assumptions to convey the essential ideas with a low-order model. It is assumed that the outside tank surfaces are insulated so that no heat flows to the environment, and that the liquid in each tank is mixed so that it can be represented by one lump. The thermal capacitance of the walls is combined with that of the liquid. The thermal capacitances of the heaters are neglected so that all the heat produced flows to the liquid.

Let w denote the product of the volumetric flow rate, density, and specific heat of the liquid, and T_d be its entrance temperature. Let C_1 and C_2 denote the thermal capacitances of the tanks and q_1' and q_2' be the heats generated by the heaters per unit time. Conservation of energy for each tank yields

$$C_1 \frac{dT_1}{dt} = w(T_d - T_1) + q_1' \tag{3.83}$$

$$C_2 \frac{dT_2}{dt} = w(T_1 - T_2) + q_2' \tag{3.84}$$

For small deviations about equilibrium, q_1' and q_2' are related linearly to u_1 and u_2, respectively, as

$$q_1' = c_3 u_1 \tag{3.85}$$

$$q_2' = c_4 u_2 \tag{3.86}$$

where the constants c_3 and c_4 include the power amplifier gains. Letting $x_1 = T_1$, $x_2 = T_2$, $\tau_1 = C_1/w$, $\tau_2 = C_2/w$, and using Eqs. (3.85) and (3.86) in Eqs. (3.83) and (3.84), respectively, with $b_1 = c_3/C_1$, and $b_2 = c_4/C_2$, we obtain

$$\dot{x}_1 = -\frac{1}{\tau_1} x_1 + b_1 u_1 + \frac{1}{\tau_1} T_d \tag{3.87}$$

$$\dot{x}_2 = \frac{1}{\tau_2} x_1 - \frac{1}{\tau_2} x_2 + b_2 u_2 \tag{3.88}$$

where the inlet temperature deviation T_d is the disturbance. These equations are in the form

$$\dot{\mathbf{x}} = \mathbf{A}\mathbf{x} + \mathbf{B}\mathbf{u} + \mathbf{B}_1 T_d$$

$$\mathbf{y} = \mathbf{C}\mathbf{x}$$

The matrices are described by

$$\mathbf{A} = \begin{bmatrix} -1/\tau_1 & 0 \\ 1/\tau_2 & -1/\tau_2 \end{bmatrix} \quad \mathbf{B} = \begin{bmatrix} b_1 & 0 \\ 0 & b_2 \end{bmatrix} \quad \mathbf{B}_1 = \begin{bmatrix} 1/\tau_1 \\ 0 \end{bmatrix} \quad \mathbf{C} = \begin{bmatrix} 1 & 0 \\ 0 & 1 \end{bmatrix} \tag{3.89}$$

The transfer-function matrices relating the outputs to the control inputs and the distrubance are given by

$$\mathbf{Y}(s) = \mathbf{C}(s\mathbf{I} - \mathbf{A})^{-1}\mathbf{B}\mathbf{U}(s) + \mathbf{C}(s\mathbf{I} - \mathbf{A})^{-1}\mathbf{B}_1 T_d \tag{3.90}$$

Now, we obtain

$$(s\mathbf{I} - \mathbf{A})^{-1} = \frac{1}{(s + 1/\tau_1)(s + 1/\tau_2)} \begin{bmatrix} s + 1/\tau_2 & 0 \\ 1/\tau_2 & s + 1/\tau_1 \end{bmatrix} \tag{3.91}$$

Substituting the results from Eqs. (3.89) and (3.91) in Eqs. (3.90), it follows that

$$\mathbf{Y}(s) = \begin{bmatrix} \dfrac{b_1}{s + 1/\tau_1} & 0 \\ \dfrac{b_1/\tau_2}{(s + 1/\tau_1)(s + 1/\tau_2)} & \dfrac{b_2}{s + 1/\tau_2} \end{bmatrix} \mathbf{U}(s) + \begin{bmatrix} \dfrac{1/\tau_1}{s + 1/\tau_1} \\ \dfrac{1/\tau_1\tau_2}{(s + 1/\tau_1)(s + 1/\tau_2)} \end{bmatrix} T_d \tag{3.92}$$

This equation is in the form

$$\mathbf{Y}(s) = \mathbf{G}_p(s)\mathbf{U}(s) + \mathbf{G}_d(s) T_d(s) \tag{3.93}$$

where $\mathbf{G}_p(s)$ is the 2×2 plant matrix and $\mathbf{G}_d(s)$ is the 2×1 disturbance matrix defined by Eq. (3.92).

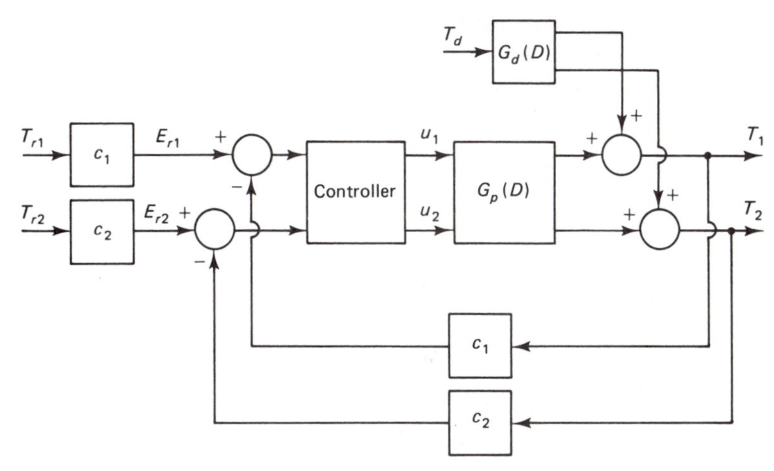

Figure 3.27 Block diagram of a multivariable system.

It is seen that the plant matrix is not diagonal. The off-diagonal zero element signifies that the input u_2 does not affect the temperature of the upstream tank. The block diagram of the control system is shown in Fig. 3.27. The design of the controller for multivariable systems is discussed in Chapter 7 with output feedback and in Chapter 8 with state feedback.

3.8 SUMMARY

In this chapter, we have discussed the various advantages and disadvantages of electrical, hydraulic, and pneumatic control systems so that a designer can use them as guidelines in formulating a conceptual design. The main aim of the chapter is to illustrate the techniques of obtaining linear mathematical models of control systems from their conceptual models. Several examples are given of modeling passive, electrical, hydraulic, and pneumatic control systems.

The emphasis has been on single-input, single-output systems, but multivariable systems are included at the end of the chapter. At this stage, however, the design of a control law for multivariable systems has been omitted. This topic will be covered in detail in Chapters 7 and 8. The examples we have considered employ a simple proportional control law, that is, the control u is proportional to the error between the command input and the controlled output. It is important to realize that a proportional control law may not yield satisfactory performance for reasons that will become obvious later in Chapter 7. Other control laws that yield improved performance are covered in Chapters 7, 8, and 9.

The linear differential equations models of control systems have been expressed both in the form of block diagrams with transfer functions and alternatively in the form of state equations. A block diagram with transfer functions yields valuable insight into the nature of control systems. A state-equations model is however more convenient for most of the analysis and synthesis techniques covered in later chapters. The mathematical models of the various examples we have studied in this chapter will be employed in later chapters to analyze the proposed conceptual designs concerning speed of response, stability, accuracy, and disturbance-rejection properties. For further examples of modeling control systems, see Raven (1978), Palm (1983), Brewer (1974), and Dorf (1980).

REFERENCES

BREWER, J. W. (1974). *Control Systems, Analysis, Design, and Simulation.* Englewood Cliffs, NJ: Prentice-Hall.

DORF, R. C. (1980). *Modern Control Systems.* 3d ed. Reading, MA: Addison-Wesley.

HERCEG, E. E. (1972). *Handbook of Measurement and Control.* Pennsauken, NJ: Schaevitz Engineering.

PALM, W. J. (1983). *Modeling, Analysis, and Control of Dynamic Systems.* New York: John Wiley & Sons.

PELLY, B. R. (1971). *Thyristor Phase-Controlled Converter and Cycloconverter.* New York: John Wiley & Sons.

RAVEN, F. H. (1978). *Automatic Control Engineering.* 3d ed. New York: McGraw-Hill.

SEN, P. C., and MACDONALD, M. L. (1978, November). "Thyristorized DC Drives with Regenerative Braking and Speed Reversal." *IEEE Trans. on Industrial Electronics and Control Instrumentation IECI-25* (4).

PROBLEMS

3.1. A dc electrical system for speed control of a machine tool spindle is shown in Fig. P3.1. The desired speed ω_r is set on an input potentiometer, which provides a reference

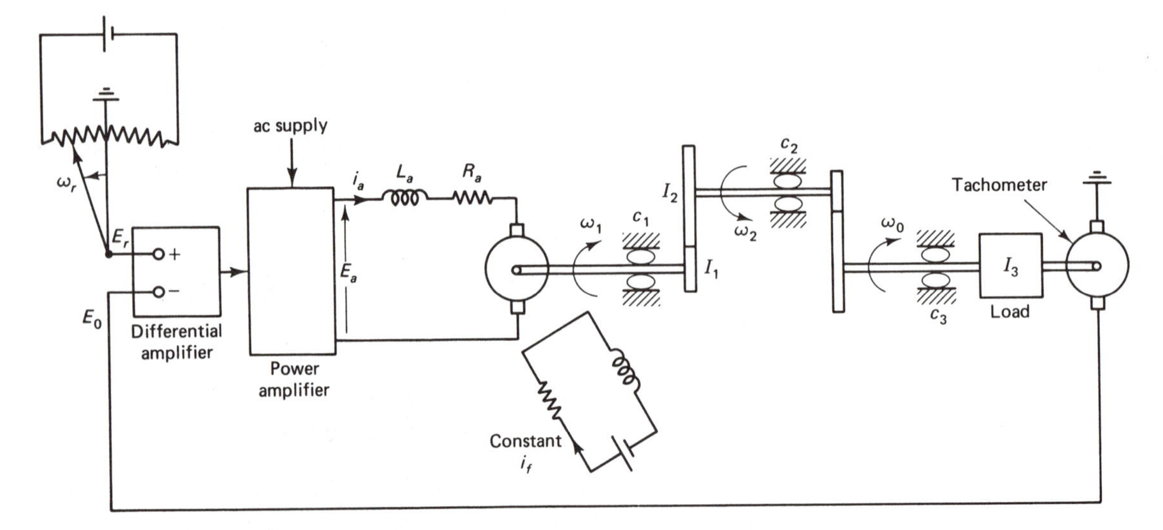

Figure P3.1 Electrical system for speed control of a spindle.

voltage E_r. The speed ω_0 of the load is measured by a tachometer, which supplies a feedback voltage E_0. The load shaft is connected to the motor shaft by gears with ratios $n_1 = \omega_2/\omega_1$ and $n_2 = \omega_0/\omega_2$. Obtain a mathematical model of the control system and express it in the form of the transfer function and the state equations.

3.2. A copying system for positioning a tool in the cross-slide direction for machining a workpiece is shown in Fig. P3.2. The desired position y_r of the tool is obtained by a stylus moving over a template and is converted to a voltage E_r by the input LVDT. The output position y_0 of the tool is sensed by the output LVDT. A differential amplifier acts as the error detector and a thyristor power amplifier supplies a voltage E_a to a dc motor. Treat the reactive force of the cutting tool as a disturbance. Obtain the transfer function and the state equations.

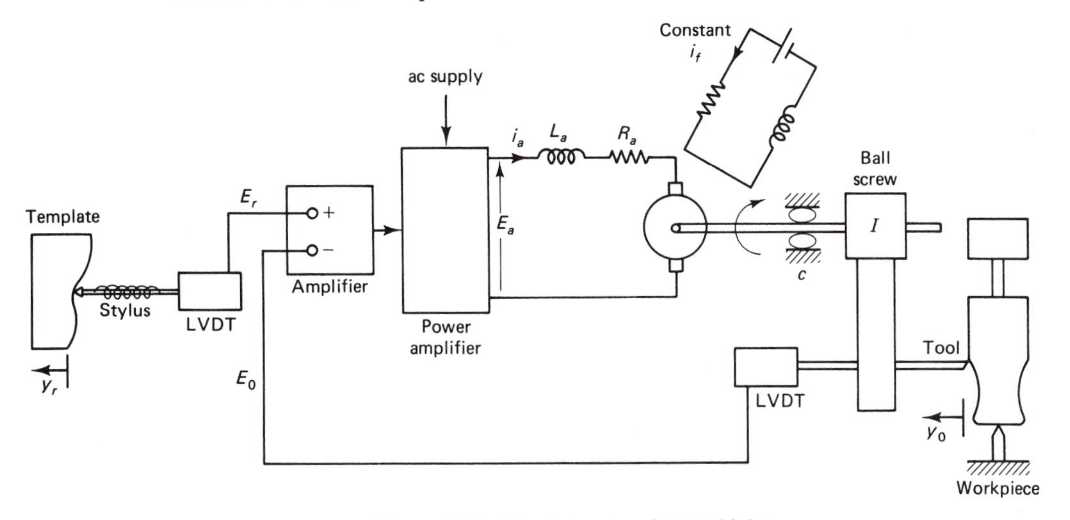

Figure P3.2 Copying system for machining.

3.3. A voltage regulator for controlling the voltage E_0 across a load resistance R_L is shown in Fig. P3.3. The desired voltage E_r is set on the input potentiometer. A differential amplifier determines the error voltage $(E_r - E_0)$ and produces a voltage $E_a = k_a(E_r - E_0)$, which is supplied to the field of the dc generator. The total load on the

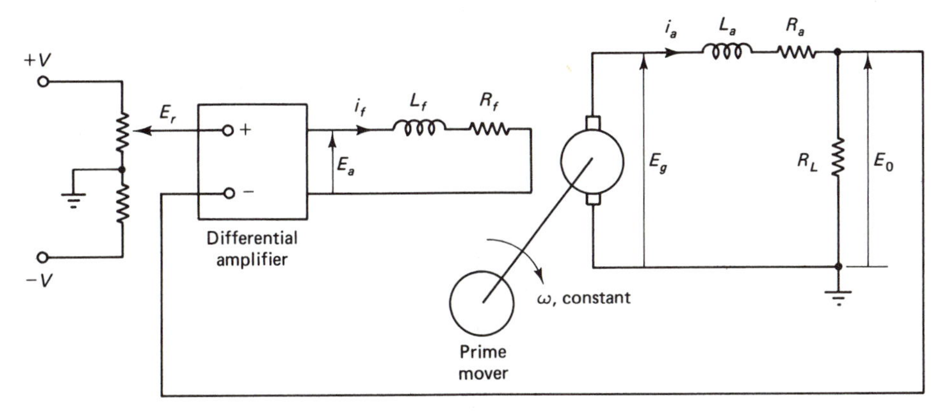

Figure P3.3 Voltage regulator system.

generator consists of its armature inductance L_a, resistance R_a, and the load resistance R_L. Obtain the transfer function and the state equations.

3.4. A system using a two-phase ac servomotor (discussed in section 2.9.2) for positioning a load is shown in Fig. P3.4. The desired position θ_r is transduced to a voltage E_r by a potentiometer and the load position θ_0 is sensed by an ac RVDT. The ac power amplifier voltage $E = k_a(E_r - E_0)$. Obtain the transfer function and the state equations.

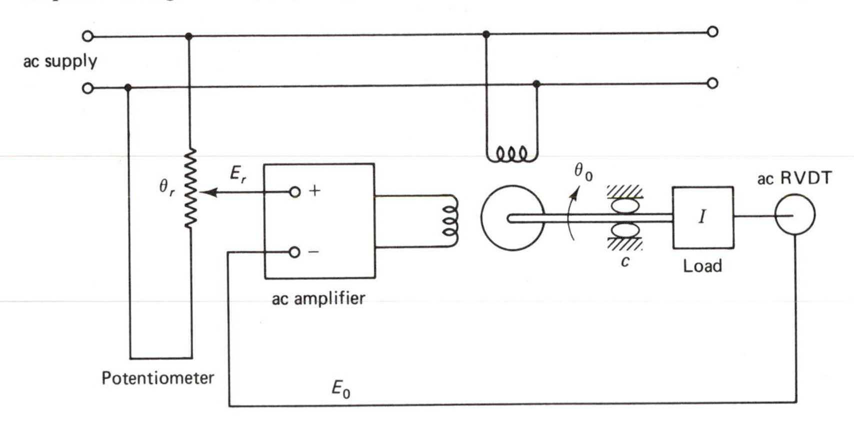

Figure P3.4 Ac load positioning system.

3.5. An electrohydraulic system for controlling the thickness of rolled steel in a steel mill is shown in Fig. P3.5. The desired change in the thickness y_r is set on a potentiometer such that $E_r = c_1 y_r$. The actual thickness is measured by a contactless sensor with signal conditioning such that $E_0 = c_2 y_0$. Include the normal force on the roller as a load disturbance. Obtain the transfer function and give the state equations.

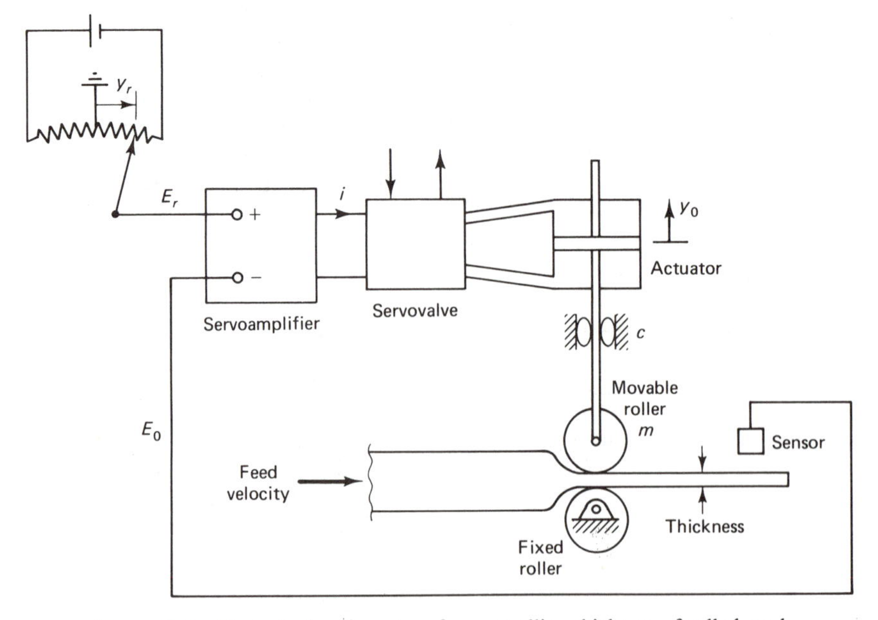

Figure P3.5 Electrohydraulic system for controlling thickness of rolled steel.

3.6. The system shown in Fig. P3.6 is for remotely positioning the rudder of an aircraft. The desired angle θ_r is set on the input potentiometer and the output angle θ_0 is sensed by a RVDT. The mass moment of inertia of the rudder about the hinge point is I and the coefficient of viscous friction moment is c. Obtain the transfer function and the state equations.

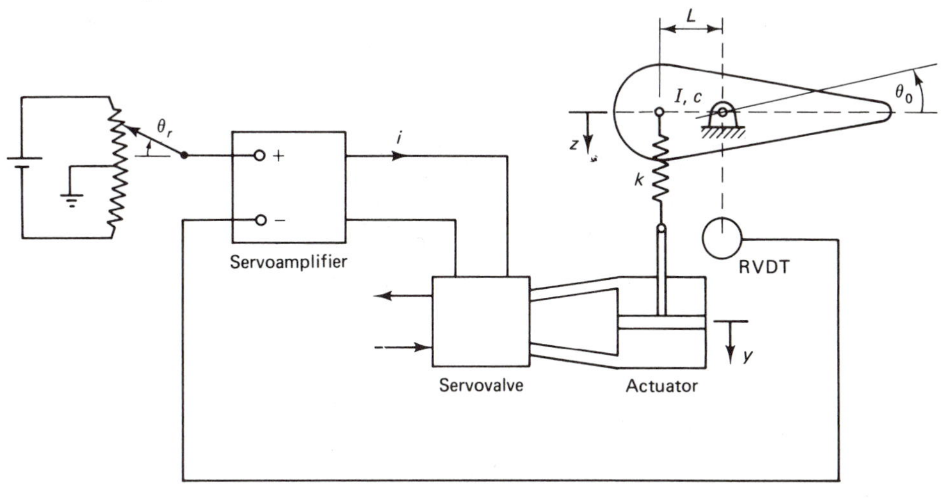

Figure P3.6 Aircraft rudder positioning system.

3.7. A active vibration-isolation system is shown in Fig. P3.7. The purpose is to maintain the displacement $y(t)$ of the mass near the desired value of $y_r = 0$ when the base on which it is mounted vibrates with displacement $w(t)$. Both w and y are measured with respect to a fixed datum and $y = z - w$. The accelerometer measures $\ddot{y}$, which is integrated twice by the signal conditioner so that $E_0 = c_1 y$. Considering the command input $y_r = 0$ and w as a disturbance input, obtain the block diagram and give the state equations.

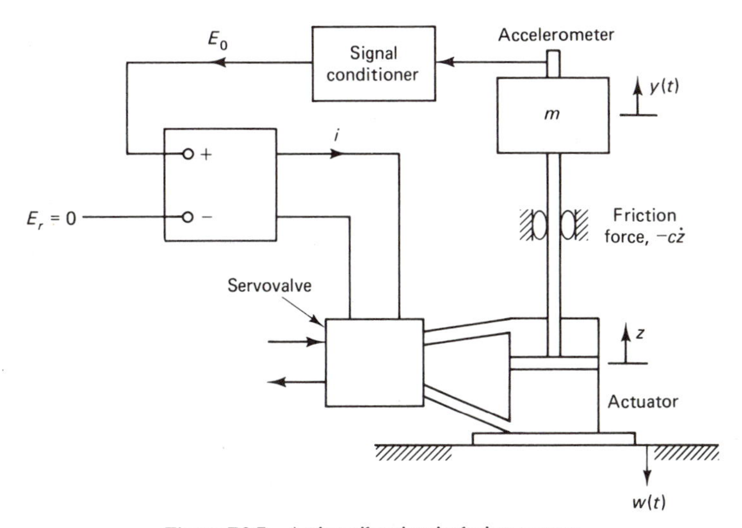

Figure P3.7 Active vibration-isolation system.

3.8. Figure P3.8 shows a positioning system to produce a desired yaw angle between the plane of a tire and a surface in a tire-testing dynamometer. The surface is simulated by a large rotating flywheel. The full output required is ±20 degrees of rotation. The desired yaw angle θ_r is set as a voltage $E_r = c_1\theta_r$. The output yaw angle θ_0 is measured by a RVDT. The system includes a servoamplifier, a servovalve, and a hydraulic motor whose shaft is connected to the load, which consists of inertia, damping torque, and a load torque T_L. Obtain the transfer function and the state equations.

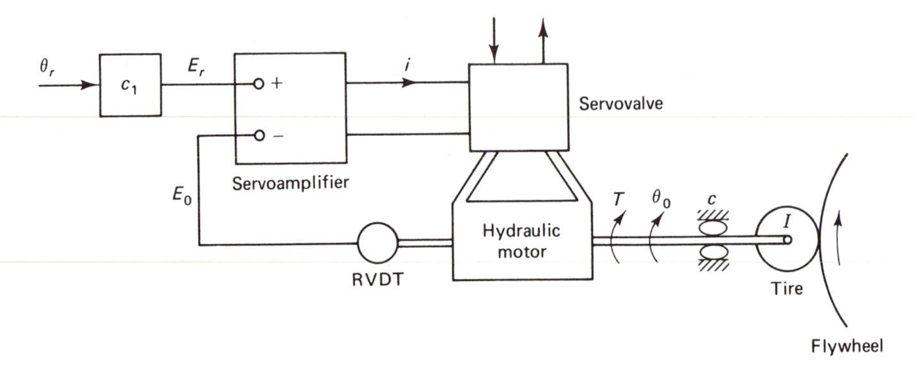

Figure P3.8 Yaw angle control system.

3.9. In a tire-testing dynamometer the control system shown in Fig. P3.9 produces a time-varying normal load on the tire corresponding to the normal load at the tire-runway interface. The runway surface is simulated by a large rotating flywheel. The desired load command F_r is obtained from a tape reader such that $E_r = c_1F_r$. The actual normal load F_0 is measured by a strain-gage load cell whose output $E_0 = c_2F_0$. The tire elasticity and damping are represented by a spring k and damper c, respectively. Obtain the transfer function and the state equations.

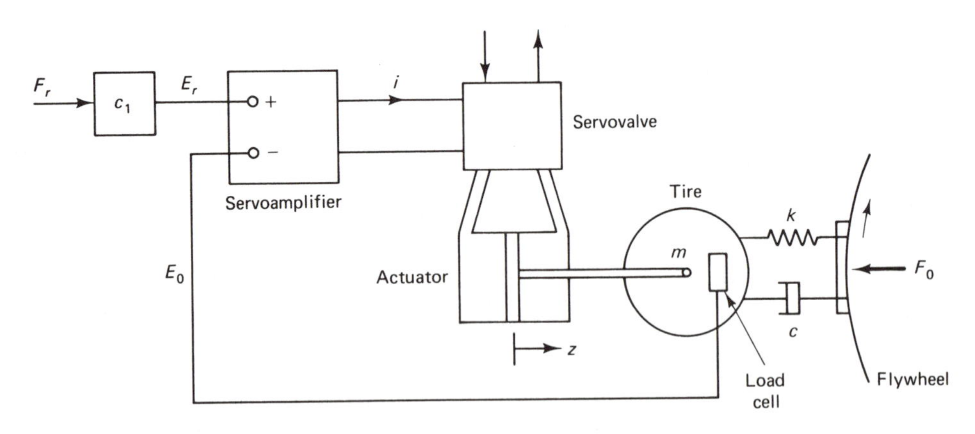

Figure P3.9 System for control of tire normal load.

3.10. The purpose of the control system shown in Fig. P3.10 is to maintain a constant pressure inside a pressurized tank. Make-up gas is admitted from a supply source through a control valve to compensate for the pressure changes caused by the time-varying outflow

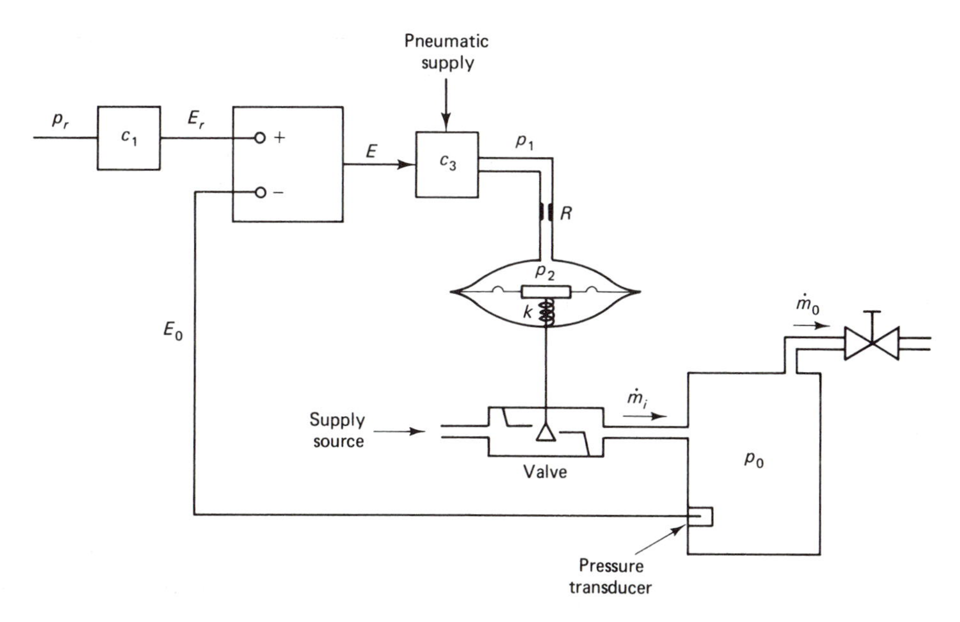

Figure P3.10 Pressure control system.

$\dot{m}_0$. The desired pressure p_r is set on a potentiometer so that $E_r = c_1 p_r$. The controlled pressure p_0 is sensed by a transducer whose output $E_0 = c_2 p_0$. For the electropneumatic transducer, assume that $p_1 = c_3 E$. Obtain the transfer function and the state equations.

3.11. A system to control the liquid level h_2 in tank 2 is shown in Fig. P3.11. The desired level is h_r and the hydrostatic pressure in the tank is sensed by a bubbler tube filled with air and connected to a bellows. A nozzle-flapper determines the error and supplies a pressure to a pneumatic actuator. Obtain the transfer function and the state equations.

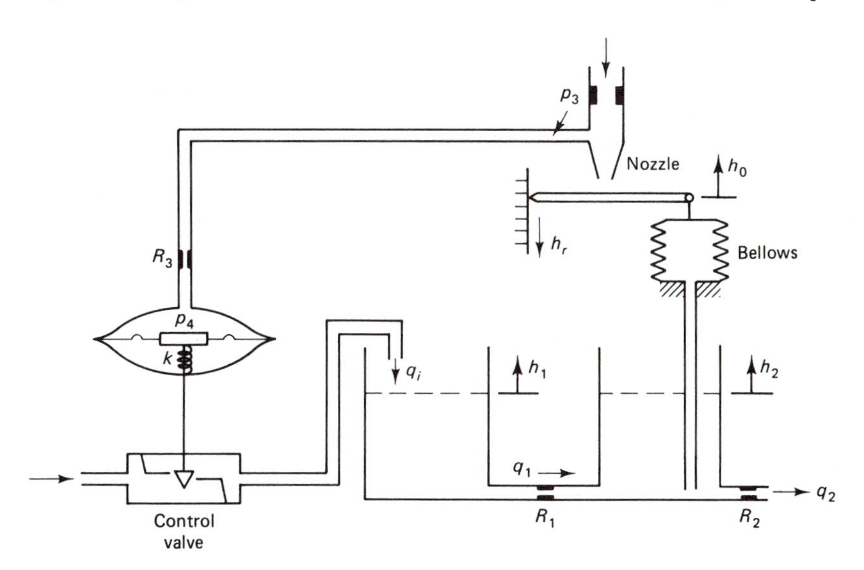

Figure P3.11 Liquid level control system.

3.12. A temperature-control system for a furnace is shown in Fig. P3.12. Here, T_1, T_2, and T_a are the temperatures of the heater, furnace, and ambient air, respectively. The furnace walls have negligible thermal capacitance and C_1 and C_2 are the capacitances of the heater and furnace, respectively. The desired temperature T_r is set on a potentiometer and the furnace temperature T_2 is sensed by a resistance thermometer connected to a Wheatstone bridge so that $E_0 = c_2 T_2$. Obtain a linear mathematical model for small deviations from equilibrium, and give the block diagram and the state equations.

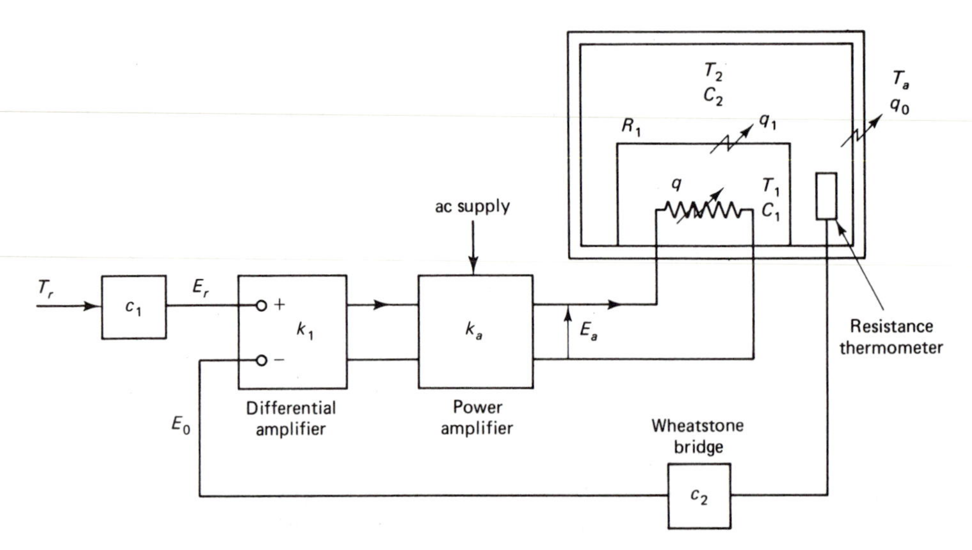

Figure P3.12 Temperature control system for a furnace.

3.13. In a heating system, air flows into and out of a chamber at constant volumetric flow rate q. A system to control the outflow temperature T_0 is shown in Fig. P3.13. The

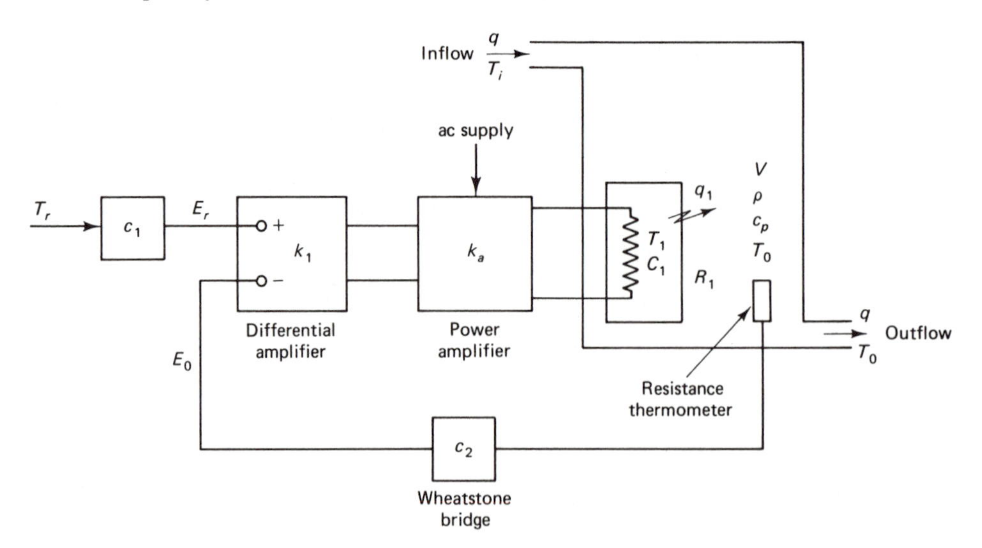

Figure P3.13 Thermal control system.

desired temperature T_r is set on a potentiometer whose output $E_r = c_1 T_r$. The temperature T_0 is sensed by a resistance thermometer with Wheatstone bridge whose output $E_0 = c_2 T_o$. The heater has temperature T_1, capacitance C_1, and thermal resistance R_1. The walls of the chamber are perfectly insulated and have negligible thermal capacitance. The air in the chamber, whose volume is V, is mixed and is at temperature T_0. Its specific heat is c_p and density is ρ. Considering the inflow temperature T_i as a disturbance, obtain a linear mathematical model, and give the block diagram and the state equations.

3.14. In a variable air-volume-type air conditioning system, the pressure of the supply air in the duct is controlled by varying the speed of the fan as shown in Fig. P3.14. The desired pressure p_r is set on a potentiometer so that $E_r = c_1 p_r$. The pressure p in the duct is sensed by a pressure transducer with signal conditioning, whose output $E_0 = c_2 p$. The armature-controlled dc motor runs the fan at variable speed ω. The characteristic curves for pressure versus flow rate of the fan are shown in Fig. P3.14(b) for different speeds ω. Obtain a linear mathematical model of the system for small deviations from equilibrium, and give the block diagram and the state equations.

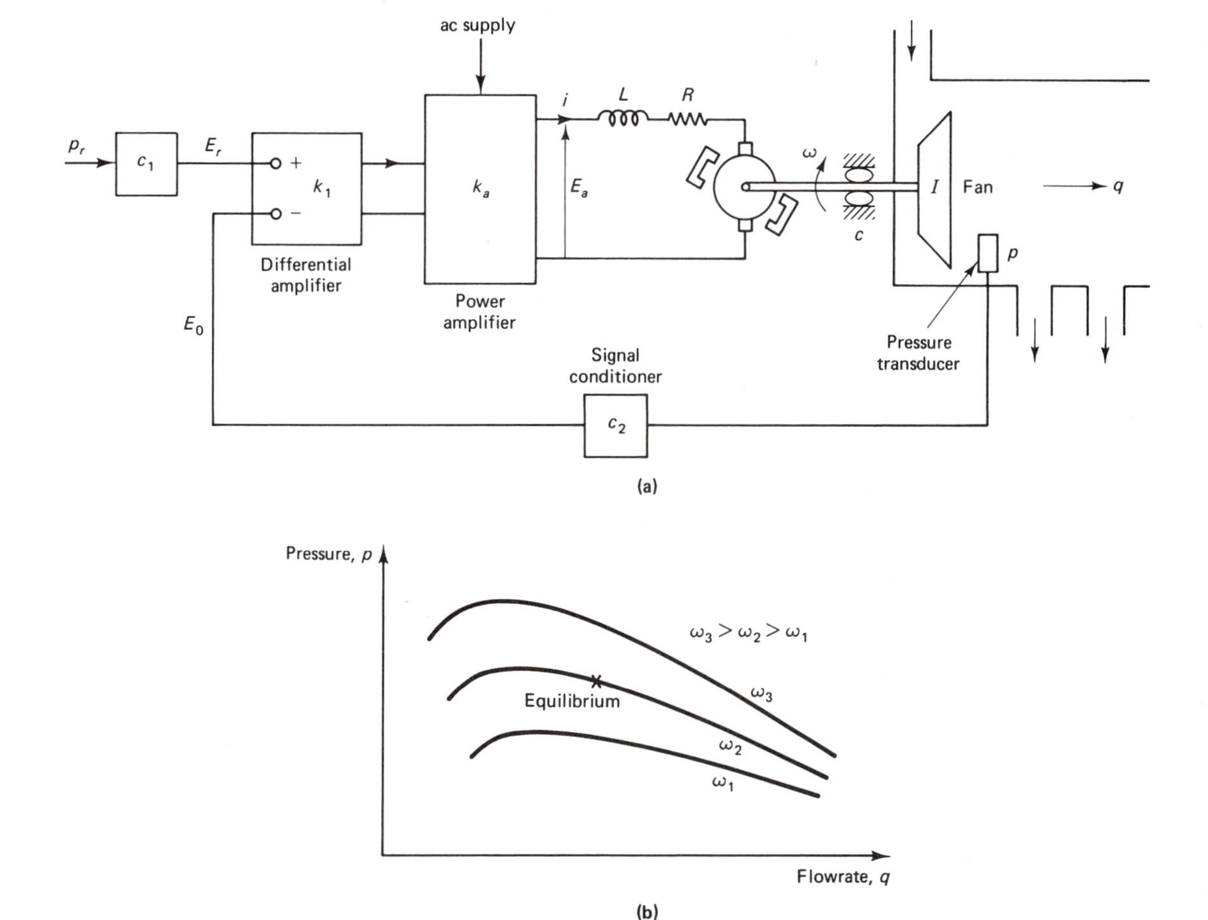

Figure P3.14 Fan speed control system.

4

Transient Response of Control Systems

4.1. INTRODUCTION

In the mathematical models of control systems developed in the previous chapter, time is the independent variable and it is of interest to study their transient behavior or time response when subjected to inputs and/or initial conditions. It is seen that control systems may be forced by command inputs and disturbances. Usually, disturbances are random functions of time and are described by their probability density functions. The stochastic response of control systems to random disturbances is beyond the scope of this book. Here, the effect of disturbances is taken into account indirectly by assigning some initial conditions to the state variables and this implies impulsively acting disturbances.

The command inputs are usually deterministic functions of time, but in some cases, they may not be known in advance. For example, for a surface-to-air or air-to-air missile the control objective is for the missile to track and home in on an enemy aircraft. The position of the aircraft is the command input but the aircraft may be performing an evasive maneuver. Hence, a basis may be established for comparing the transient performance of various systems by specifying some test signals and then studying the transient response to these test inputs.

The testing inputs commonly employed for this purpose consist of a step function, an impulse function, a ramp function, and a sinusoidal function. The performance specifications in the time domain are stated for a step function as a test input. The response to a sinusoidal test input is called frequency response, and because of its importance in its own right, it is covered separately in the next chapter. In this chapter, our objective is to obtain the time response of control systems for any general input and then consider step and impulse test inputs as special cases.

Specifically, we obtain closed-form solutions of linear, time-invariant state-equations representation of control systems. The relationship between the state-variables representation and the transfer function is established. The effect of the system parameters on the transient response is studied. Similarity transformation of state variables is included so that digital-computer-based matrix methods can be used for the closed-form solution of high-order systems.

4.2. STATE-VARIABLES REPRESENTATION

In Chapter 3, we have expressed the mathematical models of feedback control systems in the form of state equations and also obtained their block diagram representation. It has been shown that the closed-loop transfer function relating the controlled output to the command input can be obtained directly from the block diagram by using Eq. (3.8) or from the state equations by using Eq. (3.6). A system with n state variables is described by a set of n-coupled first-order differential equations. Alternatively, the system could be described by a nth-order differential equation, relating the output to the input by employing the transfer function.

We have the choice of employing either the state-variables representation or the transfer function for the transient response analysis. Here, we employ the state-variables representation because it has several advantages over a transfer function. The matrix formulation provides a suitable structure for digital computer applications. It is convenient for the analysis and design of multivariable systems. Also, it permits direct application of several techniques discussed in later chapters.

As mentioned in the previous section, we neglect the disturbance inputs for the transient response analysis and at most assign some initial conditions to the state variables to include their effects indirectly. Hence, from Eqs. (3.1) to (3.3), we obtain the following representation.

$$\text{Plant:} \qquad \dot{\mathbf{x}}(t) = \mathbf{A}\mathbf{x}(t) + \mathbf{B}\mathbf{u}(t) \tag{4.1}$$

$$\text{Control law:} \qquad \mathbf{u}(t) = \mathbf{N}\mathbf{r}(t) - \mathbf{K}\mathbf{x}(t) \tag{4.2}$$

$$\text{Output:} \qquad \mathbf{y}(t) = \mathbf{C}\mathbf{x}(t) \tag{4.3}$$

In a multivariable system where there are m controlled outputs $\mathbf{y}$, we assume that the command inputs are also m in number. Then, $\mathbf{u}$, $\mathbf{r}$, and $\mathbf{y}$ are $m \times 1$ column matrices and $\mathbf{x}$ is an $n \times 1$ column matrix. The dimensions of $\mathbf{A}$, $\mathbf{B}$, $\mathbf{N}$, $\mathbf{K}$, and $\mathbf{C}$ are $n \times n$, $n \times m$, $m \times m$, $m \times n$, and $m \times n$, respectively. In a single-input, single-output system, u, r, and y are scalars and $m = 1$, that is, $\mathbf{B}$ is an $n \times 1$ column matrix, $\mathbf{K}$ and $\mathbf{C}$ are $1 \times n$ row matrices, and N is a scalar. The state equations of the closed-loop control system are obtained by substituting for u from Eq. (4.2) in Eq. (4.1) and we get

$$\dot{\mathbf{x}} = (\mathbf{A} - \mathbf{B}\mathbf{K})\mathbf{x} + \mathbf{B}\mathbf{N}\mathbf{r} \tag{4.4}$$

Each state $\mathbf{x}$ of the system may be represented as a point in an n-dimensional Euclidean space whose coordinates are $x_1, x_2, \ldots, x_n$ as shown in Fig. 4.1. The Euclidean space E^n is a linear space that is complete and normed with an inner

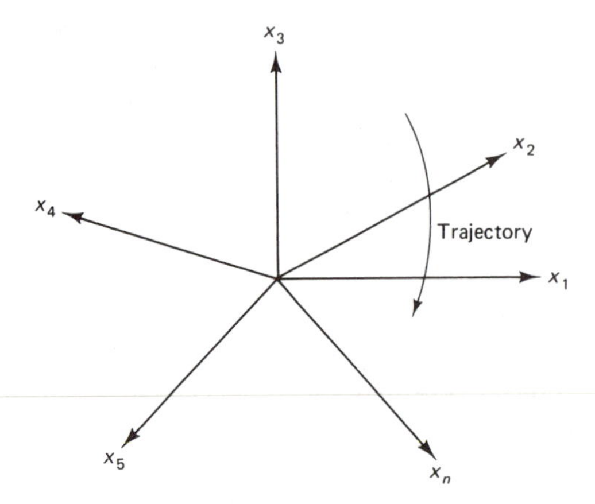

Figure 4.1 State space.

product. The norm and inner product are defined by

$$\|\mathbf{x}\| = \left(\sum_{i=1}^{n} x_i^2 \right)^{1/2} \tag{4.5}$$

$$\langle \mathbf{x}, \mathbf{x} \rangle = \sum_{i=1}^{n} x_i^2 = \|\mathbf{x}\|^2 \tag{4.6}$$

The norm is a measure of distance and the Euclidean norm is the generalization from three-dimensinal vector space to an n-dimensional space. Given the initial state $\mathbf{x}_0$ at initial time $t = t_0$ and specified input $\mathbf{r}(t)$, the state of system of Eq. (4.4) will change from $\mathbf{x}_0$ with time. The set of values that the state takes for $t > t_0$ is denoted by $\mathbf{x}(\mathbf{x}_0, t_0; \mathbf{r})$. The set of points traced out by $\mathbf{x}(\mathbf{x}_0, t_0; \mathbf{r})$ is called the state trajectory. Hence, a state trajectory is the geometrical representation in state space of a particular solution when the initial condition $\mathbf{x}_0$ and the input $\mathbf{r}(t)$ are specified. Here, time t plays the role of a parameter along the system trajectory in state space.

In some optimal control problems, the objective is to find the control $\mathbf{u}$ such that the state of the system is changed from an initial state $\mathbf{x}_0$ at time t_0 to some final state $\mathbf{x}(t_f)$ at the terminal time t_f. The initial and final states may be only partly specified, that is, the values of some of the state variables are specified at the initial time t_0 and some at the terminal time t_f. This class of problems is called the boundary value problem of ordinary differential equations. When all n initial conditions $\mathbf{x}_0$ for the system of Eq. (4.4) are specified at the initial time t_0, we have an initial value problem of ordinary differential equations. In this chapter, we study only the initial-value problem.

4.3. SOLUTION OF INITIAL-VALUE PROBLEM

Our objective is to obtain the transient response of the closed-loop control system of Eq. (4.4) with initial conditions $\mathbf{x}(t_0) = \mathbf{x}_0$ at time t_0 and subjected to the input $\mathbf{r}(t)$. Since our model is time-invariant, that is, the coefficients of matrices $\mathbf{A}$, $\mathbf{B}$, $\mathbf{K}$, $\mathbf{N}$, and $\mathbf{C}$ in Eq. (4.4) are constants, we can arbitrarily choose the initial time to be zero and set $t_0 = 0$. The first concern in whether this initial value problem has a solution and the second whether the solution is unique.

There exists a theorem that yields sufficient conditions to guarantee the existence and uniqueness of the solution of the initial value problem of ordinary differential equations that in general are nonlinear. A statement of the theorem and its proof are given by D'Souza and Garg (1984). In our case, the linear unforced system corresponding to Eq. (4.4) is

$$\dot{\mathbf{x}} = (\mathbf{A} - \mathbf{BK})\mathbf{x} \qquad \mathbf{x} = \mathbf{x}(0) \text{ at } t = 0 \tag{4.7}$$

To satisfy the conditions of the theorem, we have to find a finite Lipschitz constant k such that for any two states $\mathbf{x}_\alpha$ and $\mathbf{x}_\beta$,

$$\|(\mathbf{A} - \mathbf{BK})\mathbf{x}_\alpha - (\mathbf{A} - \mathbf{BK})\mathbf{x}_\beta\| \le k\|\mathbf{x}_\alpha - \mathbf{x}_\beta\| \tag{4.8}$$

Now, we have

$$\|(\mathbf{A} - \mathbf{BK})\mathbf{x}_\alpha - (\mathbf{A} - \mathbf{BK})\mathbf{x}_\beta\| \le \|\mathbf{A} - \mathbf{BK}\|\,\|\mathbf{x}_\alpha - \mathbf{x}_\beta\| \tag{4.9}$$

It can be shown (Bellman, 1960) that $\|\mathbf{A} - \mathbf{BK}\| \le |\lambda_{\text{max}}|$, where $|\lambda_{\text{max}}|$ is the absolute value of the maximum eigenvalue of matrix $(\mathbf{A} - \mathbf{BK})$. Hence, a Lipschitz constant can always be found such that $k \ge |\lambda_{\text{max}}|$ for any $\mathbf{x}_\alpha$ and $\mathbf{x}_\beta$ throughout the state space, i.e., globally. As a result, the linear system of Eq. (4.7) always satisfies the conditions for global existence and uniqueness of solution throughout the state space.

In the case of the forced system of Fig. (4.4), it is sufficient for the existence and uniqueness of solution that the input $\mathbf{r}(t)$ be piecewise continuous. At the points of discontinuity of $\mathbf{r}(t)$, the sufficiency conditions are not satisfied. In fact, if $\mathbf{r}(t)$ includes impulse functions (that is, Dirac delta functions), then at the time of application of the impulse input, there is a jump in one or more of the state variables, and the solution is not unique at that point. Having established the existence of a solution to Eq. (4.4), we now study the methods of getting the solution.

First, we consider the homogeneous or unforced system obtained by setting $\mathbf{r} = \mathbf{0}$ in Eq. (4.4), i.e.,

$$\dot{\mathbf{x}} = (\mathbf{A} - \mathbf{BK})\mathbf{x} \qquad \mathbf{x} = \mathbf{x}(0) \text{ at } t = 0 \tag{4.10}$$

Let an $n \times n$ matrix $\mathbf{\Phi}(t)$ be the solution of the matrix differential equation

$$\frac{d}{dt}\mathbf{\Phi}(t) = (\mathbf{A} - \mathbf{BK})\mathbf{\Phi}(t) \qquad \mathbf{\Phi}(0) = \mathbf{I} \tag{4.11}$$

where $\mathbf{\Phi}(0)$ ia the initial condition, and $\mathbf{I}$ is the identity matrix. Then the solution of Eq. (4.10) is given by

$$\mathbf{x}(t) = \mathbf{\Phi}(t)\mathbf{x}(0) \qquad t \ge 0 \tag{4.12}$$

We can verify that Eq. (4.12) is the solution of Eq. (4.10), where $\mathbf{\Phi}(t)$ satisfies Eq. (4.11) as follows. Differentiating Eq. (4.12) with respect to time, we obtain

$$\dot{\mathbf{x}} = \frac{d}{dt}[\mathbf{\Phi}(t)]\mathbf{x}(0)$$

From Eq. (4.11), it follows that

$$\dot{\mathbf{x}} = (\mathbf{A} - \mathbf{BK})\mathbf{\Phi}(t)\mathbf{x}(0)$$

and from Eq. (4.12), we get

$$\dot{\mathbf{x}} = (\mathbf{A} - \mathbf{BK})\mathbf{x}$$

which is the original equation, Eq. (4.10). We also note that the initial condition that $\mathbf{x} = \mathbf{x}(0)$ at $t = 0$ is satisfied by Eq. (4.12) since $\mathbf{\Phi}(0) = \mathbf{I}$. Since $\mathbf{\Phi}(t)$ satisfies the homogeneous equation, it defines the transition of the state from its initial value $\mathbf{x}(0)$ at the initial time $t = 0$ to a state $\mathbf{x}(t)$ at time t when the input $\mathbf{r} = \mathbf{0}$. Hence, $\mathbf{\Phi}(t)$ is called the *state transition matrix.* We now discuss some methods of obtaining the state transition matrix.

4.3.1 Methods of Determining the State Transition Matrix

Solution of the Defining Equation. This method involves the solution of the matrix differential equation, Eq. (4.11), with the initial condition $\mathbf{\Phi}(0) = \mathbf{I}$. Since $\mathbf{\Phi}(t)$ is an $n \times n$ matrix, we have to solve n^2-coupled, first-order differential equations. Hence, this method is not computationally convenient for high-order systems.

EXAMPLE 4.1

We consider the control system of Example 3.2. The state equation of the closed-loop system is given by Eq. (3.28), namely,

$$\mathbf{A} - \mathbf{BK} = \begin{bmatrix} 0 & 1 \\ 3g/2L - 3k/m & -3c/mL^2 \end{bmatrix} \tag{4.13}$$

Let the numerical values of the parameters be such that

$$\mathbf{A} - \mathbf{BK} = \begin{bmatrix} 0 & 1 \\ -3 & -4 \end{bmatrix} \tag{4.14}$$

The four first-order equations obtained from Eq. (4.11) for this closed-loop matrix are

$$\dot{\phi}_{11} = \phi_{21} \qquad \phi_{11}(0) = 1 \tag{4.15}$$

$$\dot{\phi}_{12} = \phi_{22} \qquad \phi_{12}(0) = 0 \tag{4.16}$$

$$\dot{\phi}_{21} = -3\phi_{11} - 4\phi_{21} \qquad \phi_{21}(0) = 0 \tag{4.17}$$

$$\dot{\phi}_{22} = -3\phi_{12} - 4\phi_{22} \qquad \phi_{22}(0) = 1 \tag{4.18}$$

Differentiating Eq. (4.17) with respect to time and substituting for $\dot{\phi}_{11}$ from Eq. (4.15), we obtain

$$\ddot{\phi}_{21} + 4\dot{\phi}_{21} + 3\phi_{21} = 0$$

and its solution is given by

$$\phi_{21}(t) = c_1 e^{-t} + c_2 e^{-3t} \tag{4.19}$$

Substituting this solution of $\phi_{21}(t)$ in Eq. (4.15), we get

$$\phi_{11}(t) = -c_1 e^{-t} - \tfrac{1}{3}c_2 e^{-3t} \tag{4.20}$$

The constants c_1 and c_2 are evaluated from the initial conditions. Since $\phi_{21}(0) = 0$, from Eq. (4.19) it follows that $c_1 + c_2 = 0$, and since $\phi_{11}(0) = 1$, from Eq. (4.20), $-c_1 - c_2/3 = 1$. Hence, $c_1 = -3/2$ and $c_2 = 3/2$. The other two functions ϕ_{12} and ϕ_{22} are obtained similarly from Eqs. (4.16) and (4.18). Hence, we get

$$\mathbf{\Phi}(t) = \begin{bmatrix} \tfrac{3}{2}e^{-t} - \tfrac{1}{2}e^{-3t} & \tfrac{1}{2}e^{-t} - \tfrac{1}{2}e^{-3t} \\ -\tfrac{3}{2}e^{-t} + \tfrac{3}{2}e^{-3t} & -\tfrac{1}{2}e^{-t} + \tfrac{3}{2}e^{-3t} \end{bmatrix} \tag{4.21}$$

Matrix Exponential. Consider a scalar first-order equation

$$\dot{x} = ax \qquad x = x(0) \text{ at } t = 0$$

Integrating this equation, we get

$$\int_{x(0)}^{x} \frac{dx}{x} = a\int_0^t dt$$

i.e.,

$$x(t) = x(0)\, e^{at}$$

Thus, the state transition scalar is $\phi(t) = \exp(at)$. The solution of Eq. (4.10) can be expressed similarly as

$$\mathbf{x}(t) = [e^{(\mathbf{A}-\mathbf{BK})t}]\mathbf{x}(0) \tag{4.22}$$

from which it follows that

$$\mathbf{\Phi}(t) = e^{(\mathbf{A}-\mathbf{BK})t} \tag{4.23}$$

To verify that Eq. (4.22) satisfies Eq. (4.10), we express Eq. (4.22) as

$$\mathbf{x}(t) = \left[\mathbf{I} + (\mathbf{A}-\mathbf{BK})t + \frac{1}{2!}(\mathbf{A}-\mathbf{BK})^2t^2 + \frac{1}{3!}(\mathbf{A}-\mathbf{BK})^3t^3 + \cdots\right]\mathbf{x}(0)$$

and differentiating this equation with respect to time, we get

$$\begin{aligned}\dot{\mathbf{x}}(t) &= (\mathbf{A}-\mathbf{BK})\left[\mathbf{I} + (\mathbf{A}-\mathbf{BK})t + \frac{1}{2!}(\mathbf{A}-\mathbf{BK})^2t^2 + \cdots\right]\mathbf{x}(0)\\ &= (\mathbf{A}-\mathbf{BK})\mathbf{x}(t)\end{aligned}$$

Thus, it is seen that Eq. (4.22) does indeed satisfy Eq. (4.10), and $\mathbf{x} = \mathbf{x}(0)$ at $t = 0$. Hence, the state transition matrix is also given by Eq. (4.23).

EXAMPLE 4.2

For the control system of Example 3.2, the state equation of the closed-loop system is described by by Eq. (4.13) and the numerical values given by Eq.

(4.14). Hence, we have

$$\mathbf{A} - \mathbf{BK} = \begin{bmatrix} 0 & 1 \\ -3 & -4 \end{bmatrix} \qquad (\mathbf{A} - \mathbf{BK})^2 = \begin{bmatrix} -3 & -4 \\ 12 & 13 \end{bmatrix}$$

$$(\mathbf{A} - \mathbf{BK})^3 = \begin{bmatrix} 12 & 13 \\ -39 & -40 \end{bmatrix}$$

Substituting these results in the infinite series representation of Eq. (4.23),

$$\mathbf{\Phi}(t) = \sum_{n=0}^{\infty} \frac{1}{n!}(\mathbf{A} - \mathbf{BK})^n t^n$$

$$= \begin{bmatrix} 1 - \frac{3}{2}t^2 + 2t^3 + \cdots & t - 2t^2 + \frac{13}{6}t^3 + \cdots \\ -3t + 6t^2 - \frac{13}{2}t^3 + \cdots & 1 - 4t + \frac{13}{2}t^2 - \frac{20}{3}t^3 + \cdots \end{bmatrix} \tag{4.24}$$

It can be verified that the infinite series representing each element of Eq. (4.24) converges to the corresponding element of the state-transition matrix given by Eq. (4.21). The disadvantage of employing the matrix exponential method is that the functions to which the infinite series converge are not readily apparent.

Laplace Transformation Method. The Laplace transformation of Eq. (4.10) with initial conditions $\mathbf{x}(0)$ at $t = 0$ yields

$$s\mathbf{X}(s) - \mathbf{x}(0) = (\mathbf{A} - \mathbf{BK})\mathbf{X}(s)$$

Solving for $\mathbf{X}(s)$ from the preceding equation, we obtain

$$\mathbf{X}(s) = (s\mathbf{I} - \mathbf{A} + \mathbf{BK})^{-1}\mathbf{x}(0) \tag{4.25}$$

From the inverse Laplace transformation of Eq. (4.25), it follows that

$$\mathbf{x}(t) = \mathscr{L}^{-1}[(s\mathbf{I} - \mathbf{A} + \mathbf{BK})^{-1}]\mathbf{x}(0) \qquad t \geq 0 \tag{4.26}$$

Comparing Eq. (4.26) with Eq. (4.12), we see that the state-transition matrix can also be expressed as

$$\mathbf{\Phi}(t) = \mathscr{L}^{-1}[(s\mathbf{I} - \mathbf{A} + \mathbf{BK})^{-1}] \tag{4.27}$$

It is shown in the next section that the state-transition matrix is nonsingular and hence the inverse matrix in Eqs. (4.25) or (4.27) is uniquely determined. The disadvantage of using this method is that it requires the inversion of an $n \times n$ matrix and inverse Laplace transformation.

EXAMPLE 4.3

We again consider the control system of Example 3.2, where the state equation of the closed-loop system is described by

$$\mathbf{A} - \mathbf{BK} = \begin{bmatrix} 0 & 1 \\ -3 & -4 \end{bmatrix}$$

Hence,

$$(s\mathbf{I} - \mathbf{A} + \mathbf{BK}) = \begin{bmatrix} s & -1 \\ 3 & s+4 \end{bmatrix}$$

and

$$(s\mathbf{I} - \mathbf{A} + \mathbf{BK})^{-1} = \frac{1}{(s+1)(s+3)} \begin{bmatrix} s+4 & 1 \\ -3 & s \end{bmatrix}$$

The partial fraction expansion of each element of the preceding matrix yields

$$(s\mathbf{I} - \mathbf{A} + \mathbf{BK})^{-1} = \begin{bmatrix} \dfrac{3/2}{s+1} - \dfrac{1/2}{s+3} & \dfrac{1/2}{s+1} - \dfrac{1/2}{s+3} \\ -\dfrac{3/2}{s+1} + \dfrac{3/2}{s+3} & -\dfrac{1/2}{s+1} + \dfrac{3/2}{s+3} \end{bmatrix} \tag{4.28}$$

Taking the inverse Laplace transformation of Eq. (4.28), we obtain

$$\mathbf{\Phi}(t) = \begin{bmatrix} \frac{3}{2}e^{-t} - \frac{1}{2}e^{-3t} & \frac{1}{2}e^{-t} - \frac{1}{2}e^{-3t} \\ -\frac{3}{2}e^{-t} + \frac{3}{2}e^{-3t} & -\frac{1}{2}e^{-t} + \frac{3}{2}e^{-3t} \end{bmatrix} \tag{4.29}$$

It is seen that this state-transition matrix is identical to that given by Eq. (4.21).

Matrix Diagonalization. When the order of the system exceeds three, the hand calculation of the state-transition matrix becomes very cumbersome. In this connection, it is helpful to employ a method where a digital computer can be used to provide computational assistance to obtain a closed-form expression for the state transition matrix. The method is based on diagonalizing the matrix $(\mathbf{A} - \mathbf{BK})$ when possible or reducing it to a Jordan normal form.

If the matrix is diagonalized, then the transformed state variables become uncoupled. The state-transition matrix for the transformed state variables becomes diagonal and is obtained by inspection. The state-transition matrix for the original state equations can then be obtained by a simple transformation. But the method requires the study of some preliminaries and for this reason it is covered later in this chapter.

4.3.2 Properties of the State-Transition Matrix

The state-transition matrix possesses some important properties that are useful for the study of the transient behavior. The properties are listed in the following for time-invariant systems.

1. $\mathbf{\Phi}(0) = \mathbf{I}$ (4.30)

Here, $\mathbf{I}$ is an $n \times n$ identity matrix. This property follows from the defining equation, Eqs. (4.11) or (4.12), and is also obvious from Eq. (4.23) by setting $t = 0$.

2. $\mathbf{\Phi}(t)$ is nonsingular and hence $\mathbf{\Phi}^{-1}(t)$ is unique

To verify this property, we note that the n columns of $\mathbf{\Phi}(t)$ are made up of n linearly independent solutions of Eq. (4.10). A necessary and sufficient condition for this is that the Wronskian is nonzero, i.e., the determinant

$$\det \begin{vmatrix} \phi_{11}(t) \cdots \phi_{1n}(t) \\ \cdot \quad \cdots \quad \cdot \\ \phi_{n1}(t) \cdots \phi_{nn}(t) \end{vmatrix} \neq 0 \qquad t \geq 0$$

3. $\mathbf{\Phi}(t + t_1) = \mathbf{\Phi}(t)\mathbf{\Phi}(t_1)$ (4.31)

Since,

$$\mathbf{\Phi}(t) = \exp{(\mathbf{A} - \mathbf{BK})t}$$

we have

$$\begin{aligned} \mathbf{\Phi}(t + t_1) &= \exp{(\mathbf{A} - \mathbf{BK})(t + t_1)} \\ &= \exp{(\mathbf{A} - \mathbf{BK})t} \exp{(\mathbf{A} - \mathbf{BK})t_1} \\ &= \mathbf{\Phi}(t)\mathbf{\Phi}(t_1) \end{aligned}$$

4. $\mathbf{\Phi}^{-1}(t) = \mathbf{\Phi}(-t)$ (4.32)

This property can be verified by using property 3. In Eq. (4.31), letting $t_1 = -t$, we obtain

$$\mathbf{\Phi}(0) = \mathbf{\Phi}(t)\mathbf{\Phi}(-t)$$

But $\mathbf{\Phi}(0) = \mathbf{I}$ and hence, $\mathbf{\Phi}^{-1}(t) = \mathbf{\Phi}(-t)$. As a result of this property, from Eq. (4.12) we obtain

$$\mathbf{x}(0) = \mathbf{\Phi}(-t)\mathbf{x}(t) \tag{4.33}$$

which means that knowing $\mathbf{x}(0)$, we can solve for $\mathbf{x}(t)$, or knowing $\mathbf{x}(t)$, we can obtain $\mathbf{x}(0)$ in the absence of inputs.

5. $\mathbf{\Phi}(t_2 - t_1)\mathbf{\Phi}(t_1 - t_0) = \mathbf{\Phi}(t_2 - t_0)$ for any t_0, t_1, t_2 (4.34)

To verify this property, we note that

$$\begin{aligned} \mathbf{\Phi}(t_2 - t_1)\mathbf{\Phi}(t_1 - t_0) &= \exp{(\mathbf{A} - \mathbf{BK})(t_2 - t_1)} \exp{(\mathbf{A} - \mathbf{BK})(t_1 - t_0)} \\ &= \exp{(\mathbf{A} - \mathbf{BK})(t_2 - t_0)} \\ &= \mathbf{\Phi}(t_2 - t_0) \end{aligned}$$

We see from this property that the state transition from $t = t_0$ to $t = t_2$ is equal to the transition from t_0 to t_1, and then from t_1 to t_2. Hence, the state transition can be broken up into a sequence of transitions.

4.4 FORCED RESPONSE

We have so far characterized the response of closed-loop control systems to initial conditions in the absence of inputs by the state transition matrix. We now study their response when also forced by a command input. The state equations of the closed-loop control system are described by

$$\dot{\mathbf{x}}(t) = (\mathbf{A} - \mathbf{BK})\mathbf{x}(t) + \mathbf{BNr}(t) \qquad \mathbf{x} = \mathbf{x}(0) \text{ at } t = 0 \tag{4.35}$$

$$\text{Output:} \qquad \mathbf{y}(t) = \mathbf{Cx}(t) \tag{4.36}$$

An advantage of using the state-representation model for the transient response analysis is that, in general, we let Eq. (4.35) and Eq. (4.36) represent multivariable systems. For the special case of single-input, single-output systems, we let N be a scalar, $\mathbf{B}$ an $n \times 1$ column matrix and $\mathbf{C}$ a $1 \times n$ row matrix. In the following, we obtain the solution of Eq. (4.35) by using two methods, namely, the classical method and the Laplace transform method.

4.4.1 Classical Method of Solution

The classical method of solution that we use is called the method of variation of parameters. We assume a solution in the form

$$\mathbf{x}(t) = \mathbf{\Phi}(t)\mathbf{f}(t) \tag{4.37}$$

where $\mathbf{\Phi}(t)$ is the state-transition matrix and $\mathbf{f}(t)$ is an $n \times 1$ column matrix of unknown functions of time. Differentiating Eq. (4.37) with respect to time, we get

$$\dot{\mathbf{x}} = \dot{\mathbf{\Phi}}\mathbf{f} + \mathbf{\Phi}\dot{\mathbf{f}} \tag{4.38}$$

Substituting for $\dot{\mathbf{\Phi}}$ in the preceding equation from Eq. (4.11), it follows that

$$\dot{\mathbf{x}}(t) = (\mathbf{A} - \mathbf{BK})\mathbf{\Phi}(t)\mathbf{f}(t) + \mathbf{\Phi}(t)\dot{\mathbf{f}}(t) \tag{4.39}$$

Also, substituting for $\mathbf{x}(t)$ from Eq. (4.37) in Eq. (4.35), we obtain

$$\dot{\mathbf{x}}(t) = (\mathbf{A} - \mathbf{BK})\mathbf{\Phi}(t)\mathbf{f}(t) + \mathbf{BNr}(t) \tag{4.40}$$

On comparing Eqs. (4.39) and (4.40), we see that $\mathbf{f}(t)$ satisfies the equation

$$\mathbf{\Phi}(t)\dot{\mathbf{f}}(t) = \mathbf{BNr}(t) \tag{4.41}$$

Multiplying both sides of Eq. (4.41) by $\mathbf{\Phi}^{-1}(t)$ and integrating with respect to time, we see that

$$\mathbf{f}(t) = \mathbf{f}(0) + \int_0^t \mathbf{\Phi}^{-1}(t')\mathbf{BNr}(t')\, dt' \tag{4.42}$$

where t' is the dummy of integration. Using properties of Eqs. (4.30) and (4.32) in Eq. (4.37), we get

$$\mathbf{f}(0) = \mathbf{\Phi}^{-1}(0)\mathbf{x}(0) = \mathbf{x}(0)$$

Substituting this solution of $\mathbf{f}(t)$ in Eq. (4.37), we obtain the solution of the state equation, Eq. (4.35), as

$$\mathbf{x}(t) = \mathbf{\Phi}(t)\mathbf{x}(0) + \mathbf{\Phi}(t)\int_0^t \mathbf{\Phi}^{-1}(t')\mathbf{BNr}(t')\, dt' \tag{4.43}$$

In the foregoing equation, we replace $\mathbf{\Phi}^{-1}(t')$ by $\mathbf{\Phi}(-t')$ from the property of Eq. (4.32) and then $\mathbf{\Phi}(t)\mathbf{\Phi}(-t')$ by $\mathbf{\Phi}(t - t')$ from the property of Eq. (4.31). Hence, Eq. (4.43) becomes

$$\mathbf{x}(t) = \mathbf{\Phi}(t)\mathbf{x}(0) + \int_0^t \mathbf{\Phi}(t - t')\mathbf{BNr}(t')\, dt' \qquad t \geq 0 \tag{4.44}$$

The first term on the right-hand side of the foregoing equation is the complementary solution due to the initial conditions and the second term is the particular solution due to the command input. The solution of the output can now be obtained from Eq. (4.36) as

$$\mathbf{y}(t) = \mathbf{C\Phi}(t)\mathbf{x}(0) + \mathbf{C}\int_0^t \mathbf{\Phi}(t - t')\mathbf{BNr}(t')\, dt' \qquad t \geq 0 \tag{4.45}$$

It is sometimes convenient, as in the case of computer control systems of Chapter 9, to represent the initial time by t_0, the initial state by $\mathbf{x}(t_0)$, with the input $\mathbf{r}(t)$ applied for $t \geq t_0$. In Eq. (4.44), we set $t = t_0$, premultiply the equation by $\mathbf{\Phi}^{-1}(t_0)$, which is replaced by $\mathbf{\Phi}(-t_0)$ from the property of Eq. (4.32) and obtain

$$\mathbf{x}(0) = \mathbf{\Phi}(-t_0)\mathbf{x}(t_0) - \mathbf{\Phi}(-t_0)\int_0^{t_0} \mathbf{\Phi}(t_0 - t')\mathbf{BNr}(t')\, dt' \tag{4.46}$$

Substituting for $\mathbf{x}(0)$ from Eq. (4.46) in Eq. (4.44), and setting $\mathbf{\Phi}(t)\mathbf{\Phi}(-t_0) = \mathbf{\Phi}(t - t_0)$ from the property of Eq. (4.31), we get

$$\mathbf{x}(t) = \mathbf{\Phi}(t - t_0)\mathbf{x}(t_0) - \mathbf{\Phi}(t - t_0)\int_0^{t_0} \mathbf{\Phi}(t_0 - t')\mathbf{BNr}(t')\, dt' + \int_0^t \mathbf{\Phi}(t - t')\mathbf{BNr}(t')\, dt' \tag{4.47}$$

By employing the property of Eq. (4.34), the second term on the right-hand side of Eq. (4.47) is expressed as

$$-\int_0^{t_0} \mathbf{\Phi}(t - t')\mathbf{BNr}(t')\, dt'$$

In the third term on the right-hand side of Eq. (4.47), we split the time interval from 0 to t_0 and from t_0 to t. Hence, Eq. (4.47) is now expressed as

$$\mathbf{x}(t) = \mathbf{\Phi}(t - t_0)\mathbf{x}(t_0) + \int_{t_0}^t \mathbf{\Phi}(t - t')\mathbf{BNr}(t')\, dt' \qquad t \geq t_0 \tag{4.48}$$

and

$$\mathbf{y}(t) = \mathbf{C\Phi}(t - t_0)\mathbf{x}(t_0) + \mathbf{C}\int_{t_0}^t \mathbf{\Phi}(t - t')\mathbf{BNr}(t')\, dt' \tag{4.49}$$

4.4.2 Laplace Transformation Method of Solution

Laplace transforming Eq. (4.35) with initial conditions $\mathbf{x}(0)$ at $t = 0$, we obtain

$$s\mathbf{X}(s) - \mathbf{x}(0) = (\mathbf{A} - \mathbf{BK})\mathbf{X}(s) + \mathbf{BNR}(s)$$

Solving for $\mathbf{X}(s)$,

$$\mathbf{X}(s) = (s\mathbf{I} - \mathbf{A} + \mathbf{BK})^{-1}\mathbf{x}(0) + (s\mathbf{I} - \mathbf{A} + \mathbf{BK})^{-1}\mathbf{BNR}(s) \tag{4.50}$$

From Eq. (4.27), we note that

$$\mathbf{\Phi}(t) = \mathscr{L}^{-1}[(s\mathbf{I} - \mathbf{A} + \mathbf{BK})^{-1}] \tag{4.51}$$

In the second term on the right-hand side of Eq. (4.50), we can obtain the inverse Laplace transforms of $(s\mathbf{I} - \mathbf{A} + \mathbf{BK})^{-1}$ and $\mathbf{R}(s)$ separately and then employ the convolution integral to express the results. Hence, taking the inverse Laplace transform of Eq. (4.50), we get

$$\mathbf{x}(t) = \mathbf{\Phi}(t)\mathbf{x}(0) + \int_0^t \mathbf{\Phi}(t - t')\mathbf{BNr}(t')\,dt' \qquad t \geq 0 \tag{4.52}$$

which is identical to Eq. (4.44). The solution of the output is given by Eq. (4.45). In case the initial time is t_0 instead of zero, the solution of the state is given by Eq. (4.48) and the output by Eq. (4.49).

4.4.3 State Representation and Transfer Function

The relationship between state equations and transfer functions for closed-loop control systems has been discussed in Chapter 3. Here, we review this relationship in view of our study of the state transition matrix and transient response. The state-variables formulation of closed-loop control systems is expressed by

$$\dot{\mathbf{x}}(t) = (\mathbf{A} - \mathbf{BK})\mathbf{x}(t) + \mathbf{BNr}(t) \tag{4.53}$$

$$\text{Output:} \qquad \mathbf{y}(t) = \mathbf{Cx}(t) \tag{4.54}$$

Laplace transforming Eqs. (4.53) and (4.54) with initial conditions $\mathbf{x}(0) = \mathbf{0}$, we obtain

$$s\mathbf{X}(s) = (\mathbf{A} - \mathbf{BK})\mathbf{X}(s) + \mathbf{BNR}(s) \tag{4.55}$$

$$\mathbf{Y}(s) = \mathbf{CX}(s) \tag{4.56}$$

It follows from Eqs. (4.55) and (4.56) that

$$(s\mathbf{I} - \mathbf{A} + \mathbf{BK})\mathbf{X}(s) = \mathbf{BNR}(s)$$

$$\mathbf{X}(s) = (s\mathbf{I} - \mathbf{A} + \mathbf{BK})^{-1}\mathbf{BNR}(s) \tag{4.57}$$

and

$$\mathbf{Y}(s) = \mathbf{C}(s\mathbf{I} - \mathbf{A} + \mathbf{BK})^{-1}\mathbf{BNR}(s) \tag{4.58}$$

Hence, the closed-loop transfer matrix relating the command input $\mathbf{R}(s)$ to the controlled output $\mathbf{Y}(s)$ is given by

$$\mathbf{G}_c(s) = \mathbf{C}(s\mathbf{I} - \mathbf{A} + \mathbf{BK})^{-1}\mathbf{BN} \tag{4.59}$$

In multivariable systems where m command inputs $\mathbf{r}(t)$ are used to control m controlled outputs $\mathbf{y}(t)$, $\mathbf{G}_c(s)$ is an $(m \times m)$ transfer matrix, whereas for single-input, single-output systems, $G_c(s)$ is a scalar transfer function. Letting $\hat{\mathbf{\Phi}}(s)$ denote the Laplace transform of the state-transition matrix, we note from Eqs. (4.51) that

$$\hat{\mathbf{\Phi}}(s) = (s\mathbf{I} - \mathbf{A} + \mathbf{BK})^{-1} \tag{4.60}$$

Hence, the transfer function, Eq. (4.59), can also be represented by

$$\mathbf{G}_c(s) = \mathbf{C}\hat{\mathbf{\Phi}}(s)\mathbf{BN} \tag{4.61}$$

Letting Adj ( ) and det| | denote the adjoint and determinant of a matrix, respectively, we can express Eq. (4.59) as

$$\mathbf{G}_c(s) = \frac{\mathbf{C}[\text{Adj}\,(s\mathbf{I} - \mathbf{A} + \mathbf{BK})]\mathbf{BN}}{\det|s\mathbf{I} - \mathbf{A} + \mathbf{BK}|} \tag{4.62}$$

and the block diagram is shown in Fig. 4.2.

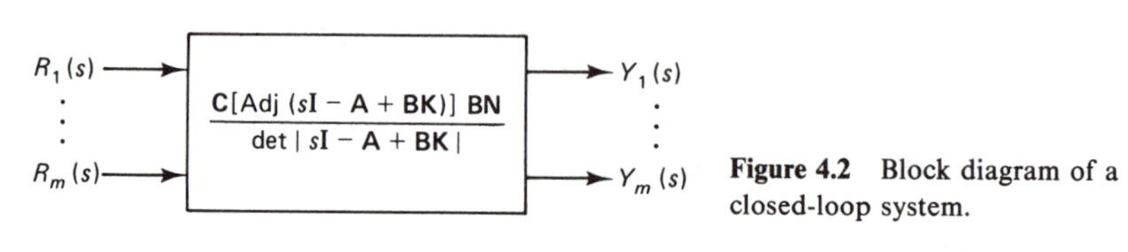

Figure 4.2 Block diagram of a closed-loop system.

The equation obtained by setting the denominator of the transfer function, Eq. (4.62), to zero is called the characteristic equation, that is, the characteristic equation of the closed-loop control system is defined by

$$\det|s\mathbf{I} - \mathbf{A} + \mathbf{BK}| = 0 \tag{4.63}$$

It can be obtained from the transfer function, Eq. (4.62), or from the state equation, Eq. (4.53). The roots of the characteristic equation play an important role in the transient response and stability of a control system.

EXAMPLE 4.4

We consider the hydraulic control system of Example 3.5, whose block diagram is shown in Fig. 3.19. It is a fourth-order system. If the natural frequency of the spool valve is high enough that it is much beyond the system bandwidth, which is discussed in Chapter 5, we can neglect the dynamics of the servovalve and represent it by a simple gain k_1. With this assumption, the order of the system is reduced to two. Neglecting the disturbance torque T_d, the block diagram of Fig. 3.19 is simplified and shown in Fig. 4.3.

The output speed ω_0 and the actuator displacement y_a are chosen as the two state variables x_1 and x_2 as shown in Fig. 4.3. The servovalve current i is the control input u, and the desired speed ω_r is the command input $r(t)$. The

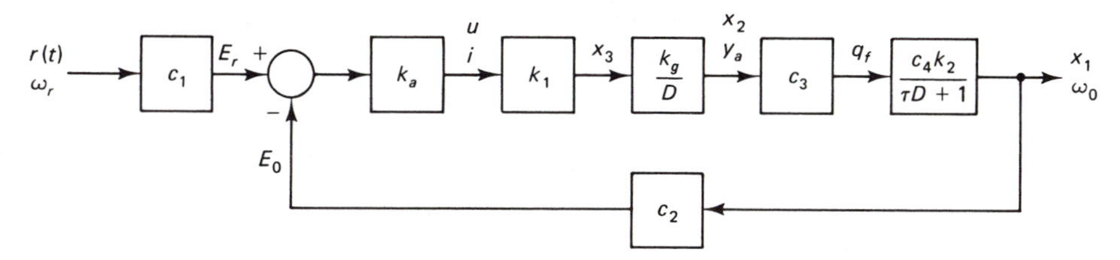

Figure 4.3 Block diagram of a simplified speed-control system.

scale factor c_1 is chosen such that $c_1 = c_2$. The state equations can be obtained from Fig. 4.3 as follows.

$$\dot{x}_1 = -\frac{1}{\tau}x_1 + \frac{1}{\tau}k_2c_3c_4x_2$$

$$\dot{x}_2 = k_1k_gu$$

Control law: $u = k_ac_1(r - x_1)$

Output: $y = x_1$

Comparing these equations to the standard form

Plant: $\dot{\mathbf{x}} = \mathbf{Ax} + \mathbf{B}u$

Control law: $u = Nr - \mathbf{Kx}$

Output: $y = \mathbf{Cx}$

we obtain

$$\mathbf{A} = \begin{bmatrix} -\dfrac{1}{\tau} & \dfrac{1}{\tau}k_2c_3c_4 \\ 0 & 0 \end{bmatrix} \qquad \mathbf{B} = \begin{bmatrix} 0 \\ k_1k_g \end{bmatrix} \tag{4.64}$$

$$N = k_ac_1 \qquad \mathbf{K} = \lfloor k_ac_1 \quad 0 \rfloor \qquad \mathbf{C} = \lfloor 1 \quad 0 \rfloor$$

The state equation of the closed-loop system is

$$\dot{\mathbf{x}} = (\mathbf{A} - \mathbf{BK})\mathbf{x} + \mathbf{B}Nr(t)$$

where

$$(\mathbf{A} - \mathbf{BK}) = \begin{bmatrix} -\dfrac{1}{\tau} & \dfrac{1}{\tau}k_2c_3c_4 \\ -k_1k_gk_ac_1 & 0 \end{bmatrix} \qquad \mathbf{B}N = \begin{bmatrix} 0 \\ k_1k_gk_ac_1 \end{bmatrix} \tag{4.65}$$

The Laplace transform of the state transition matrix is given by

$$\hat{\mathbf{\Phi}}(s) = (s\mathbf{I} - \mathbf{A} + \mathbf{BK})^{-1}$$

$$= \frac{1}{s^2 + \dfrac{1}{\tau}s + \dfrac{1}{\tau}k_1k_2k_gk_ac_1c_3c_4} \begin{bmatrix} s & \dfrac{1}{\tau}k_2c_3c_4 \\ -k_1k_gk_ac_1 & s + \dfrac{1}{\tau} \end{bmatrix} \tag{4.66}$$

The closed-loop transfer function relating the output $y(t)$ to the command input $r(t)$ is given by

$$G_c(s) = \mathbf{C}\hat{\mathbf{\Phi}}(s)\mathbf{B}N \tag{4.67}$$

Substituting for the right-hand side of Eq. (4.67) from Eqs. (4.64), (4.66), and (4.65) and multiplying the resulting matrices, we obtain

$$G_c(s) = \frac{\frac{1}{\tau} k_1 k_2 k_g k_a c_1 c_3 c_4}{s^2 + \frac{1}{\tau} s + \frac{1}{\tau} k_1 k_2 k_g k_a c_1 c_3 c_4} \tag{4.68}$$

This closed-loop transfer function can also be obtained from the block diagram of Fig. 4.3. Since $c_1 = c_2$, the block diagram of Fig. 4.3 can be

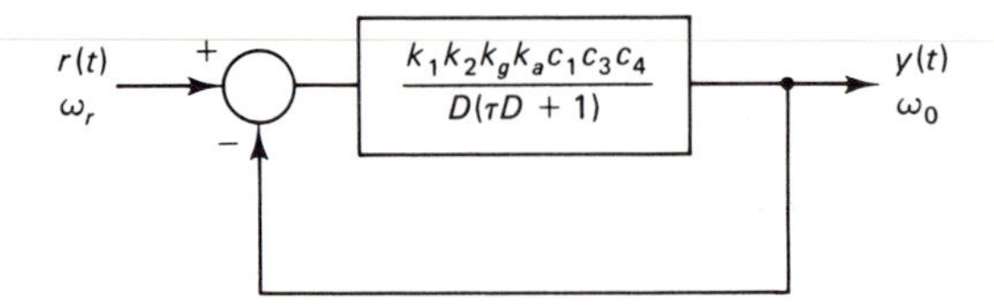

Figure 4.4 Block diagram of a control system.

represented as shown in Fig. 4.4, from which we obtain

$$\frac{y}{r} = \frac{\left(\dfrac{k_1 k_2 k_g k_a c_1 c_3 c_4}{D(\tau D + 1)}\right)}{1 + \left(\dfrac{k_1 k_2 k_g k_a c_1 c_3 c_4}{D(\tau D + 1)}\right)} \tag{4.69}$$

$$y(t) = \left(\frac{k_1 k_2 k_g k_a c_1 c_3 c_4}{\tau D^2 + D + k_1 k_2 k_g k_a c_1 c_3 c_4}\right) r(t)$$

or in the Laplace domain,

$$Y(s) = \left(\frac{k_1 k_2 k_g k_a c_1 c_3 c_4}{\tau s^2 + s + k_1 k_2 k_g k_a c_1 c_3 c_4}\right) R(s) \tag{4.70}$$

We note that the transfer function of Eq. (4.70) is identical to that of Eq. (4.68). The characteristic equation of the closed-loop control system can be obtained from Eq. (4.66) or Eq. (4.70) as

$$s^2 + \frac{1}{\tau} s + \frac{1}{\tau} k_1 k_2 k_g k_a c_1 c_3 c_4 = 0 \tag{4.71}$$

4.5 STEP AND IMPULSE RESPONSE

The forced response of control systems was analyzed in the previous section and represented in the form of a convolution integral for any arbitrary command input.

In this section, we study the transient response to specific test signals. The Heaviside step function and Dirac's delta or impulse function are chosen as the testing inputs. We show that a great deal of information about the transient behavior can be obtained by studying the location of the roots of the characteristic equation in the complex s-plane.

4.5.1 Response of First-Order Systems

Control systems of first order are rarely encountered in practice but our purpose here is to demonstrate how the value of a real root of the characteristic equation affects the transient response. We illustrate the procedure by considering an example but the results are applicable to any first-order system.

EXAMPLE 4.5

A first-order system was modeled in Example 3.1 and its state equation is given by Eq. (3.20). Neglecting the disturbance, the closed-loop state equation becomes

$$\dot{x} = -\frac{1}{\tau_1}\left(1 + \frac{Rk_1}{\rho g}\right)x + \left(\frac{Rk_1}{\tau_1 \rho g}\right)h_r \tag{4.72}$$

$$\text{Output:} \qquad y = x,$$

where the state variable x represents a change in the tank liquid level, and h_r is the command input $r(t)$. The closed-loop transfer function is given by Eq. (3.21), and letting $y = x$ and $h_r = r$, we obtain

$$Y(s) = \left(\frac{Rk_1/\rho g}{\tau_1 s + 1 + Rk_1/\rho g}\right)R(s) \tag{4.73}$$

Defining two constants a and k as

$$\frac{1}{a} = \frac{\tau_1}{1 + Rk_1/\rho g} \qquad k = \frac{k_1 R/\rho g}{1 + Rk_1/\rho g} \tag{4.74}$$

where we note that $a > 0$ and $0 < k < 1$, Eq. (4.73) can be expressed as

$$Y(s) = \left(\frac{k}{(1/a)s + 1}\right)R(s) \tag{4.75}$$

If y and r are constants, then the steady-state relationship becomes $y = kr$, and hence k is called the steady-state gain and $1/a = \tau$, where τ is the time constant of the closed-loop system. Now, Eq. (4.72) is expressed as

$$\dot{x} = -ax + kar \tag{4.76}$$

The characteristic equation is given by

$$s + a = 0 \tag{4.77}$$

and its root is $-a$, which is real and negative. The state transition matrix, which in this case is a scalar, is obtained as

$$\phi(t) = e^{-at} \tag{4.78}$$

and the solution, Eq. (4.44), is expressed for $t > 0$ by

$$y(t) = x(t) = e^{-at}x(0) + \int_0^t e^{-a(t-t')}kar(t')\, dt' \tag{4.79}$$

Unit Step Response. Let $r(t)$ be a unit Heaviside step function, i.e., $r(t) = 1$ for $t > 0$, and $r(t) = 0$ for $t < 0$. From Eq. (4.79), we obtain

$$\begin{aligned} y(t) = x(t) &= e^{-at}x(0) + ka\, e^{-at} \int_0^t e^{at'}\, dt' \\ &= e^{-at}x(0) + k(1 - e^{-at}) \\ &= e^{-t/\tau}x(0) + k(1 - e^{-t/\tau}) \end{aligned} \tag{4.80}$$

This unit step response is shown in Fig. 4.5 for $x(0) = 0$. The slope of the

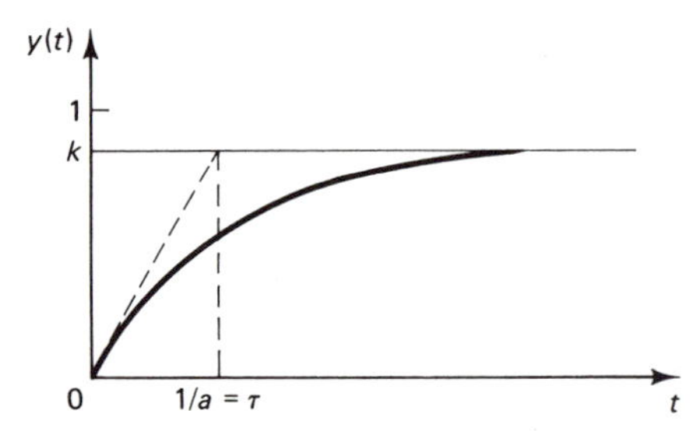

Figure 4.5 Step response of a first-order system.

tangent line at $t = 0$, shown in Fig. 4.5 for $x(0) = 0$, is obtained as

$$\begin{aligned} \frac{dy}{dt}(0) &= ka\, e^{-at}\Big|_{t=0} \\ &= ka = k/\tau \end{aligned} \tag{4.81}$$

Another important characteristic is that when $t = 1/a = \tau$, $y = k(1 - e^{-1}) = 0.632k$, i.e., when $t = \tau$, 63.2% of the response is completed. Hence, for fast response, the value of a should be large where $-a$ is the root of the characteristic equation or equivalently, the value of the time constant τ should be small. We note that $r(t) = 1$ for $t > 0$, but the final value of y is k, where we note from Eq. (4.74) that $0 < k < 1$. Hence, for the case of Example 3.1, there is a steady-state error between the command input and controlled output for a step change in the input. It will be shown in Chapter 7 that this steady-state error for step input is typical for control systems that are classified as type 0.

Unit Impulse Response. Let the initial condition as t approaches zero from the left be $x(-0)$ and $r(t)$ be a unit impulse or Dirac delta function that is defined as

$$\delta(t) = 0 \text{ for } t \neq 0 \qquad \text{and} \qquad \int_{-\infty}^{\infty} \delta(t)\, dt = 1$$

Consider a pulse of height $1/\Delta t$ and duration Δt such that its area is unity as shown in Fig. 4.6(a). When we take the limit as $\Delta t \to 0$, the unit pulse becomes the unit impulse $\delta(t)$ shown in Fig. 4.6(b).

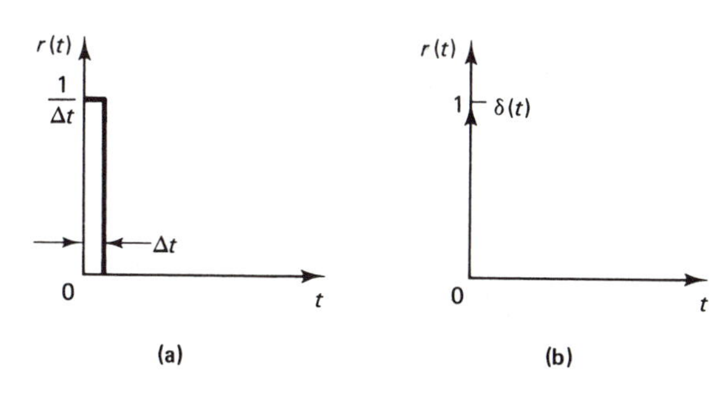

Fig. 4.6 (a) Unit pulse, and (b) unit impulse.

Substituting this result in Eq. (4.79), we obtain

$$y(t) = x(t) = e^{-at}x(-0) + ka\, e^{-at} \int_0^t e^{at'}\delta(t')\, dt' \tag{4.82}$$

Since $\delta(t') = 0$ for $t' \neq 0$ and $e^{at'} = 1$ for $t' = 0$, Eq. (4.82) becomes

$$y(t) = x(t) = [x(-0) + ka]\, e^{-at} \tag{4.83}$$

It is seen from Eq. (4.83) that when t approaches zero from the right, we get $x(+0) = x(-0) + ka$. The solution is not unique at $t = 0$. This is a consequence of the fact that the input is not piecewise continuous at $t = 0$ and the conditions for the existence and uniqueness theorem are not satisfied at $t = 0$. The solution is however unique for $t > 0$. The unit-impulse response of Eq. (4.83) is shown in Fig. 4.7 for $x(-0) = 0$. Equivalently, we can consider the initial condition at $t = +0$ as $x(+0) = x(-0) + ka$, and disregard the unit-impulse input.

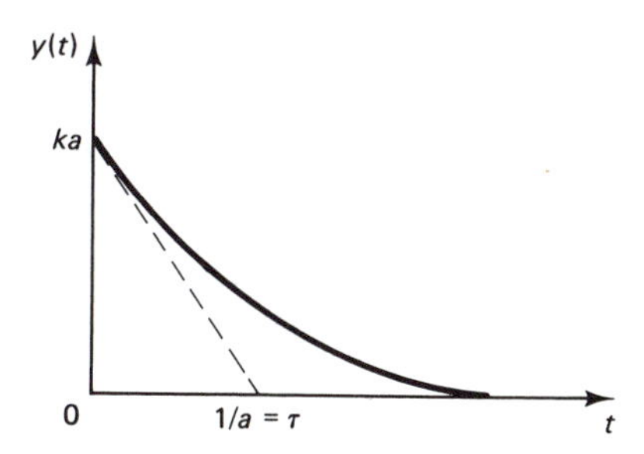

Figure 4.7 Impulse response of a first-order system.

The slope of the tangent line at $t = +0$, as shown in Fig. 4.7, is given by

$$\frac{dy}{dt}(+0) = -ka^2 = -\frac{k}{\tau^2}$$

Again, it is seen that for the impulse response to decay fast to zero, the value of a should be large or equivalently τ should be small. For this example, it is seen from Eq. (4.74) that $a > 0$. If a were such that $a < 0$, then the root

$-a$ of the characteristic equation would be real positive, and the step and impulse responses would grow out of bound with time, resulting in unstable behavior.

4.5.2 Response of Second-Order Systems

We study second-order systems to investigate the effect of the two roots of a quadratic equation on the transient response of a control system. We again illustrate the procedure by considering an example, but the results are applicable to any second-order system.

EXAMPLE 4.6

The simplified model of the hydraulic control system was studied in Example 4.4. The state equation representation of the closed-loop system is given by Eq. (4.65) and the transfer function by Eq. (4.68). In Eq. (4.68), let $(k_1k_2k_gk_ac_1c_3c_4)/\tau = \omega_n^2$ and $1/\tau = 2\zeta\omega_n$, where ω_n is the natural frequency, and ζ the damping ratio of the closed-loop control system. Then, Eq. (4.70) is expressed by

$$Y(s) = \left(\frac{\omega_n^2}{s^2 + 2\zeta\omega_n s + \omega_n^2}\right) R(s) \tag{4.84}$$

The state transition matrix is obtained by taking the inverse Laplace transform of $\hat{\mathbf{\Phi}}(s)$ in Eq. (4.66). The characteristic equation is given by

$$s^2 + 2\zeta\omega_n s + \omega_n^2 = 0 \tag{4.85}$$

and its roots are

$$s_1, s_2 = -\zeta\omega_n \pm \omega_n\sqrt{\zeta^2 - 1} \tag{4.86}$$

We consider the case where the initial conditions $x_1(0) = 0$ and $x_2(0) = 0$. The forced response is given by

$$\mathbf{x}(t) = \int_0^t \mathbf{\Phi}(t - t')\mathbf{B}Nr(t')\, dt' \tag{4.87}$$

and the output by

$$y(t) = \mathbf{C}\int_0^t \mathbf{\Phi}(t - t')\mathbf{B}Nr(t')\, dt' \tag{4.88}$$

Substituting for $\mathbf{C}$, $\mathbf{B}$, and N in Eq. (4.88), from Eq. (4.64), and multiplying the matrices, we obtain

$$y(t) = k_1k_gk_ac_1\int_0^t \phi_{12}(t - t')r(t')\, dt' \tag{4.89}$$

where $\phi_{12}(t)$ is the element in the first row and second column of the state transition matrix. From Eq. (4.66), we obtain

$$k_1 k_g k_a c_1 \hat{\phi}_{12}(s) = \frac{\omega_n^2}{s^2 + 2\zeta\omega_n s + \omega_n^2} \tag{4.90}$$

Hence, it is seen that the Laplace transformation of Eq. (4.89) yields Eq. (4.84).

Unit Step Response. Let the initial conditions be zero and $r(t)$ be a unit step.

Case 1. Overdamped case, $\zeta > 1$. The roots s_1 and s_2 of the characteristic equation as given by Eq. (4.86) are real and unequal. For this case, we can also define two time constants and express the roots as

$$s_1 = -1/\tau_1 = -\zeta\omega_n + \omega_n\sqrt{\zeta^2 - 1}$$

and

$$s_2 = -1/\tau_2 = -\zeta\omega_n - \omega_n\sqrt{\zeta^2 - 1} \tag{4.91}$$

The inverse Laplace transform of Eq. (4.90) yields

$$k_1 k_g k_a c_1 \phi_{12}(t) = \frac{\omega_n}{2\sqrt{\zeta^2 - 1}}[e^{-t/\tau_1} - e^{-t/\tau_2}] \tag{4.92}$$

Using Eq. (4.92) and letting $r(t') = 1$ in Eq. (4.89), we obtain

$$y(t) = 1 + \frac{1}{2\sqrt{\zeta^2 - 1}}[-(\zeta + \sqrt{\zeta^2 - 1})\, e^{-t/\tau_1} + (\zeta - \sqrt{\zeta^2 - 1})\, e^{-t/\tau_2}] \tag{4.93}$$

This response is shown in Fig. 4.8 with the corresponding root locations. It is seen that as $t \to \infty$, $y(t) = 1$, and since $r(t) = 1$ for $t > 0$, there is no steady-state error between the command input and the controlled output. It will be shown in Chapter 7 that there is no steady-state error for step input for control systems that are classified as type 1.

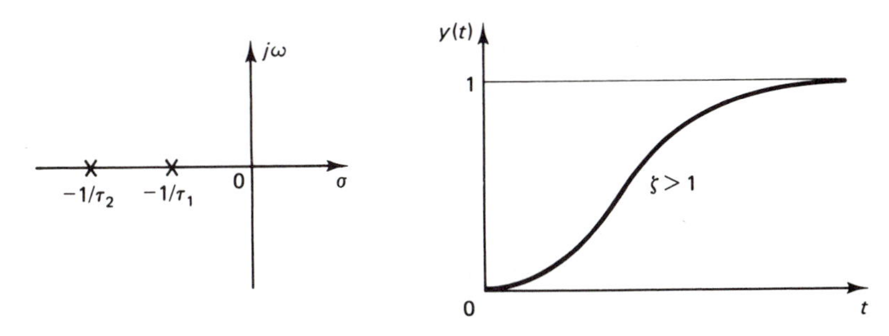

Figure 4.8 Step response of an overdamped second-order system.

Case 2. Critically damped case, $\zeta = 1$. It is seen from Eq. (4.86) that the two roots are now real, negative, and equal, that is, $s_1 = s_2 = -1/\tau = -\omega_n$. Equation (4.90) becomes

$$k_1 k_g k_a c_1 \hat{\phi}_{12}(s) = \omega_n^2/(s + \omega_n)^2 \tag{4.94}$$

and its inverse Laplace transform yields

$$k_1 k_g k_a c_1 \phi_{12}(t) = \omega_n^2 (t\, e^{-\omega_n t}) \tag{4.95}$$

Using Eq. (4.95) and letting $r(t') = 1$ in Eq. (4.89), we get

$$\begin{aligned} y(t) &= 1 - (\omega_n t + 1)\, e^{-\omega_n t} \\ &= 1 - [(t/\tau) + 1]\, e^{-t/\tau} \qquad t \geq 0 \end{aligned} \tag{4.96}$$

This response is shown in Fig. 4.9 with the corresponding repeated root location.

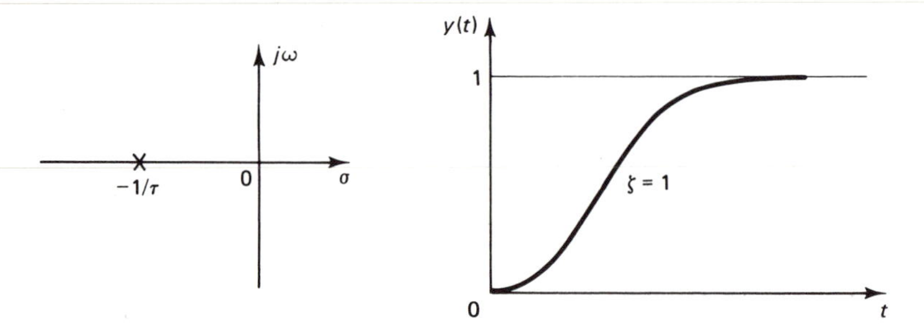

Figure 4.9 Step response of a critically damped second-order system.

Case 3. Underdamped case, $0 < \zeta < 1$. For this case, the roots are complex conjugate with a negative real part and from Eq. (4.86) are given by

$$s_1, s_2 = -\zeta\omega_n \pm j\omega_n\sqrt{1-\zeta^2} \tag{4.97}$$

The inverse Laplace transfer of Eq. (4.90) yields

$$k_1 k_g k_a c_1 \phi_{12}(t) = \left(\frac{\omega_n}{\sqrt{1-\zeta^2}}\right) e^{-\zeta\omega_n t} \sin \omega_n\sqrt{1-\zeta^2}\, t \tag{4.98}$$

Using Eq. (4.98) and letting $r(t') = 1$ in Eq. (4.89), we obtain

$$y(t) = 1 + \left(\frac{1}{\sqrt{1-\zeta^2}}\right) e^{-\zeta\omega_n t} \sin\left(\omega_n\sqrt{1-\zeta^2}\, t - \tan^{-1}\sqrt{1-\zeta^2}/-\zeta\right) \tag{4.99}$$

This response is shown in Fig. 4.10 with the corresponding root locations. It is seen that for a low value of the damping ratio, the response is fast, has a large overshoot, and is highly oscillatory before settling to its equilibrium value of unity. For $\zeta \geq 1$, there is no overshoot, but the response is slow. As seen in Fig. 4.10, the distance of the root from the origin is equal to the natural frequency ω_n and the damping ratio is given by $\zeta = \sin\theta$. The imaginary part of the root, $\omega_n\sqrt{1-\zeta^2}$ is called the damped natural frequency.

The period T of the damped oscillations shown in Fig. 4.10 is related to the damped natural frequency by

$$2\pi/T = \omega_n\sqrt{1-\zeta^2} \tag{4.100}$$

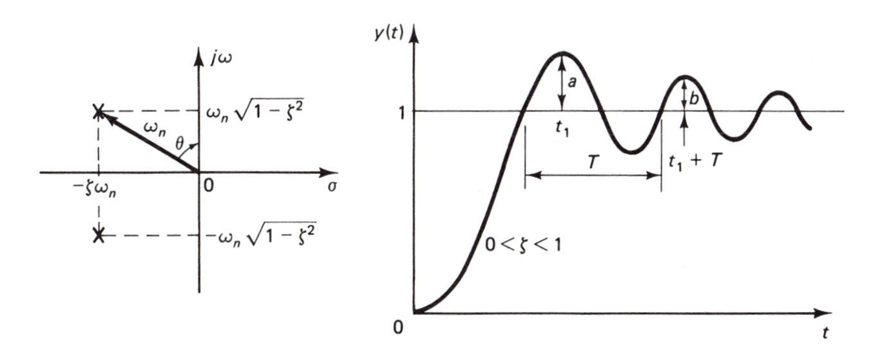

Figure 4.10 Step response of an underdamped second-order system.

Since the sine function is periodic, the ratio of the overshoots at times t_1 and $(t_1 + T)$ in Fig. 4.10 can be obtained from Eq. (4.99) as

$$\frac{a}{b} = \frac{e^{-\zeta\omega_n t}}{e^{-\zeta\omega_n(t+T)}} = e^{\zeta\omega_n T} \tag{4.101}$$

Using Eq. (4.100) in Eq. (4.101), we obtain

$$\ln\frac{a}{b} = \frac{\zeta(2\pi)}{\sqrt{1-\zeta^2}} \tag{4.102}$$

The logarithm of the ratio of any two successive overshoots separated by a period is called the logarithmic decrement. Hence, if the step response of a second-order system is available, we can identify the damping ratio from Eq. (4.102) and the natural frequency from Eq. (4.100).

Unit Impulse Response. We let the initial conditions be $x_1(-0) = 0$, $x_2(-0) = 0$, and $r(t) = \delta(t)$.

Case 1. Overdamped case. Using Eq. (4.92) and letting $r(t') = \delta(t')$ in Eq. (4.89), we obtain

$$y(t) = \left(\frac{\omega_n}{2\sqrt{\zeta^2-1}}\right)(e^{-t/\tau_1} - e^{-t/\tau_2}) \tag{4.103}$$

where τ_1 and τ_2 are defined by Eq. (4.91). This response is shown in Fig. 4.11. Differentiating Eq. (4.103) with respect to t, we obtain

$$\frac{dy}{dt} = \left(\frac{\omega_n}{2\sqrt{\zeta^2-1}}\right)\left(-\frac{1}{\tau_1}e^{-t/\tau_1} + \frac{1}{\tau_2}e^{-t/\tau_2}\right)$$

and

$$\begin{aligned}\frac{dy}{dt}(+0) &= \left(\frac{\omega_n}{2\sqrt{\zeta^2-1}}\right)\left(-\frac{1}{\tau_1} + \frac{1}{\tau_2}\right) \\ &= \omega_n^2\end{aligned} \tag{4.104}$$

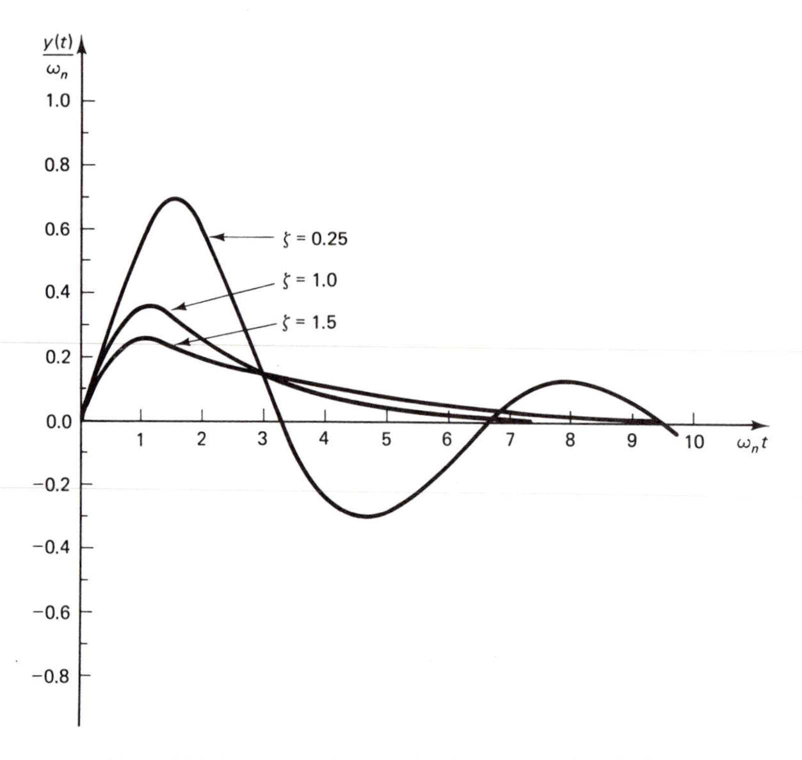

Figure 4.11 Response of a second-order system to impulse input.

For the impulse response of a second-order system, we see from Eq. (4.103) that $y(+0) = y(-0)$, but $\dot{y}(+0) \neq \dot{y}(-0)$. There is a jump in $\dot{y}$ at the time of application of the impulse and the solution of $x_2(t)$ is not unique at that point. This is consistent with the impulse-momentum principle of dynamics (D'Souza and Garg, 1984), which states that change in momentum is equal to the impulse.

Case 2. Critically damped case. Letting $r(t') = \delta(t')$ in Eq. (4.89) and using Eq. (4.95), we obtain

$$y(t) = \omega_n^2(t\, e^{-\omega_n t}) \tag{4.105}$$

This response is also shown in Fig. 4.11. Differentiating Eq. (4.105) with respect to time, we can show that $\dot{y}(+0) = \omega_n^2$.

Case 3. Underdamped case. Using Eq. (4.98) and letting $r(t') = \delta(t')$ in Eq. (4.89), we get

$$y(t) = \left(\frac{\omega_n}{\sqrt{1-\zeta^2}}\right) e^{-\zeta\omega_n t} \sin \omega_n\sqrt{1-\zeta^2}\,t \tag{4.106}$$

This response exhibits damped oscillations that decay to zero with time as shown in Fig. 4.11.

4.5.3 Response of High-Order Systems

The transient response of a high-order closed-loop control system depends on the roots of its characteristic equation, which is given by Eq. (4.63). The contribution of a real root to the response becomes obvious from our preceding study of first-order systems. Complex roots occur in conjugate pairs because the system parameters are real. The contribution of a pair of complex conjugate roots to the response can be appreciated from our study of the response of an underdamped second-order system. In fact, a great deal of information about the transient response can be gained by examining the location of the roots of the characteristic equation in the complex plane.

The impulse responses for various root locations in the complex s-plane are shown in Fig. 4.12, where the complex conjugate root locations are omitted. The

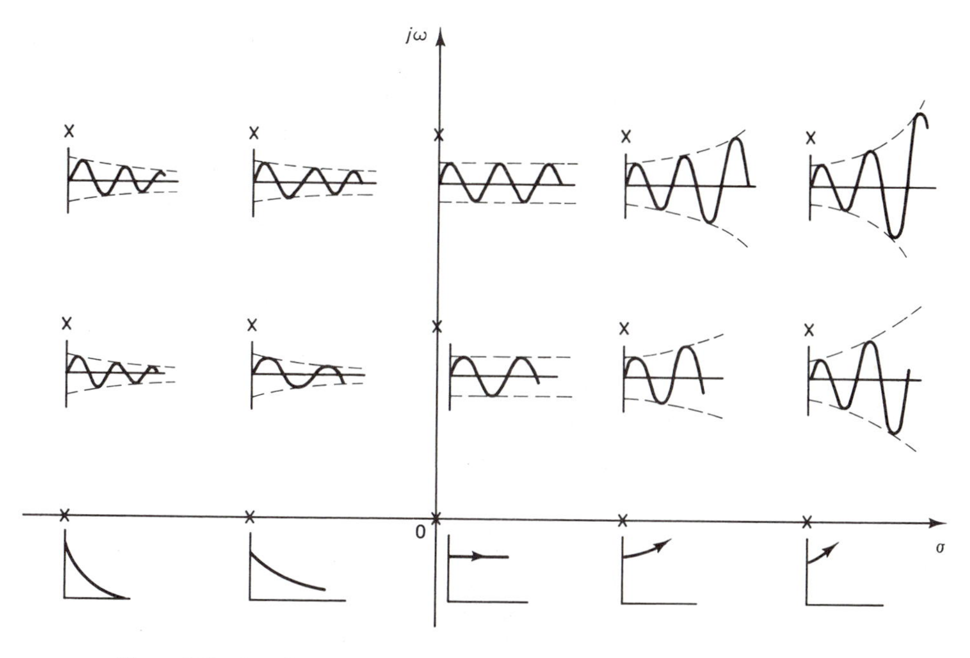

Figure 4.12 Impulse responses for various root locations in the complex plane. (The conjugate root locations are omitted.)

response due to a negative real root is an exponential decay and the farther the root is from the imaginary axis, the faster is the decay. A complex conjugate pair with a negative real part contributes a damped sinusoidal response and the natural frequency and damping ratio with the root location are as shown in Fig. 4.10. A single root at the origin contributes a constant, and a conjugate pair on the imaginary axis produces an undamped sinusoidal response.

With a positive root or a complex conjugate pair with a positive real part, the response becomes unbounded with time. Hence, it is obvious that for an asymptotically stable response, i.e., an impulse response that decays to zero with time, all roots of the characteristic equation must have a negative real part. For a fast response, all roots should be far to the left of the imaginary axis, but in this case noise and vibrations will not be filtered out. (This topic is studied in the next chapter.) A balance between speed of response and noise rejection is therefore desired.

4.5.4 Time-Domain Performance Specifications

The transient performance specifications are usually stated for the response of the control system to a unit step input as a test signal. The typical performance criteria include rise time, delay time, maximum overshoot, and settling time. These criteria are illustrated for the unit step response shown in Fig. 4.13 and are defined in the following.

1. *Rise time.* The rise time T_r is defined as the time required for the step response to rise from 10% to 90% of its final value. It is a measure of the speed of response.
2. *Delay time.* The delay time T_d is defined as the time required for the step response to reach 50% of its final value.
3. *Maximum overshoot.* The maximum overshoot is the maximum deviation of the output above its steady-state final value. The maximum overshoot may also be stated as a percentage of the final steady-state value of the response as

$$\text{Percent maximum overshoot} = \frac{\text{peak value} - \text{final value}}{\text{final value}} \times 100\%$$

4. *Settling time.* The settling time T_s is defined as the time required for the response to settle within a certain percentage of its final value. The commonly specified value is 5%, but a 2% value is sometimes used.

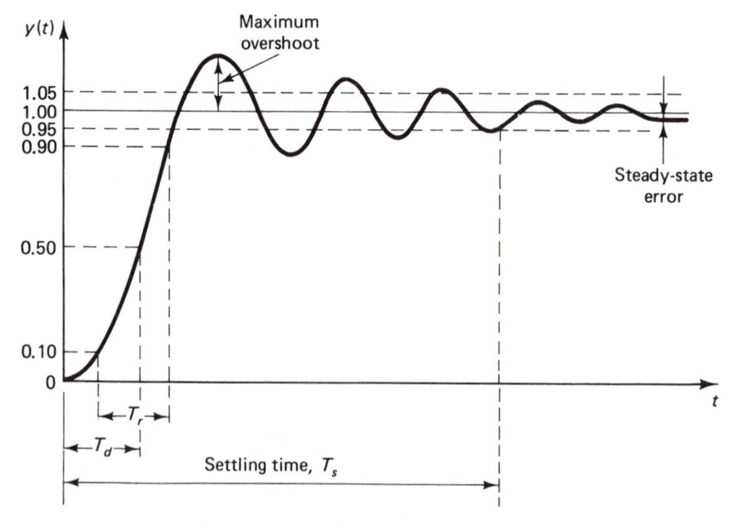

Figure 4.13 Transient performance specifications based on a step response.

The four quantities defined in the preceding are relatively easy to obtain when the step response is already plotted. However, the selection of the the roots of the characteristic equation to satisfy the specifications is not straightforward for systems of order greater than two. A first-order system or an overdamped second-order system has no overshoot or damped oscillations. In a second-order system, decreasing the damping ratio decreases the rise time, but the maximum overshoot increases and the settling time may also increase. Hence, a damping ratio $\zeta = 0.7$ is a good compromise. In robotic manipulators, overshoot is undesirable and can be hazardous; for this reason, critical damping is a good choice.

It will be shown in Chapter 7 that for a control system that is classified as type 0, there is a steady-state error between the final value of the response and the step input. For systems that are classified as type 1 or higher, there is no steady-state error for step input.

4.6 TRANSFORMATION OF STATE VARIABLES

As discussed earlier, hand calculation of the state-transition matrix becomes very cumbersome when the order of the system exceeds three. Here, matrix methods based on digital computers are useful. The method we consider yields a closed-form solution of the state-transition matrix and is distinct from the method of numerically integrating the state equations with a digital computer. Given the state equations of the closed-loop control system described by

$$\dot{\mathbf{x}} = (\mathbf{A} - \mathbf{BK})\mathbf{x} + \mathbf{BNr}(t) \qquad \mathbf{x} = \mathbf{x}(0) \text{ at } t = 0 \tag{4.107}$$

$$\text{Output: } \mathbf{y} = \mathbf{Cx} \tag{4.108}$$

we seek a linear transformation

$$\mathbf{x} = \mathbf{Px}^* \tag{4.109}$$

where $\mathbf{P}$ is an $n \times n$ constant matrix and $\mathbf{x}^*$ are the new transformed state variables. Substitution of Eq. (4.109) in Eq. (4.107) yields

$$\mathbf{Px}^* = (\mathbf{A} - \mathbf{BK})\mathbf{Px}^* + \mathbf{BNr}(t) \tag{4.110}$$

If $\mathbf{P}$ is a nonsingular matrix, we obtain

$$\mathbf{x}^* = \mathbf{P}^{-1}(\mathbf{A} - \mathbf{BK})\mathbf{Px}^* + \mathbf{P}^{-1}\mathbf{BNr}(t) \tag{4.111}$$

with initial conditions $\mathbf{x}^*(0) = \mathbf{P}^{-1}\mathbf{x}(0)$.

Letting $\mathbf{P}^{-1}(\mathbf{A} - \mathbf{BK})\mathbf{P} = \mathbf{\Lambda}$, matrix $(\mathbf{A} - \mathbf{BK})$ is said to be similar to matrix $\mathbf{\Lambda}$ and the transformation of Eq. (4.109) is called similarity transformation. We seek a nonsingular matrix $\mathbf{P}$ such that $\mathbf{\Lambda}$ is a diagonal matrix of the form

$$\mathbf{\Lambda} = \begin{bmatrix} \lambda_1 & 0 & \cdots & 0 \\ 0 & \lambda_2 & \cdots & 0 \\ & & \ddots & \\ 0 & 0 & & \lambda_n \end{bmatrix} \tag{4.112}$$

where $\lambda_1, \lambda_2, \ldots, \lambda_n$ are the eigenvalues of $(\mathbf{A}-\mathbf{BK})$, which are the same as the roots of the characteristic equation $|s\mathbf{I}-\mathbf{A}+\mathbf{BK}|=0$. The unforced system corresponding to Eq. (4.111) becomes

$$\dot{\mathbf{x}}^* = \mathbf{\Lambda}\mathbf{x}^*$$

The state variables are now uncoupled and the state-transition matrix can be obtained by inspection as

$$\mathbf{\Phi}^*(t) = \begin{bmatrix} e^{\lambda_1 t} & 0 & \cdots & 0 \\ 0 & e^{\lambda_2 t} & \cdots & 0 \\ & & \ddots & \\ 0 & 0 & & e^{\lambda_n t} \end{bmatrix} \tag{4.113}$$

The uncoupled state variables $\mathbf{x}^*$ are said to be in the normal form. The solution of Eq. (4.111) can be expressed as

$$\mathbf{x}^*(t) = \mathbf{\Phi}^*(t)\mathbf{x}^*(0) + \int_0^t \mathbf{\Phi}^*(t-t')\mathbf{P}^{-1}\mathbf{BNr}(t')\,dt' \tag{4.114}$$

The solution of the original state equation, Eq. (4.107), is obtained by using Eq. (4.109) in Eq. (4.114) as

$$\mathbf{x}(t) = \mathbf{P\Phi}^*(t)\mathbf{P}^{-1}\mathbf{x}(0) + \int_0^t \mathbf{P\Phi}^*(t-t')\mathbf{P}^{-1}\mathbf{BNr}(t')\,dt' \tag{4.115}$$

It is obvious from Eq. (4.115) that the state-transition matrix for the original state equation, Eq. (4.107), is given by

$$\mathbf{\Phi}(t) = \mathbf{P\Phi}^*(t)\mathbf{P}^{-1} \tag{4.116}$$

We now show that a matrix $(\mathbf{A}-\mathbf{BK})$ can be reduced to a diagonal matrix $\mathbf{\Lambda}$ by a similarity transformation if and only if it has a set of n linearly independent eigenvectors. Since $(\mathbf{A}-\mathbf{BK})\mathbf{P}=\mathbf{P\Lambda}$, then by partitioned matrix multiplication, it follows that $(\mathbf{A}-\mathbf{BK})\{p_i\}=\lambda_i\{p_i\}$, where $\{p_i\}$ are the columns of $\mathbf{P}$. Hence, the transformation matrix $\mathbf{P}$ has the eigenvectors of $(\mathbf{A}-\mathbf{BK})$ as its columns, i.e.,

$$\mathbf{P} = [\{v_1\}\{v_2\}\cdots\{v_n\}] \tag{4.117}$$

where $\{v_i\}$ are the eigenvectors of $(\mathbf{A}-\mathbf{BK})$. If any two or more columns of $\mathbf{P}$ are linearly dependent, then $\det\mathbf{P}=0$ and $\mathbf{P}^{-1}$ exists if and only if its columns are linearly independent. In the following, we consider two cases that may arise.

Case 1. Matrix $(\mathbf{A}-\mathbf{BK})$ has distinct eigenvalues. In this case, the roots of the characteristic equation are distinct and it can be shown that the eigenvectors of $(\mathbf{A}-\mathbf{BK})$ are linarly independent; that is, if

$$c_1\{v_1\} + c_2\{v_2\} + \cdots + c_n\{v_n\} = \{0\}$$

for constants c_i, then this is possible only if $c_1 = c_2 = \cdots = c_n = 0$. The proof is by contradiction (Ogata, 1967). In this case matrix $(\mathbf{A}-\mathbf{BK})$ can always be diagonalized.

EXAMPLE 4.7

We consider the electrical position control system of Example 3.4, where the closed-loop matrix $(\mathbf{A} - \mathbf{BK})$ is given by Eq. (3.50). The identification of parameter values is discussed in Chapter 5. Let the parameter values be such that $(\mathbf{A} - \mathbf{BK})$ becomes

$$(\mathbf{A} - \mathbf{BK}) = \begin{bmatrix} 0 & 1 & 0 \\ 0 & -3 & 5 \\ -3 & -1 & -6 \end{bmatrix}$$

$$|s\mathbf{I} - \mathbf{A} + \mathbf{BK}| = \det \begin{vmatrix} s & -1 & 0 \\ 0 & s+3 & -5 \\ 3 & 1 & s+6 \end{vmatrix} \tag{4.118}$$

$$= s^3 + 9s^2 + 23s + 15$$

$$= (s+1)(s+3)(s+5)$$

The three distinct eigenvalues are $\lambda_1 = -1$, $\lambda_2 = -3$, $\lambda_3 = -5$. The eigenvectors corresponding to the respective eigenvalues are obtained from the solution of

$$(\mathbf{A} - \mathbf{BK})\mathbf{v}_i = \lambda_i \mathbf{v}_i \tag{4.119}$$

If $\mathbf{v}_i$ is a solution of Eq. (4.119), then $c\mathbf{v}_i$ is also a solution for any constant c. Hence, only the direction of the eigenvector can be determined from Eq. (4.119) and the length is arbitrary. For $\lambda_1 = -1$, we get

$$\begin{bmatrix} 0 & 1 & 0 \\ 0 & -3 & 5 \\ -3 & -1 & -6 \end{bmatrix} \begin{bmatrix} v_{11} \\ v_{21} \\ v_{31} \end{bmatrix} = -1 \begin{bmatrix} v_{11} \\ v_{21} \\ v_{31} \end{bmatrix}$$

or

$$v_{21} = -v_{11}$$

$$-3v_{21} + 5v_{31} = -v_{21}$$

$$-3v_{11} - v_{21} - 6v_{31} = -v_{31}$$

These three equations are not linearly independent. Hence, we arbitrarily choose $v_{11} = 1$ and obtain $v_{21} = -1$ and $v_{31} = -2/5$. Similarly, we obtain the eigenvectors corresponding to the eigenvalues $\lambda_2 = -3$ and $\lambda_3 = -5$ and get

$$\{v_1\} = \begin{bmatrix} 1 \\ -1 \\ -2/5 \end{bmatrix} \qquad \{v_2\} = \begin{bmatrix} 1 \\ -3 \\ 0 \end{bmatrix} \qquad \{v_3\} = \begin{bmatrix} 1 \\ -5 \\ 2 \end{bmatrix}$$

Hence, the transformation matrix $\mathbf{P}$ and its inverse are obtained as

$$\mathbf{P} = \begin{bmatrix} 1 & 1 & 1 \\ -1 & -3 & -5 \\ -2/5 & 0 & 2 \end{bmatrix} \qquad \mathbf{P}^{-1} = \tfrac{5}{8} \begin{bmatrix} 3 & 1 & 1 \\ -2 & -6/5 & -2 \\ 3/5 & 1/5 & 1 \end{bmatrix} \tag{4.120}$$

It can be verified by using Eqs. (4.118) and (4.120) that

$$\mathbf{P}^{-1}(\mathbf{A}-\mathbf{BK})\mathbf{P} = \begin{bmatrix} -1 & 0 & 0 \\ 0 & -3 & 0 \\ 0 & 0 & -5 \end{bmatrix}$$

The state-transition matrix for the normal state variables is obtained as

$$\mathbf{\Phi}^*(t) = \begin{bmatrix} e^{-t} & 0 & 0 \\ 0 & e^{-3t} & 0 \\ 0 & 0 & e^{-5t} \end{bmatrix} \tag{4.121}$$

The state-transition matrix for the original state equations is obtained from $\mathbf{\Phi}(t) = \mathbf{P}\mathbf{\Phi}^*(t)\mathbf{P}^{-1}$. Using Eqs. (4.120) and (4.121), we get

$$\mathbf{\Phi}(t) = \frac{1}{8}\begin{bmatrix} 15e^{-t} - 10e^{-3t} + 3e^{-5t} & 5e^{-t} - 6e^{-3t} + e^{-5t} & 5e^{-t} - 10e^{-3t} + 5e^{-5t} \\ -15e^{-t} + 30e^{-3t} - 15e^{-5t} & -5e^{-t} + 18e^{-3t} - 5e^{-5t} & -5e^{-t} + 30e^{-3t} - 25e^{-5t} \\ -6e^{-t} + 6e^{-5t} & -2e^{-t} + 2e^{-5t} & -2e^{-t} + 10e^{-5t} \end{bmatrix} \tag{4.122}$$

In the preceding example, the distinct roots are all real. The case of distinct roots includes the occurrence of complex conjugate roots. However, the eigenvectors and the diagonal elements of matrix $\mathbf{\Lambda}$ will now be complex. The procedure remains the same and is illustrated by the following example.

EXAMPLE 4.8

We consider the hydraulic control system of Example 3.5, which has been studied in Examples 4.4 and 4.6. The closed-loop matrix is given by Eq. (4.65). Let the parameter values be such that

$$(\mathbf{A}-\mathbf{BK}) = \begin{bmatrix} -2 & 1 \\ -5 & 0 \end{bmatrix}$$

$$|s\mathbf{I}-\mathbf{A}+\mathbf{BK}| = \det\begin{vmatrix} s+2 & -1 \\ 5 & s \end{vmatrix} \tag{4.123}$$

$$= s^2 + 2s + 5$$

The eigenvalues are $\lambda_1 = -1 + j2$ and $\lambda_2 = -1 - j2$, which are distinct but complex conjugates. This is Case 3 of Example 4.6. The eigenvectors are obtained from the solution of

$$\begin{bmatrix} -2 & 1 \\ -5 & 0 \end{bmatrix}\mathbf{v}_i = \lambda_i \mathbf{v}_i$$

and are given by

$$\mathbf{v}_1 = \begin{bmatrix} 1 \\ 1 + j2 \end{bmatrix} \qquad \mathbf{v}_2 = \begin{bmatrix} 1 \\ 1 - j2 \end{bmatrix}$$

Hence, the transformation matrix $\mathbf{P}$ and its inverse are obtained as

$$\mathbf{P} = \begin{bmatrix} 1 & 1 \\ 1 + j2 & 1 - j2 \end{bmatrix} \qquad \mathbf{P}^{-1} = \frac{1}{4j}\begin{bmatrix} -1 + j2 & 1 \\ 1 + j2 & -1 \end{bmatrix} \tag{4.124}$$

The state-transition matrix for the normal state variables is

$$\mathbf{\Phi}^*(t) = \begin{bmatrix} e^{(-1+j2)t} & 0 \\ 0 & e^{(-1-j2)t} \end{bmatrix} \tag{4.125}$$

The state-transition matrix for the original state equations is obtained from $\mathbf{\Phi}(t) = \mathbf{P}\mathbf{\Phi}^*(t)\mathbf{P}^{-1}$. Using Eqs. (4.124) and (4.125), we get

$$\mathbf{\Phi}(t) = e^{-t}\begin{bmatrix} \cos 2t - \frac{1}{2}\sin 2t & \frac{1}{2}\sin 2t \\ -\frac{5}{2}\sin 2t & \cos 2t + \frac{1}{2}\sin 2t \end{bmatrix} \tag{4.126}$$

Case 2. Matrix $(\mathbf{A} - \mathbf{BK})$ does not have distinct eigenvalues. As stated earlier, any $n \times n$ matrix can be diagonalized if and only if it has a set of n linearly independent eigenvectors. Hence, in this case, the requirement for diagonalization is that for each multiple eigenvalue λ_i of multiplicity m_i there must exist m_i linearly independent eigenvectors corresponding to λ_i. If this condition is not satisfied, a matrix that does not have distinct eigenvalues is called degenerate and cannot be diagonalized. The transformation matrix defined by Eq. (4.117) would be singular. The simplest form to which a degenerate matrix can be reduced is called the Jordan normal form.

A Jordan normal matrix has the form

$$\mathbf{\Lambda} = \begin{bmatrix} \lambda_1 & \alpha_1 & 0 & 0 & \cdots & 0 \\ 0 & \lambda_2 & \alpha_2 & 0 & \cdots & 0 \\ \cdot & \cdot & \cdot & \cdot & \cdots & \alpha_{n-1} \\ 0 & 0 & 0 & 0 & \cdots & \lambda_n \end{bmatrix} \tag{4.127}$$

where all elements below the principal diagonal are zero, the diagonal elements λ_i are the eigenvalues, and all elements above the principal diagonal are zero except possibly those elements that are adjacent to two equal diagonal elements depending on the degeneracy, i.e., α_i are either zero or one. The Jordan normal state variables are not completely uncoupled, but the state equations are simplified so that the state transition matrix can be obtained without much difficulty.

The transformation matrix $\mathbf{P}$ is given by Eq. (4.117), but $\mathbf{v}_1, \mathbf{v}_2, \ldots, \mathbf{v}_n$ are determined as follows. Since $\mathbf{P\Lambda} = (\mathbf{A} - \mathbf{BK})\mathbf{P}$, we get

$$[\mathbf{v}_1 \quad \mathbf{v}_2 \cdots \mathbf{v}_n]\mathbf{\Lambda} = (\mathbf{A} - \mathbf{BK})[\mathbf{v}_1 \quad \mathbf{v}_2 \cdots \mathbf{v}_n]$$

where $\mathbf{\Lambda}$ is the Jordan normal matrix given by Eq. (4.127). Hence, we get

$$\begin{aligned} \lambda_1 \mathbf{v}_1 &= (\mathbf{A} - \mathbf{BK})\mathbf{v}_1 \\ \alpha_1 \mathbf{v}_1 + \lambda_2 \mathbf{v}_2 &= (\mathbf{A} - \mathbf{BK})\mathbf{v}_2 \\ \alpha_2 \mathbf{v}_2 + \lambda_3 \mathbf{v}_3 &= (\mathbf{A} - \mathbf{BK})\mathbf{v}_3 \\ &\vdots \\ \alpha_{n-1}\mathbf{v}_{n-1} + \lambda_n \mathbf{v}_n &= (\mathbf{A} - \mathbf{BK})\mathbf{v}_n \end{aligned} \tag{4.128}$$

where α_i is either zero or one, depending on the degeneracy of $(\mathbf{A} - \mathbf{BK})$.

EXAMPLE 4.9

Let the closed-loop matrix be given by

$$(\mathbf{A} - \mathbf{BK}) = \begin{bmatrix} -1 & 2 & -1 \\ 0 & -1 & 0 \\ 0 & 0 & -1 \end{bmatrix} \tag{4.129}$$

The characteristic equation becomes

$$\det |s\mathbf{I} - \mathbf{A} + \mathbf{BK}| = (s+1)(s+1)(s+1) = 0 \tag{4.130}$$

Here, the eigenvalue -1 is repeated thrice and $\lambda_1 = \lambda_2 = \lambda_3 = -1$. Using the first equation of Eq. (4.128), we get

$$\begin{bmatrix} -1 & 2 & -1 \\ 0 & -1 & 0 \\ 0 & 0 & -1 \end{bmatrix} \begin{bmatrix} v_{11} \\ v_{21} \\ v_{31} \end{bmatrix} = -1 \begin{bmatrix} v_{11} \\ v_{21} \\ v_{31} \end{bmatrix} \qquad \text{i.e.,} \qquad \begin{aligned} 2v_{21} - v_{31} &= 0 \\ 0 &= 0 \\ 0 &= 0 \end{aligned}$$

Hence, we find two linearly independent eigenvectors corresponding to this eigenvalue and they are given by

$$\mathbf{v}_1 = \begin{bmatrix} 1 \\ 0 \\ 0 \end{bmatrix} \qquad \mathbf{v}_3 = \begin{bmatrix} 0 \\ 1 \\ 2 \end{bmatrix}$$

The degeneracy of $(\mathbf{A} - \mathbf{BK})$ is one and its Jordan normal form is

$$\mathbf{\Lambda} = \begin{bmatrix} -1 & 1 & 0 \\ 0 & -1 & 0 \\ 0 & 0 & -1 \end{bmatrix} \tag{4.131}$$

We determine $\mathbf{v}_2$ from the second equation of Eq. (4.128), namely.

$$\mathbf{v}_1 - \mathbf{v}_2 = (\mathbf{A} - \mathbf{BK})\mathbf{v}_2$$

Solving for $\mathbf{v}_2$, we get

$$\mathbf{v}_2 = \begin{bmatrix} 0 \\ 1 \\ 1 \end{bmatrix}$$

The transformation matrix $\mathbf{P}$ and its inverse are given by

$$\mathbf{P} = \begin{bmatrix} 1 & 0 & 0 \\ 0 & 1 & 1 \\ 0 & 1 & 2 \end{bmatrix} \qquad \mathbf{P}^{-1} = \begin{bmatrix} 1 & 0 & 0 \\ 0 & 2 & -1 \\ 0 & -1 & 1 \end{bmatrix}$$

It can be verified that $\mathbf{P}^{-1}(\mathbf{A} - \mathbf{BK})\mathbf{P} = \mathbf{\Lambda}$, where $\mathbf{\Lambda}$ is the Jordan normal matrix of Eq. (4.131).

When matrix $(\mathbf{A} - \mathbf{BK})$ is real and symmetric, it can always be diagonalized, even if it has multiple eigenvalues, by using an orthogonal transformation matrix. We omit this case because closed-loop symmetric matrix $(\mathbf{A} - \mathbf{BK})$ is rarely encountered for control systems and the reader may consult the technique given by D'Souza and Garg (1984). The method discussed in this section involves the determination of the eigenvalues and eigenvectors of a matrix, and standard computer programs are available for this purpose (Smith, et al., 1970).

4.7. SUMMARY

In this chapter, we have studied the transient response of closed-loop control systems to standard test inputs by employing their state-variables representation. The techniques in general are applicable to multivariable control systems. It is shown that when the inputs are piecewise continuous, the linear time-invariant model has a unique solution. The time response of a system is characterized by its state-transition matrix and several methods are given for its determination. The dynamic response due to initial conditions and forcing functions has been obtained by using the state-transition matrix and convolution integral.

We have discussed the relationship between the state representation and transfer function. The effect on the transient response of the location of the roots of the characteristic equation in the complex plane has been studied. The time-domain performance specifications, such as rise time, maximum overshoot, and settling time, have been discussed. The last section covers a method that is suitable for a closed-form solution of the state-transition matrix of high-order systems by using computer-based techniques. In this method, the state-transition matrix is obtained by similarity transformation and matrix diagonalization.

REFERENCES

BELLMAN, R. (1960). *Introduction to Matrix Analysis.* New York: McGraw-Hill.

D'SOUZA, A. F., and GARG, V. K. (1984). *Advanced Dynamics: Modeling and Analysis.* Englewood Cliffs, NJ: Prentice-Hall.

KUO, B. C. (1982). *Automatic Control Systems.* 4th ed. Englewood Cliffs, NJ: Prentice-Hall.

OGATA, K. (1967). *State Space Analysis of Control Systems.* Englewood Cliffs, NJ: Prentice-Hall.

SMITH, B. T., BOYLE, J. M., GARBOW, B. S., IKEBE, Y., KLEMA, V. C., and MOLER, C. B. (1970). *Matrix Eigensystem Routines—EISPACK Guide.* New York: Springer-Verlag.

WIBERG, D. M. (1971). *State Space and Linear Systems.* Schaum's Outline Series. New York: McGraw-Hill.

PROBLEMS

4.1. For the closed-loop system of Example 3.2, let

$$(\mathbf{A} - \mathbf{BK}) = \begin{bmatrix} 0 & 1 \\ -4 & -4 \end{bmatrix}$$

Obtain the state-transition matrix by using the following methods.

(a) Method of solution of the defining equation, Eq. (4.11).

(b) Matrix exponential method of Eq. (4.23).

(c) Laplace transformation method of Eq. (4.27).

4.2. For the closed-loop system of Example 3.2, let

$$(\mathbf{A} - \mathbf{BK}) = \begin{bmatrix} 0 & 1 \\ -13 & -4 \end{bmatrix}$$

Obtain the state-transition matrix by using the following methods.

(a) Method of solution of the defining equation, Eq. (4.11).

(b) Laplace transformation method of Eq. (4.27).

4.3. For the state-transition matrix of Problem 4.1., verify the properties given by Eqs. (4.31), (4.32), and (4.34) by obtaining the right- and left-hand sides of those equations and showing that they are equal to one another.

4.4. For the state-transition matrix of Problem 4.2., verify the properties given by Eqs. (4.31), (4.32), and (4.34) by obtaining the right- and left-hand sides of those equations and showing that they are equal to one another.

4.5. For the electrical control system of Example 3.4, whose closed-loop matrix is given by Eq. (3.50) and other matrices by Eq. (3.49), let

$$(\mathbf{A} - \mathbf{BK}) = \begin{bmatrix} 0 & 1 & 0 \\ 0 & -3 & 5 \\ -3 & -1 & -6 \end{bmatrix} \qquad \mathbf{B} = \begin{bmatrix} 0 \\ 0 \\ 10 \end{bmatrix}$$

$$\mathbf{C} = \lfloor 1 \quad 0 \quad 0 \rfloor \qquad N = 0.3$$

Obtain the closed-loop transfer function of Eq. (4.62), relating the output θ_0 to the input θ_r.

4.6. For the electrical control system of Example 3.3, whose closed-loop matrix is given by Eq. (3.42) and other matrices by Eq. (3.41), let

$$(\mathbf{A} - \mathbf{BK}) = \begin{bmatrix} -0.5 & 5 & 0 \\ -0.2 & -2 & 10 \\ -0.6 & 0 & -3 \end{bmatrix} \qquad \mathbf{B} = \begin{bmatrix} 0 \\ 0 \\ 2 \end{bmatrix}$$

$$\mathbf{C} = \lfloor 1 \quad 0 \quad 0 \rfloor \qquad N = 0.3$$

Obtain the closed-loop transfer function of Eq. (4.62), relating the output speed ω_0 to the input ω_r.

4.7. The system matrices of the hydraulic control system of Example 4.4 are given by Eqs. (4.64) and (4.65). Let

$$(\mathbf{A} - \mathbf{BK}) = \begin{bmatrix} -2 & 0.4 \\ -25 & 0 \end{bmatrix} \qquad \mathbf{B}N = \begin{bmatrix} 0 \\ 25 \end{bmatrix}$$

(a) By using Eq. (4.62) show that the closed-loop transfer function of Eq. (4.69) is given by

$$y(t) = \left(\frac{10}{D^2 + 2D + 10}\right) r(t)$$

(b) For $t < 0$, r is a constant and equal to 50 rad/s, and the system is in steady state. Evaluate the initial states $x_1(0)$ and $x_2(0)$.

(c) At $t = 0$, r is brought to zero as a step function. Determine the response $y(t)$. (*Hint*: Use $y = \mathbf{C}\mathbf{\Phi}(t)\mathbf{x}(0)$.)

4.8. The parameter values of the hydraulic control system of Example 4.4 are given in Problem 4.7. But now let $x_1(0) = 0$, $x_2(0) = 0$, and $r(t)$ be a unit step. Obtain the response by using Eq. (4.89).

4.9. Solve Problem 4.8 but instead of using Eq. (4.89), use a digital-computer program (see Appendix D). Give a plot of response $y(t)$ versus time. Evaluate the rise time, maximum overshoot, 5% settling time, and steady-state error.

4.10. The block diagram of the hydraulic control system of Example 4.4 is shown in Fig. 4.4. Let $\tau = 0.1$ second and $k_1 k_2 k_g k_a c_1 c_3 c_4 = k$. From Eq. (4.71), determine the natural frequency and damping ratio as functions of k. The unit step response for a second-order underdamped system is given by Eq. (4.99) for zero initial conditions. Using Eq. (4.99), obtain the value of k such that the maximum overshoot is 15%. For this value of k, determine the rise time. (*Hint*: First, differentiate Eq. (4.99) with respect to t and set it equal to zero to obtain the time at which the maximum overshoot occurs.)

4.11. The experimental unit step response of Eq. (4.84) with zero initial conditions is shown in Fig. P4.11. Determine the values of the natural frequency and the damping ratio.

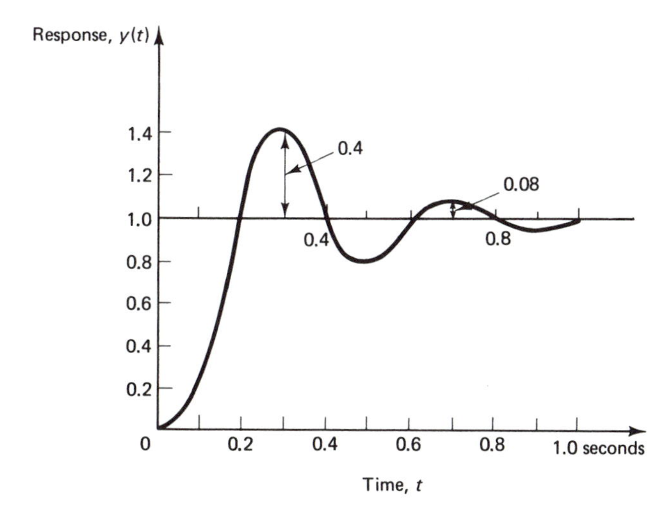

Figure P4.11 Experimental unit step response.

4.12. Consider the first-order control system whose transfer function is given by Eq. (4.75). When the initial condition is zero and the input $r(t)$ is a step function of magnitude 5, the experimentally obtained response is described by $y(t) = 4.5(1 - e^{-12t})$. Determine the closed-loop transfer function, relating output y to input r.

4.13. In a second-order closed-loop control system, when the input is a unit impulse and all initial conditions are zero, the experimentally obtained response is described by $y(t) = 12.5\, e^{-6t} \sin 8t$. Determine the transfer function, relating output y to input r.

4.14. Obtain the response of the system of Problem 4.5 when the input is a unit step and all initial conditions are zero. Verify your analytical solution by obtaining the response using a digital-computer program.

4.15. Obtain the response of the system of Problem 4.6 when the input is a unit step and all initial conditions are zero. Verify your analytical solution by obtaining the response using a digital computer program.

4.16. In a third-order closed-loop control system, let

$$(\mathbf{A} - \mathbf{BK}) = \begin{bmatrix} -3 & 1 & 0 \\ 2 & -3 & 2 \\ 0 & 1 & -3 \end{bmatrix}$$

Obtain the state-transition matrix by first determining the similarity-transformation matrix $\mathbf{P}$ that uncouples the state variables and then using Eqs. (4.113) and (4.116).

5

Frequency Response

5.1 INTRODUCTION

The preceding chapter treated the transient response of control systems. This chapter is concerned with frequency response when the test input is a sinusoidal signal. A control system during its normal operation may not encounter any sinusoidal signal as a command input. But a sinusoidal signal is an important testing input because frequency response is a very useful tool for the analysis and design of control systems. Frequency response is defined as the steady-state response of a system to sinusoidal input. By steady-state response, we mean the response that remains after all transients due to the initial conditions have decayed to zero. It is therefore implied that the system is asymptomatically stable or equivalently that all roots of the characteristic equation have negative real parts.

For linear time-invariant models that are asymptotically stable, we show that the response is also sinusoidal with the same frequency as the input. The steady-state output differs from the sinusoidal input only in the amplitude and phase angle. The output-input amplitude ratio and the phase angle between the output and input sinusoids are obtained from the frequency response function, which is the transfer function with s replaced by $j\omega$, where ω is the frequency. For the frequency domain analysis, it is therefore more convenient to employ the transfer function, or transfer function matrix for multivariable systems, instead of the state-variables representation.

In this chapter, the frequency-response functions are represented graphically by logarithmic plots called Bode diagrams. The applications of Bode diagrams to the analysis and design of control systems are then studied. The applications include filter characteristics and noise rejection properties, identification of the unknown transfer function of a system and parameter evaluation from experimental frequency response, and performance specifications in the frequency domain. Bode diagrams are also useful for the analysis of linear forced vibrations, but this topic is beyond the scope of this book.

5.2 FREQUENCY-RESPONSE FUNCTION

For frequency-response analysis, it is more convenient to use the scalar transfer function for single-input, single-output systems or transfer function matrix for multivariable systems rather than the state-variables formulation. The reason for this preference will become clear from what follows. Neglecting the disturbance, the state-variables formulation of a closed-loop control system is expressed as shown in Eqs. (3.1) to (3.3) by

$$\text{Plant:} \qquad \dot{\mathbf{x}} = \mathbf{A}\mathbf{x} + \mathbf{B}\mathbf{u} \tag{5.1}$$

$$\text{Control law:} \qquad \mathbf{u} = \mathbf{N}\mathbf{r} - \mathbf{K}\mathbf{x} \tag{5.2}$$

$$\text{Output:} \qquad \mathbf{y} = \mathbf{C}\mathbf{x} \tag{5.3}$$

These equations can be combined and expressed in the Laplace domain assuming zero initial conditions as

$$\mathbf{Y}(s) = \mathbf{C}(s\mathbf{I} - \mathbf{A} + \mathbf{B}\mathbf{K})^{-1}\mathbf{B}\mathbf{N}\mathbf{R}(s) \tag{5.4}$$

Hence, the closed-loop transfer function matrix relating the output $\mathbf{Y}(s)$ to the command input $\mathbf{R}(s)$ is given as shown in Eqs. (4.61) and (4.62) by

$$\mathbf{G}_c(s) = \frac{\mathbf{C}[\text{Adj}\,(s\mathbf{I} - \mathbf{A} + \mathbf{B}\mathbf{K})]\mathbf{B}\mathbf{N}}{\det|s\mathbf{I} - \mathbf{A} + \mathbf{B}\mathbf{K}|} \tag{5.5}$$

$$= \mathbf{C}\hat{\mathbf{\Phi}}(s)\mathbf{B}\mathbf{N} \tag{5.6}$$

where $\hat{\mathbf{\Phi}}(s)$ is the Laplace transform of the state-transition matrix, and the subscript of $\mathbf{G}_c$ indicates the closed-loop transfer function. For single-input, single-output system, $\mathbf{G}_c(s)$ is a scalar. For the frequency response of a closed-loop transfer function, we let $r(t)$ be a sinusoidal testing signal, i.e., $r(t) = a \sin \omega t$, instead of a command input and employ the transfer function, Eq. (5.5).

However, we also find it necessary to conduct frequency-response analysis of the open-loop system, Eqs. (5.1) and (5.3), before synthesizing the control law, Eq. (5.2). For this case, we have

$$\mathbf{Y}(s) = \mathbf{C}(s\mathbf{I} - \mathbf{A})^{-1}\mathbf{B}\mathbf{U}(s) \tag{5.7}$$

Hence, the open-loop transfer matrix relating the output $\mathbf{Y}(s)$ to the control input $\mathbf{U}(s)$ is given by

$$\mathbf{G}_o(s) = \frac{\mathbf{C}[\text{Adj}\,(s\mathbf{I} - \mathbf{A})]\mathbf{B}}{\det|s\mathbf{I} - \mathbf{A}|} \tag{5.8}$$

where the subscript of $\mathbf{G}_o$ indicates the open-loop transfer function. For frequency-response analysis of open-loop systems, we let $u(t)$ be a sinusoidal testing signal, i.e., $u(t) = a \sin \omega t$, and use the transfer function, Eq. (5.8). In this chapter, we employ the notation $\mathbf{G}_c(s)$ when treating specifically the closed-loop transfer function Eq. (5.5), the notation $\mathbf{G}_o(s)$ when treating specifically the open-loop transfer function, Eq. (5.8), and omit the subscript when the transfer function may be either open loop or closed loop.

Let a scalar transfer function be described by

$$G(s) = \frac{b_m s^m + b_{m-1} s^{m-1} + \cdots + b_0}{a_n s^n + a_{n-1} s^{n-1} + \cdots + a_1 s + a_0} \qquad n > m \tag{5.9}$$

The characteristic equation is given by

$$a_n s^n + a_{n-1} s^{n-1} + \cdots + a_0 = 0 \tag{5.10}$$

which is identical in Eq. (5.5) to $\det|s\mathbf{I} - \mathbf{A} + \mathbf{BK}| = 0$ or in Eq. (5.8) to $\det |s\mathbf{I} - \mathbf{A}| = 0$. In the time domain, we have

$$y(t) = \left(\frac{b_m D^m + b_{m-1} D^{m-1} + \cdots + b_0}{a_n D^n + a_{n-1} D^{n-1} + \cdots + a_0}\right) a \sin \omega t \tag{5.11}$$

We now prove that for this sinusoidal input, when all the roots of the characteristic equation, Eq. (5.10), have negative real parts, in steady state, $y(t) = a|G(j\omega)| \sin(\omega t + \phi)$, where $\phi = \measuredangle G(j\omega)$. Taking the Laplace transformation of Eq. (5.11) and including initial conditions, we obtain

$$Y(s) = \left(\frac{p(s)}{a_n s^n + a_{n-1} s^{n-1} + \cdots + a_0}\right) + \left(\frac{b_m s^m + \cdots + b_0}{a_n s^n + a_{n-1} s^{n-1} + \cdots + a_0}\right)\left(\frac{a\omega}{s^2 + \omega^2}\right)$$

where $p(s)$ is a polynomial in s of order $n - 1$ or less due to the initial conditions. If Eq. (5.10) has only distinct roots $-\lambda_1, \ldots, -\lambda_n$, then partial fraction expansion yields

$$Y(s) = \left(\frac{c_1}{s + \lambda_1} + \cdots + \frac{c_n}{s + \lambda_n}\right) + \left(\frac{k_1}{s + \lambda_1} + \cdots + \frac{k_n}{s + \lambda_n} + \frac{k_{n+1}}{s - j\omega} + \frac{k_{n+2}}{s + j\omega}\right) \tag{5.12}$$

where c_i and k_i are constants of partial fraction expansion. Since all roots of the characteristic equation have negative real parts, $c_i e^{-\lambda_i t} \to 0$ and $k_i e^{-\lambda_i t} \to 0$ as t approaches infinity. If Eq. (5.10) has a repeated root $-\lambda_i$ of multiplicity q, then $y(t)$ will include such terms as $t^h e^{-\lambda_i t}$ (where $h = 0, 1, \ldots, q - 1$), which approach zero as t approaches infinity since the real part of the root is negative. Hence, whether Eq. (5.10) has distinct or repeated roots, the steady-state response as t approaches infinity is obtained from Eq. (5.12) as

$$y(t) = \mathcal{L}^{-1}\left(\frac{k_{n+1}}{s - j\omega} + \frac{k_{n+2}}{s + j\omega}\right) \tag{5.13}$$

where the symbol $\mathscr{L}^{-1}$ denotes the inverse Laplace transformation. Now

$$k_{n+1} = \lim_{s \to j\omega} \left[(s - j\omega)G(s)\left(\frac{a\omega}{(s - j\omega)(s + j\omega)}\right)\right]$$

$$= a\left(\frac{G(j\omega)}{2j}\right)$$

$$= \frac{a}{2j}|G(j\omega)|\, e^{j\phi} \qquad \phi = \measuredangle G(j\omega).$$

$$k_{n+2} = \frac{a}{-2j}|G(j\omega)|\, e^{-j\phi}$$

Hence, Eq. (5.13) yields the steady-state value of $y(t)$ as

$$y(t) = a|G(j\omega)|\left(\frac{e^{j(\omega t+\phi)} - e^{-j(\omega t+\phi)}}{2j}\right)$$

$$= a|G(j\omega)| \sin(\omega t + \phi) \tag{5.14}$$

Hence, it is seen, as illustrated in Fig. 5.1, that the steady-state value of the output is also sinusoidal with the same frequency as the input. With y_0 as the amplitude of the output, from Eq. (5.14), the amplitude ratio $y_0/a = |G(j\omega)|$ and the phase angle between the output and input is $\phi = \measuredangle G(j\omega)$. The function $G(j\omega)$, which is obtained by replacing the operator D by $j\omega$ in $G(D)$ or by replacing s by $j\omega$ in $G(s)$, is called the *frequency-response function* or *harmonic-response function.* The condition for the frequency-response function to exist such that Eq. (5.14) is valid is that all the roots of the characteristic equation have negative real parts. A method for checking whether this condition has been satisfied, without actually determining the roots, is the Routh criterion, which is discussed in Chapter 6.

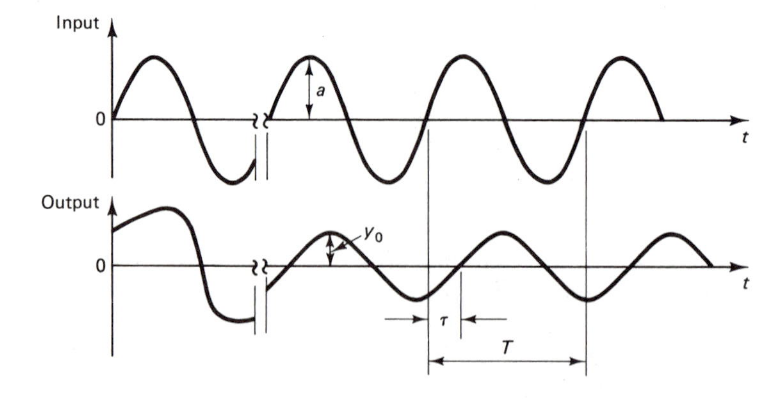

Period = T

Phase lag, $\phi = \frac{\tau}{T}(360°)$

Figure 5.1 Sinusoidal input and response.

The reason for this restriction is obvious from the preceding development and Fig. 4.12. If any root of Eq. (5.10) has a positive real part, the response will grow to infinity with time and steady state will not result. Where Eq. (5.10) has a pair of complex conjugate roots on the imaginary axis in the s-plane and the remaining roots have negative real parts, the response will contain terms with two frequencies: the natural frequency ω_n and the forcing frequency ω. When $\omega = \omega_n$, resonance will result. If Eq. (5.10) has a double root $\lambda = 0$ at the origin and the remaining roots have negative real parts, the response contains such terms as $t^h(h = 0, 1)$ and it will grow to infinity with time. Let us consider the case where Eq. (5.10) has a single root $\lambda = 0$ at the origin and the remaining roots have negative real parts. From Eq. (5.12), the steady-state response for this case becomes

$$y(t) = \alpha + a|G(j\omega)| \sin(\omega t + \phi) \tag{5.15}$$

where α is a constant and Eq. (5.14) differs from Eq. (5.15) by this additive constant.

Also, the reason for the restriction can be explained by considering the Fourier transform. The transfer function $G(s)$ results from the Laplace transformation of the differential equation whereas the frequency response function $G(j\omega)$ results from the Fourier transform. The Fourier transform of a function $y(t)$ is defined by

$$Y(j\omega) = \mathscr{F}[y(t)] = \int_{-\infty}^{\infty} y(t)\, e^{-j\omega t}\, dt \tag{5.16}$$

A sufficient condition for its existence, such that the integral will converge, is given by

$$\int_{-\infty}^{\infty} |y(t)|\, dt < \infty$$

which implies that $y(t) \to 0$ as $t \to \infty$. This condition is more restrictive than that for the existence of a Laplace transform. As seen in Appendix A, a Laplace transform of an unstable function exists provided that it does not grow with time faster than an exponential function. Since in a Laplace transformation $s = \sigma + j\omega$, we choose σ to make the integral converge and the transfer function $G(s)$ exists for unstable systems. But in a Fourier transform, $s = j\omega$ and its existence requires asymptotic stability of the system.

The result given by Eq. (5.14) for single-input, single-output systems can be extended easily to multivariable systems. For the purpose of illustration, we consider a multivariable system with two inputs and two outputs, such as the mechanical system of Example 3 in Chapter 2, whose transfer function is shown in the block diagram of Fig. 2.14, or the temperature-control system of Example 3.8, whose transfer function matrix is given by Eq. (3.92). Let the transfer function matrix be described by

$$\begin{bmatrix} Y_1(s) \\ Y_2(s) \end{bmatrix} = \begin{bmatrix} \dfrac{N_{11}(s)}{\Delta(s)} & \dfrac{N_{12}(s)}{\Delta(s)} \\ \dfrac{N_{21}(s)}{\Delta(s)} & \dfrac{N_{22}(s)}{\Delta(s)} \end{bmatrix} \begin{bmatrix} U_1(s) \\ U_2(s) \end{bmatrix} \tag{5.17}$$

where $N_{ik}(s)$ is the numerator polynomial in s, $u_1 = a_1 \sin \omega_1 t$, and $u_2 = a_2 \sin(\omega_2 t + \psi)$ are the two sinusoidal inputs, and ψ is the phase angle between them. We have

$$Y_1(s) = \left(\frac{N_{11}(s)}{\Delta(s)}\right) U_1(s) + \left(\frac{N_{12}(s)}{\Delta(s)}\right) U_2(s)$$

$$Y_2(s) = \left(\frac{N_{21}(s)}{\Delta(s)}\right) U_1(s) + \left(\frac{N_{22}(s)}{\Delta(s)}\right) U_2(s) \tag{5.18}$$

First, we have to check whether all roots of the characteristic equation $\Delta(s) = 0$ have negative real parts. Assuming that this condition is satisfied, we let $s = j\omega_1$ or $j\omega_2$ and express the steady-state frequency response as

$$y_1(t) = \left|\frac{N_{11}(j\omega_1)}{\Delta(j\omega_1)}\right| a_1 \sin(\omega_1 t + \phi_{11}) + \left|\frac{N_{12}(j\omega_2)}{\Delta(j\omega_2)}\right| a_2 \sin(\omega_2 t + \psi + \phi_{12})$$

$$y_2(t) = \left|\frac{N_{21}(j\omega_1)}{\Delta(j\omega_1)}\right| a_1 \sin(\omega_1 t + \phi_{21}) + \left|\frac{N_{22}(j\omega_2)}{\Delta(j\omega_2)}\right| a_2 \sin(\omega_2 t + \psi + \phi_{22}) \tag{5.19}$$

where

$$\phi_{ik} = \measuredangle \frac{N_{ik}(j\omega_k)}{\Delta(j\omega_k)} \qquad i = 1, 2;\ k = 1, 2$$

The extension of this procedure to multivariable systems with any number of inputs and outputs is very straightforward.

EXAMPLE 5.1

The development described in the preceding is applicable to the analysis of linear vibrations of multivariable systems where the sinusoidal input is the exciting force instead of a testing signal. In this example, we consider a vibration absorber as a special case of the system of Example 2.3, whose transfer function matrix is shown in the block diagram of Fig. 2.14. In Fig. 2.12, let m_2 be the mass of a machine mounted on a frame with suspension stiffness k_2 and damping coefficient c_2, and subject to a sinusoidal exciting force $F_2 = a_2 \sin \omega_2 t$. In order to absorb the forced vibrations, a small mass m_1 is attached to m_2 through a spring of stiffness k_1 and no damping, i.e., $c_1 = 0$ and the force $F_1 = 0$.

In the transfer matrix given by Eq. (2.46), we set $c_1 = 0$ and $F_1 = 0$ and obtain

$$\begin{bmatrix} Y_1(s) \\ Y_2(s) \end{bmatrix} = \begin{bmatrix} \dfrac{m_2 s^2 + c_2 s + k_1 + k_2}{\Delta(s)} & \dfrac{k_1}{\Delta(s)} \\ \dfrac{k_1}{\Delta(s)} & \dfrac{m_1 s^2 + k_1}{\Delta(s)} \end{bmatrix} \begin{bmatrix} 0 \\ F_2(s) \end{bmatrix} \tag{5.20}$$

where $\Delta(s)$ is obtained from Eq. (2.47) by setting $c_1 = 0$ as

$$\Delta(s) = m_1 m_2 s^4 + m_1 c_2 s^3 + (m_1 k_1 + m_1 k_2 + m_2 k_1)s^2 + c_2 k_1 s + k_1 k_2 \tag{5.21}$$

It can be checked by the application of the Routh criterion, which is discussed in the next chapter, that all roots of the characteristic equation $\Delta(s) = 0$ have negative real parts. This implies that the equilibrium about which the vibrations occur is asymptotically stable. Hence, the frequency-response function matrix is obtained by substituting $j\omega$ for s in the transfer function matrix given by Eq. (5.20). Since

$$Y_1(s) = \left(\frac{k_1}{\Delta(s)}\right) F_2(s)$$

$$Y_2(s) = \left(\frac{m_1 s^2 + k_1}{\Delta(s)}\right) F_2(s)$$

and $F_2(t) = a_2 \sin \omega_2 t$, for steady-state vibrations, we obtain

$$y_1(t) = \left|\frac{k_1}{\Delta(j\omega_2)}\right| a_2 \sin(\omega_2 t + \phi_{12}) \tag{5.22}$$

$$y_2(t) = \left|\frac{-m_1\omega_2^2 + k_1}{\Delta(j\omega_2)}\right| a_2 \sin(\omega_2 t + \phi_{22}) \tag{5.23}$$

where

$$\phi_{12} = \measuredangle \frac{k_1}{\Delta(j\omega_2)} \qquad \phi_{22} = \measuredangle \frac{-m_1\omega_2^2 + k_1}{\Delta(j\omega_2)}$$

The vibration absorber is tuned such that $k_1/m_1 = \omega_2^2$. It then follows from Eq. (5.23) that $y_2 = 0$. Hence, the machine of mass m_2 on which the exciting force is acting does not vibrate at all. What vibrates is the vibration absorber of mass m_1. Substituting $s = j\omega_2$ and $k_1/m_1 = \omega_2^2$ in Eq. (5.21), we get $\Delta(j\omega) = -k_1^2$ and $\phi_{12} = -\pi$. Hence, the steady-state vibration of the absorber becomes

$$y_1 = (a_2/k_1) \sin(\omega_2 t - \pi) \tag{5.24}$$

The spring force acting on mass m_2 is $k_1 y_1 = -a_2 \sin \omega_2 t$, which is equal and opposite to the exciting force. The net force acting on the machine m_2 is thus zero and hence it does not vibrate at all. The assumption that mass m_1 can be attached to mass m_2 with only a spring without any damping is only an idealization. In practice, the damping coefficient c_1 can be made very small, but it will not be zero, and hence mass m_2 will have some small amplitude vibrations. Also, since the absorber is tuned to the forcing frequency, it is useful only in cases where the exciting frequency is constant.

5.3 BODE DIAGRAMS

The preceding section has shown that when the input is sinusoidal and the frequency-response function $G(j\omega)$ exists, the steady-state response of linear systems is also sinusoidal with the same frequency as the input. The output/input amplitude ratio in steady state is given by $|G(j\omega)|$ and the phase angle between the output and the input by $\phi = \measuredangle G(j\omega)$. Hence, a plot of $|G(j\omega)|$ and $\measuredangle G(j\omega)$ versus the frequency ω is very useful for analysis and design in the frequency domain.

Several alternative methods may be used for graphical portrayal of the frequency-response function. One method is to present the magnitude and phase angle of $G(j\omega)$ as a polar plot with frequency as a parameter. Because of its general usefulness, we employ a graphical presentation in the form of a logarithmic plot of the frequency-response function. The logarithmic plots are called Bode diagrams in honor of Herman Bode who used them in his studies of feedback amplifiers at the Bell Telephone Laboratories.

A Bode diagram employs semilogarithmic paper with the frequency ω in rad/s along the logarithmic coordinate as shown in Fig. 5.2(a). Two sets of axes are used along the linear rectangular coordinate. One axis shows $\log|G(j\omega)|$ or $20\log|G(j\omega)|$, and the other $\phi(\omega) = \measuredangle G(j\omega)$ in degrees. The logarithm is to the base 10 and it is more common to employ $20\log|G(j\omega)|$ in decibels (dB) rather

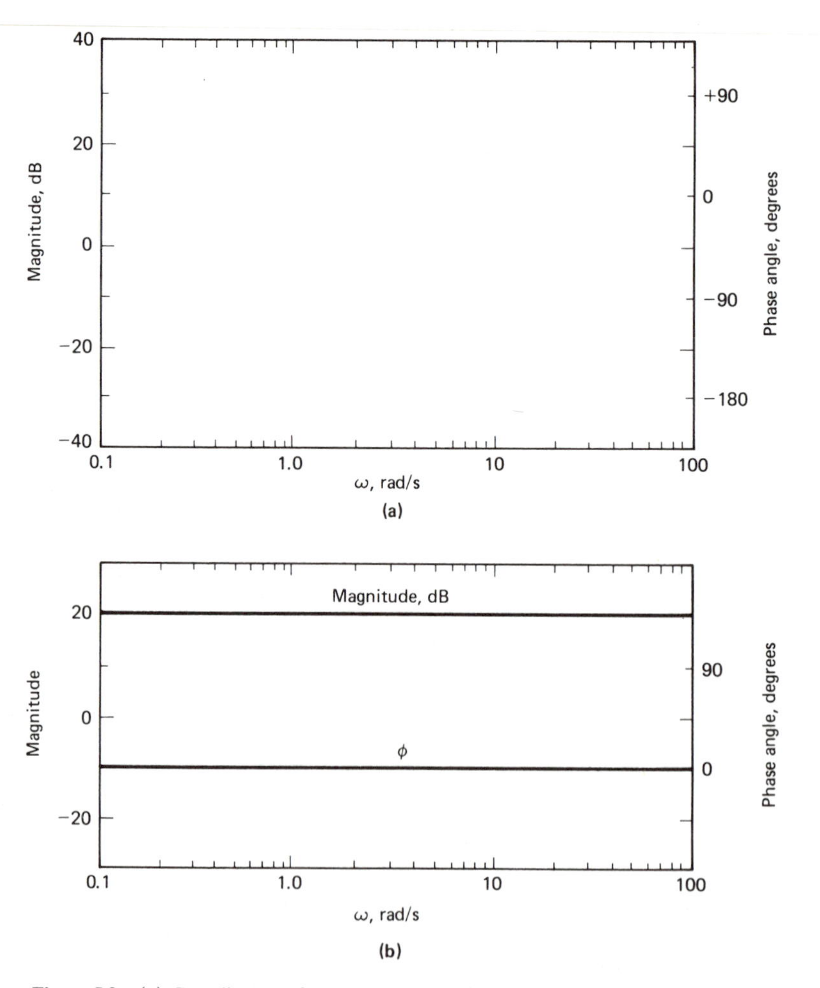

Figure 5.2 (a) Coordinates of a Bode diagram. (b) Bode diagram for constant k_g.

than $\log|G(j\omega)|$. The Bode diagram consists of two plots, one with $20\log|G(j\omega)|$ in decibels versus $\log\omega$, and the other with the phase of $G(j\omega)$ in degrees versus $\log\omega$. The range of the scales shown in Fig. 5.2(a) depend on the particular frequency-response function.

To explain the reasons for the choice of these coordinates, we consider a transfer function given by

$$G(s) = \frac{k_g(\tau_a s + 1)\cdots}{s(\tau_1 s + 1)\left(\dfrac{s^2}{\omega_n^2} + \dfrac{2\zeta}{\omega_n}s + 1\right)\cdots} \tag{5.25}$$

In the theory of complex variables, values of s that make the denominator of $G(s)$ equal to zero are called the poles of $G(s)$, and the values of s that make the numerator of $G(s)$ equal to zero are called its zeros. The poles of $G(s)$ are identical to the roots of the characteristic equation.

We note that in Eq. (5.25), $s = 0$ is a pole of $G(s)$. It is seen that if we disregard the constant α in Eq. (5.15), then a single root at the origin is the only exception we can permit to the restriction that all roots of the characteristic equation have negative real parts. Assuming that all remaining poles of $G(s)$ have negative real parts, we substitute $j\omega$ for s in Eq. (5.25) and obtain the frequency-response function. It follows that

$$\begin{aligned}
20\log|G(j\omega)| &= 20\log k_g + 20\log|j\tau_a\omega + 1| + \cdots \\
&\quad - 20\log|j\omega| - 20\log|j\tau_1\omega + 1| \\
&\quad - 20\log|(1 - \omega^2/\omega_n^2) + j2\zeta\omega/\omega_n| - \cdots \\
&= 20\log k_g + \tfrac{20}{2}\log(\tau_a^2\omega^2 + 1) + \cdots \\
&\quad - 20\log\omega - \tfrac{20}{2}\log(\tau_1^2\omega^2 + 1) \\
&\quad - \tfrac{20}{2}\log[(1 - \omega^2/\omega_n^2)^2 + (2\zeta\omega/\omega_n)^2] - \cdots
\end{aligned} \tag{5.26}$$

$$\begin{aligned}
\phi &= \measuredangle G(j\omega) \\
&= \measuredangle k_g + \measuredangle(j\tau_a\omega + 1) + \cdots \\
&\quad - \measuredangle j\omega - \measuredangle(j\tau_1\omega + 1) - \measuredangle[(1 - \omega^2/\omega_n^2) + j2\zeta\omega/\omega_n] - \cdots \\
&= \tan^{-1}\tau_a\omega + \cdots - 90^\circ - \tan^{-1}\tau_1\omega - \tan^{-1}\left(\frac{2\zeta\omega/\omega_n}{1 - \omega^2/\omega_n^2}\right) - \cdots
\end{aligned} \tag{5.27}$$

Additional product terms of $G(s)$ in Eq. (5.25) would simply produce similar terms in the expressions for the magnitude and phase angle in Eq. (5.26) and (5.27), respectively.

The reasons for the choice of the coordinates for the Bode diagram now become clear. By employing a logarithmic coordinate for the frequency, we can accurately represent a large range of frequencies. Usually, a semilog paper of four cycles is adequate to represent the frequency response of control systems. The advantage of plotting the logarithm of the magnitude is that the multiplicative factors in Eq. (5.25) have been converted into additive terms in Eq. (5.26). Also, it will be shown that with this choice of coordinates, the magnitude plot can be approximated by straight-line segments. However, log-log graph paper is not used because the phase angle in Eq. (5.27) has been already expressed with additive terms without using logarithms.

The plotting of Bode diagrams is facilitated by first studying the plots of individual factors. The four different kinds of factors that may occur in a transfer function are the following:

1. Gain constant, k_g.
2. Pole at the origin $(j\omega)^{-1}$ or zeros at the origin $(j\omega)^p$.
3. Real simple pole or zero, $(j\tau\omega + 1)^{\pm 1}$.
4. Quadratic poles or zeros, $[(1 - \omega^2/\omega_n^2) + j2\zeta\omega/\omega_n]^{\pm 1}$.

where $p = 1, 2, \ldots$ In the following, we first study the plotting of Bode diagrams for each of these individual factors and then utilize them to draw the Bode diagram for any general frequency response function.

Gain constant, k_g

In this case, $20 \log k_g$ = constant and $\measuredangle k_g = 0$, since k_g is a real positive constant. The Bode diagram is shown in Fig. 5.2(b) for $k_g = 10$. Both the magnitude and phase angle curves are straight lines of zero slope and do not depend on the frequency.

Pole at the origin $(j\omega)^{-1}$ or zeros at the origin $(j\omega)^p$

First, we consider a pole at the origin. We have

$$20 \log |1/j\omega| = -20 \log \omega \qquad \text{and} \qquad \measuredangle 1/j\omega = -90° \tag{5.28}$$

Both the magnitude and phase angle curves are straight lines and the phase angle is not dependent on the frequency. The Bode diagram is shown in Fig. 5.3. As the

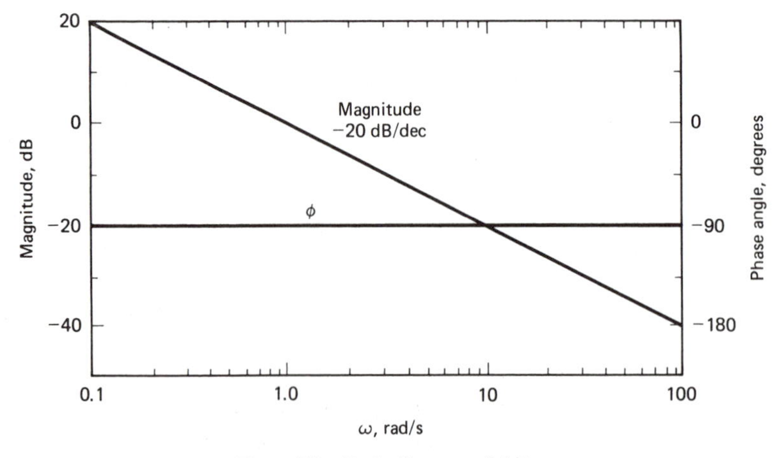

Figure 5.3 Bode diagram of $1/j\omega$.

frequency is increased from 0.1 to 1 rad/s, the magnitude curve decreases from 20 to 0 dB. Hence, the slope of the magnitude curve is −20 dB/decade. A decade is the distance on the frequency coordinate from any value of ω to 10ω. Thus, the distance from $\omega = 3$ to $\omega = 30$ rad/s is a decade. A slope of −20 dB/decade is equivalent to a slope of −6 dB/octave. An octave is the distance on the frequency coordinate from any value of ω to 2ω. It is common for Bode diagrams to state the slope per decade rather than per octave.

We now consider zeros at the origin, $(j\omega)^p$ for $p = 1, 2, \ldots$. We have $20 \log |(j\omega)^p| = 20p \log \omega$ and $\measuredangle (j\omega)^p = p(90°)$. The slope of the magnitude curve, which becomes a straight line, is $20p$ dB/decade and the phase angle is not dependent on the frequency. The Bode diagram is shown in Fig. 5.4 for $p = 2$, that is, a double zero at the origin.

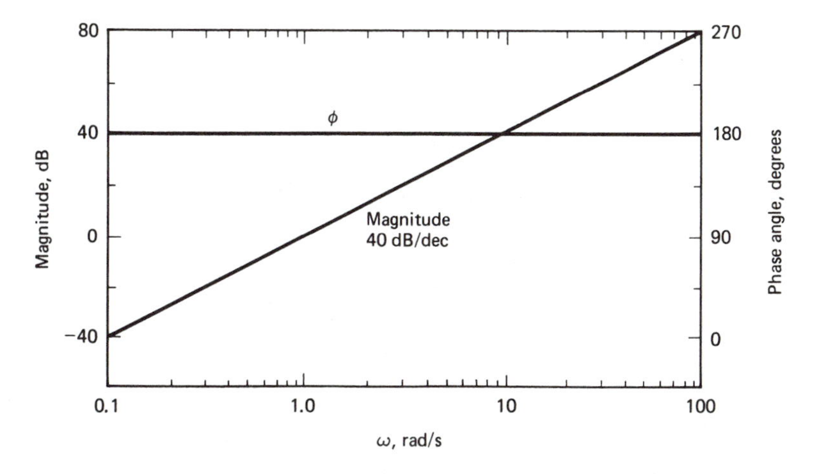

Figure 5.4 Bode diagram of $(j\omega)^2$.

Real Simple Pole or Zero, $(j\tau\omega + 1)^{\pm 1}$

We first consider a real pole and obtain

$$20 \log \left| \frac{1}{j\tau\omega + 1} \right| = -\frac{20}{2} \log (\tau^2\omega^2 + 1) \tag{5.29}$$

$$\phi = \measuredangle \frac{1}{j\tau\omega + 1} = -\tan^{-1} \tau\omega \tag{5.30}$$

The expression given by Eq. (5.29) can be approximated by two asymptotes. For low frequencies, such that $\omega \ll 1/\tau$, we approximate $\tau^2\omega^2 + 1$ by 1 and Eq. (5.29) becomes $-10 \log 1 = 0$ dB. For high frequencies, such that $\omega \gg 1/\tau$, we approximate $\tau^2\omega^2 + 1$ by $\tau^2\omega^2$ and Eq. (5.29) becomes

$$20 \log \left| \frac{1}{j\tau\omega + 1} \right| \approx -\frac{20}{2} \log \tau^2\omega^2 = -20 \log \tau\omega$$

which is a straight line with a slope of −20 dB/decade. Hence, the logarithmic magnitude curve given by Eq. (5.29) can be approximated by two straight lines called the low-frequency and high-frequency asymptotes shown in Fig. 5.5. The low-frequency asymptote has a slope of 0 dB/decade. The two asymptotes intersect at $\omega = 1/\tau$, which is called the corner frequency or break frequency.

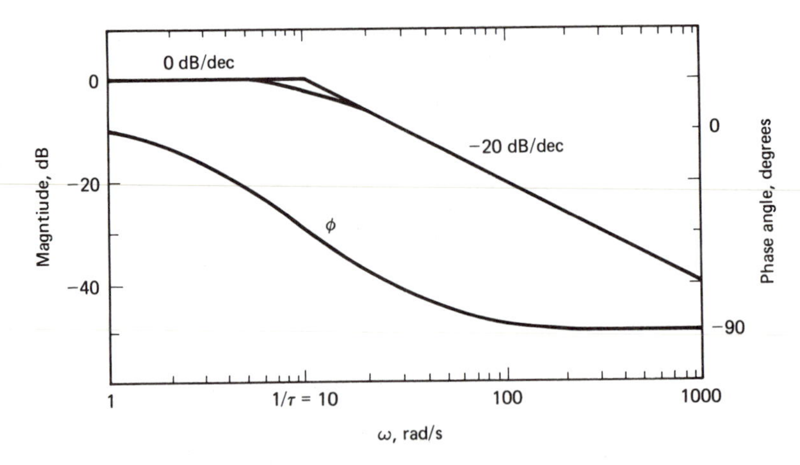

Figure 5.5 Bode diagram of $(j\tau\omega + 1)^{-1}$ for $\tau = 0.1$ second.

The worst error between the exact and asymptotic representation is at the corner frequency $\omega = 1/\tau$, where the actual value is $-10 \log 2 = -3$ dB, which is the worst error. Hence, Eq. (5.29) can be plotted by first drawing the two asymptotes and joining them by a smooth symmetrical curve passing through a point -3 dB below the corner. The phase angle of Eq. (5.30) has the values of 0° at $\omega = 0$, $-45°$ at the corner frequency $\omega = 1/\tau$, and $-90°$ as $\omega \to \infty$. A straight-line approximation to the phase angle curve can be made by drawing a line from 0° at one decade below the corner frequency to $-90°$ at one decade above the corner frequency. The maximum error between the straight-line approximation and the exact curve is less than 6°. This approximation provides a useful means of readily determining the form of the phase angle curve. A Bode diagram of Eqs. (5.29) and (5.30) is shown in Fig. 5.5 for $\tau = 0.1$ s.

For the case of real zero $(j\tau\omega + 1)$, we have

$$20 \log |j\tau\omega + 1| = \tfrac{20}{2} \log (\tau^2\omega^2 + 1) \tag{5.31}$$

$$\phi = \measuredangle(j\tau\omega + 1) = \tan^{-1} \tau\omega \tag{5.32}$$

On comparing Eq. (5.31) with Eq. (5.29) and Eq. (5.32) with Eq. (5.30), we note that the only difference is a change in the sign. Hence, a Bode diagram of Eqs. (5.31) and (5.32) can be drawn by extending the preceding analysis for the case of a real pole. The low-frequency asymptote, which has a slope of 0 dB/decade, intersects the high-frequency asymptote, which has a slope of 20 dB/decade, at the corner frequency $\omega = 1/\tau$. The exact curve passes through a point 3 dB above the corner. The phase angle varies from 0° at $\omega = 0$ to 90° at $\omega = \infty$, and it is 45° at the corner frequency. A Bode diagram of Eqs. (5.31) and (5.32) is shown in Fig. 5.6 for $\tau = 0.1$ second.

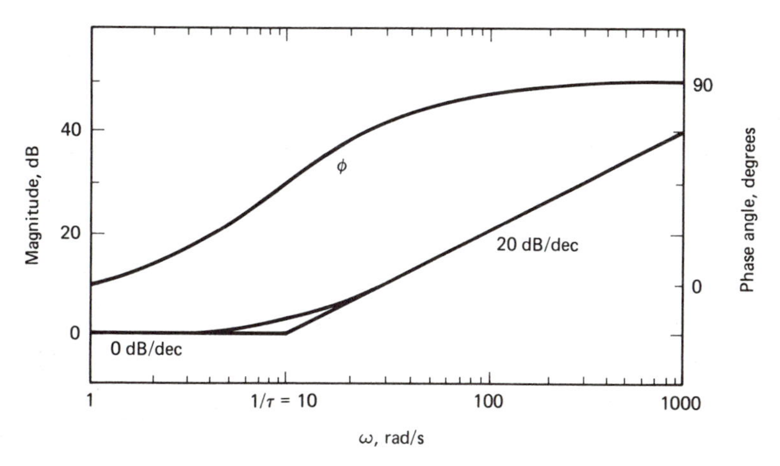

Figure 5.6 Bode diagram of $(j\tau\omega + 1)$ for $\tau = 0.1$ second.

Quadratic poles or zeros

We now consider quadratic poles when the transfer function contains a factor

$$\frac{1}{1 - (\omega/\omega_n)^2 + j2\zeta(\omega/\omega_n)}$$

The logarithmic magnitude and phase angle of this factor are given by

$$\text{Logarithmic magnitude} = -\tfrac{20}{2}\log\left[(1 - \omega^2/\omega_n^2)^2 + (2\zeta\omega/\omega_n)^2\right] \tag{5.33}$$

$$\text{Phase angle, } \phi = -\tan^{-1}\left(\frac{2\zeta\omega/\omega_n}{1 - \omega^2/\omega_n^2}\right) \tag{5.34}$$

At low frequencies, such that $\omega/\omega_n \ll 1$, we approximate the argument of the logarithm in Eq. (5.33) by 1 and hence it becomes $-(20/2)\log 1 = 0$ dB. Thus, the low-frequency asymptote is a straight line with a slope of 0 dB/decade. At high frequencies, such that $\omega/\omega_n \gg 1$, we approximate the argument of the logarithm in Eq. (5.33) by $(\omega/\omega_n)^4$ by neglecting the other terms as insignificant and obtain

$$\text{Logarithmic magnitude} = -\tfrac{20}{2}\log(\omega/\omega_n)^4 = -40\log(\omega/\omega_n) \tag{5.35}$$

This expression represents the high-frequency asymptote and it is a straight line with a slope of -40 dB/decade. The low- and high-frequency asymptotes intersect at $\omega = \omega_n$, which is the corner frequency. When $\omega = \omega_n$, the exact value of the logarithmic magnitude can be obtained from Eq. (5.33) as $-20\log 2\zeta$, which depends on the damping ratio. Hence, the exact-magnitude plot may differ strikingly from the asymptotes in the vicinity of the corner frequency. The asymptotes and the exact-magnitude plots are shown in Fig. 5.7 for $\omega_n = 10$ and several values of ζ. When $\zeta = 0.05$, the magnitude is 20 dB above the corner, and when $\zeta = 1$, it is 6 dB below the corner.

We are interested only in those cases where $\zeta \le 1$. When $\zeta > 1$, the quadratic factor has two real and unequal poles that belong to the previous case. For $\zeta = 1$, we get a double pole, i.e., the roots of the quadratic term are real, negative, and

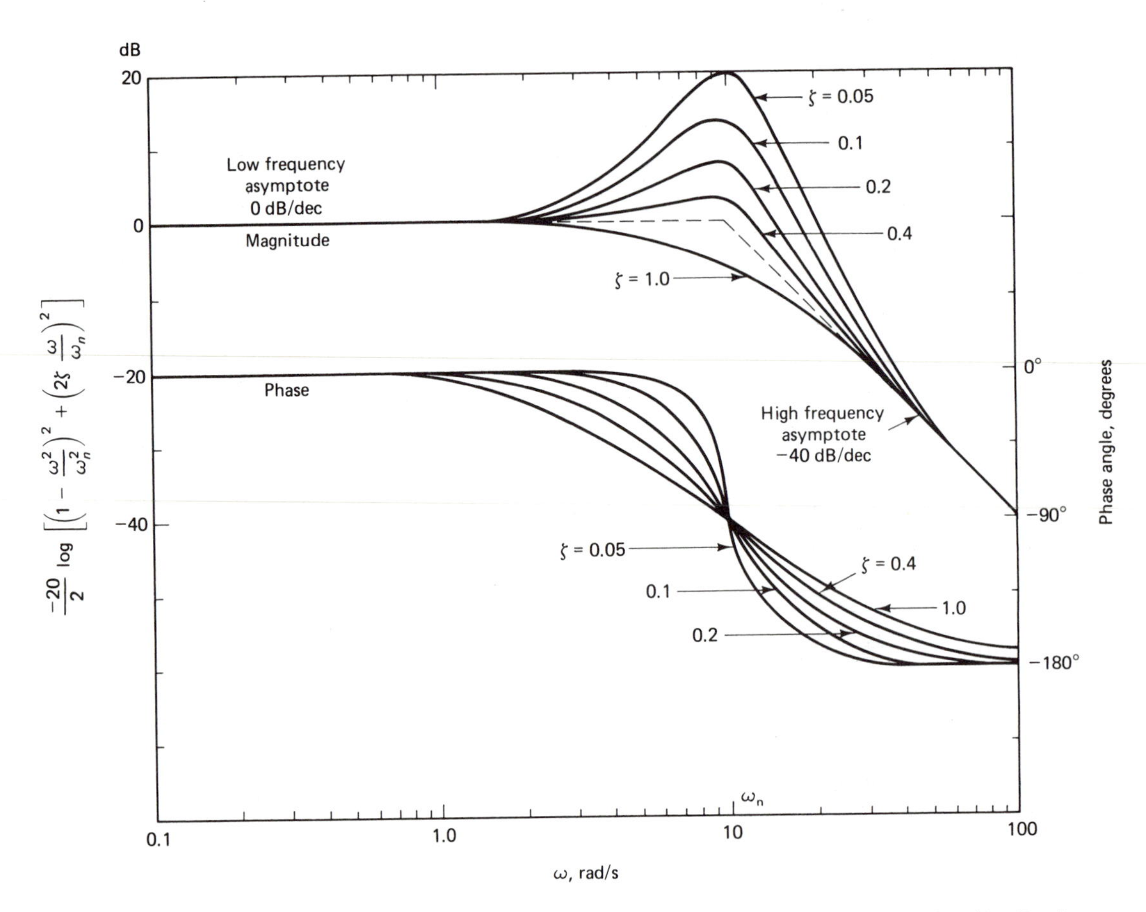

Figure 5.7 Bode diagram of $1/(1 - \omega^2/\omega_n^2 + j2\zeta\omega/\omega_n)$ for $\omega_n = 10$ rad/s. (A. Frank D'Souza/Vijay K. Garg, *Advanced Dynamics: Modeling and Analysis*, © 1984, p. 258. Reprinted by permission of Prentice-Hall Inc., Englewood Cliffs, New Jersey.)

equal. Hence, to draw the magnitude curve of Eq. (5.33), we first locate the corner frequency, draw the asymptotes, and then make a correction in the vicinity of the corner frequency by using Fig. 5.7 for the corresponding damping ratio. The phase angle curve varies from 0°, when $\omega \to 0$, to $-180°$, when $\omega \to \infty$, and it is $-90°$ at the corner frequency. In the vicinity of the corner frequency, the phase angle depends on the damping ratio as shown in Fig. 5.7.

For the case of quadratic zeros, the signs in Eqs. (5.33) and (5.34) are changed from negative to positive, and we can draw the Bode diagram by extending the preceeding analysis for the case of quadratic poles. The low-frequency asymptote, which has a slope of 0 dB/decade, intersects the high-frequency asymptote, which has a slope of 40 dB/decade, at the corner frequency $\omega = \omega_n$. At the corner frequency, the magnitude is $20 \log 2\zeta$. When $\zeta = 0.05$, the magnitude is 20 dB below the corner, and when $\zeta = 1$, it is 6 dB above the corner. The phase angle varies from 0°, when $\omega \to 0$, to 180°, when $\omega \to \infty$, and it is 90° at the corner frequency.

The Bode diagram of a transfer function $G(s)$ containing several poles and zeros can be drawn by adding the plots of each of its individual factors. In fact, an asymptotic curve can be plotted easily from the knowledge of the initial slope, the corner frequencies, and the subsequent slopes in the order of increasing frequency as illustrated by the following examples.

EXAMPLE 5.2

We consider the hydraulic speed-control system of Example 3.5, whose block diagram is shown in Fig. 3.19. The state-variables formulation is given by Eqs. (3.60) and (3.61). A Bode diagram is to be drawn for the open-loop transfer function relating the control input u to the output ω_0. Instead of using Eq. (5.8), we can obtain the transfer function directly from the block diagram of Fig. 3.19 as

$$G_0(s) = \frac{k}{s(\tau s + 1)(s^2/\omega_n^2 + 2\zeta s/\omega_n + 1)} \tag{5.36}$$

where the constant $k = k_1 k_g c_3 c_4 k_2$. Let the parameter values be such that

$$G_0(s) = \frac{3.163}{s(0.1s + 1)(s^2/200^2 + 2(0.3)(s/200) + 1)} \tag{5.37}$$

The factors of this transfer function consist of a gain constant k, a pole at the origin $s = 0$, a pole at $s = -10$, and a pair of quadratic poles with natural frequency $\omega_n = 200$, and damping ratio $\zeta = 0.3$. Since there is a single pole at the origin and the remaining poles have negative real parts, we substitute $s = j\omega$ in Eq. (5.37) and obtain the frequency-response function as

$$G_0(j\omega) = \frac{3.163}{j\omega(j0.1\omega + 1)(1 - \omega^2/200^2 + j2(0.3)\omega/200)} \tag{5.38}$$

Hence, we obtain

$$\begin{aligned} 20 \log |G_0(j\omega)| &= 20 \log 3.163 - 20 \log \omega - \tfrac{20}{2} \log (0.1^2\omega^2 + 1) \\ &\quad - \tfrac{20}{2} \log [(1 - \omega^2/200^2)^2 + (2(0.3)\omega/200)^2] \end{aligned} \tag{5.39}$$

$$\measuredangle G_0(j\omega) = -90° - \tan^{-1} 0.1\omega - \tan^{-1}\left(\frac{2(0.3)\omega/200}{1 - \omega^2/200^2}\right) \tag{5.40}$$

Let us consider the contribution of the individual terms in Eq. (5.39). The first term $20 \log 3.163 = 10$ dB is a line labeled 1 in Fig. 5.8. The second term $-20 \log \omega$ is a straight line of slope -20 dB/decade with a value of 0 dB at $\omega = 1$ rad/s and is labeled 2. The third term is represented by two asymptotes of slopes 0 and -20 dB/decade, respectively, with the corner frequency at $\omega = 1/\tau = 10$ rad/s and is labeled 3 in Fig. 5.8. The fourth term is represented by two asymptotes of slopes 0 and -40 dB/decade, respectively, with the corner frequency at $\omega = \omega_n = 200$ rad/s and labeled 4 in Fig. 5.8. A correction is shown since the damping ratio $\zeta = 0.3$.

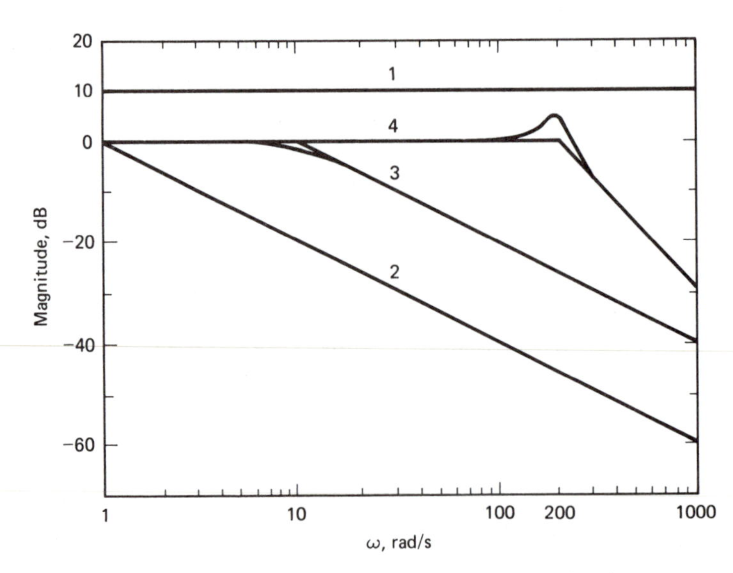

Figure 5.8 Magnitude plots of the terms in Eq. (5.39).

The magnitude curve for Eq. (5.39) can be drawn simply by adding the contributions of the individual terms in Fig. 5.8 and is shown in Fig. 5.9. However, this extra labor is not needed. The magnitude curve of Fig. 5.9 can be drawn directly from the starting value, initial slope, the corner frequencies, and the subsequent slopes in order of increasing frequency. At $\omega = 1$ rad/s,

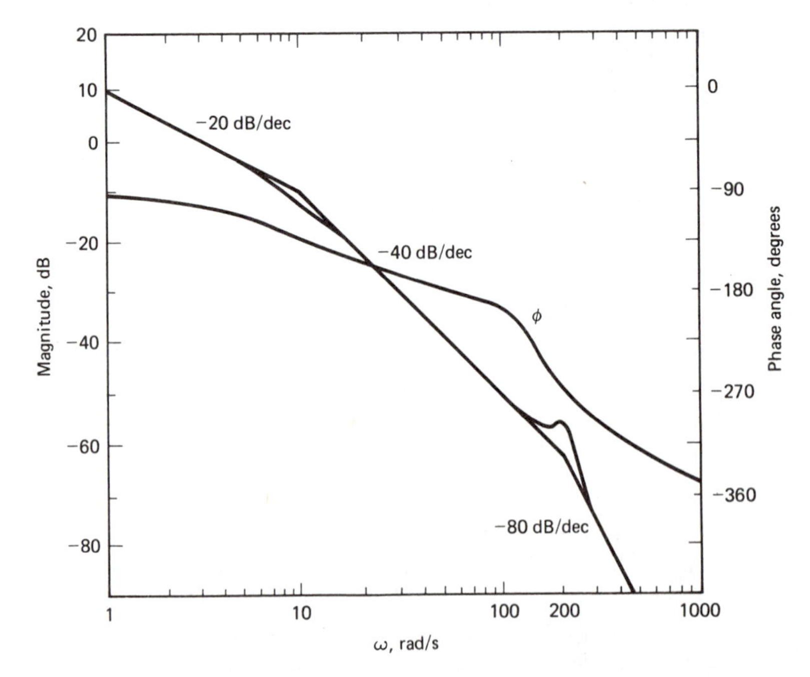

Figure 5.9 Bode diagram for Eq. (5.38).

the starting value from Eq. (5.39) is 10 dB. The initial slope, i.e., the slope before the first break occurs, is −20 dB/decade and the corner frequencies are 10 and 200 rad/s. The scale in Fig. 5.8 is chosen so that the breaks can be represented by the diagram. The phase angle of Eq. (5.40) varies from −90°, as $\omega \to 0$, to −360°, as $\omega \to \infty$.

EXAMPLE 5.3

The transfer-function matrix of the two-input, two-output system of Example 2.3 is shown in the block diagram of Fig. 2.14. A special case of this system is the vibration absorber studied in Example 5.1. From Fig. 2.14, we obtain

$$Y_1(s) = \left(\frac{m_2 s^2 + (c_1 + c_2)s + k_1 + k_2}{\Delta(s)}\right) F_1(s) + \left(\frac{c_1 s + k_1}{\Delta(s)}\right) F_2(s) \qquad (5.41)$$

where $\Delta(s)$ is given by Eq. (2.47). We wish to study the steady-state response $y_1(t)$ when $F_2(t)$ is sinusoidal. Hence, a Bode diagram is to be drawn for the transfer-function matrix element

$$G_{12}(s) = \frac{c_1 s + k_1}{\Delta(s)} \qquad (5.42)$$

It can be shown by using the Routh criterion, discussed in Chapter 6, that all poles of $G_{12}(s)$, i.e., roots of $\Delta(s) = 0$, have negative real parts. After dividing the polynomial $\Delta(s)$ of Eq. (2.47) throughout by $k_1 k_2$, factoring the quartic into two quadratic factors, and defining $\tau = c_1/k_1$, we express Eq. (5.42) as

$$G_{12}(s) = \frac{(1/k_2)(\tau s + 1)}{(s^2/\omega_{n1}^2 + 2\zeta_1 s/\omega_{n1} + 1)(s^2/\omega_{n2}^2 + 2\zeta_2 s/\omega_{n2} + 1)} \qquad (5.43)$$

Let the parameter values be given by $k_2 = 0.1$, $\tau = 25$ seconds, $\omega_{n_1} = 0.2$ rad/s, $\zeta_1 = 0.4$, $\omega_{n2} = 1.5$ rad/s, and $\zeta_2 = 0.05$. With these parameter values, after noting that $20 \log 1/k_2 = 20$ dB, we obtain

$$\begin{aligned} 20 \log |G_{12}(j\omega)| = 20 &+ \tfrac{20}{2} \log (25^2\omega^2 + 1) \\ &- \tfrac{20}{2} \log [(1 - \omega^2/0.2^2)^2 + (2(0.4)\omega/0.2)^2] \\ &- \tfrac{20}{2} \log [(1 - \omega^2/1.5^2)^2 + (2(0.05)\omega/1.5)^2] \end{aligned} \qquad (5.44)$$

$$\measuredangle G_{12}(j\omega) = \tan^{-1} 25\omega - \tan^{-1}\left(\frac{2(0.4)\omega/0.2}{1 - \omega^2/0.2^2}\right) - \tan^{-1}\left(\frac{2(0.05)\omega/1.5}{1 - \omega^2/1.5^2}\right) \qquad (5.45)$$

The frequency scale of Fig. 5.10 is chosen to include all the corner frequencies. At $\omega = 0.01$ rad/s, the starting value from Eq. (5.44) is 20 dB. The initial slope, before the first break occurs, is 0 dB/decade. The first break occurs at $\omega = 1/\tau = 0.04$ rad/s and the slope becomes +20 dB/decade. The second break occurs at $\omega = 0.2$ rad/s, where the net slope changes to (20 − 40) = −20 dB/decade. The third break occurs at $\omega = 1.5$ rad/s, where the slope changes to (−20 − 40) = −60 dB/decade. The magnitude curve of Fig.

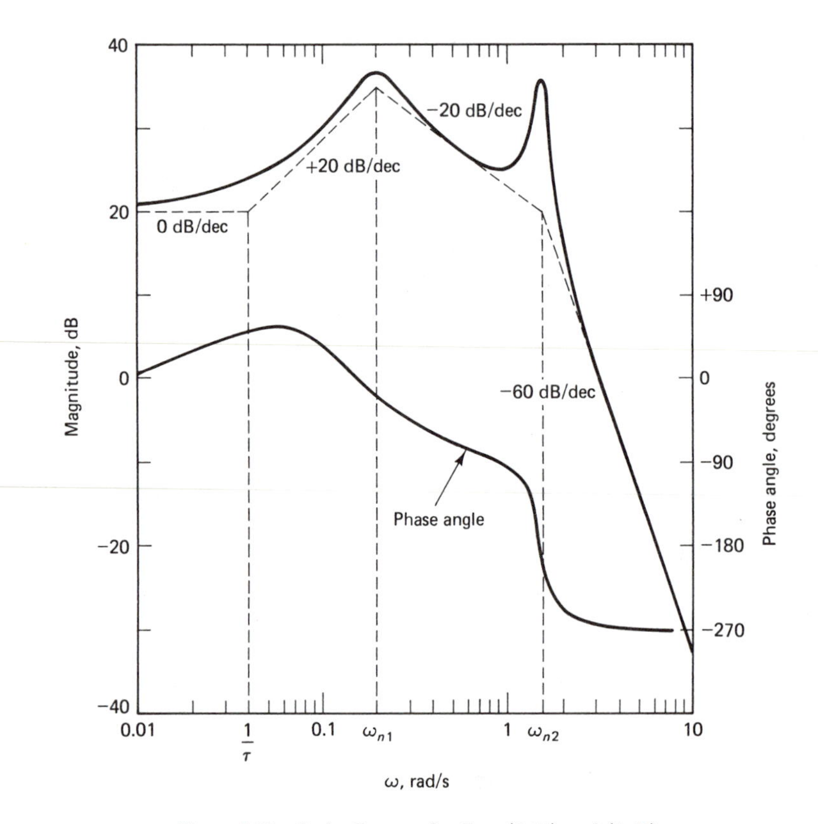

Figure 5.10 Bode diagram for Eqs. (5.44) and (5.45).

5.10 is drawn only by using this information, and corrections due to the damping ratios are made as shown. The phase angle is plotted from Eq. (5.45) and it varies from 0°, as $\omega \to 0$, to $-270°$, as $\omega \to \infty$.

5.4 FILTER CHARACTERISTICS

We have seen that for steady-state frequency response, the amplitude of the output is equal to $a|G(j\omega)|$, where a is the amplitude of the input sinusoid and $G(j\omega)$ is the frequency-response function. Hence, when $|G(j\omega)| > 1$, the input amplitude is magnified, and when $|G(j\omega)| < 1$, it is attenuated by the system dynamics. If $|G(j\omega)| \ll 1$ in a certain range of frequencies, the input amplitude is completely attenuated and the system does not respond in that range of frequencies. The system acts as a filter of those frequencies. The filter characteristics have important applications for the design of control systems and are discussed in the following.

Filter of Low Frequencies. A transfer function that attenuates the amplitude of low-frequency input is called a filter of low frequencies or a high-pass filter. It is noted that the distinction between low and high frequencies is dependent on the particular application; frequencies that are considered high for an automobile suspension may be low for a microwave system.

EXAMPLE 5.4

Let the frequency-response function of a system be given by

$$G(j\omega) = \frac{0.1(j\omega)}{j0.01\omega + 1} \tag{5.46}$$

We get

$$20 \log |G(j\omega)| = 20 \log 0.1 + 20 \log \omega - \tfrac{20}{2} \log (0.01^2\omega^2 + 1) \tag{5.47}$$

$$\measuredangle G(j\omega) = 90° - \tan^{-1} 0.01\omega \tag{5.48}$$

The magnitude curve for the expression of Eq. (5.47) is shown in Fig. 5.11.

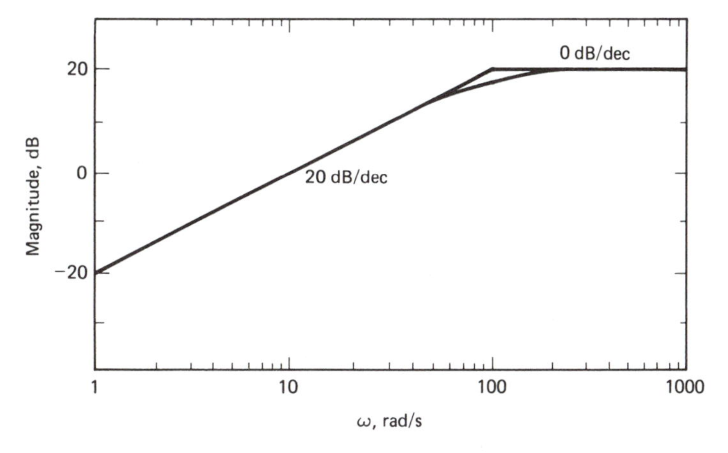

Figure 5.11 Low-frequency filter.

It is seen that this transfer function amplifies the amplitude of frequencies above 10 rad/s and attenuates the amplitude of those below 10 rad/s. At $\omega = 0.1$ rad/s, the attenuation factor corresponding to −40 dB is 0.01. Hence, this transfer function is a filter of low frequencies.

Filter of High Frequencies. A transfer function that attenuates the amplitude of high-frequency inputs is called a filter of high frequencies or a low-pass filter. The filtering of unwanted high-frequency noise and vibrations requires careful consideration. For example, the suspension system of an automobile is required to filter out the high-frequency random undulations of the road surface and not to transmit them to the sprung mass.

EXAMPLE 5.5(a)

The Bode diagram of the open-loop transfer function of the hydraulic speed-control system of Example 3.5 has been considered in Example 5.2. The frequency-response function is given by Eq. (5.38) and the Bode diagram is shown in Fig. 5.9. It is seen that the amplitude of all frequencies above 3 rad/s is attenuated and the attenuation increases with frequency. At the frequency $\omega = \omega_n = 200$ rad/s, where the quadratic break occurs in Fig. 5.9, the magnitude has a value of nearly -70 dB, which corresponds to an attenuation factor of 0.0003.

Such high frequencies are filtered out and we can discard the quadratic factor in Eq. (5.36) without affecting the accuracy of the transfer function. The order of the transfer function is thus reduced from fourth to second as in Example 4 in Chapter 4. Hence, in a transfer function, the factors with time constants and natural frequencies, which contribute a break in the magnitude curve at frequencies that are filtered out by the remaining factors, can be discarded. The order of the system is thus reduced without affecting the dynamic performance. For example, we have simplified Eq. (2.160) to Eq. (2.161).

Intermediate-Pass Filter. A transfer function that filters out both low- and high-frequency inputs but transmits inputs in the intermediate-frequency range is called an intermediate-pass filter.

EXAMPLE 5.5(b)

The transfer function relating the displacement output to the pressure input of the pneumatic device of Example 2.9 is given by Eq. (2.105). This transfer function is

$$G(s) = \frac{k_g s}{(\tau_1 s + 1)(\tau_2 s + 1)} \tag{5.49}$$

Let the numerical values be $k_g = 10$, $\tau_1 = 1$, and $\tau_2 = 0.1$. The frequency-

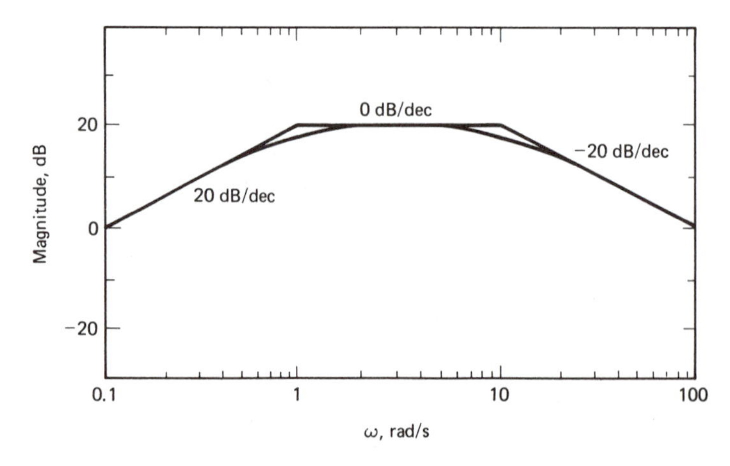

Figure 5.12 Intermediate-pass filter.

response function becomes

$$G(j\omega) = \frac{10(j\omega)}{(j\omega + 1)(j0.1\omega + 1)} \tag{5.50}$$

Its magnitude is given by

$$20 \log |G(j\omega)| = 20 + 20 \log \omega - \tfrac{20}{2} \log (\omega^2 + 1) - \tfrac{20}{2} \log (0.1^2\omega^2 + 1) \tag{5.51}$$

The magnitude curve is shown in Fig. 5.12. It is seen that the amplitudes of frequencies in the range from 0.1 to 100 rad/s are amplified. The amplitudes of frequencies outside that range are attenuated.

Frequency-Domain Specification of Instrumentation. The frequency-domain specification of sensors, recorders, and other instruments used to measure and/or record dynamic variables is usually given by stating that the frequency-response curve of the magnitude is flat up to a certain maximum frequency. This signifies that up to that maximum frequency, the dynamics of the instrument or recorder itself can be neglected and its input and output related by only a gain constant. This gain constant can be determined from the steady-state calibration and then the instrument can be used for measurement of dynamic variables.

Beyond the maximum stated frequency, the dynamics of the instrument itself cannot be neglected and its input and output are related by a transfer function that is no longer a gain constant. Hence, the steady-state calibration becomes invalid and the output will be phase shifted from the input and the input amplitude multiplied by the value of $|G(j\omega)|$, where $G(j\omega)$ is the frequency-response function of the instrument. Hence, care is required in the selection of instrumentation and recording equipment to ensure that their frequency response is suitable for the measurement of dynamic variables under consideration. For example, strip chart recorders usually have their frequency response magnitude curve flat up to about 50 Hz, whereas an oscillograph has a flat frequency response up to some kHz range.

EXAMPLE 5.6

The transfer function of the U-tube manometer of Example 2.6 is given by Eq. (2.80) and is shown in the block diagram of Fig. 2.26. Let the parameter values be such that this transfer function becomes

$$G(s) = \frac{0.1}{s^2/5^2 + 2(0.4)s/5 + 1} \tag{5.52}$$

We get

$$20 \log |G(j\omega)| = 20 \log 0.1 - \tfrac{20}{2} \log [(1 - \omega^2/5^2)^2 + (2(0.4)\omega/5)^2] \tag{5.53}$$

The magnitude curve is shown in Fig. 5.13. The corner frequency is the natural frequency $\omega_n = 5$ rad/s, and because the damping ratio $\zeta = 0.4$, the magnitude curve is flat up to about 1 rad/s. The steady-state gain calibration of the manometer is not valid for measurement of pressure with frequency

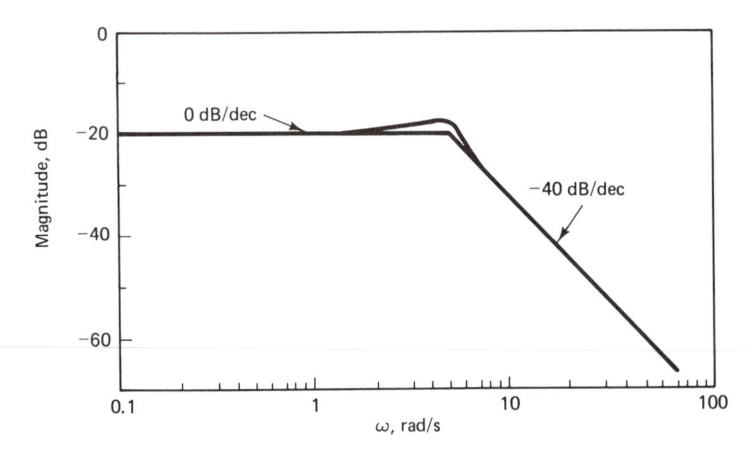

Figure 5.13 Frequency response of a manometer.

content above 1 rad/s. For the measurement of a transient pressure signal that changes fast with time, we need to select some other instrument, such as a strain-gage pressure transducer.

5.5 PERFORMANCE SPECIFICATIONS IN THE FREQUENCY DOMAIN

The transient performance specifications of control systems, such as rise time, maximum overshoot, and settling time, have been discussed in Chapter 4. We now give the performance specifications of control systems in the frequency domain in terms of peak resonance and bandwidth. We consider the closed-loop transfer function $G_c(s)$ of a control system defined by Eq. (5.5). It is assumed for asymptotic stability that all poles of $G_c(s)$ have negative real parts, that is, all roots of the characteristic equation $\det|s\mathbf{I} - \mathbf{A} + \mathbf{BK}| = 0$ from Eq. (5.5) have negative real parts. Hence, $G_c(j\omega)$ is the closed-loop frequency response function. A typical plot of $20 \log |G_c(j\omega)|$ is shown in Fig. 5.14.

Peak Resonance M_p. The peak resonance M_p is defined as the maximum value $|G_c(j\omega)|$ shown in Fig. 5.14. It gives an indication of the relative stability of the closed-loop control system. In the design of control systems, it is not sufficient that all roots of the characteristic equation have negative real parts so that asymptotic stability is ensured. If the roots are located too close to the left of the imaginary axis in the s-plane, then due to wear and tear and changes in the temperature and other conditions, the roots might migrate to the right half of the s-plane, thus giving rise to instability.

A large value of M_p usually corresponds to a large value of the maximum overshoot of the step response. The commonly specified optimum value of M_p is in the range from 1.1 to 1.5, corresponding to the maximum value of $20 \log |G_c(j\omega)|$, lying in the range from 0.83 to 3.52 dB.

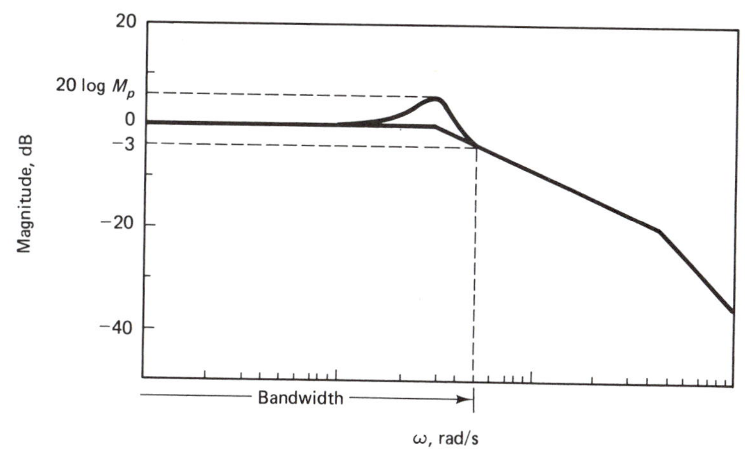

Figure 5.14 Typical magnitude plot of a control system.

Bandwidth. The cutoff frequency ω_c is defined as the frequency at which $20 \log |G_c(j\omega)|$ falls below -3 dB, which corresponds to the magnitude $|G_c(j\omega)|$ dropping below 0.707. It is an indication of the filter characteristics of the system. The bandwidth is defined as the spread between zero frequency and the cutoff frequency, as shown in Fig. 5.14. In some systems, there may be two cutoff frequencies. In that case, the bandwidth is the spread between the two frequencies. For example, in Fig. 5.12 there are two cutoff frequencies and the bandwidth is the spread from about 0.07 to slightly above 100 rad/s.

A control system may be subjected to high-frequency electrical noise, vibrations, hydraulic and pneumatic pressure fluctuations, and other disturbances. In case the bandwidth is very high, these unwanted disturbances will be transmitted around the loop. An optimum bandwidth is one that transmits the signals unattenuated but filters out the unwanted noise. The bandwidth is also a measure of the transient response. A large bandwidth corresponds to the roots of the characteristic equation that are far to the left of the imaginary axis in the complex s-plane and yields a fast rise time. A small bandwidth yields a time response that is generally slow and sluggish.

EXAMPLE 5.7

The transient response of the first-order control system of Example 3.1 has been studied in Example 4.5. The closed-loop transfer function given by Eq. (4.75) is

$$G_c(s) = \frac{k}{(1/a)s + 1} \tag{5.54}$$

where the time constant $\tau = 1/a$ and the gain constant k are defined by Eq. (4.74). The root of the characteristic equation is $-a$. Letting $k = 0.95$, we get

$$20 \log |G_c(j\omega)| = -0.45 - \tfrac{20}{2} \log (\omega^2/a^2 + 1) \tag{5.55}$$

The magnitude curve is shown in Fig. 5.15.

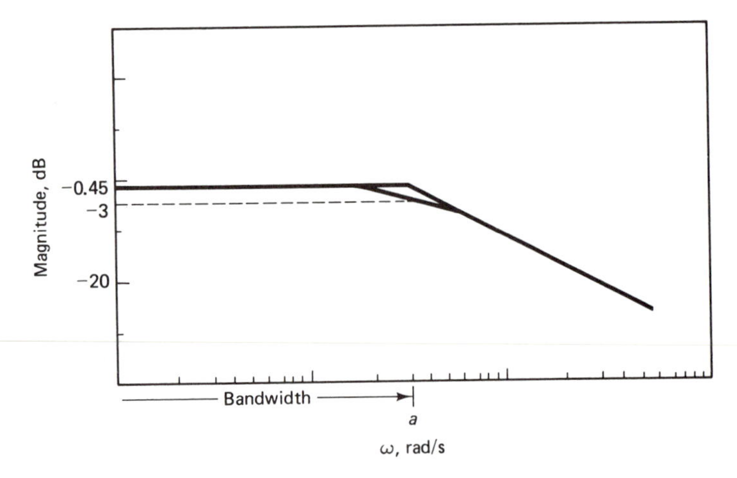

Figure 5.15 First-order control system.

It is seen that this control system does not have a resonance peak and the maximum value of $|G_c(j\omega)|$ is 0.95, which corresponds to −0.45 dB. The bandwidth is close to a rad/s. If the bandwidth is increased by increasing the value of a, it is seen from Fig. 4.5 that the transient response has a fast rise time. On the other hand, if the bandwidth is decreased by decreasing the value of a, the transient response of Fig. 4.5 becomes sluggish.

EXAMPLE 5.8

The transient response of the simplified model of the hydraulic control system of Example 3.5 has been investigated in Example 4.6. The closed-loop transfer function is obtained from Eq. (4.84) as

$$G_c(s) = \frac{1}{s^2/\omega_n^2 + 2\zeta s/\omega_n + 1} \tag{5.56}$$

This second-order transfer function has been studied earlier and its Bode diagram is shown in Fig. 5.7. It is seen from this figure that the peak resonance value M_p depends on the damping ratio ζ and that the bandwidth is close to the natural frequency. Their exact values can be obtained as follows. We have

$$M = \frac{1}{[(1 - \omega^2/\omega_n^2)^2 + (2\zeta\omega/\omega_n)^2]^{1/2}} \tag{5.57}$$

For maximum value of M, we set $dM/d\omega = 0$ and obtain

$$4(\omega/\omega_n)^3 - 4(\omega/\omega_n) + 8\zeta^2\omega/\omega_n = 0$$

The two positive roots of this equation are

$$\omega = 0 \tag{5.58}$$

$$\omega = \omega_n\sqrt{1 - 2\zeta^2} \qquad \zeta \le 0.707 \tag{5.59}$$

Substituting for ω from Eq. (5.59) in Eq. (5.57) and simplifying the resulting equation, we obtain

$$M_p = \frac{1}{2\zeta\sqrt{1-\zeta^2}} \tag{5.60}$$

which shows that M_p is a function of ζ only for $\zeta < 0.707$. For $\zeta \geq 0.707$, solution of Eq. (5.58) is the valid one and $M_p = 1$. To obtain the bandwidth frequency, we set the right-hand side of Eq. (5.57) equal to 0.707 and solve for ω. The solution yields the cutoff frequency as

$$\omega_c = \omega_n[(1-2\zeta^2) + \sqrt{4\zeta^4 - 4\zeta^2 + 2}]^{1/2} \tag{5.61}$$

The bandwidth is the spread from zero frequency to this cutoff frequency. The bandwidth is directly proportional to ω_n. For a fixed value of ω_n, both the bandwidth and the resonance peak M_p increase as ζ decreases. This frequency response can be compared with the unit step response of the system studied in Example 4.5. The maximum overshoot of the unit step response depends upon ζ only. The rise time increases with ζ, whereas the bandwidth decreases with the increase of ζ for a fixed ω_n.

5.6 MODELING FROM EXPERIMENTAL FREQUENCY RESPONSE

The mathematical models of control systems are usually obtained by making several assumptions. It is therefore desirable to validate the models experimentally and at the same time identify the parameter values, such as gain, time constants, natural frequencies, and damping ratios. Sometimes we encounter difficulties in obtaining the transfer function of a system by analytical means. Experimental frequency response is a powerful and practical method of validating a linear model and identifying the parameter values or even developing a linear model solely from experimental results.

A sinusoidal test signal $a \sin \omega t$ is used as input. The output of a nonlinear system usually will contain harmonics. For stable linear systems, the output is $b \sin(\omega t + \phi)$, where $b/a = |G(j\omega)|$ and $\phi = \measuredangle G(j\omega)$. Here, $G(s)$ is the transfer function relating the input and output. A device called a frequency-response analyzer measures the amplitude ratio b/a and the phase angle ϕ as the frequency is swept from a very low value to some maximum value at which the input is completely filtered out. A Bode diagram is obtained by plotting $20 \log b/a = 20 \log |G(j\omega)|$ and $\phi = \measuredangle G(j\omega)$ versus the frequency. Some frequency-response analyzers are equipped with a plotter for machine plotting the Bode diagram. Earlier, we studied the drawing of a Bode diagram for a given transfer function. Now, we have the inverse problem where given an experimentally obtained plot of a Bode diagram, we have to identify the transfer function.

The experimental frequency response is usually conducted on the open-loop system, Eq. (5.7), to identify the transfer function $G_o(s)$ of Eq. (5.8) with

$u = a \sin \omega t$ and $y = b \sin(\omega t + \phi)$. Since the analysis and design are performed before constructing the control system, the open-loop transfer function $G_o(s)$ is used in synthesizing the control law of the closed-loop system as shown in later chapters. Also, experimentally obtained Bode diagrams are usually available for many control components from their manufacturers. Experimental frequency response may then be conducted only on the existing part of the system, usually the system to be controlled. However, when the open-loop transfer function $G_o(s)$ is not stable, experimental frequency response may be conducted on the closed-loop system of Eq. (5.4) to identify the transfer function $G_c(s)$ of Eq. (5.5).

The transfer function is identified from the magnitude curve, and the phase angle curve is used for a check of the results. First, we determine whether $G(s)$ has a pole or zeros at the origin from the initial slope of the magnitude curve. The initial slope refers to the slope before the first corner frequency. If $G(s)$ has a pole at the origin, the initial slope will be -20 dB/decade. If there is a zero at the origin of order p, the initial slope will be $p(20)$ dB/decade. Next, the corners are drawn to the smooth experimental magnitude curve to obtain the time constants and natural frequencies. For a simple pole, the slope at the corner frequency changes by -20 DB/decade and for a simple zero by 20 dB/decade. For quadratic poles, the slope at the corner frequency changes by -40 dB/decade and for quadratic zeros by 40 dB/decade. The gain constant is determined last as shown in the following examples. The phase angle curve is then used for a check of the results.

EXAMPLE 5.9

The open-loop transfer function of a control system is given by

$$Y(s) = G_o(s)U(s) \tag{5.62}$$

An experimental frequency response is conducted to identify $G_o(s)$. The system is subjected to a sinusoidal test input $u(t) = a \sin \omega t$ and the frequency ω is varied. The output is given by $y(t) = b \sin(\omega t + \phi)$. Figure 5.16 shows the plots of $20 \log b/a$ and ϕ versus ω and the transfer function is to be identified from this diagram. The initial slope remains unchanged for frequencies lower than those shown in Fig. 5.16.

The initial slope of the magnitude curve is -20 dB/decade, which indicates that $G_o(s)$ has a pole at the origin. Two corners can be drawn to the experimental data as shown in Fig. 5.16. The first corner frequency is 10 rad/s, where the slope changes from -20 to -40 dB/decade, i.e., the change is -20 dB/decade. Hence, $G_o(s)$ has a first-order term on the denominator with a time constant $\tau_1 = 1/10 = 0.1$ second. The next corner frequency is 100 rad/s, where the change in the slope is -20 dB/decade. Thus, $G_o(s)$ has another first-order term on the denominator with a time constant $\tau_2 = 1/100 = 0.01$ second. Hence, we get

$$G_o(s) = \frac{k}{s(0.1s + 1)(0.01s + 1)} \tag{5.63}$$

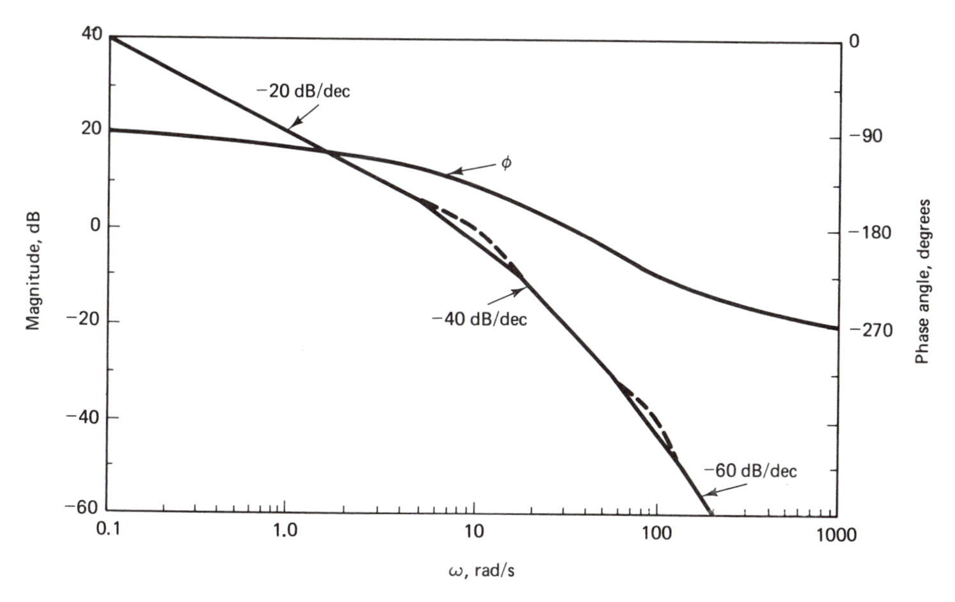

Figure 5.16 Experimental frequency response of an open-loop system.

In order to determine the gain constant, we note that

$$20 \log |G_0(j\omega)| = 20 \log k - 20 \log \omega - \tfrac{20}{2} \log (0.1^2\omega^2 + 1) - \tfrac{20}{2} \log (0.01^2\omega^2 + 1) \tag{5.64}$$

We now select a value of ω, obtain the value of the left-hand side of Eq. (5.64) from the Bode diagram, and solve for k. The computation is simplified if we select $\omega = 1$ rad/s and we obtain $20 = 20 \log k$. Hence, $k = 10$. The unknown transfer function and the parameter values are now identified as

$$G_0(s) = \frac{10}{s(0.1s + 1)(0.01s + 1)} \tag{5.65}$$

For this $G_o(s)$, we have

$$\measuredangle G_o(j\omega) = -90° - \tan^{-1} 0.1\omega - \tan^{-1} 0.01\omega \tag{5.66}$$

It is seen that the phase angle plot of Fig. 5.16 satisfies Eq. (5.66) and this fact serves as a check that the transfer function of Eq. (5.65) has been identified correctly.

EXAMPLE 5.10

The open-loop transfer function of a control system is given by Eq. (5.62). The experimental frequency response is shown in Fig. 5.17. The unknown transfer function is to be identified from this diagram.

The initial slope of the magnitude curve is −20 dB/decade, which indicates that $G_o(s)$ has a pole at the origin. Three corners can be drawn to the

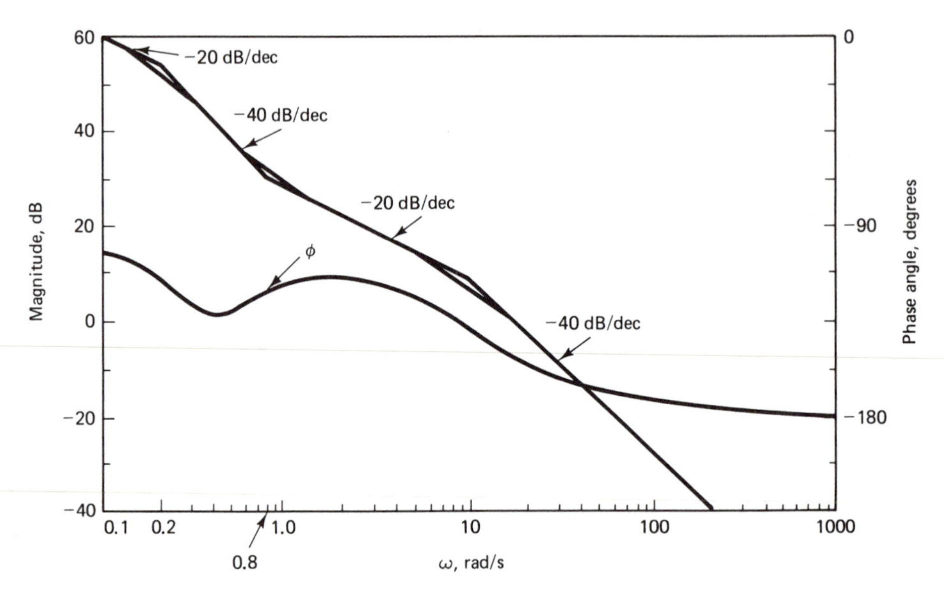

Figure 5.17 Experimental frequency response of an open-loop system.

experimental data as shown in Fig. 5.17. The first corner frequency is 1/5 rad/s, where the slope changes by −20 dB/decade, which indicates a first-order term on the denominator with time constant of 5 seconds. The next corner frequency is 1/1.25 rad/s, where the slope changes by +20 dB/decade, which indicates a first-order term on the numerator with time constant 1.25 seconds. The last corner frequency is 10 rad/s, where the slope changes by −20 dB/decade. Hence, $G_o(s)$ has a first-order term on the denominator with time constant 0.1 second. We, therefore, obtain

$$G_o(s) = \frac{k(1.25s + 1)}{s(5s + 1)(0.1s + 1)} \tag{5.67}$$

To determine the gain constant k, we note that

$$20 \log |G_o(j\omega)| = 20 \log k + \tfrac{20}{2} \log (1.25^2\omega^2 + 1) - 20 \log \omega - \tfrac{20}{2} \log (5^2\omega^2 + 1) - \tfrac{20}{2} \log (0.1^2\omega^2 + 1) \tag{5.68}$$

Now we select $\omega = 0.1$ rad/s and find from Fig. 5.17 that for this frequency, the left-hand side of Eq. (5.68) is 60 dB. Hence,

$$60 = 20 \log k + 0.07 + 20 - 0.97 - 0$$

i.e., $20 \log k = 40.9$ or $k = 110.9$. The unknown transfer function is now identified as

$$G_o(s) = \frac{110.9(1.25s + 1)}{s(5s + 1)(0.1s + 1)} \tag{5.69}$$

For this $G_o(j\omega)$, we obtain

$$\measuredangle G_o(j\omega) = \tan^{-1} 1.25\omega - 90° - \tan^{-1} 5\omega - \tan^{-1} 0.1\omega \tag{5.70}$$

It can be verified that the phase angle plot of Fig. 5.17 does indeed satisfy Eq. (5.70) as a check.

EXAMPLE 5.11

An experimental frequency response is conducted to identify the closed-loop transfer function, Eq. (5.5), of a control system. For this purpose, in Eq. (5.4) $r(t)$ is a test signal $a \sin \omega t$ and the output is given by $y = b \sin(\omega t + \phi)$. The plots of $20 \log b/a$ and ϕ versus ω are given in Fig. 5.18. Identify the closed-loop transfer function.

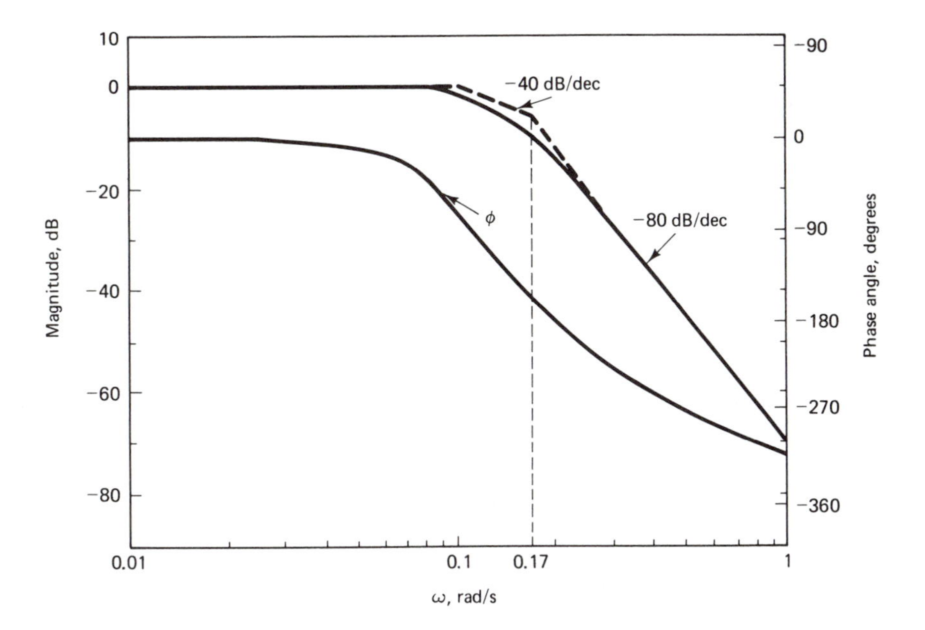

Figure 5.18 Experimental frequency response of a closed-loop system.

The initial slope of the magnitude curve in Fig. 5.18 is 0 dB/decade, which indicates that $G_c(s)$ has neither a pole nor a zero at the origin. Two corners can be fitted to the experimental data as shown in the figure. The first corner frequency is 0.1 rad/s, where the slope changes from 0 to −40 dB/decade, which indicates a quadratic factor on the denominator. The second corner frequency is 0.17 rad/s, where the slope changes from −40 to −80 dB/decade. Hence, the transfer function has another quadratic factor on the denominator. The natural frequencies of the two quadratic factors are the corner frequencies, which are 0.1 and 0.17 rad/s, respectively. Inspection of the magnitude plot at the corner frequencies and comparison with Fig. 5.7 yield the damping ratios as 0.6 and 0.8, respectively. Hence, we obtain

$$G_c(s) = \frac{k}{(s^2/0.1^2 + 2(0.6)s/0.1 + 1)(s^2/0.17^2 + 2(0.8)(s/0.17) + 1)} \tag{5.71}$$

The gain constant k can be identified from the procedure of the previous two examples and it is found that $k = 1$. For this $G_c(j\omega)$, we obtain

$$\measuredangle\, G_c(j\omega) = -\tan^{-1}\left(\frac{2(0.6)\omega/0.1}{1 - \omega^2/0.1^2}\right) - \tan^{-1}\left(\frac{2(0.8)\omega/0.17}{1 - \omega^2/0.17^2}\right) \qquad (5.72)$$

It can be verified that the phase angle plot of Fig. 5.18 does indeed satisfy Eq. (5.72).

The procedure for conducting experimental frequency response requires test equipment that is capable of operating in the range from very low frequencies to frequencies that are higher than the bandwidth frequency of the system to be identified. In most systems, the order of the denominator of the transfer function is higher than the order of its numerator, and the amplitude ratio is attenuated at high frequencies.

The output signal levels then diminish at the higher end of the spectrum and get corrupted by noise. This difficulty can be alleviated by using test equipment that has good noise-rejection capability.

A simple setup for conducting experimental frequency response is shown in Fig. 5.19. A signal generator supplies a sinusoidal signal at variable frequency to the system to be identified. The system output is measured by a sensor/transducer and both the sinusoidal input and output signals are displayed on a recorder. In case the recorder is a dual-channel strip chart recorder, the phase angle can be measured as illustrated in Fig. 5.1. If a dual-trace oscilloscope is employed, the phase angle can be obtained from the Lissajous figure.

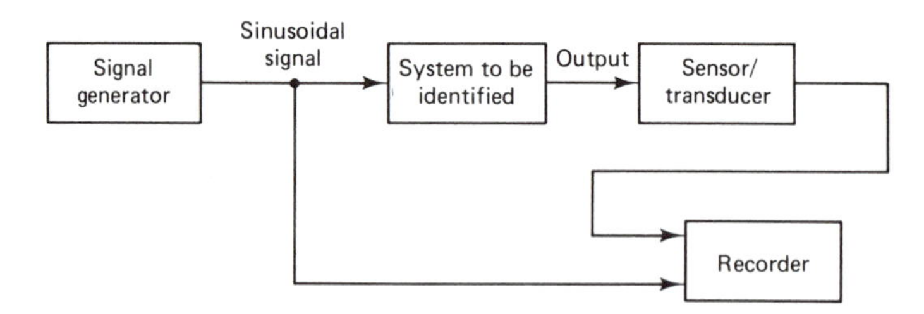

Figure 5.19 Experimental setup for frequency response.

Specialized types of equipment, which can be very sophisticated but expensive, are available commercially. These are called frequency-response analyzers. They can compute the magnitude ratio and phase angle and store the results for plotting. The signal generators are usually incorporated into the analyzers.

5.7 SUMMARY

In this chapter, we have studied the frequency response of control systems when the input is a sinusoidal testing signal. It is shown that when a linear, asymptotically stable system is subjected to a sinusoidal input, the steady-state output is also sinusoidal with the same frequency but different amplitude and there is a phase angle between the input and output. The ratio of the amplitude of the output to that of the input is given by the magnitude of the frequency-response function. The phase angle is equal to the angle of the frequency-response function.

The Bode diagram is a very useful graphical representation of the magnitude and angle of the frequency-response function versus frequency. Computer programs are available (Melsa and Jones, 1973) for calculating the frequency response of a transfer function and for providing a graphical output of the Bode diagram.* We have considered several applications of the Bode diagrams, including filter characteristics of transfer functions, and frequency-domain specifications of instrumentation and closed-loop control systems. Bode diagrams are also useful for the analysis of linear vibrations (D'Souza and Garg, 1984), but this topic is beyond our scope.

It is shown that experimental frequency response is a simple but powerful method for modeling linear stable systems. It can be used for identifying an unknown transfer function and/or validating an analytically obtained model and evaluating the parameter values. This provides one motivation for using parameters in the form of time constants, natural frequencies, damping ratios, and gain rather than the physical parameters of a system. Experimental frequency response is often undertaken in practice and several frequency-response analyzers are available for this purpose.

REFERENCES

DORF, R. C. (1980). *Modern Control Systems.* 3d ed. Reading, MA: Addison-Wesley.

D'SOUZA, A. F., and GARG, V. K. (1984). *Advanced Dynamics: Modeling and Analysis.* Englewood Cliffs, NJ: Prentice-Hall.

MELSA, J. L., and JONES, S. K. (1973). *Computer Programs for Computational Assistance in the Study of Linear Control Theory.* New York: McGraw-Hill.

RAVEN, F. H. (1978). *Automatic Control Engineering.* 3d ed. New York: McGraw-Hill.

PROBLEMS

5.1. The response T of a thermometer for measuring the input temperature T_1 is described by

$$2dT/dt + T = T_1$$

When T_1 is changing sinusoidally at frequency 1.2 rad/s, the amplitude of T is 2° C. Determine the amplitude of T_1.

5.2. In a seismograph shown in Fig. P5.2, u is the displacement of the ground with respect to a fixed datum and y is measured relative to u.

*See the references listed in Appendix D.

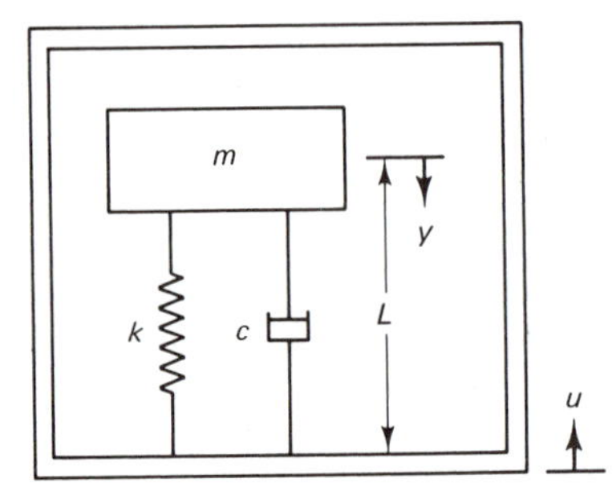

Figure P5.2 Seismograph.

(**a**) Show that

$$y = \left(\frac{mD^2}{mD^2 + cD + k}\right)u$$

(**b**) When $m = 20$ kg, $c = 8$ kg/s, $k = 80$ kg/s^2, and $u = 0.1 \sin 4t$, $y = y_0 \sin (4t + \phi)$. Determine y_0 and ϕ.

5.3. A machine of mass m is mounted on the floor with a foundation spring of stiffness k and a damper with coefficient c as shown in Fig. P5.3. It is subjected to a sinusoidal input force $F = f_0 \sin \omega t$. Obtain the transfer function $G(D)$ relating the force F_T transmitted to the floor to the input force F. Obtain the amplitude a of the transmitted force F_T when $f_0 = 200$ newtons, $\omega/\omega_n = 10$, and $\zeta = 0.05$, where ω_n and ζ are the natural frequency and the damping ratio, respectively.

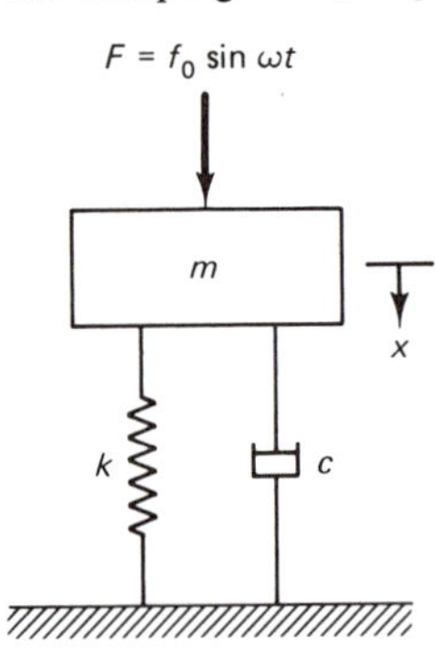

Figure P5.3 Machine mounted on a foundation.

5.4. Draw the Bode diagram for the following transfer functions. Also, comment on their filtering characteristics.

(**a**) $G(s) = \dfrac{100s(0.1s + 1)}{(s^2 + 0.18s + 1)(s + 1)}$

(**b**) $G(s) = \dfrac{10(0.1s + 1)}{0.01s^2 + 0.02s + 1}$

(**c**) $G(s) = \dfrac{100s}{(0.5s + 1)(5s + 1)}$

(**d**) $G(s) = \dfrac{10}{(s^2 + 1.2s + 1)(0.01s^2 + 0.02s + 1)}$

5.5. Draw the Bode diagram for the transfer function and parameter values of Problem 5.2 for $0.1 \le \omega \le 100$ rad/s. Give the frequency range for which the dynamics of the seismograph can be neglected. How can this range be increased?

5.6. Draw the Bode diagram for the transfer function of Problem 5.3. Use dimensionless frequency ω/ω_n and give the plots for $0.1 \leq \omega/\omega_n \leq 100$ for the following two cases of the damping ratio:
(**a**) $\zeta = 0.05$
(**b**) $\zeta = 0.5$
Comment on the difference in the high-frequency filter characteristics for these two values of the damping ratio.

5.7. For the hydraulic control system whose closed-loop transfer function is given in Problem 4.7, draw the Bode diagram and obtain the values of the peak resonance M_p and the bandwidth. Verify your answers by using Eqs. (5.60) and (5.61).

5.8. For the numerical values of the matrices of the control system of Example 3.3 given in Problem 4.6, obtain the closed-loop transfer function. Draw the Bode diagram for this transfer function and obtain values of the peak resonance M_p and the bandwidth. Is the value of M_p satisfactory? (*Hint*: The denominator of this transfer function can be factored as $(s + 5.033)(s^2 + 0.467s + 7.153)$.)

5.9. The asymptotic magnitude curves for two transfer functions are shown in Fig. P5.9. The slopes are maintained for frequencies lower and higher than those shown in the figure. Determine the transfer function of each system.

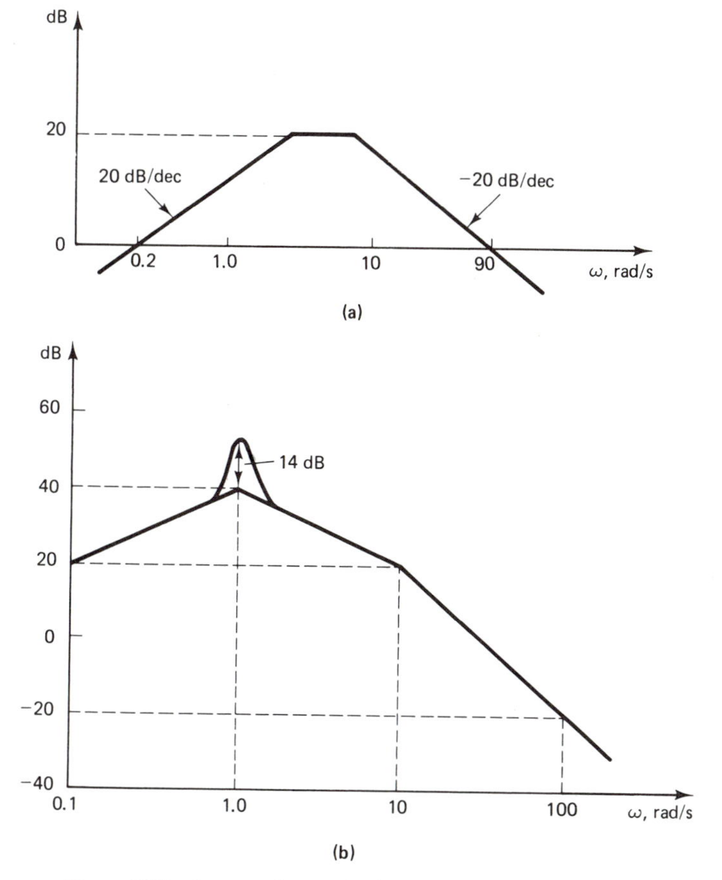

Figure P5.9 Asymptotic magnitude curve of a transfer function.

5.10. An experimentally determined magnitude curve of the Bode diagram has the following characteristics:

(1) Initial slope of -20 dB/decade at low frequencies.
(2) A break upwards so that the slope becomes 0 dB/decade at $\omega = 10$ rad/s at which frequency the magnitude is -10 dB.
(3) A break downwards at $\omega = 200$ rad/s so that the slope becomes -20 dB/decade and is maintained at higher frequencies.

Sketch the asymptotic magnitude curve and identify the transfer function represented by these data. The phase angle approaches $-90°$ both for very low and very high frequencies.

5.11. An experimentally determined magnitude curve of the Bode diagram has the following characteristics:

(1) Initial slope of 0 dB/decade and initial magnitude of 20 dB at low frequencies.
(2) A break downwards at $\omega = 0.5$ rad/s so that the slope becomes -20 dB/decade.
(3) A break upwards at $\omega = 2$ rad/s so that the slope becomes 0 dB/decade.
(4) A break downwards at $\omega = 80$ rad/s so that the slope becomes -20 dB/decade and is maintained at higher frequencies.

The phase angle approaches 0° at low frequencies and -90 at very high frequencies. Sketch the asymptotic magnitude curve and identify the transfer function represented by these data.

5.12. Experimental frequency-response data of a system are given in the following. The slopes are maintained for frequencies lower and higher than those given. Plot the Bode diagram and identify the transfer function.

Frequency (rad/s)	Magnitude Ratio (dB)	Phase angle (degrees)
0.1	20	−97
0.3	10	−110
0.6	3	−128
0.9	−2	−138
1	−3	−146
2	−13	−175
5	−31	−214
8	−42	−231
12	−51	−242
16	−58	−250
30	−74	−261

5.13. Plot the Bode diagram and identify the transfer function as for Problem 5.12 but for the following data.

Frequency (rad/s)	Magnitude ratio (dB)	Phase angle (degrees)
0.1	40	−92
0.2	34	−95
0.5	25	−100
1	20	−108
2	14	−126
3	10	−138
5	2	−160
10	−9	−190
20	−23	−220
30	−32	−235
40	−40	−243
50	−46	−248
100	−64	−258

5.14. Plot Bode diagram and identify the transfer function as for Problem 5.12 but for the following data.

Frequency (rad/s)	Magnitude ratio (dB)	Phase angle (degrees)
0.1	20	−11
0.2	19.6	−22
0.4	17	−37
0.6	15.5	−42
0.8	13.5	−46
1	12	−47
2	8	−45
3	7	−43
5	6	−44
7	5.5	−48
10	4	−55
20	1	−68
40	−6	−78
70	−10.5	−85

6

Stability and Root Locus

6.1 INTRODUCTION

We have seen in Chapter 4 that the transient response of a control system, whose model is described by linear, ordinary differential equations with constant coefficients, is governed by the roots of the characteristic equation. In particular, if any root has a positive real part, the response is unstable and grows out of bound. Among the performance specifications of control systems, the most important requirement is stability. An unstable control system will fail to function properly and in some cases, instability can cause a catastrophic failure. In some cases, nonlinearities, such as saturation, that we have neglected, limit the growth of the unstable response and the control system exhibits nonlinear oscillations called limit cycles.

The performance specifications demand not only a stable response but also a certain margin of stability as a factor of safety to account for parameter variations and to ensure stability under all operating conditions. The maximum overshoot of the step-response specifications of Chapter 4 and the peak resonance of the frequency-domain specifications of Chapter 5 depend on the margin of stability.

In this chapter, we first give definitions of stability and show that for control systems described by linear, ordinary differential equations with constant coefficients, stability depends only on the characteristic equation. The requirement for stability is that all roots of the characteristic equation have negative real parts. We study the Routh stability criterion, which is a very simple method of determining stability without actually finding the roots of the characteristic equation.

The locations of the roots of the characteristic equation in the complex plane are very important for the dynamic response. The locations determine the stability and the margin of stability. Frequently, one or more parameters are to be adjusted to obtain suitable root locations. It is therefore necessary to determine how the roots of the characteristic equation move in the complex plane as the parameters are varied. This is the objective of the root-locus method studied in this chapter. The root-locus method is also useful for the study of the sensitivity of the roots of the characteristic equation to parameter variations.

6.2 DEFINITIONS OF STABILITY

There are several concepts of stability, such as stability in the sense of Lagrange, Poincaré, Lyapunov, boundedness of response, and input-output stability. In nonlinear systems, these different concepts of stability may not lead to identical results. What is considered as stable in one sense may be unstable in another. In our case, where the models of control systems are described by linear, constant-parameter, ordinary differential equations, the distinctions among the different concepts of stability get blurred. In the following, we consider two definitions of stability. The first definition is in the sense of Lyapunov and the second is in the sense of bounded input-bounded output (BIBO). We show that for the mathematical models that we employ, the stability condition is independent of the input and is solely determined by the roots of the characteristic equation.

The mathematical models of closed-loop control systems have been formulated in Chapter 3 in the state-variables representation as

$$\dot{\mathbf{x}}(t) = (\mathbf{A} - \mathbf{BK})\mathbf{x}(t) + \mathbf{BNr}(t) + \mathbf{B}_1\mathbf{v}(t) \tag{6.1}$$

where $\mathbf{r}(t)$ is the comand input and $\mathbf{v}(t)$ is a disturbance. The technique of modeling and identifying parameter values from experimental frequency response has been studied in Chapter 5. In this chapter, we assume that Eq. (6.1) represents a known mathematical model.

6.2.1 Stability in the Sense of Lyapunov

An important concept of stability is in the sense of Lyapunov. This concept of stability does not admit any inputs and the control system is disturbed only by the initial conditions, which may be caused by impulsive changes in the command inputs or disturbances at the initial time. Hence, with $\mathbf{r}(t) = 0$ and $\mathbf{v}(t) = 0$ in Eq. (6.1), the homogeneous state equation is

$$\dot{\mathbf{x}}(t) = (\mathbf{A} - \mathbf{BK})\mathbf{x}(t) \qquad \text{with initial conditions } \mathbf{x}(0) \text{ at } t = 0 \tag{6.2}$$

For equilibrium, we have $\dot{\mathbf{x}} = \mathbf{0}$ and from Eq. (6.2), the equilibrium state is given by

$$\mathbf{x} = (\mathbf{A} - \mathbf{BK})^{-1}\mathbf{0} \tag{6.3}$$

Assuming that $(\mathbf{A} - \mathbf{BK})$ is nonsingular, the only equilibrium state of Eq. (6.2) is obtained from Eq. (6.3) as $\mathbf{x}_e = \mathbf{0}$, which is the origin of the state space. It is the equilibrium about which we have linearized the differential equations, and the state variables represent perturbations about this equilibrium.

If the response $\mathbf{x}(t)$, when subjected to the initial state $\mathbf{x}(0)$, does not stray too far from $\mathbf{x}(0)$ and returns to the equilibrium state $\mathbf{0}$ as time tends to infinity, then this equilibrium state is said to be asymptotically stable in the sense of Lyapunov. However, a formal definition applicable to Eq. (6.2) can be stated as follows. More general definitions applicable to nonlinear, time-varying parameter systems are given by D'Souza and Garg (1984).

The equilibrium state $\mathbf{0}$ of Eq. (6.2) is said to be asymptotically stable if for every $\epsilon > 0$, there exists a $\delta > 0$, where δ may depend on ϵ, such that $\|\mathbf{x}(0)\| \le \delta$ implies that $\|\mathbf{x}(t)\| < \epsilon$ for all $t \ge 0$ and $\lim_{t\to\infty} \|\mathbf{x}(t)\| = 0$. The equilibrium state is said to be unstable if there is a $\epsilon > 0$ for which no $\delta > 0$ can be found, such that $\|\mathbf{x}(0)\| \le \delta$ implies that $\|\mathbf{x}(t)\| < \epsilon$. Here, $\|\mathbf{x}(t)\|$ represents the norm of the state vector in euclidean space, that is,

$$\|\mathbf{x}(t)\| = \left[\sum_{i=1}^{n} x_i^2(t) \right]^{1/2} \tag{6.4}$$

These definitions are illustrated in Fig. 6.1 for a second-order system. In Fig. 6.1(a), when the equilibrium state is asymptotically stable, for every ϵ we can find a δ, such that if the initial state $\mathbf{x}(0)$ is inside a circle of radius δ, the response is inside a circle of radius ϵ, and it decays to the origin as time approaches infinity. In Fig. 6.1(b), which represents an unstable equilibrium, for a given ϵ we can find no δ such that if the initial state $\mathbf{x}(0)$ is inside a circle of radius δ, the response is inside a circle of radius ϵ. For higher-order systems, where $n > 2$, the circles become hyperspheres of radius δ and ϵ, respectively.

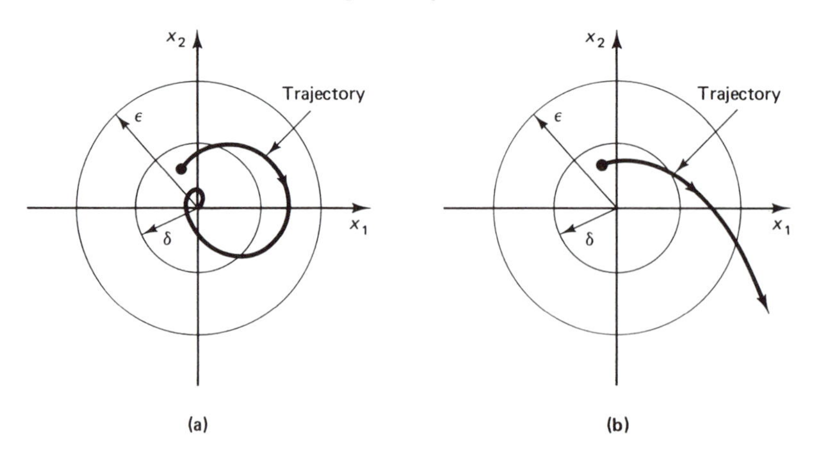

Figure 6.1 (a) Asymptotically stable equilibrium, and (b) unstable equilibrium.

In the following, we show that the preceding definition of asymptotic stability is satisfied when all roots of the characteristic equation have negative real parts. If the characteristic equation has a root with a positive real part, then the equilibrium state is unstable. As shown in Chapter 4, the response of Eq. (6.2) is given by

$$\mathbf{x}(t) = \mathbf{\Phi}(t)\mathbf{x}(0) \tag{6.5}$$

where $\mathbf{\Phi}(t)$ is the state-transition matrix. Taking the norm of both sides of Eq.

(6.5), we obtain

$$\|\mathbf{x}(t)\| = \|\mathbf{\Phi}(t)\mathbf{x}(0)\| \tag{6.6a}$$

and it follows that

$$\|\mathbf{x}(t)\| \le \|\mathbf{\Phi}(t)\| \, \|\mathbf{x}(0)\| \tag{6.6b}$$

Since for asymptotic stability $\lim_{t\to\infty} \|\mathbf{x}(t)\| = 0$, from Eqs. (6.6a) and (6.6b), we obtain the condition

$$\lim_{t\to\infty} \|\mathbf{\Phi}(t)\| = 0 \qquad \text{i.e., } \lim_{t\to\infty} \phi_{ij}(t) = 0 \tag{6.7}$$ [1]

for $i, j = 1, 2, \ldots, n$, where ϕ_{ij} is an element of the state-transition matrix. As given by Eq. (4.27), the state-transition matrix can be represented by

$$\begin{aligned}\mathbf{\Phi}(t) &= \mathscr{L}^{-1}[(s\mathbf{I} - \mathbf{A} + \mathbf{BK})^{-1}] \\ &= \mathscr{L}^{-1}\left[\frac{\text{Adj}\,(s\mathbf{I} - \mathbf{A} + \mathbf{BK})}{\det|s\mathbf{I} - \mathbf{A} + \mathbf{BK}|}\right]\end{aligned} \tag{6.8}$$

The characteristic equation is given by $\det|s\mathbf{I} - \mathbf{A} + \mathbf{BK}| = 0$. It is seen from the transient response studied in Chapter 4 that to satisfy Eq. (6.7), all roots of the characteristic equation must have negative real parts. If the characteristic equation has a root with a positive real part, the equilibrium is unstable.

6.2.2 Bounded-Input, Bounded-Output Stability

A system is said to be bounded-input, bounded-output (BIBO) stable, if its output is bounded for any bounded input. For stability, it does not matter whether the input is a command signal or a disturbance. Hence, we can set $\mathbf{v} = 0$ in Eq. (6.1). The output is given by $\mathbf{y} = \mathbf{Cx}$. A system is called BIBO stable if given

$$\|\mathbf{r}(t)\| \le \delta < \infty \qquad \text{for all } t \ge 0 \tag{6.9}$$

implies that

$$\|\mathbf{y}(t)\| \le \epsilon < \infty \qquad \text{for all } t \ge 0 \tag{6.10}$$

We show that BIBO stability again leads to the requirement that all roots of the characteristic equation have negative real parts. For systems described by linear time-invariant models, stability is independent of the inputs. As shown by Eq. (4.45), the output with initial conditions $\mathbf{x}(0) = \mathbf{0}$ is given by

$$\mathbf{y}(t) = \mathbf{C}\int_0^t \mathbf{\Phi}(t - t')\mathbf{BNr}(t')\,dt' \tag{6.11}$$

Since Eq. (6.10) must be valid for all time, we can let $t \to \infty$ and making use of the property of the convolution integral, express Eq. (6.11) as

$$\mathbf{y}(\infty) = \mathbf{C}\int_0^\infty \mathbf{\Phi}(t')\mathbf{BNr}(t - t')\,dt' \tag{6.12}$$

Taking the norm of both sides of Eq. (6.12), we obtain

$$\|\mathbf{y}(\infty)\| = \left\|\mathbf{C}\int_0^\infty \mathbf{\Phi}(t')\mathbf{BNr}(t - t')\,dt'\right\|$$

and making use of Eq. (6.9), it follows that

$$\|\mathbf{y}(\infty)\| \le \delta\left\|\mathbf{C}\int_0^\infty \mathbf{\Phi}(t')\mathbf{BN}\,dt'\right\| \tag{6.13}$$

If the output is to be bounded, then

$$\left\|\mathbf{C}\int_0^\infty \mathbf{\Phi}(t')\mathbf{BN}\,dt'\right\| < \infty \tag{6.14}$$

The condition for this integral to converge and be finite is that

$$\lim_{t\to\infty} \|\mathbf{\Phi}(t)\| = 0 \qquad \text{i.e., } \lim_{t\to\infty} \phi_{ij}(t) = 0 \qquad (6.15)^2$$

for $i, j = 1, \ldots, n$. From Eq. (6.8), we conclude that for BIBO stability, all roots of the characteristic equation must have negative real parts.

6.3 ROUTH CRITERION

The stability of a control system described by a linear time-invariant model can be investigated by finding the roots of the characteristic equation. However, the determination of the roots of a high-order polynomial usually requires a digital computer. Since the computer program uses a numerical algorithm, obtaining the region of stability in the parameter space is a time-consuming process. A simple method of stability analysis, without actually finding the exact location of the roots of the characteristic equation, is based on the Routh criterion.

The Routh criterion is an algebraic method that indicates whether or not all roots of the characteristic equation have negative real parts without actually finding the roots. In case the stability condition is not satisfied, the method also indicates the number of roots that lie in the right half of the s-plane and on the imaginary axis, that is, the number of roots that have positive and zero real parts. The technique was developed independently by Hurwitz and Routh in the 1890s and for this reason, it is also called the Routh–Hurwitz criterion.

We have seen earlier that the characteristic equation of a closed-loop control system, Eq. (6.1), is given by

$$\det|s\mathbf{I} - \mathbf{A} + \mathbf{BK}| = a_n s^n + a_{n-1}s^{n-1} + \cdots + a_1 s + a_0 = 0 \tag{6.16}$$

where n is the order of the system, and a_i are constant coefficients. A necessary condition for all the roots of Eq. (6.16) to have negative real parts is that all the coefficients have the same sign and that a_i $(i = n - 1, \ldots, 0)$ be nonzero. When this requirement is not satisfied, we can tell from inspection of Eq. (6.16) that it is unstable and further use of the Routh criterion will also indicate instability.

However, this requirement is not a sufficient condition and we employ the Routh criterion for stability analysis.

The criterion was stated by Hurwitz in terms of determinants as follows. A necessary and sufficient condition for all roots of Eq. (6.16) to have negative real parts is that the Hurwitz determinants $D_1, D_2, \ldots, D_n$ must all be positive.

The Hurwitz determinants of Eq. (6.16) are defined by

$$D_1 = a_{n-1} \qquad D_2 = \begin{vmatrix} a_{n-1} & a_{n-3} \\ a_n & a_{n-2} \end{vmatrix} \qquad D_3 = \begin{vmatrix} a_{n-1} & a_{n-3} & a_{n-5} \\ a_n & a_{n-2} & a_{n-4} \\ 0 & a_{n-1} & a_{n-3} \end{vmatrix}$$

$$D_n = \begin{vmatrix} a_{n-1} & a_{n-3} & a_{n-5} & \cdots & 0 \\ a_n & a_{n-2} & a_{n-4} & \cdots & \cdot \\ 0 & a_{n-1} & a_{n-3} & \cdots & \cdot \\ 0 & a_n & a_{n-2} & \cdots & \cdot \\ 0 & 0 & a_{n-1} & \cdots & \cdot \\ \vdots & \vdots & \vdots & \cdots & \cdot \\ 0 & 0 & 0 & \cdots & a_0 \end{vmatrix} \tag{6.17}$$

where the coefficients with negative indices are replaced by zeros and n determinants must be considered for an nth-order system. There is labor involved in evaluating determinants for high-order systems and for this reason we employ the more convenient array formulation of Routh. The coefficients of the characteristic equation are arranged in the following Routh array.

$$\begin{array}{cccccc l} a_n & a_{n-2} & a_{n-4} & a_{n-6} & \cdots & 0 & \text{first row, } s^n \\ a_{n-1} & a_{n-3} & a_{n-5} & a_{n-7} & \cdots & 0 & \text{second row, } s^{n-1} \\ b_1 & b_2 & b_3 & b_4 & \cdots & 0 & \\ c_1 & c_2 & c_3 & \cdots & & & \\ \cdot & \cdot & \cdot & & & & \\ d_1 & d_2 & 0 & & & & \\ e_1 & e_2 & 0 & & & & \\ f_1 & 0 & 0 & & & & n^{\text{th}} \text{ row, } s^1 \\ g_1 & 0 & 0 & & & & (n+1)^{\text{th}} \text{ row, } s^0 \end{array}$$

The first two rows of this array are obtained from the coefficients of the characteristic equation. The remaining rows are calculated by employing an algorithm. The row of b terms is obtained as follows:

$$b_1 = \frac{a_{n-1}a_{n-2} - a_n a_{n-3}}{a_{n-1}}$$

$$b_2 = \frac{a_{n-1}a_{n-4} - a_n a_{n-5}}{a_{n-1}} \tag{6.18}$$

$$b_3 = \frac{a_{n-1}a_{n-6} - a_n a_{n-7}}{a_{n-1}}$$

By dropping down a row, the same pattern is used to obtain the c terms as

$$c_1 = \frac{b_1 a_{n-3} - a_{n-1} b_2}{b_1}$$
$$c_2 = \frac{b_1 a_{n-5} - a_{n-1} b_3}{b_1} \tag{6.19}$$

This process is continued until it is terminated at the $(n+1)^{\text{th}}$ row, where n is the order of the system. The Routh criterion states that a necessary and sufficient condition for all roots of the characteristic equation to have negative real parts is that all elements of the first column of the array have the same sign. Furthermore, the number of changes of sign of the elements in the first column of the array is equal to the number of roots of the characteristic equation with positive real parts.

The conditions on the Hurwitz determinants are in fact equivalent to the conditions on the elements of the first column of the Routh array. It can be shown that

$$a_{n-1} = D_1,\, b_1 = D_2/D_1,\, c_1 = D_3/D_2, \ldots, g_1 = D_n/(D_{n-1}) \tag{6.20}$$

The first two equations of Eq. (6.20) can be proved by inspection. The proof of the remaining equations requires some computations. It therefore follows that if all the Hurwitz determinants are positive, the elements in the first column of the Routh array also have the same sign. We employ the Routh array because it is computationally simpler. Several examples are now presented to illustrate the criterion. A proof of the criterion is given by Routh (1955).

EXAMPLE 6.1

The block diagram of a hydraulic control system is shown in Fig. 3.19. The characteristic equation of this fourth-order system can be obtained from Eq. (3.61). Let the parameter values be such that the characteristic equation becomes

$$\det|s\mathbf{I} - \mathbf{A} + \mathbf{BK}| = s^4 + 3s^3 + s^2 + 9s + 12 = 0 \tag{6.21}$$

It is noted that since all coefficients have the same sign and none is zero, inspection of Eq. (6.21) does not reveal anything about its stability. The Routh array is

1	1	12	First row
3	9	0	Second row
b_1	b_2	0	
c_1	0	0	
d_1	0	0	Fifth row

where

$$b_1 = \frac{3(1) - 9(1)}{3} = -2 \qquad b_2 = \frac{3(12) - 0(1)}{3} = 12$$

$$c_1 = \frac{b_1(9) - b_2(3)}{b_1} = 27 \qquad d_1 = \frac{c_1 b_2 - 0(b_1)}{c_1} = 12$$

Hence, the Routh array can be written as

$$\begin{array}{rrr} 1 & 1 & 12 \\ 3 & 9 & 0 \\ -2 & 12 & 0 \\ 27 & 0 & 0 \\ 12 & 0 & 0 \end{array}$$

Since there are two changes in sign in the first column of the array, the characteristic equation, Eq. (6.21), has two roots with a positive real part; and the control system of Fig. 3.19 is unstable for these values of the parameters.

EXAMPLE 6.2

A mechanical, passive, feedback control system has been discussed in Example 3.2. Its block diagram is shown in Fig. 3.6 and the characteristic equation as given by Eq. (3.28) is

$$\det|s\mathbf{I} - \mathbf{A} + \mathbf{BK}| = s^2 + \frac{3c}{mL^2}s - \frac{3g}{2L} + \frac{3k}{m} = 0 \tag{6.22}$$

We determine the minimum value of the spring constant k for asymptotic stability. The Routh array for this second-order system can be constructed as

$$\begin{array}{ccl} 1 & 3k/m - 3g/2L & \text{First row} \\ 3c/mL^2 & 0 & \text{Second row} \\ 3k/m - 3g/2L & 0 & \text{Third row} \end{array}$$

Since all parameters have positive values, there will be no change in sign of the elements in the first column when $3k/m - 3g/2L > 0$. Hence, for asymptotic stability, the spring constant must be such that $k > mg/2L$. It is seen that the three elements in the first column are the three coefficients of the characteristic equation, Eq. (6.22). Hence, a necessary and sufficient condition for asymptotic stability of a second-order system is that the three coefficients of the characteristic equation be nonzero and have the same sign. The sufficient condition is valid only for the special case of a first- or second-order system.

EXAMPLE 6.3

We consider the dc electrical position-control system of Example 3.4. Its characteristic equation is obtained from Eq. (3.50) as

$$\det|s\mathbf{I} - \mathbf{A} + \mathbf{BK}| = s^3 + \left(\frac{1}{\tau_1} + \frac{1}{\tau_2}\right)s^2 + \left(\frac{1}{\tau_1\tau_2}\right)\left(1 + \frac{k_t k_b}{R_a c}\right)s + \left(\frac{k_t k_r k_a c_1}{\tau_1 \tau_2 R_a c}\right) = 0 \tag{6.23}$$

Let τ_1 and k_a be free parameters, $\tau_2 = 2$, and other parameter values be such that Eq. (6.23) becomes

$$s^3 + \left(\frac{1}{\tau_1} + \frac{1}{2}\right)s^2 + \frac{1}{\tau_1}s + \frac{k_a}{2\tau_1} = 0 \tag{6.24}$$

We determine the region of asymptotic stability in the parameter space of τ_1 and k_a, where $\tau_1 > 0$ and $k_a > 0$. Rewriting Eq. (6.24) as

$$2\tau_1 s^3 + (2 + \tau_1)s^2 + 2s + k_a = 0$$

we obtain the Routh array as

$$\begin{array}{cc} 2\tau_1 & 2 \\ 2 + \tau_1 & k_a \\ 2 - \left(\dfrac{2\tau_1 k_a}{2 + \tau_1}\right) & 0 \\ k_a & \end{array}$$

Since $\tau_1 > 0$ and $k_a > 0$, there will be no change in sign of the elements in the first column when

$$2 - \left(\frac{2\tau_1 k_a}{2 + \tau_1}\right) > 0 \qquad \text{i.e., } k_a < (1/\tau_1) + 1 \tag{6.25}$$

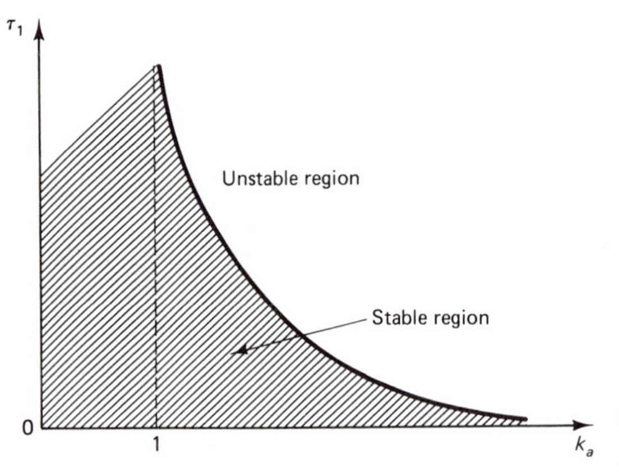

Figure 6.2 Region of asymptotic stability in parameter space.

The region of asymptotic stability in parameter space is shown by the shaded region of Fig. 6.2. It is noted that the determination of this region by actually evaluating the roots of the cubic equation would be a time-consuming task.

In high-order systems, it is convenient to employ a computer program to obtain the coefficients of the characteristic equation instead of calculating the determinant on the left-hand side of Eq. (6.16). If the determinant in Eq. (6.16) is expanded, it is found that the coefficient a_{n-1} is equal to the negative sum of the diagonal elements of $(\mathbf{A} - \mathbf{BK})$. The sum of the diagonal elements of a matrix is called the trace of the matrix.

Denoting the trace of $(\mathbf{A} - \mathbf{BK})^m$ by T_m, a useful recursive formula for the coefficients of the characteristic equation is obtained as

$$a_n = 1$$

$$a_{n-1} = -T_1$$

$$a_{n-2} = -\tfrac{1}{2}(a_{n-1}T_1 + T_2)$$

$$\vdots$$

$$a_0 = -\frac{1}{n}(a_1 T_1 + a_2 T_2 + \cdots + a_{n-1}T_{n-1} + T_n)$$

This result is known as Bôcher's formula (DeRusso, et al., 1965) and is useful for the computer determination of the coefficients of the characteristic equation.[3]

For example, let

$$(\mathbf{A} - \mathbf{BK}) = \begin{bmatrix} 0 & 1 \\ -3 & -4 \end{bmatrix}$$

Then,

$$a_1 = -T_1 = -(0 - 4) = 4$$

$$(\mathbf{A} - \mathbf{BK})^2 = \begin{bmatrix} -3 & -4 \\ 12 & 13 \end{bmatrix}$$

Thus,

$$T_2 = (-3 + 13) = 10$$

and

$$a_0 = -\tfrac{1}{2}[4(-4) + 10] = 3$$

The characteristic equation is obtained as

$$s^2 + 4s + 3 = 0$$

Special cases

The appearance of a zero in the first column before the termination of the array requires special considerations. In this case, if we use the algorithm indicated by Eqs. (6.18) and (6.19), the determination of the subsequent row will involve division by zero. This difficulty can be resolved by the following procedure. There are two special cases that must be treated separately.

Case 1. There is a zero in the first column and some of the other elements of that row are nonzero. We replace the zero element in the first column of the array by an arbitrary, small positive number ϵ, such that $0 < \epsilon \ll 1$ and complete the array.

EXAMPLE 6.4

Let the characteristic equation of a control system be given by

$$|s\mathbf{I} - \mathbf{A} + \mathbf{BK}| = s^5 + 2s^4 + 2s^3 + 4s^2 + 6s + 8 = 0 \qquad (6.26)$$

The Routh array is obtained as

1	2	6	First row
2	4	8	
$\epsilon \to 0$	2	0	
c_1	8	0	
d_1	0	0	
8	0	0	Sixth row

The zero in the first column and third row is replaced by ϵ, and we obtain

$$c_1 = 4 - 4/\epsilon$$

$$d_1 = 2 - \frac{8\epsilon}{c_1} = \frac{8\epsilon - 8 - 8\epsilon^2}{4\epsilon - 4}$$

Since ϵ is a small positive number, it follows that $c_1 < 0$ and $d_1 \to 2$ as $\epsilon \to 0$. There are two changes in sign of the elements of the first column and hence Eq. (6.26) has two roots with positive real parts, thus indicating instability.

Case 2. There is a zero in the first column and all other elements of that row are zero. This case occurs when the characteristic equation has roots that are symmetrically located about the origin of the s-plane. For example, the characteristic equation has factors such as $(s + \sigma)(s - \sigma)$, which yield real roots with opposite signs, and/or factors such as $(s + j\omega)(s - j\omega)$, which yield a pair of imaginary roots.

The equation formed by using the coefficients of the row just above the all-zero row is called the auxiliary equation. The symmetrical roots of the characteristic equation can be obtained from the solution of the auxiliary equation. We divide the characteristic equation by the auxiliary equation to obtain the reduced-order polynomial. Since some of the roots of the characteristic equation are known from the solution of the auxiliary equation, the Routh criterion is then applied to the reduced-order polynomial to investigate the locations of the remaining roots.

EXAMPLE 6.5

Let the parameter values of the hydraulic control system of Example 6.1 be such that the characteristic equation becomes

$$|s\mathbf{I} - \mathbf{A} + \mathbf{BK}| = s^4 + 4s^3 + 7s^2 + 16s + 12 = 0 \tag{6.27}$$

The Routh array is formed as

1	7	12	First row, s^4
4	16	0	Second row, s^3
3	12	0	Third row, s^2
0	0	0	Fourth row, s^1

The fourth row consists of all-zero elements and hence there is difficulty in obtaining the fifth row. The auxiliary equation is obtained from the row preceding the row of zeros, whch in this case is the third row. Hence the auxiliary equation becomes $3s^2 + 12 = 0$, i.e., $s^2 + 4 = 0$, which has imaginary roots $\pm j2$. To examine the remaining roots, we divide the characteristic equation, Eq. (6.27), by the auxiliary equation and obtain the reduced-order equation

$$\left(\frac{1}{s^2+4}\right)|s\mathbf{I} - \mathbf{A} + \mathbf{BK}| = s^2 + 4s + 3 = 0 \tag{6.28}$$

The Routh array for Eq. (6.28) is obtained as

$$\begin{matrix} 1 & 3 \\ 4 & 0 \\ 3 & 0 \end{matrix}$$

Hence, the roots of Eq. (6.28) have negative real parts. We have now established that the characteristic equation, Eq. (6.27), has two imaginary roots and its remaining roots have negative real parts. Hence, the control system is on the borderline of stability and is not asymptotically stable. It is BIBO unstable.

6.4 RELATIVE STABILITY BY THE ROUTH CRITERION

It is seen that the Routh criterion is a quick and easy method of stability analysis without actually determining the roots of the characteristic equation. When the application of the Routh criterion indicates asymptotic stability, it is desirable to further determine the distance of the closest root, or roots, to the left of the imaginary axis in the s-plane. In case a root is located too close to the left of the imaginary axis, then there is a danger that during service it might migrate to the right of the imaginary axis due to parameter variations, and thus give rise to instability. The performance specifications regarding stability usually demand a certain margin of stability as a safety factor.

The Routh criterion can be easily extended to study the relative stability. Let the application of Routh criterion to the characteristic equation indicate asymptotic stability. In that case, we wish to determine if the characteristic equation has any roots to the right of a line parallel to the imaginary axis in the complex s-plane and passing through $s = -\sigma_1$ as shown in Fig. 6.3(a).

A new characteristic equation is created by shifting the imaginary axis of the s-plane to the left by $-\sigma_1$. For this purpose, we replace s by $(p - \sigma_1)$ in the characteristic equation and obtain a new characteristic equation as a polynomial in p. The line passing through $s = -\sigma_1$ in the s-plane now becomes the imaginary axis of the new p-plane as shown in Fig. 6.3(b). We now apply the Routh criterion to the new characteristic equation in p. The number of roots lying to the right of the vertical line passing through $s = -\sigma_1$ is equal to the number of roots lying to the right of the imaginary axis in the p-plane.

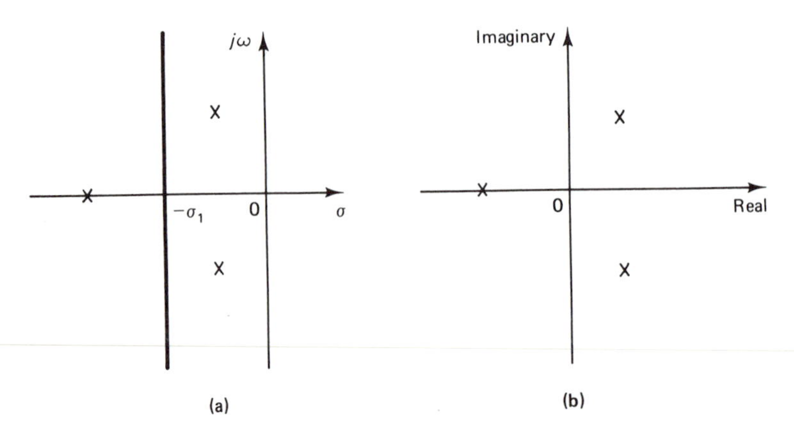

Figure 6.3 Root locations in (a) the s-plane, and (b) the p-plane.

EXAMPLE 6.6

Let the parameter values of the electrical position-control system of Example 3.4 be such that its characteristic equation becomes

$$s^3 + 5s^2 + 11s + 15 = 0 \tag{6.29}$$

We obtain the Routh array as

$$\begin{array}{cc} 1 & 11 \\ 5 & 15 \\ 8 & 0 \\ 15 & 0 \end{array}$$

All elements of the first column have the same sign and hence the control system is asymptotically stable. We now wish to determine if the characteristic equation has any roots to the right of a line parallel to the imaginary axis and passing through $s = -2$. We replace s by $(p - 2)$ in Eq. (6.29) and obtain a new characteristic equation in p as

$$(p-2)^3 + 5(p-2)^2 + 11(p-2) + 15 = 0$$

i.e.,

$$p^3 - p^2 + 3p + 5 = 0 \tag{6.30}$$

The coefficients of Eq. (6.30) do not have the same sign and hence it is unstable. The Routh array is obtained as

$$\begin{array}{cc} 1 & 3 \\ -1 & 5 \\ 8 & \\ 5 & \end{array}$$

There are two changes in sign of the elements in the first column, and hence Eq. (6.30) has two roots located to the right of the imaginary axis of the p-plane. This implies that Eq. (6.29) has two roots located to the right of the line passing through $s = -2$ in the s-plane. We can show by determining the roots of Eq. (6.29) that it can be expressed as

$$(s+3)(s+1+j2)(s+1-j2) = 0 \tag{6.31}$$

and its roots are given by $s = -3, -1 \pm j2$. After replacing s by $(p-2)$, we obtain

$$(p-2+3)(p-2+1+j2)(p-2+1-j2) = 0 \tag{6.32}$$

This equation is identical to Eq. (6.30), and its roots are given by $p = -1$, $1 \pm j2$. It is thus seen that Eq. (6.30) has two roots with positive real parts, and we conclude that Eq. (6.29) has two roots located to the right of the line passing through $s = -2$ in the s-plane.

6.5 THE ROOT-LOCUS METHOD

The transient response and stability of a control system is dependent on the location of the roots of the closed-loop characteristic equation. It is therefore important to investigate how the roots move in the complex s-plane as a parameter is varied. For this purpose, a graphical procedure called the root-locus method was introduced by Evans in 1948 (Evans, 1954) and is frequently used in practice. The method also provides a measure of the sensitivity of the roots of the characteristic equation to variations in the parameter under consideration. Very often, only a sketch of the root loci is sufficient to make design decisions. The root-locus method can be used for both single-input, single-output and multivariable systems. The method can also be extended to the case where more than a single parameter varies by using root-locus contours.

The root-locus method as originally developed by Evans refers to a single-input, single-output, closed-loop control system whose mathematical model is expressed as shown in the block diagram of Fig. 6.4. Its open-loop transfer function is given by

$$G_o(s) = \frac{k(s+z_1)(s+z_2)\cdots(s+z_m)}{(s+p_1)(s+p_2)\cdots(s+p_n)} \tag{6.33}$$

where k is the gain of $G_o(s)$, and $n \geq m$ for causal systems.

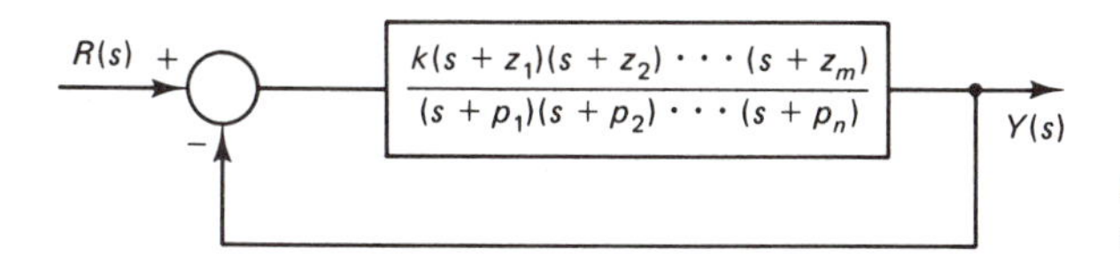

Figure 6.4 Block diagram of a control system.

Here, $-p_1, -p_2, \ldots, -p_n$ are called the poles of $G_o(s)$ because when s takes on any of these values, the denominator of $G_o(s)$ becomes zero. When s takes on any of the values $-z_1, -z_2, \ldots, -z_m$, the numerator of $G_o(s)$ becomes zero and for this reason they are called the zeros of $G_o(s)$. The poles and zeros of $G_o(s)$ may be real or occur in complex conjugate pairs. The closed-loop transfer function can be expressed as

$$\begin{aligned} Y(s) &= \left(\frac{G_o(s)}{1 + G_o(s)}\right) R(s) \\ &= \frac{k(s + z_1)(s + z_2) \cdots (s + z_m)}{(s + p_1)(s + p_2) \cdots (s + p_n) + k(s + z_1)(s + z_2) \cdots (s + z_m)} R(s) \end{aligned} \tag{6.34}$$

The characteristic equation is obtained from Eq. (6.34) as

$$(s + p_1)(s + p_2) \cdots (s + p_n) + k(s + z_1)(s + z_2) \cdots (s + z_m) = 0 \tag{6.35}$$

The root loci are the paths of the roots of the characteristic equation, Eq. (6.35), in the complex s-plane as k is varied from zero to infinity, that is, $0 \le k < \infty$. When k takes on negative values, that is, $-\infty < k \le 0$, the loci are referred to as complementary root loci. In this text, we restrict k such that $0 \le k < \infty$, and refer to Kuo (1982) for the complementary root locus. The characteristic equation, Eq. (6.35), can be expressed as

$$\frac{k(s + z_1)(s + z_2) \cdots (s + z_m)}{(s + p_1)(s + p_2) \cdots (s + p_n)} = -1 \tag{6.36}$$

i.e.,

$$G_o(s) = -1$$

Since the roots in general may be complex, Eq. (6.36) in fact provides two equations that must be satisfied for a value of s to be a root of the characteristic equation. The real and imaginary parts on both sides of Eq. (6.36) must be equal to each other, separately. In polar form, we can separately equate the magnitudes and phase angles on both sides of Eq. (6.36). After noting that $k \ge 0$, we get the following two conditions.

Angle condition:

$$\begin{gathered} \measuredangle(s + z_1) + \cdots + \measuredangle(s + z_m) - \measuredangle(s + p_1) - \cdots - \measuredangle(s + p_n) = \pm i\pi \\ i = 1, 3, 5, \ldots \end{gathered} \tag{6.37}$$

Magnitude condition:

$$\frac{|s + z_1|\,|s + z_2| \cdots |s + z_m|}{|s + p_1|\,|s + p_2| \cdots |s + p_n|} = \frac{1}{|k|} = \frac{1}{k} \tag{6.38}$$

The angle condition does not depend on the value of k, and the root locus is constructed by finding all points in the s-plane that satisfy Eq. (6.37). Then the values of k along the loci for a given root location are determined from the magnitude condition, Eq. (6.38). The construction of the root locus is basically a graphical procedure that uses some rules of construction arrived at analytically.

The root-locus method was originally developed for determining the loci of the roots of the characteristic equation of the single-input, single-output control system of Fig. 6.4 as k is varied from zero to infinity. The method is readily applicable to multivariable systems and for drawing the root locus when a parameter other than the gain is varied from zero to infinity. The characteristic equation of a multivariable system is represented as given in Eq. (6.16) by

$$\det|s\mathbf{I} - \mathbf{A} + \mathbf{BK}| = s^n + a_{n-1}s^{n-1} + \cdots + a_2s^2 + a_1s + a_0 = 0 \tag{6.39}$$

where we have assumed without loss of generality that $a_n = 1$. Since $a_n \neq 0$, we can always divide the equation throughout by a_n and represent it in the form of Eq. (6.39). Suppose that all parameters except a_2 are fixed at their nominal values and the root locus is required as a_2 is varied from zero to infinity. We rewrite Eq. (6.39) in the form

$$\frac{a_2s^2}{s^n + a_{n-1}s^{n-1} + \cdots + a_3s^3 + a_1s + a_0} + 1 = 0 \tag{6.40}$$

Factoring the denominator of Eq. (6.40) by using a computer program if necessary, and letting $a_2 = k$, where $0 \leq k < \infty$, we get

$$\frac{ks^2}{(s+p_1)(s+p_2)\cdots(s+p_n)} = -1 \tag{6.41}$$

This equation is in the standard form of Eq. (6.36) with $z_1 = 0$ and $z_2 = 0$ as the two zeros and $-p_1, \ldots, -p_n$ as the poles. In case the parameter to be varied does not appear solely as a coefficient a_i of the characteristic equation, Eq. (6.39), the parameter needs to be isolated. The general procedure is to rearrange the characteristic equation, Eq. (6.39), if necessary, so that the parameter to be varied appears as a multiplying factor in the form

$$D(s) + kN(s) = 0$$

i.e.,

$$kN(s)/D(s) = -1 \tag{6.42}$$

where $D(s)$ and $N(s)$ are polynomials in s with constant coefficients. Then factoring $D(s)$ and $N(s)$, if necessary, we can express Eq. (6.42) in the standard form of Eq. (6.36). With $P(s) = kN(s)/D(s)$, Eq. (6.42) may be expressed as

$$P(s) = -1 \tag{6.43}$$

where $P(s)$ now plays the role of the transfer function $G_o(s)$ of Eq. (6.36). In the angle and magnitude conditions of Eqs. (6.37) and (6.38), we now interpret $-z_1, \ldots, -z_m$ as the zeros of $P(s)$ and $-p_1, \ldots, -p_n$ as the poles of $P(s)$.

EXAMPLE 6.7

Let the characteristic equation of a closed-loop control system be given by

$$\det|s\mathbf{I} - \mathbf{A} + \mathbf{BK}| = s^4 + (8 + 2\alpha)s^3 + (27 + 5\alpha)s^2 + 44s + 24 = 0 \tag{6.44}$$

The root locus of this characteristic equation is to be drawn as the parameter

α is varied from zero to infinity. Rearranging Eq. (6.44), we get

$$s^4 + 8s^3 + 27s^2 + 44s + 24 + \alpha s^2(2s + 5) = 0$$

This equation can be expressed as

$$\frac{2\alpha s^2(s + 2.5)}{s^4 + 8s^3 + 27s^2 + 44s + 24} + 1 = 0$$

Letting $k = 2\alpha$ and factoring the denominator of $P(s)$, we obtain

$$\frac{ks^2(s + 2.5)}{(s + 1)(s + 3)(s + 2 - j2)(s + 2 + j2)} = -1 \tag{6.45}$$

Here, $P(s)$ has three zeros at $s = 0$, $s = 0$, and $s = -2.5$, and four poles. The root locus can now be drawn from the angle condition, Eq. (6.37), and the values of k, and hence those of α, along the loci can be obtained from the magnitude condition, Eq. (6.38).

6.6 PROCEDURE FOR CONSTRUCTION OF THE ROOT LOCUS

The root locus is constructed by using certain rules that are now developed in the following. As discussed earlier, the characteristic equation of a control system is expressed in the form

$$P(s) + 1 = 0 \tag{6.46}$$

where

$$P(s) = \frac{kN(s)}{D(s)} = \frac{k(s + z_1)(s + z_2) \cdots (s + z_m)}{(s + p_1)(s + p_2) \cdots (s + p_n)} \tag{6.47}$$

Here, $N(s)$ and $D(s)$ are polynomials in s of order m and n, respectively. It is assumed that $n \geq m$. The values $-z_1, -z_2, \ldots, -z_m$ are the zeros of $P(s)$ and $-p_1, -p_2, \ldots, -p_n$ are its poles. We are interested in drawing the loci of the roots of the characteristic equation given by Eq. (6.35) as k varies from zero to infinity.

Rule 1. Number of Loci. The number of branches of the root loci is equal to the number of the roots given by the order of the characteristic equation. Since it is assumed that $n \geq m$, the characteristic equation has n roots and hence there are n loci.

EXAMPLE 6.8

For the system of Example 6.7, we note from Eq. (6.45) that $n = 4$ and $m = 3$. Hence, the characteristic equation is of fourth order and there are four branches of the root loci.

Rule 2. Origin of Loci. The loci originate when $k = 0$ at the poles of $P(s)$. This rule becomes obvious from the characteristic equation, Eq. (6.35). When $k = 0$, the roots of the characteristic equation are $-p_1, -p_2, \ldots, -p_n$.

EXAMPLE 6.9

Consider the system of Example 6.7. The characteristic equation is obtained from Eq. (6.45) as

$$(s+1)(s+3)(s+2-j2)(s+2+j2)+ks^2(s+2.5)=0 \qquad (6.48)$$

The loci originate when $k = 0$ at the poles of $P(s)$, namely, at -1, -3, $-2+j2$, and $-2-j2$.

Rule 3. Termination of Loci. When $k \to \infty$, m loci terminate at the m zeros of $P(s)$ and $(n-m)$ loci terminate at ∞ along asymptotes. Referring to the magnitude condition, Eq. (6.38), we note that when $k \to \infty$, the left-hand side approaches zero. This corresponds to s approaching $-z_i$ for $i = 1, 2, \ldots, m$. When $n > m$, we note that the left-hand side of Eq. (6.38) also approaches zero when $s \to \infty$. We may then consider $P(s)$ as having m finite zeros and $(n-m)$ zeros at infinity and restate this rule as follows. When $k \to \infty$, the loci terminate at the zeros of $P(s)$.

EXAMPLE 6.10

We again consider the system of Example 6.7. When $k \to \infty$, three loci terminate at the finite zeros of $P(s)$, which are 0, 0, and -2.5, and one locus terminates at infinity.

Rule 4. Symmetry of the Root Locus. The root locus is symmetric about the real axis of the complex s-plane. Since the values of the parameters of the characteristic equation, Eq. (6.39), are real, complex roots always occur in conjugate pairs. Hence, the root locus is symmetric about the real axis.

Rule 5. Location of Locus on the Real Axis. The root locus exists on a section of the real axis only if the total number of real poles and zeros of $P(s)$ to the right of that section is odd. The complex poles and zeros of $P(s)$ do not affect the existence of root loci on the real axis. This rule is shown to be true from an examination of the angle condition, Eq. (6.37). In Fig. 6.5, let filled-in circles denote the locations of the poles and open circles the locations of the zeros of $P(s)$.

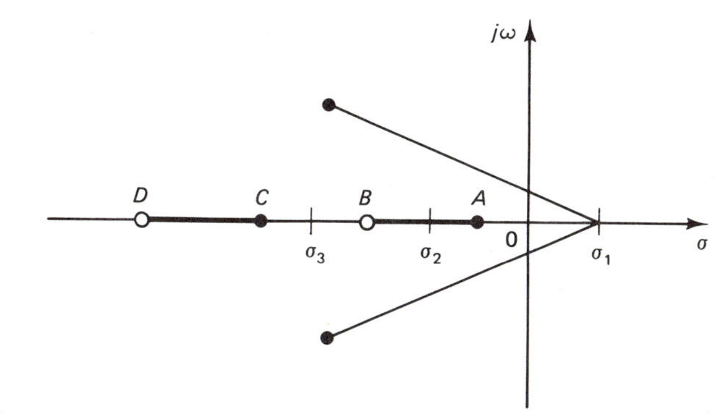

Figure 6.5 Locations of poles and zeros in the s-plane.

We wish to investigate if a value of $s = \sigma_1$ on the real axis is a root. It is seen that the angles of the vectors drawn from the pair of complex conjugate poles to σ_1 sum up to zero. This is also true for any pair of complex conjugate zeros when s lies on the real axis. Hence, when s is on the real axis, the only contribution to the angle condition, Eq. (6.37), is from the real poles and zeros of $P(s)$.

When $s = \sigma_1$, in Fig. 6.5, we note that each of the vectors drawn from the real poles and zeros of $P(s)$ to σ_1 contributes zero degrees. Hence, the angle condition, Eq. (6.37), is not satisfied and there is no locus to the right of A on the real axis. Suppose $s = \sigma_2$ in Fig. 6.5 is a root. Then the vector drawn from the real pole to the right of σ_2 to σ_2 contributes an angle of 180°. Each of the other vectors drawn from the real poles and zeros to the left of σ_2 contributes zero degrees. Hence, the angle condition is satisfied and there is a locus between A and B on the real axis.

Suppose $s = \sigma_3$ is a root. Both the pole at A and the zero at B contribute 180°. Both the pole at C and zero at D contribute 0°. The left-hand side of the angle condition, Eq. (6.37), becomes zero and hence it is not satisfied; and, there is no locus between B and C on the real axis, whereas there is a locus between C and D. These observations show that for a real value of $s = \sigma$ to be on the root locus, the total number of real poles and zeros to the right of σ must be odd.

Rule 6. Angles of Asymptotes. As $s \to \infty$, the $(n - m)$ loci that do not terminate at the finite zeros of $P(s)$ approach infinity along asymptotes. The angles that the asymptotes make with the real axis are given by

$$\measuredangle s = \frac{\pm i\pi}{n - m} \qquad i = 1, 3, 5, \ldots \tag{6.49}$$

As $s \to \infty$, we can let $\measuredangle(s + z_i) \approx \measuredangle s$ for $i = 1, 2, \ldots, m$ and $\measuredangle(s + p_i) \approx \measuredangle s$ for $i = 1, 2, \ldots, n$. Hence, from the angle condition, Eq. (6.37), as $s \to \infty$, we obtain

$$m(\measuredangle s) - n(\measuredangle s) = \pm i\pi$$

or

$$\measuredangle s = \frac{\pm i\pi}{n - m} \qquad i = 1, 3, 5, \ldots$$

EXAMPLE 6.11

We refer to the system of Example 6.7. Since $n = 4$ and $m = 3$, the angle that the asymptote makes with the real axis is

$$\measuredangle s = \frac{\pm i180°}{4 - 3} = 180°$$

We obtain only one distinct value of the angle and the asymptote lies on the real axis. This is in conformity with the results that as $k \to \infty$, three loci terminate at the zeros and one locus tends to infinity along an asymptote.

Rule 7. Intersection of Asymptotes. The point where the asymptotes intersect the real axis is given by

$$\sigma_c = \frac{\sum_{i=1}^{n} - p_i - \sum_{i=1}^{m} - z_i}{n - m} = \frac{\sum \text{poles} - \sum \text{zeros}}{n - m} \tag{6.50}$$

where the numerator is the sum of the finite poles minus the sum of the finite zeros of $P(s)$. Consider the case where $P(s)$ has three poles at $-p_1$, $-p_2$, and $-p_3$, and

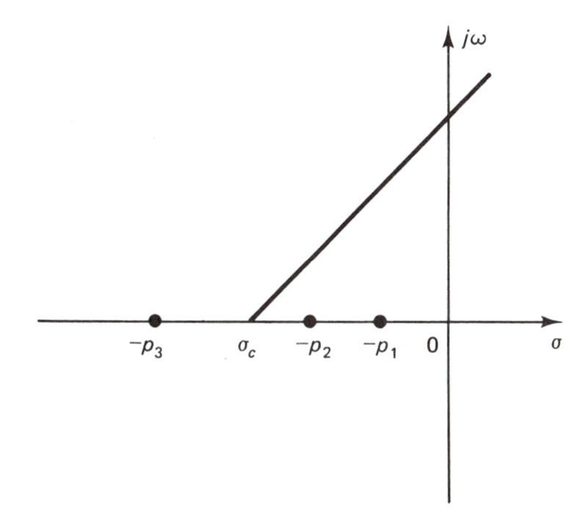

Figure 6.6 Intersection of the asymptote with the real axis.

no finite zeros. Using Eq. (6.50), the point σ_c in Fig. 6.6 where the asymptote cuts the real axis is given by

$$\sigma_c = \frac{-p_1 - p_2 - p_3}{3} \tag{6.51}$$

Thus, σ_c is the centroid or the average value of the poles from the imaginary axis. Now, multiplying the factors in Eq. (6.47), we obtain

$$P(s) = \frac{k(s^m + b_{m-1}s^{m-1} + \cdots + b_0)}{s^n + a_{n-1}s^{n-1} + \cdots + a_0} \tag{6.52}$$

where

$$-a_{n-1} = -p_1 - p_2 - \cdots - p_n \tag{6.53}$$

and

$$-b_{m-1} = -z_1 - z_2 - \cdots - z_m \tag{6.54}$$

Dividing the numerator by the denominator of Eq. (6.52), we get

$$P(s) = \frac{k}{s^{n-m} + (a_{n-1} - b_{m-1})s^{n-m-1} + \cdots}$$

As $s \to \infty$, we neglect all but the first two terms and obtain

$$P(s) \approx \frac{k}{s^{n-m-1}(s + a_{n-1} - b_{m-1})} \tag{6.55}$$

This expression has $(n - m - 1)$ poles at the origin and one pole at $-a_{n-1} + b_{m-1}$. Thus, as $s \to \infty$, $P(s)$ is reduced to a form containing only poles with no finite zeros. The centroid is obtained from Eq. (6.55) as

$$\sigma_c = \frac{-a_{n-1} + b_{m-1}}{n - m}$$

and from Eqs. (6.53) and (6.54), it follows that

$$\sigma_c = \frac{\sum_{i=1}^{n} - p_i - \sum_{i-1}^{m} - z_i}{n - m}$$

EXAMPLE 6.12

For the system of Example 6.7, using Eqs. (6.45) and (6.50), we get

$$\sigma_c = \frac{(-1 - 3 - 2 + j2 - 2 - j2) - (-0 - 0 - 2.5)}{4 - 3}$$

$$= -5.5$$

Rule 8. Breakaway and Break-in Points. When two poles of $P(s)$ on the real axis are connected by a locus, the loci approach each other as k increases until they meet and then depart from the real axis at a point called the breakaway point as shown in Fig. 6.7. This figure also shows that the locus can also enter the real axis

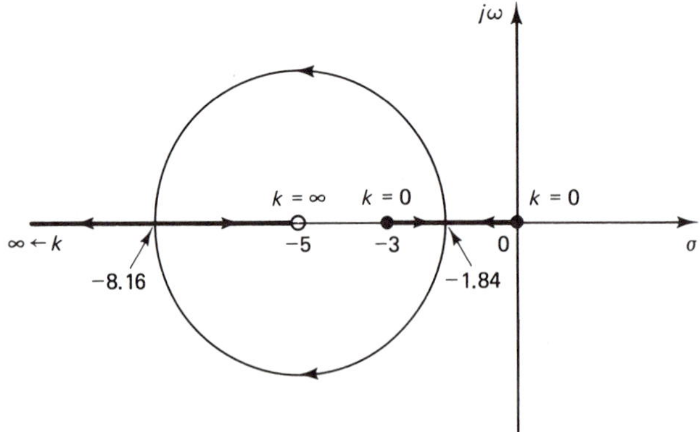

Figure 6.7 Breakaway and break-in points.

at a point that is called the break-in point. It is seen from Fig. 6.7 that at a breakaway point, k attains a local maximum value for the two roots to lie on the real axis. When k exceeds this local maximum value, the two roots become complex conjugate. Similarly, for a break-in point, k attains a local minimum for the two roots to lie on the real axis.

The locations of the breakaway and break-in points are determined from the condition that $dk/ds = 0$. Expressing Eq. (6.43) in the form of Eq. (6.42), this condition is expressed as

$$\frac{dk}{ds} = -\frac{d}{ds}\frac{D(s)}{N(s)} = 0 \tag{6.56}$$

or

$$-N(s)\frac{d}{ds}D(s) + D(s)\frac{d}{ds}N(s) = 0 \tag{6.57}$$

We now have to determine the roots of this polynomial equation to locate the breakaway and break-in points. It is possible that there exist multiple breakaway or break-in points. The breakaway and break-in points are obtained from the real roots of Eq. (6.57) and the knowledge of the location of loci on the real axis.

EXAMPLE 6.13

Determine the breakaway and break-in points when $P(s)$ is given by

$$P(s) = \frac{k(s+5)}{s(s+3)}$$

This $P(s)$ has one zero at $s = -5$ and two poles at $s = 0$ and $s = -3$. The location of the locus on the real axis is obtained from Rule 5 and is shown in Fig. 6.7. From Eq. (6.56), we get

$$\frac{dk}{ds} = -\frac{d}{ds}\frac{s(s+3)}{s+5} = 0$$

or

$$-\frac{(s+5)(2s+3) - s(s+3)}{(s+5)^2} = 0$$

Setting the numerator of this equation to zero,

$$s^2 + 10s + 15 = 0$$

Its roots are $s = -1.84$ and -8.16. From the knowledge of the location of the locus on the real axis in Fig. 6.7, $s = -1.84$ is a breakaway point and $s = -8.16$ is a break-in point.

Rule 9. Angles of Departure and Arrival. The angle of departure of the locus at $k = 0$ from a complex pole and the angle of arrival of the locus at $k = \infty$ at a complex zero are determined from the application of the angle condition, Eq. (6.37). We let s be a point on the root locus infinitesimally close to the complex pole or zero under consideration and apply the angle condition. This procedure is illustrated later in Example 6.14.

Summary of procedure

The root locus can be drawn by employing the rules given in the preceding. In addition, use can be made of a device called the Spirule (available from The Spirule

Company, 9728 El Venado Drive, Whittier, CA 90603). It is a protractor with a hinged arm that facilitates the addition of angles. A trial value of s is selected and tested to determine if it is on the root locus by summing the angles from the poles and zeros with the Spirule. When the angle condition, Eq. (6.37), is satisfied, the trial point is on the root locus. Also, computer programs are available (Melsa and Jones, 1973) that can generate and plot the root locus.* We now summarize the steps used in drawing the root locus and then illustrate their use by examples.

1. Obtain the characteristic equation such that the parameter k to be varied appears as a multiplying factor in the form $1 + P(s) = 1 + kN(s)/D(s) = 0$. Obtain the angle and magnitude conditions.
2. Locate the poles and zeros of $P(s)$ in the s-plane and mark them with closed circles and open circles, respectively.
3. Determine the number of loci from Rule 1.
4. Obtain the location of the root locus on the real axis from Rule 5.
5. Determine the angles of the asymptotes from Rule 6.
6. Obtain the intersection of the asymptotes with the real axis from Rule 7.
7. Find the breakaway and break-in points (if any) from Rule 8.
8. Obtain the angles of departure from complex poles and the angles of arrival at the complex zeros (if any) from Rule 9.
9. For any selected value of the root on the root locus, obtain the value of the parameter k from the magnitude condition.
10. If the locus crosses the imaginary axis, determine the corresponding value of k. The Routh criterion may be employed for this purpose.

It is noted that Steps 7 and 8 may be omitted when they are not necessary. For example, if $P(s)$ has no complex poles or zeros, Step 8 is omitted.

EXAMPLE 6.14

A pneumatic temperature-control system has been studied in Example 3.7 and its block diagram is given in Fig. 3.25. With the disturbance T_a at zero and $c_1 = c_2$, the block diagram is shown in Fig. 6.8(a), where the gain $k_a = c_1 k_1 k_2 c_5 k_g$. Let the parameter values, identified from experimental frequency response if necessary, be given by $\tau_1 = 1/6$, $\tau = 1/2$, $\omega_n = 5$ rad/s, and $\zeta = 0.8$. The block diagram is now rearranged as shown in Fig. 6.8(b) so that the poles and zero of the open-loop transfer function become obvious. The parameter k to be varied is defined as $k = k_a \omega_n^2 \tau_1/\tau$.

1. The characteristic equation of the closed-loop control system is given by

$$(s + 2)(s^2 + 8s + 25) + k(s + 6) = 0 \tag{6.58}$$

which can be expressed in the form

$$1 + \frac{k(s + 6)}{(s + 2)(s^2 + 8s + 25)} = 0 \tag{6.59}$$

*See also references listed in Appendix D.

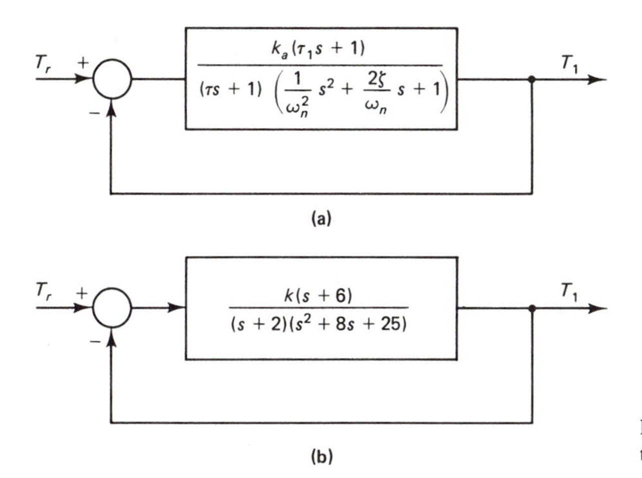

Figure 6.8 Block diagrams of a temperature-control system.

The angle condition is stated as

$$\measuredangle(s+6) - \measuredangle(s+2) - \measuredangle(s+4-j3) - \measuredangle(s+4+j3) = \pm i\pi$$

$$i = 1, 3, 5, \ldots \quad (6.60)$$

and the magnitude condition as

$$\frac{|s+6|}{|s+2||s+4-j3||s+4+j3|} = \frac{1}{k} \quad (6.61)$$

2. From Eq. (6.59), we note that there are three poles at -2, $-4+j3$, $-4-j3$, and one zero at -6. These are marked by closed circles and open circles, respectively, in Fig. 6.9.

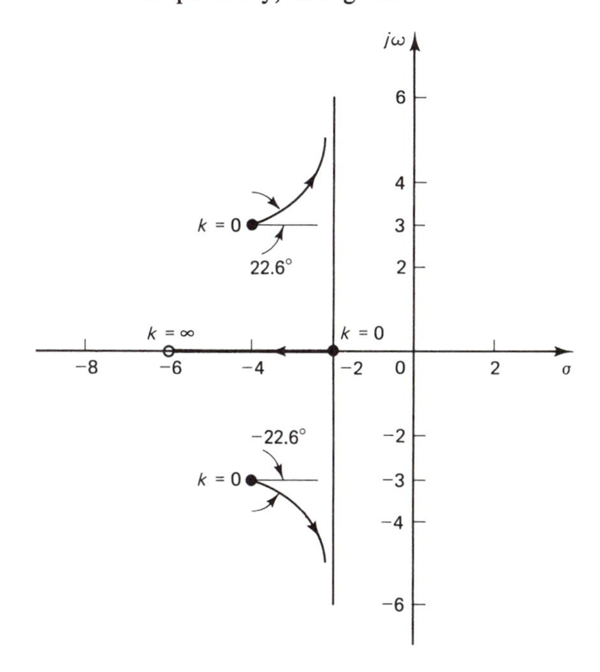

Figure 6.9 The root-locus plot for the system of Fig. 6.8(b).

3. From Rule 1, we observe that there are three separate branches of the root loci. When $k = 0$, the loci originate at the poles (Rule 2). When $k = \infty$, one locus terminates at the finite zero at -6 and two loci terminate at ∞ along asymptotes (Rule 3).
4. Employing Rule 5, we see that there is a root locus on the real axis only between $s = -2$ and $s = -6$.
5. The angles of asymptotes are obtained from Rule 6. From Eq. (6.49), we get

$$\measuredangle s = \frac{\pm i\pi}{3 - 1} = +90° \text{ and } -90° \tag{6.62}$$

We obtain only two distinct angles for the two asymptotes.

6. The intersection of the asymptotes with the real axis is obtained from Rule 7. From Eq. (6.50), we get

$$\sigma_c = \frac{(-2 - 4 + j3 - 4 - j3) - (-6)}{3 - 1} = -2 \tag{6.63}$$

7. There are no breakaway or break-in points.
8. The angle of departure from the complex pole at $-4 + j3$ is obtained from Rule 9. We let s be a point on the root locus infinitesimally close

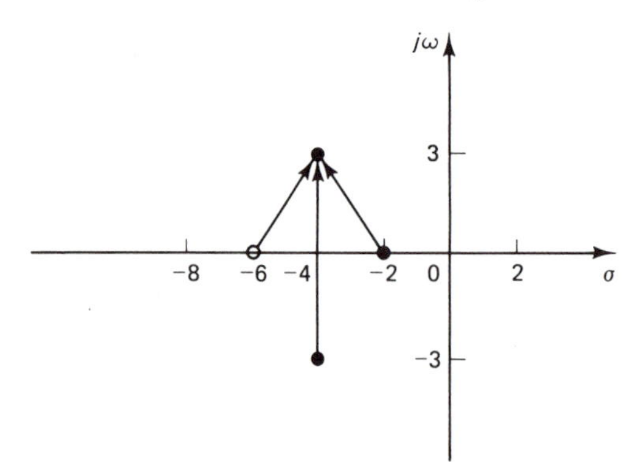

Figure 6.10 The angle of departure from a complex pole.

to $-4 + j3$ as shown in Fig. 6.10. We draw vectors to this point from the poles and zero. We obtain

$$\measuredangle(s + 6) = \tan^{-1} 3/2 = 56.3°$$

$$\measuredangle(s + 2) = \pi - \tan^{-1} 3/2 = 123.7°$$

$$\measuredangle(s + 4 + j3) = 90°$$

Substituting these values in the angle condition, Eq. (6.60), we get

$$56.3° - 123.7° - \measuredangle(s + 4 - j3) - 90° = \pm i180° \qquad i = 1, 3, 5, \ldots$$

The solution is given by $\measuredangle(s + 4 - j3) = 22.6°$ and is shown in Fig. 6.9. The angle of departure of the locus from the complex conjugate pole at $-4 - j3$ is $-22.6°$.

9. It is seen from Fig. 6.9 that $s = -3$ is a root of the characteristic equation for a value of k that can be determined from the magnitude condition. Letting $s = -3$ in Eq. (6.61), we obtain $k = 10/3$.
10. We observe from Fig. 6.9 that the root locus does not cross the imaginary axis for $0 \le k < \infty$. Hence, the control system of Fig. 6.8(b) is asymptotically stable for any positive value of k.

Suppose that a value of k is to be selected such that the complex conjugate roots of the characteristic equation exhibit a damping ratio of 0.707. As shown by Fig. 4.10, we draw a line at an angle $\theta = \sin^{-1} 0.707 = 45°$ to the imaginary axis and find the value of k where it intersects the locus.

EXAMPLE 6.15

An electrical position-control system was studied in Example 3.4 and its block diagram is shown in Fig. 3.15. We set the disturbance torque T_d to zero and $c_1 = c_2$. Let the parameter values be such that its block diagram can be

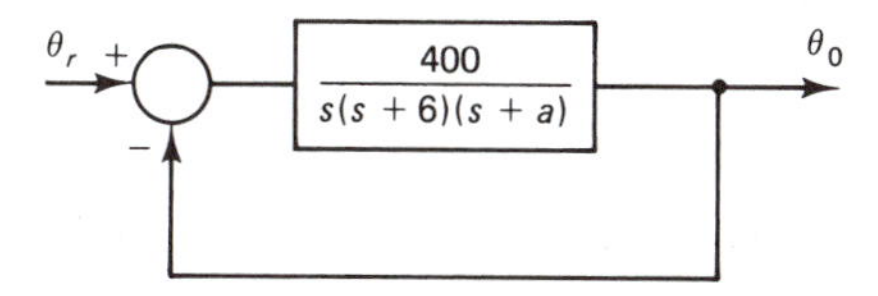

Figure 6.11 Block diagram of a position-control system.

represented by Fig. 6.11, where a is a free parameter. We draw the root locus for this system as the parameter a is varied from zero to infinity.

1. The characteristic equation of the closed-loop control system is given by

$$s(s+6)(s+a) + 400 = 0 \tag{6.64}$$

which can be rearranged as

$$-(s+a) = \frac{400}{s(s+6)}$$

$$-a = \frac{400}{s(s+6)} + s = \frac{s^3 + 6s^2 + 400}{s(s+6)}$$

$$1 + \frac{as(s+6)}{s^3 + 6s^2 + 400} = 0$$

or

$$1 + \frac{as(s+6)}{(s+10)(s-2-j6)(s-2+j6)} = 0 \tag{6.65}$$

The parameter a now appears as a multiplying factor in the desired form.

The angle condition is stated as

$$\measuredangle s + \measuredangle(s+6) - \measuredangle(s+10) - \measuredangle(s-2-j6) - \measuredangle(s-2+j6) = \pm i\pi$$
$$i = 1, 3, 5, \ldots \qquad (6.66)$$

and the magnitude condition as

$$\frac{|s||s+6|}{|s+10||s-2-j6||s-2+j6|} = \frac{1}{a} \qquad (6.67)$$

2. From Eq. (6.67), we note that there are three poles at -10, $2+j6$, $2-j6$, and two zeros at 0 and -6. These are marked by closed circles and open circles in Fig. 6.12.
3. From Rule 1, we observe that there are three separate branches of the root loci. When $a = 0$, the loci originate at the poles (Rule 2). When $a = \infty$, two loci terminate at the finite zeros at 0 and -6 and one locus terminates at ∞ along an asymptote (Rule 3).
4. From Rule 5, we observe that there are loci on the real axis between 0 and -6 and between -10 and $-\infty$.
5. The angle of asymptote is obtained from Rule 6. From Eq. (6.49), we get

$$\measuredangle s = \frac{\pm i\pi}{3-2} = 180^\circ \qquad (6.68)$$

We obtain only one distinct angle for the single asymptote.
6. The asymptote is the real axis and hence the point of intersection of the asymptote with the real axis is not meaningful.
7. The break-in point where the loci from the two complex conjugate poles enter the real axis is obtained from Rule 8. From Eq. (6.57), we get

$$-s(s+6)\frac{d}{ds}(s^3+6s^2+400) + (s^3+6s^2+400)\frac{d}{ds}[s(s+6)] = 0$$

or

$$2s^4 + 15s^3 + 6s^2 + 728s + 2400 = 0 \qquad (6.69)$$

This quartic polynomial has four roots, but from Fig. 6.12, the only admissible root is the one between 0 and -6. This root is obtained as -3.03, which is the break-in point.
8. The angle of departure from the complex pole at $2+j6$ is obtained from Rule 9. We let s be a point on the root locus infinitesimally close to $2+j6$. We obtain

$$\measuredangle s = \tan^{-1} 6/2 = 71.57^\circ$$

$$\measuredangle(s+6) = \tan^{-1} 6/8 = 36.87^\circ$$

$$\measuredangle(s+10) = \tan^{-1} 6/12 = 26.57^\circ$$

$$\measuredangle(s-2+j6) = 90^\circ$$

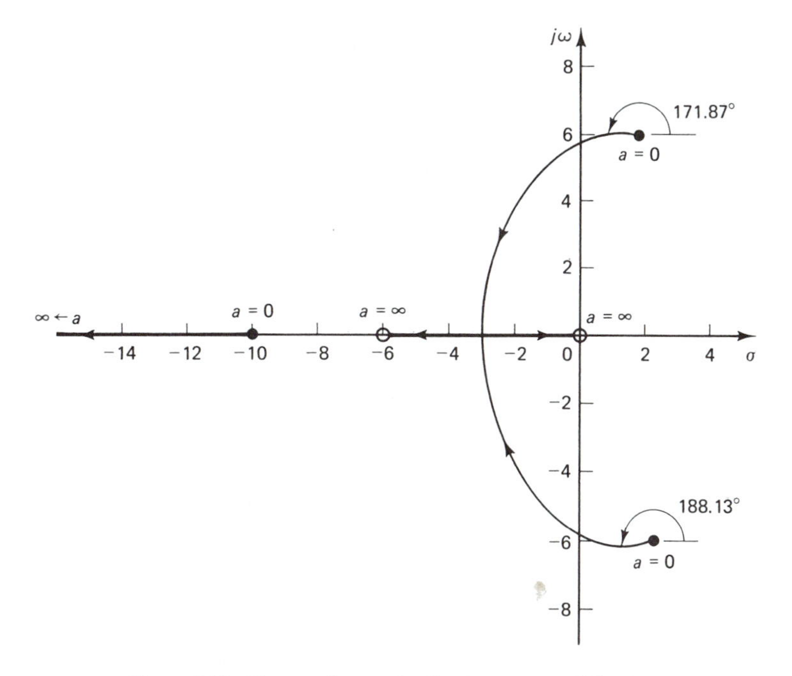

Figure 6.12 The root-locus plot for the system of Fig. 6.11.

Substituting these values in the angle condition, Eq. (6.66), we get

$$71.57° + 36.87° - 26.57° - \measuredangle(s - 2 - j6) - 90° = \pm i\pi \qquad i = 1, 3, 5, \ldots$$

Hence, $\measuredangle(s - 2 - j6) = -188.13°$ or $171.87°$, which is shown in Fig. 6.12.

9. For any value of the root selected from the root locus, the value of a can be obtained from the magnitude condition, Eq. (6.67).
10. The value of a at which the locus crosses the imaginary axis can be obtained from the Routh criterion. The characteristic equation, Eq. (6.64), is obtained as

$$s^3 + (6 + a)s^2 + 6as + 400 = 0$$

The Routh array is given by

$$\begin{array}{cc} 1 & 6a \\ 6 + a & 400 \\ 6a - 400/(6 + a) & 0 \\ 400 & \end{array}$$

For roots on the imaginary axis, we get $6a - 400/(6 + a) = 0$ (see special case 2), from which we obtain $a = 5.7$. Hence, for asymptotic stability of the system of Fig. 6.11, we need $a > 5.7$.

6.7 SENSITIVITY OF ROOTS TO PARAMETER VARIATIONS

The parameters of a control system can change with the environment and with age. The parameter values, therefore, may not remain constant throughout the entire operating life of the system. Also, the parameter values are not known exactly. One of the reasons for using feedback in control systems is to reduce the sensitivity of the system to parameter variations. One definition of sensitivity deals with the change in the closed-loop transfer function as a parameter is varied and is called transmittance sensitivity or system sensitivity.

Since the roots of the characteristic equation play an important role in transient response and stability, it is more useful to employ the definition of root sensitivity that deals with the variations in the roots of the characteristic equation due to a parameter change. The root sensitivity is defined as (Dorf, 1980)

$$s_{i,a} = \frac{\partial s_i}{\partial \ln a} = \left(\frac{\partial s_i}{\partial a}\right) a \tag{6.70}$$

where s_i is the ith root of the characteristic equation and a is the nominal value of the parameter that is varied. The partial derivative is evaluated at the nominal values of the parameters. Hence, the root sensitivity is equal to the partial derivative of the root with respect to the parameter, multiplied by the nominal value of the parameter. Since in high-order systems, the roots cannot be expressed as explicit functions of the parameters, the sensitivity coefficients can be obtained only for specific parameter values. The change in the ith root due to small variations in several parameters from their nominal values can be expressed as

$$\Delta s_i = \sum_{j=1}^{k} \frac{\partial s_i}{\partial a_j} \Delta a_j = \sum_{j=1}^{k} s_{i,a_j} \frac{\Delta a_j}{a_j} \tag{6.71}$$

where s_{i,a_j} is the sensitivity of the ith root to parameter a_j.

The sensitivity of a complex root is also complex, relating to the individual sensitivities of the real and imaginary parts. In general, it is desirable to keep the sensitivity to a small magnitude. A control system that is relatively insensitive to parameter variations and has good disturbance-rejection properties, which are discussed in Chapter 7, is called robust.

The root sensitivity can be evaluated by using the root-locus method. However, when several parameters are to be varied, the root-locus method becomes very cumbersome for this purpose and a method that uses machine computation becomes useful. We consider a method (Rudisill, 1974) that is quite suitable for machine computation of the partial derivatives of the roots with respect to the parameters. The characteristic equation of a closed-loop system is described by

$$|s\mathbf{I} - \mathbf{A} + \mathbf{BK}| = 0 \tag{6.72}$$

Let $s_1, s_2, \ldots, s_n$ be the roots of the characteristic equation for nominal values of the parameters. The right and left eigenvectors are obtained from the solution of

$$(\mathbf{A} - \mathbf{BK})\mathbf{v} = s_i\mathbf{v} \tag{6.73}$$

$$\mathbf{u}'(\mathbf{A} - \mathbf{BK}) = s_i\mathbf{u}' \tag{6.74}$$

where $\mathbf{u}'$ denotes the transpose of $\mathbf{u}$. The roots of the characteristic equation and the eigenvectors can be obtained by employing a matrix eigenvalues and eigenvectors computer program. Taking the partial derivative of Eq. (6.73) with respect to the parameter a, we obtain

$$\frac{\partial}{\partial a}(\mathbf{A}-\mathbf{BK})\mathbf{v}+(\mathbf{A}-\mathbf{BK})\frac{\partial \mathbf{v}}{\partial a}=\frac{\partial s_i}{\partial a}\mathbf{v}+s_i\frac{\partial \mathbf{v}}{\partial a} \tag{6.75}$$

Premultiplying Eq. (6.75) by $\mathbf{u}'$ and then employing Eq. (6.74), it follows that

$$\frac{\partial s_i}{\partial a}=\frac{\mathbf{u}'\dfrac{\partial}{\partial a}(\mathbf{A}-\mathbf{BK})\mathbf{v}}{\mathbf{u}'\mathbf{v}} \tag{6.76}$$

This equation can also be obtained by performing similar operations on Eq. (6.74). The sensitivity of the root s_i to the parameter a can then be obtained by using Eq. (6.76) in Eq. (6.70).

EXAMPLE 6.16

We consider the passive, mechanical, feedback control system of Example 3.2 whose stability has been investigated in Example 6.2. Let the nominal values of the parameters in Eq. (3.27) be given by

$$3g/2L-3k/m=-0.5, 3c/mL^2=1.5c, \text{ and } c=2/3$$

We investigate the root sensitivity of this system to the damping coefficient c. For these parameter values, Eq. (3.27) becomes

$$\mathbf{A}-\mathbf{BK}=\begin{bmatrix}0 & 1\\ -0.5 & -1.5c\end{bmatrix} \tag{6.77}$$

For the nominal value of $c=2/3$, the characteristic equation is obtained as

$$|s\mathbf{I}-\mathbf{A}+\mathbf{BK}|=s^2+s+0.5=0 \tag{6.78}$$

Its roots are given by $s_1=-0.5+j0.5$ and $s_2=-0.5-j0.5$. The right eigenvector corresponding to root s_1 is obtained from the solution of

$$\begin{bmatrix}0 & 1\\ -0.5 & -1\end{bmatrix}\begin{bmatrix}v_{11}\\ v_{12}\end{bmatrix}=(-0.5+j0.5)\begin{bmatrix}v_{11}\\ v_{12}\end{bmatrix} \tag{6.79}$$

and is given by

$$\mathbf{v}=\begin{bmatrix}1\\ -0.5+j0.5\end{bmatrix} \tag{6.80}$$

The left eigenvector corresponding to s_1 is obtained from

$$\lfloor u_{11} \quad u_{12}\rfloor\begin{bmatrix}0 & 1\\ -0.5 & -1\end{bmatrix}=(-0.5+j0.5)\lfloor u_{11} \quad u_{12}\rfloor \tag{6.81}$$

and is given by

$$\mathbf{u}'=\lfloor 0.5+j0.5 \quad 1\rfloor \tag{6.82}$$

The partial derivative of Eq. (6.77) with respect to c yields

$$\frac{\partial}{\partial c}(\mathbf{A}-\mathbf{BK}) = \begin{bmatrix} 0 & 0 \\ 0 & -1.5 \end{bmatrix} \tag{6.83}$$

Substituting the results from Eqs. (6.80), (6.82), and (6.83) in Eq. (6.76), we obtain

$$\frac{\partial s_1}{\partial c} = \frac{0.75 - j0.75}{j} = -0.75 - j0.75 = 1.06\angle -135^\circ \tag{6.84}$$

The sensitivity of root s_1 to parameter c for $c = 2/3$ is obtained from Eq. (6.70) as

$$s_{1,c} = (2/3)(1.06)\angle -135^\circ = 0.707\angle -135^\circ \tag{6.85}$$

The angle -135° gives the direction in which the root moves in the complex s-plane for a positive change in c. For a negative change in c, the direction is given by $180^\circ - 135^\circ = 45^\circ$. For a small change Δc in parameter c, the change in root s_1 can be obtained from Eq. (6.71) as

$$\Delta s_1 = s_{1,c}\frac{\Delta c}{c} \tag{6.86}$$

and for a 10% positive change, we obtain $\Delta s_1 = 0.0707\angle -135^\circ$.

The root-locus method can also be used to find the sensitivity of the roots to parameter c. For the nominal values of the parameters, the characteristic equation of Eq. (6.77) is expressed as

$$\begin{aligned} |s\mathbf{I} - \mathbf{A} + \mathbf{BK}| &= s^2 + 1.5cs + 0.5 \\ &= s^2 + 1.5\left(\frac{2}{3} \pm \Delta c\right)s + 0.5 = 0 \end{aligned} \tag{6.87}$$

In the standard form of root locus, we have

$$\frac{(\pm 1.5\Delta c)s}{s^2 + s + 0.5} + 1 = 0$$

or

$$\frac{(\pm 1.5\Delta c)s}{(s + 0.5 - j0.5)(s + 0.5 + j0.5)} + 1 = 0 \tag{6.88}$$

A negative change $-\Delta c$ requires the drawing of a complementary root locus and it is omitted here. The root locus can be drawn for a positive change Δc. The angle and magnitude conditions are expressed by

$$\measuredangle s - \measuredangle(s + 0.5 - j0.5) - \measuredangle(s + 0.5 + j0.5) = \pm i\pi \qquad i = 1, 3, 5, \ldots \tag{6.89}$$

$$\frac{|s|}{|s + 0.5 - j0.5||s + 0.5 + j0.5|} = \frac{1}{1.5\Delta c} \tag{6.90}$$

where $1.5\Delta c$ now plays the role of the parameter k. There are two poles at $-0.5 + j0.5$ and $-0.5 - j0.5$, and one zero at $s = 0$. These are marked by closed circles and an open circle respectively, in Fig. 6.13.

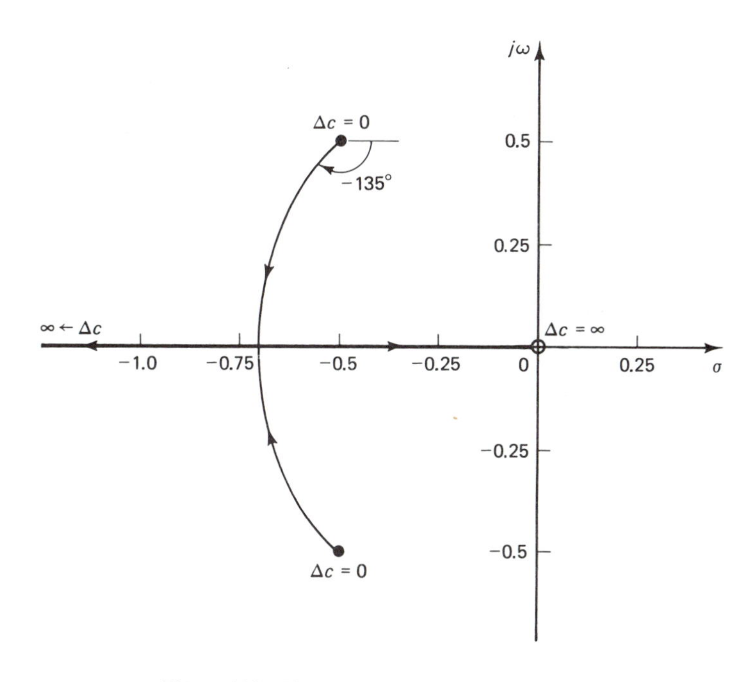

Figure 6.13 The root locus for the parameter c.

The angle of departure from the complex pole at $-0.5 + j0.5$ is obtained from Rule 9. Letting s be a point on the root locus infinitesimally close to $-0.5 + j0.5$ and employing the angle condition of Eq. (6.89), we obtain

$$135° - \measuredangle(s + 0.5 - j0.5) - 90° = \pm i180° \qquad i = 1, 3, 5, \ldots$$

where $\measuredangle s = 135°$ and $\measuredangle(s + 0.5 + j0.5) = 90°$. Hence, we get

$$\measuredangle(s + 0.5 - j0.5) = -135° \tag{6.91}$$

and it is precisely the angle obtained in Eq. (6.85). The break-in point where the loci from the two complex conjugate poles enter the real axis is obtained from Rule 8. From Eq. (6.57), we get

$$-s(2s + 1) + (s^2 + s + 0.5) = 0$$

The admissible root of this equation is -0.707, which is the break-in point. The complete root locus is shown in Fig. 6.13 from which we can obtain the values of the roots for any positive value of Δc. For a negative change in c near $\Delta c = 0$, the root at $-0.5 + j0.5$ moves in the direction of 45°.

6.8 SUMMARY

In this chapter, we have defined the concepts of stability in the sense of Lyapunov and in the sense of bounded-input-bounded-output (BIBO). For systems described by linear, ordinary differential equations with constant coefficients, the distinctions among the different concepts of stability get blurred and the requirement for stability is that all the roots of the characteristic equation have negative real parts. We have studied the Routh stability criterion, which is a very quick and simple method of determining whether or not the stability conditions are satisfied without actually obtaining the roots of the characteristic equation. The Routh criterion is also useful for finding the region of stability in the parameter space. It is shown that the Routh criterion can be easily extended to investigate relative stability.

The root-locus method provides a graphical view of the movement of the roots of the characteristic equation in the complex s-plane as one or more parameters are varied. An approximate sketch is often sufficient to obtain information concerning stability. The root-locus method can also provide useful information regarding the sensitivity of the roots to parameter variations. Since the method employs the system characteristic equation, it is equally applicable to single-input, single-output and multivariable systems. However, the selection of multiple gains to obtain desired root locations by the root-locus method is a laborious process. This topic is covered in detail in Chapters 8 and 9 and involves state feedback.

REFERENCES

DERUSSO, P. M., ROY, R. J., and CLOSE, C. M. (1965). *State Variables for Engineers.* New York: Wiley.

DORF, R. C. (1980). *Modern Control Systems.* 3d ed. Reading, MA: Addison-Wesley.

D'SOUZA, A. F., and GARG, V. K. (1984). *Advanced Dynamics: Modeling and Analysis.* Englewood Cliffs, NJ: Prentice-Hall.

EVANS, W. R. (1954). *Control System Dynamics.* New York: McGraw-Hill.

KUO, B. C. (1982). *Automatic Control Systems.* 4th ed. Englewood Cliffs, NJ: Prentice-Hall.

MELSA, J. L., and JONES, S. K. (1973). *Computer Programs for Computational Assistance in the Study of Linear Control Theory.* New York: McGraw-Hill.

RAVEN, F. H. (1978). *Automatic Control Engineering.* 3d ed. New York: McGraw-Hill.

ROUTH, E. J. (1955). *Dynamics of a System of Rigid Bodies.* New York: Dover.

RUDISILL, C. S. (1974, May). "Derivatives of Eigenvalues and Eigenvectors for a General Matrix." *AIAA Journal* 12(5), 721–722.

PROBLEMS

6.1. A closed-loop control system is described by

$$y = \left(\frac{45}{(D+5)(D^2+9)}\right) r$$

Giving reasons, state whether this system is BIBO stable or unstable. Give a particular input $r(t)$ that will cause the output $y(t)$ to be unbounded. (*Hint*: Try $r = \sin 3t$.)

6.2. A system that has bounded state variables for all bounded inputs is called bounded-input-bounded-state stable (BIBS). Show that if the system described by Eq. (6.1) is asymptotically stable, then it is also BIBS stable.

6.3. Using the Routh criterion, investigate the stability of the control systems whose characteristic equations are given by the following. For unstable systems, determine how many roots have a positive real part.

(a) $s^3 + 3s^2 + 10s + 100 = 0$

(b) $s^3 + 9s^2 + 40s + 100 = 0$

(c) $s^4 + 7s^3 + 70s^2 + 150s + 500 = 0$

(d) $s^4 + 3s^3 + 20s^2 + 80s + 200 = 0$

6.4. For the electrical control system of Example 3.3 whose closed-loop matrix is given by Eq. (3.42), let the parameter values be such that

$$\mathbf{A} - \mathbf{BK} = \begin{bmatrix} -0.5 & 5 & 0 \\ -0.2 & -2 & 10 \\ -k_1 & 0 & -3 \end{bmatrix}$$

Find the range of k_1 for stability, where $k_1 > 0$.

6.5. For the pneumatic temperature-control system of Example 3.7, whose system matrices are given by Eq. (3.82), let the parameter values be such that

$$\mathbf{A} - \mathbf{BK} = \begin{bmatrix} -0.5 & 0.5 & 0.3 \\ 0.4 & -0.8 & 0 \\ -k_1 & 0 & -0.2 \end{bmatrix}$$

Find the range of k_1 for stability, where $k_1 > 0$.

6.6. The block diagam of the hydraulic machining-control system of Example 3.6 is shown in Fig. 3.22. Now, the negative feedback is replaced by positive feedback. With $c_1 = c_2$ and $F_d = 0$, the positive feedback control system is shown in Fig. P6.6, where $k > 0$.

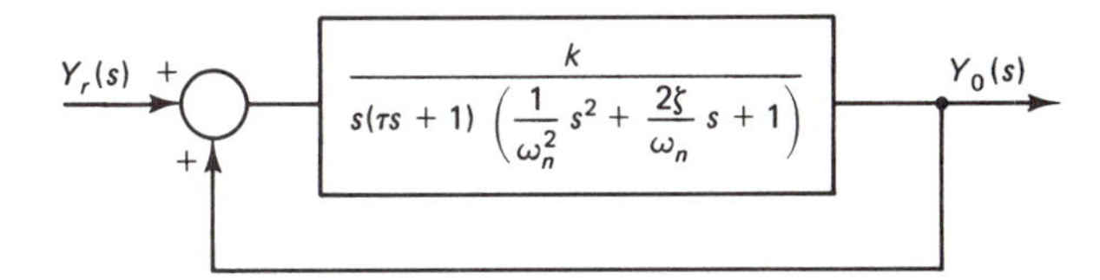

Figure P6.6 Block diagram of a hydraulic machining system.

Show that the characteristic equation can be represented in the form

$$a_4 s^4 + a_3 s^3 + a_2 s^2 + s - k = 0$$

where a_4, a_3, and a_2 are positive constants and are related to the system parameters. Using the Routh criterion, investigate the stability for $k > 0$. Comment on the stabilizing or destabilizing effect of replacing negative feedback by positive feedback.

6.7. The block diagram of the hydraulic control system of Example 4.4 is shown in Fig. 4.3. Now, k_a is replaced by $k_p + k_i/s$, where $k_p > 0$ and $k_i > 0$. This is the proportional plus integral control law discussed in Chapter 7. The other parameter values are such that the block diagram of the system is shown in Fig. P6.7. Construct a parameter plane with k_p versus k_i, and show the stable and unstable regions.

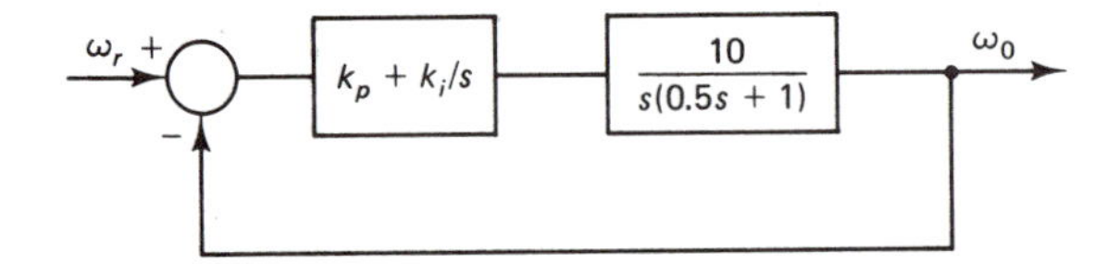

Figure P6.7 Block diagram of a hydraulic control system.

6.8. Using the Routh criterion, investigate the stability of the closed-loop control systems whose characteristic equations are given by the following special cases.

(a) $s^3 + 5s^2 + 4s + 20 = 0$

(b) $s^4 + 5s^3 + 8s^2 + 10s + 12 = 0$

(c) $s^5 + 3s^4 + 3s^3 + 9s^2 + s + 1 = 0$

6.9. Determine how many roots of the characteristic equation have real parts greater than -1 for the following cases.

(a) System of Problem 6.4 with $k_1 = 0.8$

(b) System of Problem 6.5 with $k_1 = 10$

6.10. Sketch the root locus, giving all the steps, for the closed-loop system shown in Fig. P6.10 as k varies from zero to infinity for the following cases. Discuss how the stability depends on k.

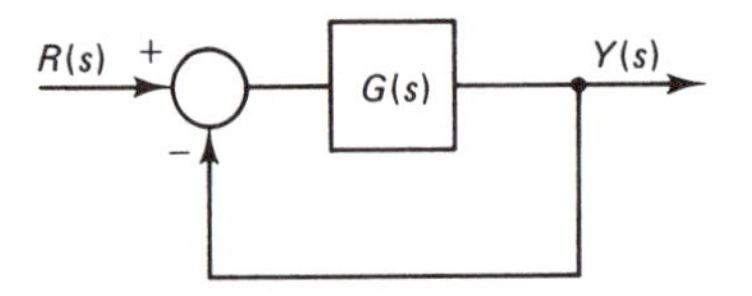

Figure P6.10 Closed-loop system.

(a) $G(s) = \dfrac{k(s+3)}{s(s+1)}$

(b) $G(s) = \dfrac{k}{s(s+2)(s+3)}$

(c) $G(s) = \dfrac{k(s+1)}{(s+3)(s^2+8s+20)}$

(d) $G(s) = \dfrac{k(s+1)}{s(s+3)(s+5)^2}$

(e) $G(s) = \dfrac{k(s+2)}{s^2(s+3)}$

(f) $G(s) = \dfrac{k(s+2)}{s^2 - 4s + 20}$

6.11. Draw the root locus for the roots of the characteristic equation of the system of Problem 6.4 as k_1 is varied from zero to infinity. Comment on the dependence of stability on k_1. (*Hint*: When arranged in the form of Eq. (6.43), the poles of $P(s)$ are -3, $-1.25 \pm j0.66$.)

6.12. Draw the root locus for the roots of the characteristic equation of the system of Problem 6.5 as k_1 is varied from zero to infinity. Comment on the dependence of stability on k_1. (*Hint*: When arranged in the form of Eq. (6.43), the poles of $P(s)$ are -0.178, -1.122, -0.2.)

6.13. For the system of Problem 6.7, let $k_i = 0.5k_p$. Draw the root locus for this system as $0 < k_p < \infty$. Obtain the value of k_p such that the damping ratio of the dominant quadratic is 0.5.

6.14. In Fig. P6.10, let

$$G(s) = \frac{10(\tau s + 1)}{s^2 + 8s + 10}$$

Draw the root locus as τ varies from zero to infinity.

6.15. Using the root locus, determine the sensitivity of the real root of the characteristic equation of the system of Problem 6.4 to the parameter k_1. The nominal value of k_1 is 0.5. Consider a change in k_1 around 20%.

6.16. Employing the root locus, determine the sensitivity of the real root of the characteristic equation of the system of Problem 6.5 to the parameter k_1. The nominal value of k_1 is 10. Consider a change in k_1 around 20%.

6.17. Solve Problem 6.15 but, instead of using the root locus, use a computer program and Eqs. (6.70) and (6.76).

6.18. Solve Problem 6.16 but, instead of using the root locus, use a computer program and Eqs. (6.70) and (6.76).

Footnotes

[1] It can be proved that this condition is also necessary by assuming otherwise and arriving at a contradiction that $\|\mathbf{x}(t)\| \to 0$ as $t \to \infty$.

[2] It can be proved that this condition is also necessary by assuming otherwise and arriving at a contradiction that $\|\mathbf{y}(\infty)\|$ is bounded.

[3] See Programs 1 and 2, Appendix D.

7

Controller Design with Output Feedback

7.1. INTRODUCTION

The central element of a feedback control system is the controller whose function is to implement a control law. In Chapter 3, we have obtained the mathematical models of several control systems that use output feedback. The controlled output is measured and compared to the command input to obtain an error. This error $e(t)$ is then used to obtain a control signal $u(t)$ that is supplied to the actuator to correct the error. The objective of this chapter is to design a controller with output feedback to implement a *control law* that is a functional relationship between the error $e(t)$ and the control signal $u(t)$. Controller design with state feedback is studied in the next chapter.

The control law plays an important role in determining the accuracy of a control system in following a command. The error results from either a change in the command input or a disturbance due to a load change or other reasons. The steady-state error is obtained from the response that remains in a stable system after all the transients have decayed to zero. In a regulator, accuracy may be expressed in terms of acceptable steady-state error after a change in the set point or the occurrence of a disturbance. In a servomechanism, we require an acceptable tracking accuracy.

In order to analyze steady-state errors, it is useful to classify the control systems into *system types*. The system type number is determined from the forward-path transfer function, which combines the transfer function of the control law, actuator, and the controlled process. The selection of an appropriate control law is dependent to a large extent on the system type that results.

In this chapter, we begin our study with the analysis of steady-state errors, tracking accuracy, and disturbance-rejection properties of control systems. This information is then used to design a suitable controller. First, the three basic system types are discussed and steady-state errors are analyzed for each type. Then the basic control laws are presented and their commonly encountered combinations are studied. The task finally involves the design of the components of the controller, the determination of the manner in which they should be connected, the selection of the gains, and compensation.

7.2 SYSTEM TYPES

The steady-state error between the command input and the controlled output depends on the system type and on the input or disturbance. For the classification of control systems into system types, it is more appropriate to express the mathematical model in the form of a block diagram with transfer functions rather than in the state-variables formulation. The block diagram of a single-input, single-output, closed-loop control system is shown in Fig. 7.1, where $R(s)$ is the command input, $Y(s)$ the controlled output, and $V(s)$ is a disturbance input.

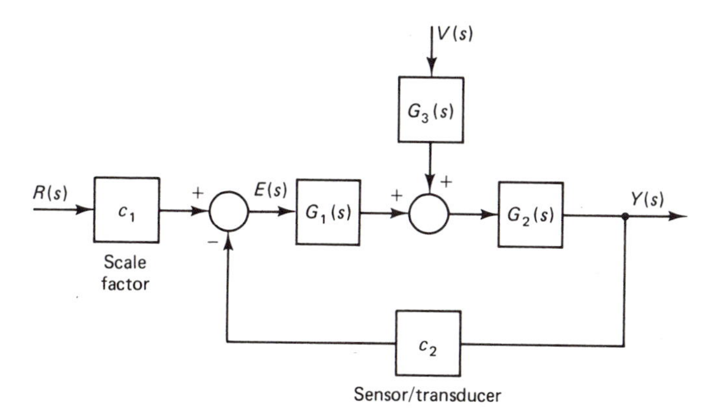

Figure 7.1 Block diagram of a closed-loop control system.

In this figure, $G_1(s)$ represents the transfer function of the control law and actuator, $G_2(s)$ the transfer function of the controlled process, c_1 the input scale factor, and c_2 the gain constant of the sensor/transducer. The reference input $r(t)$ is scaled by a scale factor c_1 such that $c_1r(t)$ and the sensor output $c_2y(t)$ are the same dimensionally and are scaled to the same level, that is, $c_1 = c_2 = c$. We can absorb this constant c into the transfer function $G_1(s)$ and without loss of generality represent the block diagram of Fig. 7.1 as a unity-feedback control system of Fig. 7.2.

As an example, consider the position-control system of Example 3.4 whose schematic diagram is shown in Fig. 3.10 and whose block diagram is shown in Fig. 3.15. Note that the reference input is the desired angular position θ_r and the controlled output is the angular position θ_0. The sensor output is $E_0 = c_2\theta_0$. Hence, θ_r is scaled such that $E_r = c_1\theta_r$, where $c_1 = c_2$. The error voltage thus becomes $e = E_r - E_0 = c(\theta_r - \theta_0)$, where $c = c_1 = c_2$.

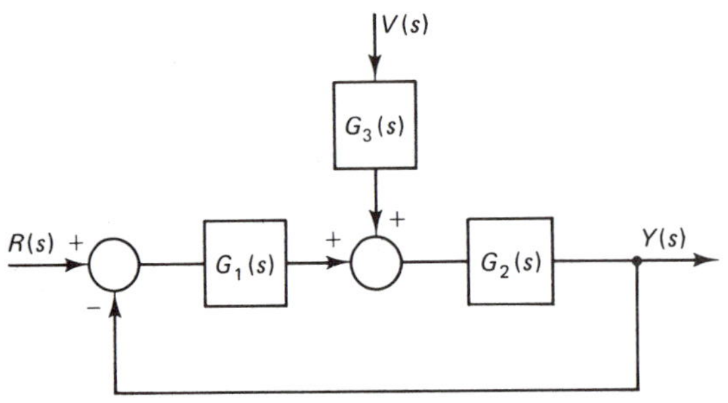

Figure 7.2 Unity-feedback control system.

We now develop an expression for the error. Since we are dealing with a linear system, we obtain the output from Fig. 7.2 by superposition as

$$Y(s) = \left(\frac{G_1(s)G_2(s)}{1 + G_1(s)G_2(s)}\right) R(s) + \left(\frac{G_2(s)G_3(s)}{1 + G_1(s)G_2(s)}\right) V(s) \tag{7.1}$$

The preceding transfer functions have been derived earlier in Chapter 3 and given by Eqs. (3.9) and (3.11). We first consider the case where the disturbance $v(t) = 0$, but the control input $r(t) \neq 0$. The Laplace transform of the error in this case is given by

$$\begin{aligned} E(s) &= R(s) - Y(s) = R(s) - \left(\frac{G_1G_2}{1 + G_1G_2}\right) R(s) \\ &= \left(\frac{1}{1 + G_1(s)G_2(s)}\right) R(s) \end{aligned} \tag{7.2}$$

Next, we consider the case of a regulator when the set point has not been changed so that $r(t) = 0$, but the disturbance $v(t) \neq 0$. In this case, the Laplace transform of the error becomes

$$\begin{aligned} E(s) &= 0 - Y(s) \\ &= -\left(\frac{G_2(s)G_3(s)}{1 + G_1(s)G_2(s)}\right) V(s) \end{aligned} \tag{7.3}$$

We define the control system type from the form of $G_1(s)G_2(s)$. In general, let

$$G_1(s)G_2(s) = \frac{k(\tau_a s + 1) \cdots (\tau_m s + 1)}{s^i(\tau_1 s + 1) \cdots (\tau_n s + 1)} \tag{7.4}$$

The type of control system refers to the order of the pole at $s = 0$ of the forward-path transfer function $G_1(s)G_2(s)$, that is, the number of pure integrators in the forward path. In Eq. (7.4), when $i = 0$, the control system is called type 0; when $i = 1$, it is called type 1; and when $i = 2$, it is called type 2. System types higher than 2 are generally not employed because of problems with stability. The steady-state error depends on the system type and on the command input and disturbance. The steady-state error is defined as the limit of the error as time tends to infinity, that is, $e_{ss} = \lim e(t)$ as $t \to \infty$. The steady-state error has significance only for control systems that are asymptotically stable.

In the following four illustrative examples, we classify into system types some of the control systems whose mathematical models were derived in Chapter 3.

EXAMPLE 7.1

A passive, mechanical control system for regulating the liquid level in a tank has been studied in Example 3.1. We note from the block diagram of Fig. 3.4(b) that

$$G_1(s) = k_1 \qquad G_2(s) = \frac{R/\rho g}{\tau_1 s + 1} \qquad G_3(s) = 1 \tag{7.5}$$

Hence, we obtain

$$G_1(s)G_2(s) = \frac{k}{\tau_1 s + 1} \qquad \text{where } k = k_1 R/\rho g \tag{7.6}$$

This system does not have any pure integrator in the forward path and is therefore classified as type 0.

EXAMPLE 7.2

In Example 3.5, we have obtained a mathematical model of an electrohydraulic control system for the speed control of a prime mover. In the block diagram of Fig. 3.19, we let $c_1 = c_2 = c$ and absorb this constant in the forward-path transfer function so that it can be represented as a unity-feedback system. It follows that

$$G_1(s) = \frac{k_3}{s\left(\dfrac{1}{\omega_n^2}s^2 + \dfrac{2\zeta}{\omega_n}s + 1\right)} \qquad G_2(s) = \frac{k_2}{\tau s + 1} \qquad G_3(s) = -c_5 \tag{7.7}$$

where we let $k_3 = ck_a k_1 k_g c_3 c_4$. Letting $k = k_2 k_3$, we get

$$G_1(s)G_2(s) = \frac{k}{s\left(\dfrac{1}{\omega_n^2}s^2 + \dfrac{2\zeta}{\omega_n}s + 1\right)(\tau s + 1)} \tag{7.8}$$

This system has one pure integrator in the forward path and is therefore classified as type 1.

EXAMPLE 7.3

An electrohydraulic control system for machining a workpiece has been considered in Example 3.6. In the block diagram of Fig. 3.22, $c_1 = c_2 = c$, and we represent it as a unity-feedback system. We obtain

$$G_1(s) = \frac{k_3}{\dfrac{1}{\omega_n^2}s^2 + \dfrac{2\zeta}{\omega_n}s + 1} \qquad G_2(s) = \frac{k_g}{s(\tau s + 1)} \qquad G_3(s) = -k_2 \tag{7.9}$$

where we have let $k_3 = ck_ak_1$. Letting $k = k_3k_g$, we get

$$G_1(s)G_2(s) = \frac{k}{s\left(\dfrac{1}{\omega_n^2}s^2 + \dfrac{2\zeta}{\omega_n}s + 1\right)(\tau s + 1)} \tag{7.10}$$

This system is therefore type 1 as in Example 7.2. However, in this system, the integrator is located in the transfer function $G_2(s)$, that is, after the disturbance enters the system as seen in Fig. 3.22 and not in the transfer function $G_1(s)$ as in Fig. 3.19 of Example 7.2. Hence, even though the systems of Examples 7.2 and 7.3 can both be classified as type 1, they differ in their disturbance-rejection properties and this difference will be clarified in the next section.

EXAMPLE 7.4

A pneumatic system for regulating the temperature has been modeled in Example 3.7. In the block diagram of Fig. 3.25, we again let $c_1 = c_2 = c$, and represent it as a unity-feedback system. From Fig. 3.25, it is seen that

$$G_1(s) = \frac{ck_1k_2c_5}{\tau s + 1} \qquad G_2(s) = \frac{k_g(\tau_1 s + 1)}{\dfrac{1}{\omega_n^2}s^2 + \dfrac{2\zeta}{\omega_n}s + 1}$$
$$G_3(s) = \frac{1/k_g}{\tau_1 s + 1} \tag{7.11}$$

Letting $k = ck_1k_2c_5k_g$, the forward-path transfer function is obtained as

$$G_1(s)G_2(s) = \frac{k(\tau_1 s + 1)}{(\tau s + 1)\left(\dfrac{1}{\omega_n^2}s^2 + \dfrac{2\zeta}{\omega_n}s + 1\right)} \tag{7.12}$$

This control system is therefore classified as type 0.

7.3 STEADY-STATE ERRORS

The steady-state error is a measure of the control system accuracy in tracking a command input or in rejecting a disturbance in the form of a load change. As mentioned earlier, the steady-state errors depend on the system type and on the inputs, which include command signals and disturbances. A method of determining the steady-state error is to obtain the solution of $e(t)$ either from Eq. (7.2) or Eq. (7.3), depending on the input, and then to take the limit of $e(t)$ as time tends to infinity. This labor can be avoided by using the final-value theorem of Laplace transformation and is the main reason for conducting this analysis with transfer functions in the Laplace domain.

According to this theorem, the steady-state error is obtained from

$$e_{ss} = \lim_{t\to\infty} e(t) = \lim_{s\to 0} sE(s) \tag{7.13}$$

provided $sE(s)$ has no poles on the imaginary axis and in the right half of the s-plane. This restriction is required for the validity of the final-value theorem. From Eqs. (7.1) and (7.4), the characteristic equation of the closed-loop system is obtained as

$$s^i(\tau_1 s + 1)\cdots(\tau_n s + 1) + k(\tau_a s + 1)\cdots(\tau_m s + 1) = 0 \tag{7.14}$$

It can be seen from Eqs. (3.6) and (3.7) that this characteristic equation is identical to the one obtained from state-variables formulation, namely,

$$\det|s\mathbf{I} - \mathbf{A} + \mathbf{BK}| = 0 \tag{7.15}$$

Hence, for the validity of the results of the final-value theorem, the roots of the closed-loop characteristic equation, Eq. (7.14), must all have negative real parts, that is, the closed-loop system must be asymptotically stable. It is therefore important to check the asymptotic stability of the closed-loop system first by the Routh criterion before employing the final-value theorem.

The command inputs to a servomechanism may not be known in advance. Also, the disturbance inputs and load changes usually cannot be predicted. Hence, three deterministic test signals, namely, step, ramp, and parabolic, are used as inputs to facilitate the steady-state error analysis.

7.3.1 Steady-State Errors Due to a Step Input

The step input represents an instantaneous jump in the command input or disturbance to a new constant value. We have

$$\begin{aligned} r(t) \text{ or } v(t) &= a \qquad \text{i.e., } R(s) \text{ or } V(s) = a/s \qquad t > 0 \\ &= 0 \qquad\qquad\qquad\qquad\qquad\quad = 0 \qquad t < 0 \end{aligned} \tag{7.16}$$

We assume that the closed-loop system is asymptotically stable so that we can employ the final-value theorem. For a step change in the command input, from Eqs. (7.2) and (7.16), we obtain

$$\begin{aligned} e_{ss} &= \lim_{t\to\infty} e(t) = \lim_{s\to 0} s\left(\frac{1}{1 + G_1(s)G_2(s)}\right)\frac{a}{s} \\ &= \frac{a}{1 + G_1(0)G_2(0)} \end{aligned} \tag{7.17}$$

For a step change in the disturbance, from Eqs. (7.3) and (7.16), we get

$$\begin{aligned} e_{ss} &= \lim_{t\to\infty} e(t) = \lim_{s\to 0} s\left(\frac{-G_2(s)G_3(s)}{1 + G_1(s)G_2(s)}\right)\frac{a}{s} \\ &= -\frac{aG_2(0)G_3(0)}{1 + G_1(0)G_2(0)} \end{aligned} \tag{7.18}$$

Type 0 System. For a type 0 system, in Eq. (7.4), we have $i = 0$ and $G_1(0)G_2(0) = k$. Hence, Eqs. (7.17) and (7.18) are given, respectively, by

$$e_{ss} = \frac{a}{1+k} \tag{7.19}$$

and

$$e_{ss} = \frac{-aG_2(0)G_3(0)}{1+k} \tag{7.20}$$

where $G_2(0)G_3(0)$ is a finite constant. Hence, a type 0 system has a finite steady-state error for a step input and it is referred to as the steady-state offset. This has been mentioned earlier in the time-domain performance specifications of Chapter 4 and shown in Fig. 4.13. The value of the gain k for a type 0 system is also called the position error constant and is denoted by k_p.

Type 1 System. For a type 1 system, in Eq. (7.4), we have $i = 1$ and $G_1(0)G_2(0) = \infty$. Hence, Eqs. (7.17) and (7.18) are given, respectively, by

$$e_{ss} = \frac{a}{1+\infty} = 0 \tag{7.21}$$

and

$$e_{ss} = \frac{-aG_2(0)G_3(0)}{1+\infty} \tag{7.22}$$

If the integrator is located in $G_1(s)$, the value of e_{ss} in Eq. (7.22) is zero, but if it is located in $G_2(s)$, then Eq. (7.22) yields a finite nonzero value of e_{ss}. This behavior is illustrated by the following Examples 7.7 and 7.8.

Type 2 System. For a type 2 system, in Eq. (7.4), we have $i = 2$ and $G_1(0)G_2(0) = \infty$. Hence in Eq. (7.17), $e_{ss} = 0$, and if not more than one of the integrators is located in $G_2(s)$, then e_{ss} of Eq. (7.18) is also zero. If $G_2(s)$ contains one of the two integrators, then

$$G_2(0)G_3(0) = \lim_{s\to 0} k_2/s$$

and from Eq. (7.18), we obtain

$$e_{ss} = \lim_{s\to 0} \frac{-ak_2}{s(1+k/s^2)} = \frac{-ak_2}{\infty} = 0 \tag{7.23}$$

This result is also true when both of the integrators are located in $G_1(s)$ such that $G_2(0)G_3(0) = k_2$.

The typical responses of type 0 and higher type systems to a step change of magnitude a in the command input are shown in Figs. 7.3(a) and (b).

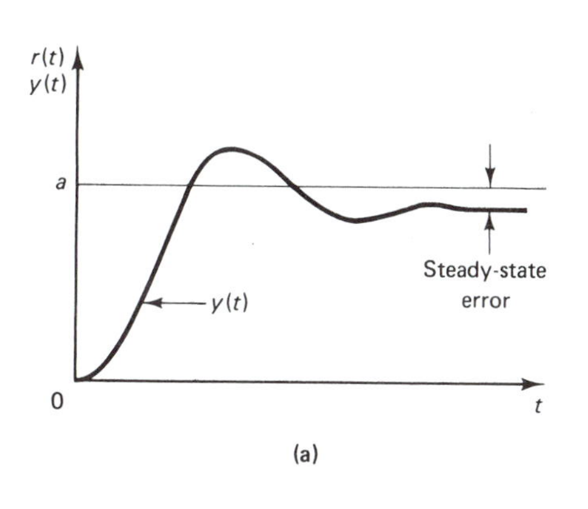

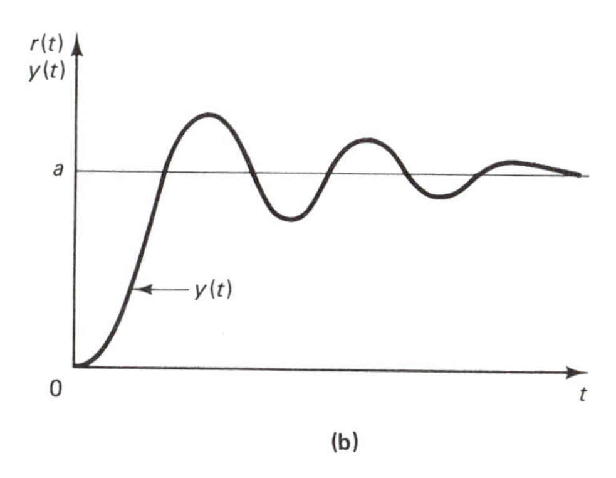

Figure 7.3 Step response of (a) a type 0 system, and (b) of a type 1 or higher system.

EXAMPLE 7.5

In Example 7.1, it is seen that the liquid-level regulator of Example 3.1 is a type 0 system. Let the set point be unchanged, but the disturbance flow q_d be a unit step of magnitude a. The characteristic equation of the closed-loop system is obtained from Eq. (3.21) as

$$\tau_1 s + 1 + k_1 R/\rho g = 0$$

Its root is negative and hence the system is asymptotically stable so that we can employ the results of the final-value theorem. From Eqs. (7.5) and (7.6), we obtain

$$G_1(0)G_2(0) = k \qquad G_2(0) = R/\rho g \qquad G_3(0) = 1$$

Substituting these results in Eq. (7.20), we get the steady-state error due to the disturbance as

$$e_{ss} = -\frac{aR/\rho g}{1+k} \tag{7.24}$$

In steady state, the liquid level rises above its desired value by an amount $aR/\rho g(1 + k)$. The physical reason for this error can be explained from the schematic diagram of Fig. 3.3. When the set point is unchanged but the disturbance flow q_d is a step, the flow-control valve opening must decrease so that the inflow q_1 is reduced and the liquid level is maintained. But it is seen from Fig. 3.3 that for the valve opening to decrease, the float must rise, and this implies that the liquid level must rise above its desired value.

EXAMPLE 7.6

We consider the pneumatic temperature-control system of Example 3.7. We have seen in Example 7.4 that this system is type 0, and the transfer functions are given by Eqs. (7.11) and (7.12). It is assumed that the closed-loop system is asymptotically stable. Referring to the block diagram of Fig. 3.25, let the change in the disturbance temperature T_a be zero, but the setpoint temperature

T_r be changed by a step of magnitude a. From Eq. (7.12), we obtain $G_1(0)G_2(0) = k$, and from Eq. (7.17), it follows that $e_{ss} = a/(1 + k)$. This implies that after the transients have decayed, T_1 settles to a steady-state value of $ak/(1 + k)$.

Now, let the set point be unchanged so that $T_r = 0$ and the ambient temperature increases by a step of magnitude a. From Eq. (7.11), we obtain $G_2(0) = k_g$ and $G_3(0) = 1/k_g$. Substituting these results in Eq. (7.18), the steady-state error is $-a/(1 + k)$. This implies that the controlled temperature T_1 has increased above its desired value by $a/(1 + k)$. A physical reason for this error can be given as in the previous example. The preceding results are not valid when the control system is unstable.

EXAMPLE 7.7

The electrohydraulic speed control system of Example 3.5 has been classified as type 1 in Example 7.2, and its transfer functions are given by Eqs. (7.7) and (7.8). The characteristic equation is obtained from Eq. (7.8) as

$$s\left(\frac{1}{\omega_n^2}s^2 + \frac{2\zeta}{\omega_n}s + 1\right)(\tau s + 1) + k = 0 \tag{7.25}$$

We assume that all four roots of this equation have negative real parts and employ the final-value theorem. In the block diagram of Fig. 3.19, let the load torque T_d be zero and the command input ω_r be a step. We have $G_1(0)G_2(0) = \infty$, and when there is a step change in the command input, the steady-state error $e_{ss} = 0$. Next, let $\omega_r = 0$, but the disturbance load torque T_d be a step. The integrator is located in $G_1(s)$, and from Eq. (7.7), we get $G_2(0)G_3(0) = -c_5k_2$. Using this result in Eq. (7.22), we see that when there is a step load change, the steady-state error $e_{ss} = 0$.

The reason why there is no steady-state error in this system for a step change in the input can be explained from the schematic diagram of Fig. 3.16. When there is a step increase in the load torque, the fuel-control valve opening must increase so that more fuel is supplied to the prime mover to maintain the speed at its desired value. The piston of the actuator can be in equilibrium at any location within the cylinder, that is, it has a floating equilibrium that is a characteristic of an integrator. Hence, after the transients have decayed, the piston arrives at a new equilibrium inside the cylinder and thus increases the fuel-valve opening.

EXAMPLE 7.8

The electrohydraulic control system of Example 3.6 for machining a workpiece has been classified in Example 7.3 as a type 1 system, and its transfer functions are given by Eqs. (7.9) and (7.10). It is assumed that this control system is asymptotically stable. In the block diagram of Fig. 3.22, let the change in the disturbance thrust force $F_d = 0$ and y_r be a step input. It can be observed from Eq. (7.10) that $G_1(0)G_2(0) = \infty$, and for a step change in the command input, we see from Eq. (7.21) that the steady-state error $e_{ss} = 0$.

Now, let $y_r = 0$, but the disturbance thrust force F_d be a step. The integrator is located in $G_2(s)$ after the disturbance enters the system. Substituting for the transfer functions from Eqs. (7.9) and (7.10) in (7.18), we obtain the steady-state error as

$$\begin{aligned} e_{ss} &= \lim_{s\to 0} \frac{ak_2k_g/s}{1 + k/s} \\ &= ak_2k_g/k \end{aligned} \tag{7.26}$$

Hence, there is a finite steady-state error for a step change in the disturbance. On comparing this system with that of the previous Example 7.7, we note that both systems are type 1 and have zero steady-state error for a step change in the command input. But, unlike the system of Example 7.7, this system has a finite, nonzero steady-state error for a step change in the disturbance. The reason for this behavior is that in the system of Example 7.7, the integrator is located in $G_1(s)$, whereas in this system, it is located in $G_2(s)$, that is, after the disturbance enters the system in the forward path.

7.3.2. Steady-State Errors Due to a Ramp Input

This test signal permits us to study how a system will track or follow a command input or respond to a disturbance that changes linearly with time. The input is given by

$$\begin{aligned} r(t) \text{ or } v(t) &= at \qquad \text{i.e., } R(s) \text{ or } V(s) = a/s^2 \qquad t \geq 0 \\ &= 0 \qquad\qquad\qquad\qquad\qquad\qquad\quad = 0 \qquad t < 0 \end{aligned} \tag{7.27}$$

We assume that the control system is asymptotically stable and use the final-value theorem. When the command input is a ramp, the steady-state error is obtained from Eq. (7.2) as

$$\begin{aligned} e_{ss} &= \lim_{s\to 0} s \left(\frac{1}{1 + G_1(s)G_2(s)}\right) \frac{a}{s^2} \\ &= \lim_{s\to 0} \frac{a}{s[1 + G_1(0)G_2(0)]} \end{aligned} \tag{7.28}$$

and when the disturbance is a ramp, from Eq. (7.3) we get

$$\begin{aligned} e_{ss} &= \lim_{s\to 0} \left(\frac{s[-G_2(s)G_3(s)]}{1 + G_1(s)G_2(s)}\right) \frac{a}{s^2} \\ &= \lim_{s\to 0} \frac{-aG_2(0)G_3(0)}{s[1 + G_1(0)G_2(0)]} \end{aligned} \tag{7.29}$$

Type 0 System. For a type 0 system, $G_1(0)G_2(0) = k$ and $G_2(0)G_3(0)$ is also finite. Hence, when the command input or disturbance is a ramp, it follows from Eqs. (7.28) and (7.29) that the steady-state error $e_{ss} = \infty$. Hence, a type 0 system will not follow a ramp input as shown in Fig. 7.4.

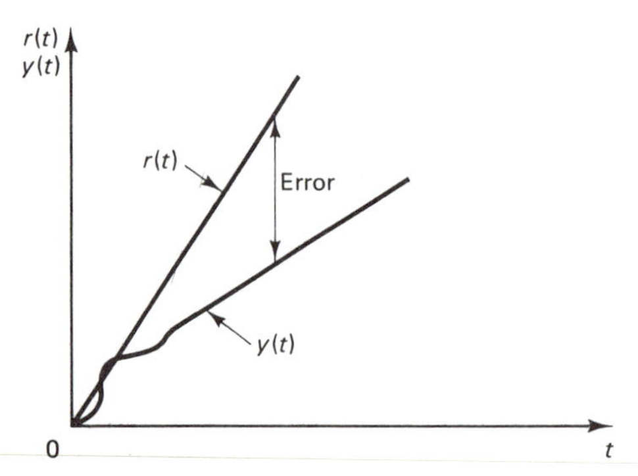

Figure 7.4 Ramp response of a type 0 system.

Type 1 System. For a type 1 system, when the command input is a ramp, it follows from Eq. (7.28) that

$$e_{ss} = \lim_{s \to 0} \frac{a}{s(1 + k/s)} = \frac{a}{k} \tag{7.30}$$

When the disturbance is a ramp and the integrator is located in $G_1(s)$, from Eq. (7.29), we obtain

$$e_{ss} = \frac{-aG_2(0)G_3(0)}{k} \tag{7.31}$$

which is a finite nonzero value. But if the integrator is located in $G_2(s)$, then $e_{ss} = \infty$. A type 1 system will follow a ramp command input but with a finite steady-state error as shown in Fig. 7.5. The value of the gain k, for a type 1 system is called the velocity error constant and is denoted by k_v.

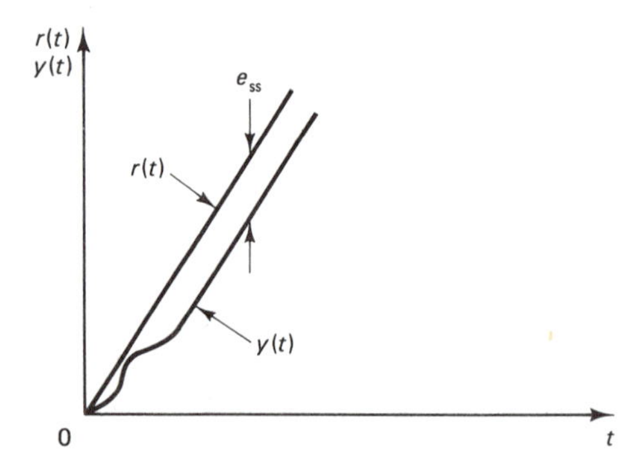

Figure 7.5 Ramp response of a type 1 system.

Type 2 System. For a type 2 system, when the command input is a ramp, it follow from Eq. (7.28) that

$$e_{ss} = \lim_{s \to 0} \frac{a}{s(1 + k/s^2)} = 0 \tag{7.32}$$

Similarly, when the disturbance is a ramp, from Eq. (7.29) we get $e_{ss} = 0$ when $G_1(s)$ contains both the integrators or e_{ss} has a finite nonzero value when $G_2(s)$ has one of the two integrators. A type 2 system will follow a ramp command input with zero steady-state error.

7.3.3 Steady-State Errors Due to a Parabolic Input

This test signal is used to study how a system would follow a command input or respond to a disturbance that is a parabolic function of time. In this case, the input is given by

$$\begin{aligned} r(t) \text{ or } v(t) &= \tfrac{1}{2}at^2 \qquad \text{i.e., } R(s) \text{ or } V(s) = a/s^3 \qquad t \geq 0 \\ &= 0 \qquad\qquad\qquad\qquad\qquad\qquad = 0 \qquad t < 0 \end{aligned} \tag{7.33}$$

We use the final-value theorem when the system is asymptotically stable. When the command input is parabolic, the steady-state error is obtained frm Eq. (7.2) as

$$\begin{aligned} e_{ss} &= \lim_{s \to 0} s \left(\frac{1}{1 + G_1(s)G_2(s)} \right) \frac{a}{s^3} \\ &= \lim_{s \to 0} \frac{a}{s^2[1 + G_1(0)G_2(0)]} \end{aligned} \tag{7.34}$$

When the disturbance is parabolic, from Eq. (7.3) we get

$$e_{ss} = \lim_{s \to 0} \frac{-aG_2(0)G_3(0)}{s^2[1 + G_1(0)G_2(0)]} \tag{7.35}$$

According to the procedure used earlier, we can draw the following conclusions from Eq. (7.34) regarding the steady-state error when the command input is a parabolic function.

Type 0 system: $e_{ss} = \infty$

Type 1 system: $e_{ss} = \infty$

Type 2 system: $e_{ss} = a/k$

Hence, type 0 and type 1 systems will not track a parabolic command input as shown in Fig. 7.6. A type 2 system will follow a parabolic input but with a finite error as shown in Fig. 7.7. The value of the gain k for a type 2 system is called the acceleration error constant and is also denoted by k_a.

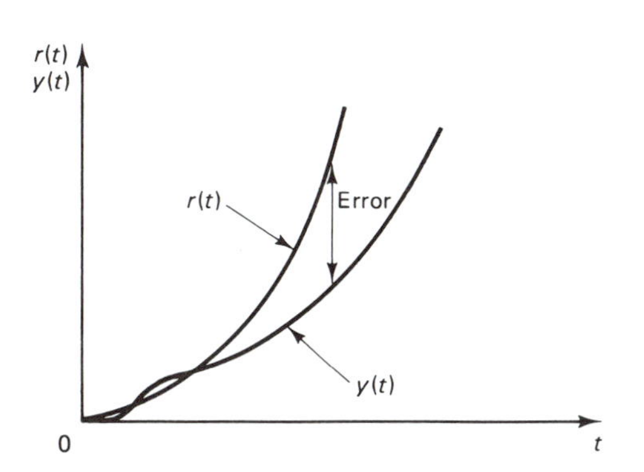

Figure 7.6 Response of type 0 and type 1 systems to parabolic input.

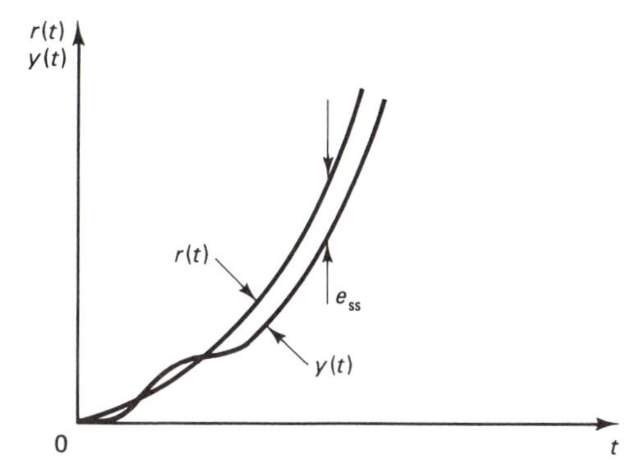

Figure 7.7 Response of a type 2 system to parabolic input.

A summary of the error analysis for asymptotically stable systems is given in Table 7.1 for the three inputs. It should be noted that it is assumed that the control systems are asymptotically stable. A similar table can be constructed for the steady-state errors due to the disturbance, depending on the total number of integrators in the forward path and their locations before or after the entrance of the disturbance. It is seen that the tracking accuracy and disturbance-rejection property are improved as the type number is increased. However, each additional integrator in the forward path tends to destabilize the control system. A compromise is therefore necessary and type 1 systems are commonly employed.

TABLE 7.1. SUMMARY OF STEADY-STATE ERRORS

	Steady-State Errors		
System type	Step input $r(t) = a$	Ramp input $r(t) = at$	Parabolic input $r(t) = \frac{1}{2}at^2$
0	$a/1 + k_p$	∞	∞
1	0	a/k_v	∞
2	0	0	a/k_a

EXAMPLE 7.9

The electrohydraulic control system of Example 3.6 for machining a workpiece is classified as type 1 in Example 7.3 and its steady-state errors for step changes in the command input and in the disturbance are analyzed in Example 7.8. Different control laws are discussed in the next section. In Fig. 3.22, let the proportional control law $i = k_a(E_r - E_0)$ be modified to an integral control law where $i = (k_a/s)(E_r - E_0)$.

The modified control system is now type 2. Its characteristic equation is obtained from Eq. (7.10) by replacing the s term on the denominator by s^2 and is given by

$$s^2\left(\frac{1}{\omega_n^2}s^2 + \frac{2\zeta}{\omega_n}s + 1\right)(\tau s + 1) + k = 0 \qquad (7.36)$$

On expanding this equation, we obtain a fifth-order polynomial whose s term coefficient is zero. Hence, the Routh array will indicate that the modified control system is unstable. For an unstable system, the steady-state error is meaningless and the final-value theorem is not applicable. Hence, the steady-state errors given in Table 7.1 for a type 2 system are not valid for this modified system.

7.4 CONTROL LAWS

In this section, we introduce the basic concepts of controller design to implement a control law. We then consider the physical realization of controllers by analog devices. An error signal $e(t)$ is the input to the controller that employs an appropriate control law to obtain its output $u(t)$. The controller output signal then becomes the control input to an actuator. In the following, we describe the three linear control laws and their combinations commonly used in practice.

Proportional Control Law. In this control law, the output of the controller is simply related to its input by a proportional constant. Hence, a proportional control law is described by

$$u(t) = k_p e(t) \tag{7.37}$$

where k_p is the proportional gain. In the examples given in Chapter 3, a proportional control law has been employed for simplicity. The output $u(t)$ depends only on the magnitude of the error at that time and a large corrective action is produced when the error is large at the current time. The main advantage of a proportional control law is that it is simple to implement with an amplifier. Its main disadvantage is that if the system has no pure integrator in the forward path, the resulting control system will be type 0 with steady-state errors as given in Table 7.1.

Integral Control Law. In this control law, the output of the controller is proportional to the time integral of the error and we have

$$u(t) = k_i \int_0^t e(t)\, dt \qquad \text{i.e., } U(s) = (k_i/s)E(s) \tag{7.38}$$

where k_i is the integral gain. The output is proportional to the accumulation of the past error and a large corrective action is produced when the integal of the error is large at the current time. Integral control is also called reset action.

One effect of integral control is that it increases the order of the system by one. The main advantage of integral control is that it increases the system type by one. Therefore, the steady-state error of the system without integral control is improved to that of one type higher. For example, if without integral control the original system is type 0, then it is converted to type 1 by integral control. Its main disadvantage is that integral control adds a pole at the origin of the s-plane to the open-loop transfer function and this has a destabilizing effect. This effect can be seen from Rule 7 of the root-locus method presented in the previous chapter. The addition of a pole at the origin of the s-plane moves the centroid closer to the imaginary axis.

Derivative Control Law. In this case, the controller output is proportional to the time derivative of the error and ideally we have

$$u(t) = k_d\, de/dt \qquad \text{i.e., } U(s) = k_d s E(s) \tag{7.39}$$

where k_d is the gain of the derivative controller. When the slope of $e(t)$ is large at current time, the magnitude of $e(t)$ will increase in the future. Hence, a derivative control law provides a large corrective action in anticipation before the error becomes large. Derivative control law is also called rate action.

The main advantage of derivative control law is that it predicts the growth of the error ahead of time and provides a corrective action before a large error actually occurs. However, if the error is a constant, then $u = 0$ and no corrective action is taken at all even when the error is large. Hence, the main disadvantage of derivative control is that it is insensitive to slowly varying errors and allows drift. For this reason, it is not used alone, but is combined with other control laws.

The transfer function of the ideal differentiator of Eq. (7.39) has a first-order numerator and zero-order denominator and hence is noncausal. For this reason, there are difficulties in implementing a derivative control law. An operational amplifier could be used for differentiation, but it is not done in practice because it causes amplification of the noise. Hence, a derivative control law is usually approximated by

$$U(s) = \left(\frac{k_d s}{\tau s + 1}\right) E(s) \tag{7.40}$$

where the time constant τ is very small. The Bode diagram of the transfer function of Eq. (7.40) shows that it acts like an ideal differentiator for frequencies up to about $\omega = 1/\tau$. For frequencies higher than $1/\tau$, the magnitude curve has a slope of 0 dB/decade instead of the 20-dB/decade slope of a differentiator and hence the high-frequency signals are not differentiated. For this reason, τ must be chosen to be appropriately small.

PID Family of Control Laws. The advantages of the three preceding control laws can be combined without their disadvantages in a single three-mode proportional plus integral plus derivative (PID) control law in the form

$$U(s) = \left(k_p + \frac{k_i}{s} + \frac{k_d s}{\tau s + 1}\right) E(s) \tag{7.41}$$

where τ is small. The effect of each of the three modes is additive. The values of the constants k_p, k_i, and k_d are chosen to satisfy the performance specifications.

Other combinations are also commonly employed. If the system already has an integrator in the forward path outside the controller, then proportional plus derivative (PD) control law may be adequate. Also, from considerations of cost, the derivative action may be omitted and proportional plus integral (PI) controllers are frequently used in practice.

In the preceding development, it is assumed that the controller operates within the linear range of the system. With proportional control law, when there is a large change in the set point, the large initial error can cause the controller or other components, such as the actuator or valve, to saturate. When the error decreases, a proportional controller will operate within its linear range.

In addition, a control law that includes an integral mode can exhibit a nonlinear behavior called reset windup. If there is a large initial error, the integrator does not respond immediately, but it integrates the error and produces a large signal some time later. The integral control action continues to increase as long as the error has not changed sign even though the controlled output may be close to its desired value. If the error remains large for a sufficiently long time, the integral mode can cause saturation even when the current value of the error is small. Consequently, a large overshoot can result.

The full PID control law of Eq. (7.41) raises the order of the control system by two. The integral mode and the practical derivative mode are each responsible for raising the order by one. Hence, two additional state variables must be defined for the state-variables representation. If an ideal derivative mode were employed, that is, $\tau = 0$ then Eq. (7.41) would raise the order only by one. The integral mode in Eq. (7.41) improves the system type by one. The PID family of control laws can be implemented by an analog controller, consisting of components that are electronic, mechanical, hydraulic, pneumatic or their combinations. The implementation of control laws by a digital computer is covered in Chapter 9.

EXAMPLE 7.10

We consider the mechanical liquid-level control system of Example 3.1. It is seen from Fig. 3.4(b) that the control law is proportional. We have classified this system as type 0 in Example 7.1 and hence it exhibits the steady-state errors given in Table 7.1 for that type. A hydromechanical controller that implements a PID control law for this system is shown in Fig. 7.8.

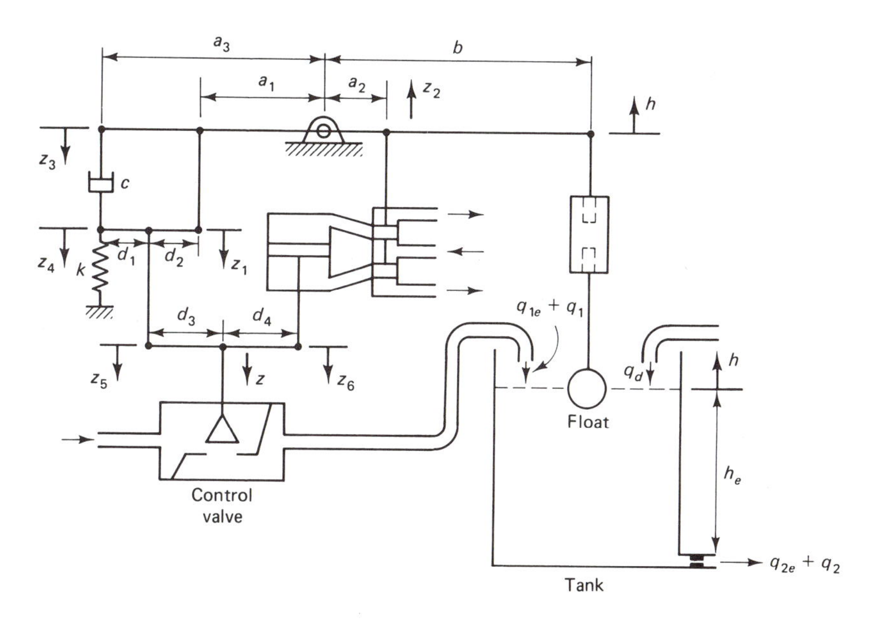

Figure 7.8 PID controller for a liquid-level control system.

The displacement z_1 is related to h by

$$z_1 = (a_1/b)h \tag{7.42}$$

and the displacement z_2 of the spool valve by

$$z_2 = (a_2/b)h \tag{7.43}$$

The load on the actuator is negligible and as shown in Eq. (2.148), we obtain

$$z_6 = (k_1/D)z_2$$

and using Eq. (7.43),

$$z_6 = \left(\frac{k_1 a_2}{b}\right)\left(\frac{1}{D}\right) h \tag{7.44}$$

Equating the damper force to the spring force, we get

$$c(\dot{z}_3 - \dot{z}_4) = kz_4$$

or

$$\begin{aligned} z_4 &= \left(\frac{cD}{cD + k}\right) z_3 \\ &= \left(\frac{\tau D}{\tau D + 1}\right)\left(\frac{a_3}{b}\right) h \end{aligned} \tag{7.45}$$

where $\tau = c/k$ and $z_3 = (a_3/b)h$. The valve movement z is obtained as

$$z = \left(\frac{d_4}{d_3 + d_4}\right) z_5 + \left(\frac{d_3}{d_3 + d_4}\right) z_6 \tag{7.46}$$

where

$$z_5 = \left(\frac{d_1}{d_1 + d_2}\right) z_1 + \left(\frac{d_2}{d_1 + d_2}\right) z_4 \tag{7.47}$$

Substituting for z_5 in Eq. (7.46) from Eq. (7.47) and then using Eqs. (7.42), (7.44), and (7.45), we obtain

$$\begin{aligned} z = &\left(\frac{d_4}{d_3 + d_4}\right)\left(\frac{d_1}{d_1 + d_2}\right)\left(\frac{a_1}{b}\right) h + \left(\frac{d_3}{d_3 + d_4}\right)\left(\frac{k_1 a_2}{b}\right)\left(\frac{1}{D}\right) h \\ &+ \left(\frac{d_4}{d_3 + d_4}\right)\left(\frac{d_2}{d_1 + d_2}\right)\left(\frac{\tau a_3}{b}\right)\left(\frac{D}{\tau D + 1}\right) h \end{aligned} \tag{7.48}$$

This equation can be expressed as

$$z = k_p h + \left(\frac{k_i}{D}\right) h + k_d \left(\frac{D}{\tau D + 1}\right) h \tag{7.49}$$

where the gains k_p, k_i and k_d are obtained by comparing the corresponding terms in Eqs. (7.48) and (7.49). The linearized equation for the flow-control valve is

$$q_1 = -c_1 z \tag{7.50}$$

and the mathematical model of the tank has been obtained in Example 3.1. The block diagram may now be completed as shown in Fig. 7.9.

Thus, the hydromechanical controller implements a PID control law. The time constant τ must be chosen to be small to extend the frequency range of the derivative mode. After summing up the three control actions, we can see that the system is now type 1. The first order of the original system of Example 3.1 has now been raised to the third order. Hence, two additional

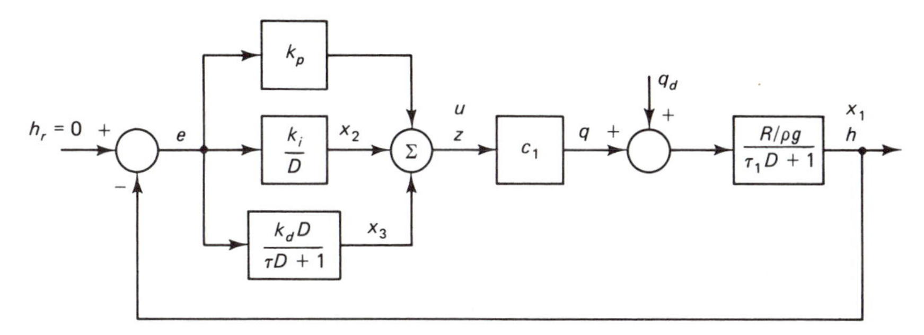

Figure 7.9 Block diagram of a liquid-level control system.

state variables must be defined as shown in Fig. 7.9 for the state-variables representation.

Noting that the set point has not been changed, that is, $h_r = 0$, we obtain the state equations as follows.

$$\begin{aligned}
\dot{x}_1 &= -\left(\frac{1}{\tau_1}\right) x_1 + \left(\frac{Rc_1}{\tau_1 \rho g}\right) u + \left(\frac{R}{\tau_1 \rho g}\right) q_d \\
\dot{x}_2 &= -k_i x_1 \\
\dot{x}_3 &= -\left(\frac{1}{\tau}\right) x_3 - \left(\frac{k_d}{\tau}\right) \dot{x}_1 \\
&= \left(\frac{k_d}{\tau\tau_1}\right) x_1 - \left(\frac{1}{\tau}\right) x_3 - \left(\frac{k_d R c_1}{\tau\tau_1 \rho g}\right) u - \left(\frac{k_d R}{\tau\tau_1 \rho g}\right) q_d
\end{aligned} \tag{7.51}$$

where in the last equation, we have substituted for $\dot{x}_1$ from the first equation. We also have

$$u = -k_p x_1 + x_2 + x_3$$

The preceding equations can be expressed in the standard form

$$\dot{\mathbf{x}} = \mathbf{A}\mathbf{x} + \mathbf{B}u + \mathbf{B}_1 v \qquad u = -\mathbf{K}\mathbf{x}$$

where

$$\mathbf{A} = \begin{bmatrix} -1/\tau_1 & 0 & 0 \\ -k_i & 0 & 0 \\ k_d/\tau\tau_1 & 0 & -1/\tau \end{bmatrix} \qquad \mathbf{B} = \begin{bmatrix} Rc_1/\tau_1\rho g \\ 0 \\ -k_d R c_1/\tau\tau_1\rho g \end{bmatrix}$$
$$\mathbf{K} = \lfloor k_p \quad -1 \quad -1 \rfloor \qquad \mathbf{B}_1 = \begin{bmatrix} R/\tau_1\rho g \\ 0 \\ -k_d R/\tau\tau_1\rho g \end{bmatrix} \tag{7.52}$$

The closed-loop system can now be expressed as

$$\dot{\mathbf{x}} = (\mathbf{A} - \mathbf{B}\mathbf{K})\mathbf{x} + \mathbf{B}_1 v$$

and its characteristic equation is given by

$$\det|s\mathbf{I} - \mathbf{A} + \mathbf{B}\mathbf{K}| = 0$$

7.5 IMPLEMENTATION OF CONTROL LAWS WITH ELECTRONIC CONTROLLERS

In electrical control systems, such as those of Examples 3.3 and 3.4, and in electrohydraulic systems, as in Examples 3.5 and 3.6, a convenient method of implementing an appropriate control law is by means of an electronic controller using operational amplifiers. An operational amplifier is shown symbolically in Fig. 7.10. All voltages are with respect to the ground, but in the subsequent diagrams the ground connection is not shown according to practice.

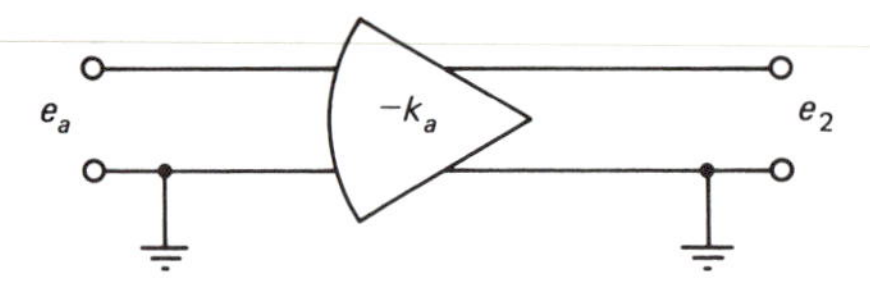

Figure 7.10 Operational amplifier.

The output voltage $e_2(t)$ of the operational amplifier is related to its input voltage $e_a(t)$ by

$$e_2 = -k_a e_a \tag{7.53}$$

where the gain k_a of the amplifier is very large and is of the order of 10^5 to 10^8. It is seen that there is also a sign reversal. The operational amplifier can be made to perform the desired operations by connecting suitable impedances Z_1 in the forward path and Z_2 in the feedback path as shown in Fig. 7.11.

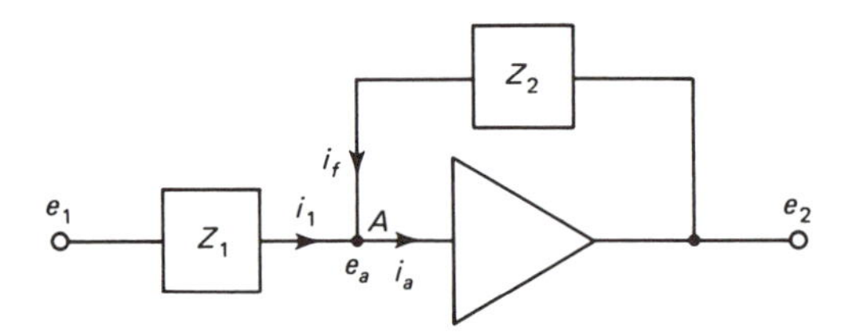

Figure 7.11 Operational amplifier with impedances.

Employing Kirchhoff's current law to the node A in Fig. 7.11, we obtain

$$i_1 + i_f = i_a \tag{7.54}$$

The operational amplifier has a high internal impedance and hence it draws a negligible current i_a, which is of the order of 10^{-9} ampere. Neglecting i_a in Eq. (7.54), we get

$$i_1 + i_f = 0 \tag{7.55}$$

The currents can be expressed by

$$i_1 = \frac{e_1 - e_a}{Z_1} \qquad i_f = \frac{e_2 - e_a}{Z_2} \tag{7.56}$$

We also have $e_2 = -k_a e_a$. The maximum value of e_2 is limited to a value that may range from 10 to 100 volts, depending on the operational amplifier, and k_a is of the order of 10^5 to 10^8. Hence, $e_a \approx 0$ and Eq. (7.56) becomes

$$i_1 = e_1/Z_1 \qquad i_f = e_2/Z_2$$

Substituting these results in Eq. (7.55), we obtain

$$e_2/e_1 = -Z_2/Z_1 \tag{7.57}$$

This is the basic relationship employed to implement a control law as described in the following. We note the sign reversal in Eq. (7.57).

Proportional controller

To obtain a proportional control law, we let Z_1 and Z_2 be two resistances R_1 and R_2, respectively. We then get

$$e_2/e_1 = -R_2/R_1 \tag{7.58}$$

where e_1 is the error voltage and the proportional gain $k_p = R_2/R_1$. This gain can be made adjustable by using a potentiometer in series with one of the resistances. The sign in Eq. (7.58) can be changed by using another amplifier in series with $Z_1 = Z_2 = R$ as shown in Fig. 7.12. The second amplifier merely changes the sign of the input voltage without affecting its magnitude and is called an inverter.

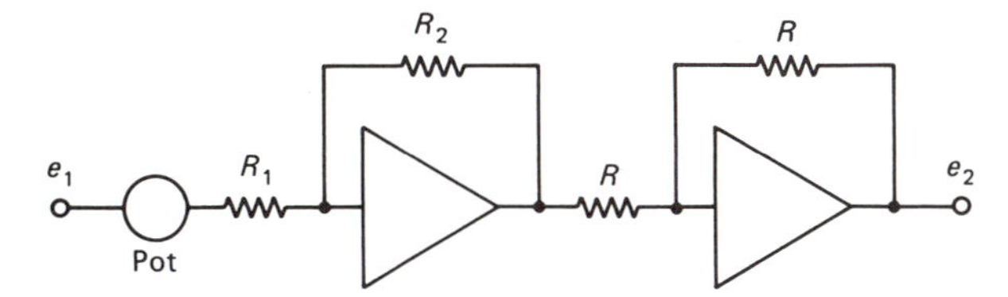

Figure 7.12 Proportional controller.

Integral controller

An integral control law can be obtained by connecting a resistance R in the forward path and a capacitance C in the feedback path as shown in Fig. 7.13. Then, $Z_1 = R$ and $Z_2 = 1/CD$, and from Eq. (7.57)

$$\frac{e_2}{e_1} = -\frac{1}{RCD} \qquad \text{i.e., } e_2 = -\frac{1}{RC}\int e_1\, dt \tag{7.59}$$

where e_1 is the error voltage and the integral gain $k_i = 1/RC$. Again, this gain can be made adjustable by using a potentiometer in series with the resistance and an inverter can be used to change the sign as shown in Fig. 7.13.

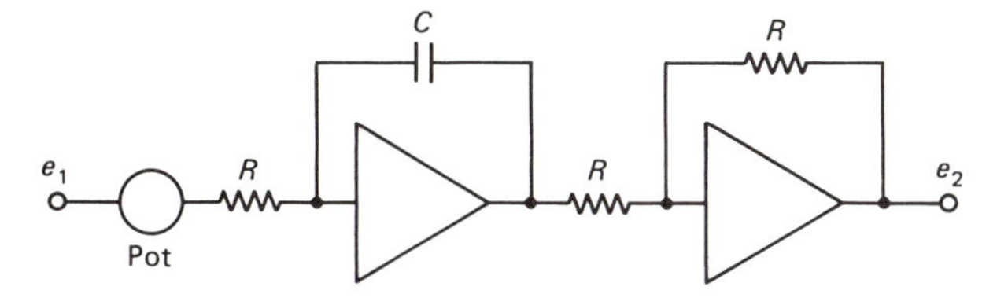

Figure 7.13 Integral controller.

Derivative controller

A derivative control law can be obtained in theory as shown in Fig. 7.14. In this case, we have $Z_1 = 1/CD$ and $Z_2 = R_2$, and from Eq. (7.57)

$$e_2/e_1 = -R_2CD \qquad \text{i.e., } e_2 = -R_2C\, de_1/dt \tag{7.60}$$

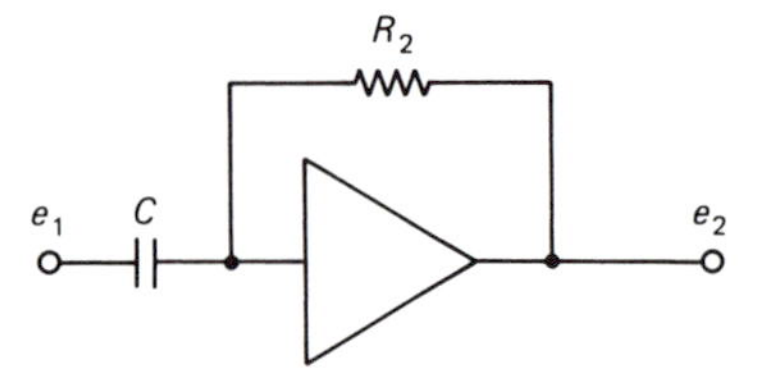

Figure 7.14 Ideal derivative controller.

The Bode diagram of a pure differentiator shows that its magnitude curve has a slope of 20 dB/decade, which implies amplification at high frequencies. A signal is usually corrupted by noise, which has high-frequency components. A differentiator will, therefore, amplify the noise. For this reason, an operational amplifier is not used in practice for differentiation, but a modification shown in Fig. 7.15 is used.

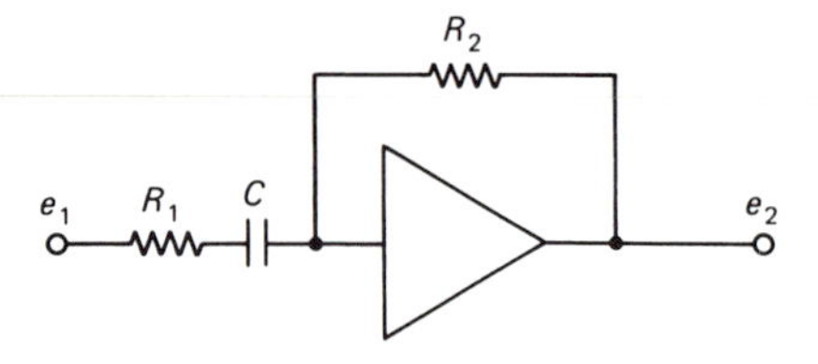

Figure 7.15 Practical derivative controller.

From Fig. 7.15, we get

$$Z_1 = R_1 + \frac{1}{CD} = \frac{R_1CD + 1}{CD} \qquad Z_2 = R_2$$

and from Eq. (7.57), it follows that

$$e_2 = -\left(\frac{R_2CD}{R_1CD + 1}\right) e_1 = -\left(\frac{k_dD}{\tau D + 1}\right) e_1 \tag{7.61}$$

where the derivative gain $k_d = R_2C$. The time constant τ must be chosen to be sufficiently small so that Eq. (7.61) behaves as a differentiator for frequencies up to about $\omega = 1/\tau$.

PID family of control laws

Many electronic controllers employed in industry have provision for combining the three basic control modes to obtain a PID family of control laws, depending on the characteristics of the plant to be controlled. The implementation of the PI control law is shown in Fig. 7.16. From this figure, we obtain

$$Z_1 = R_1 \qquad Z_2 = R_2 + 1/CD$$

$$\frac{Z_2}{Z_1} = \frac{R_2CD + 1}{R_1CD} = \frac{R_2}{R_1} + \frac{1}{R_1CD}$$

Letting $k_p = R_2/R_1$ and $k_i = 1/R_1C$,

$$e_2 = -\left(k_p + \frac{k_i}{D}\right) e_1 \tag{7.62}$$

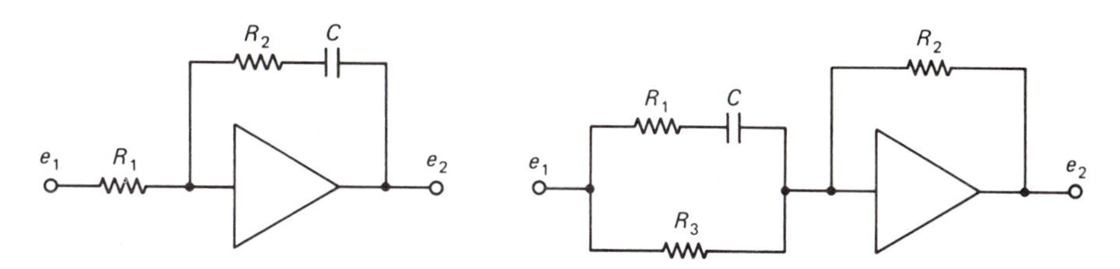

Figure 7.16 PI controller.

Figure 7.17 Practical PD controller.

A practical PD control law can be implemented as shown in Fig. 7.17. To obtain Z_1, we note that R_1 and C are in series and their combination is in parallel with R_3. Hence,

$$Z_1 = \frac{1}{\dfrac{CD}{R_1CD+1} + \dfrac{1}{R_3}} \qquad Z_2 = R_2$$

$$\frac{Z_2}{Z_1} = -\frac{R_2(R_3CD + R_1CD + 1)}{R_3(R_1CD+1)} = -\left(\frac{R_2}{R_3} + \frac{R_2CD}{R_1CD+1}\right)$$

Letting $k_p = R_2/R_3$, $k_d = R_2C$, and $\tau = R_1C$, we obtain

$$e_2 = -\left(k_p + \frac{k_dD}{\tau D + 1}\right) e_1 \tag{7.63}$$

A PID control law can be obtained by joining in parallel the PI controller of Fig. 7.16 and the practical derivative controller of Fig. 7.15. An additional amplifier is used in Fig. 7.18 as an adder/inverter. We get

$$e_2 = \left(k_p + \frac{k_i}{D} + \frac{k_dD}{\tau D + 1}\right) e_1 \tag{7.64}$$

It should be noted that there are several other possible combinations for implementing a PID control law.

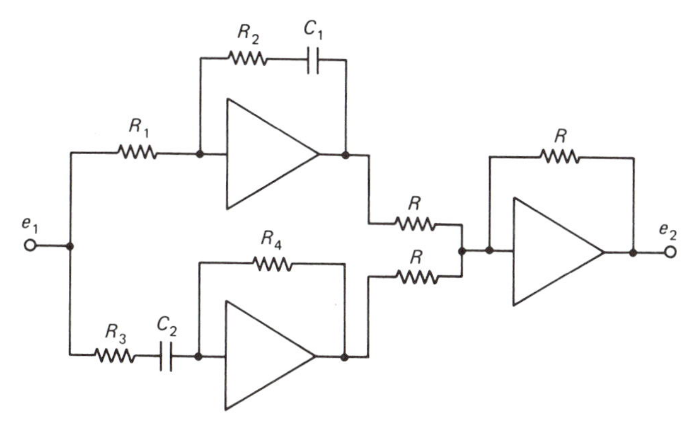

Figure 7.18 Practical PID controller.

EXAMPLE 7.11

The mathematical model of an electrical position-control system is obtained in Example 3.4. From the block diagram of Fig. 3.15, we see that the system is type 1, but the integrator is located after the disturbance enters the system in the forward path. It is required to replace the proportional controller, which is the amplifier shown in Fig. 3.10, by a PI controller. Since the control system is electrical, it is convenient to implement the PI control law by using the controller of Fig. 7.16.

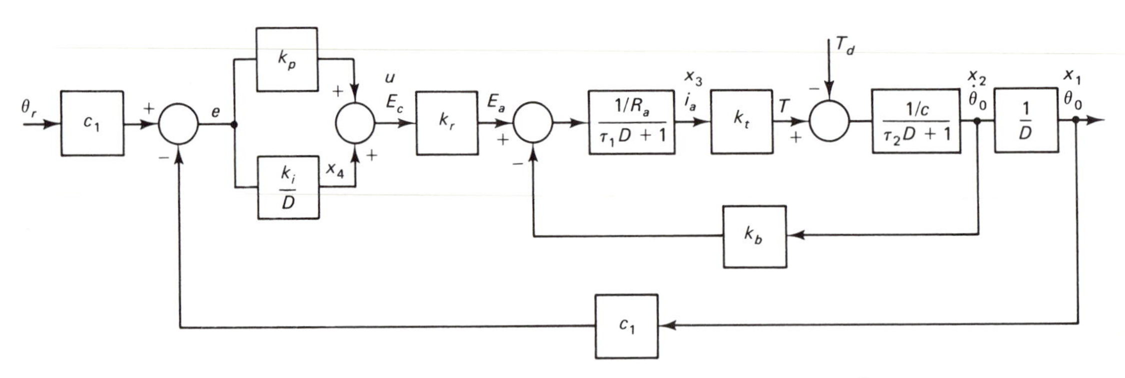

Figure 7.19 Position-control system with PI controller.

The block diagram of Fig. 3.15 is modified as shown in Fig. 7.19. The modified system is now type 2. The order of the system is raised from three to four and an additional state variable x_4 is defined as shown in Fig. 7.19. The state equations are obtained from this figure as follows:

$$\begin{aligned}
\dot{x}_1 &= x_2 \\
\dot{x}_2 &= -\left(\frac{1}{\tau_2}\right) x_2 + \left(\frac{k_t}{\tau_2 c}\right) x_3 - \left(\frac{1}{\tau_2 c}\right) T_d \\
\dot{x}_3 &= -\left(\frac{k_r k_p c_1}{\tau_1 R_a}\right) x_1 - \left(\frac{k_b}{\tau_1 R_a}\right) x_2 - \left(\frac{1}{\tau_1}\right) x_3 + \left(\frac{k_r}{\tau_1 R_a}\right) x_4 + \left(\frac{k_r k_p c_1}{\tau_1 R_a}\right) \theta_r \\
\dot{x}_4 &= -k_i c_1 x_1 + k_i c_1 \theta_r \\
u &= -k_p c_1 x_1 + x_4 + k_p c_1 \theta_r
\end{aligned} \tag{7.65}$$

The preceding equations can be expressed in the standard form

$$\dot{\mathbf{x}} = (\mathbf{A} - \mathbf{BK})\mathbf{x} + \mathbf{B}N\theta_r + B_1 T_d$$

where

$$(\mathbf{A} - \mathbf{BK}) = \begin{bmatrix} 0 & 1 & 0 & 0 \\ 0 & -1/\tau_2 & k_t/\tau_2 c & 0 \\ -k_r k_p c_1/\tau_1 R_a & -k_b/\tau_1 R_a & -1/\tau_1 & k_r/\tau_1 R_a \\ -k_i c_1 & 0 & 0 & 0 \end{bmatrix}$$

$$\mathbf{B}N = \begin{bmatrix} 0 \\ 0 \\ k_r k_p c_1/\tau_1 R_a \\ k_i c_1 \end{bmatrix} \qquad B_1 = \begin{bmatrix} 0 \\ -1/\tau_2 c \\ 0 \\ 0 \end{bmatrix}$$

$$v = T_d \text{ and } r = \theta_r \tag{7.66}$$

The closed-loop system can now be expressed as

$$\dot{\mathbf{x}} = (\mathbf{A} - \mathbf{BK})\mathbf{x} + \mathbf{B}Nr + \mathbf{B}_1 v \tag{7.67}$$

7.6 ADJUSTMENT OF CONTROLLER GAINS

After a controller has been selected to implement a desired control law, the plant and controller configurations are fixed. The problem then is to adjust the controller gains to satisfy the performance specifications, such as margin of stability, transient response, and bandwidth. The number of gains to be adjusted depends on the control law. A full PID controller has three gains to be tuned, a PI or PD controller has two, whereas a P or I controller has a single gain to be adjusted.

Trial-and-error procedures may sometimes be necessary to obtain the values of the gains that satisfy the performance specifications. After selecting the values of the gains, we can analyze the performance by using some of the techniques discussed in the previous chapters, such as the Routh criterion for checking the stability margin, transient response for time-domain specifications, and the Bode diagram for bandwidth. During the course of the design, the specifications may have to be revised to effect a compromise.

An analytical approach for this compromise is to compute the gains that minimize a performance index, which is expressed as a function of the error. We define the error between the command input $r(t)$ and the controlled output $y(t)$ as $e(t) = r(t) - y(t)$. Choosing $r(t)$ as a step input and letting all initial conditions be zero, we obtain an expression for the error as

$$\begin{aligned} E(s) &= R(s) - Y(s) = [1 - G_c(s)]R(s) \\ &= \frac{1}{s}[1 - G_c(s)] \end{aligned} \tag{7.68}$$

where $G_c(s)$ is the closed-loop transfer function and $R(s) = 1/s$. The commonly employed four choices for a performance index are the following.

Integral of absolute error (IAE),

$$I = \int_0^\infty |e(t)|\, dt \tag{7.69}$$

Integral of square error (ISE),

$$I = \int_0^\infty e^2(t)\, dt \tag{7.70}$$

Integral of time multiplied by absolute value of error (ITAE),

$$I = \int_0^\infty t|e(t)|\, dt \tag{7.71}$$

Integral of time multiplied by squared error (ITSE),

$$I = \int_0^\infty te^2(t)\, dt \tag{7.72}$$

The ISE criterion penalizes large errors. The ITAE and ITSE performance indices emphasize errors occurring later in the response. The closed-loop system must be stable and the error decay to zero with time so that the integrals converge. The optimum transfer functions for the ITAE performance index are given by Graham and Lathrop (1953) and those for ISE by Newton, Gould, and Kaiser (1957). The choice of the performance index is not limited to those given by Eqs. (7.69) to (7.72) and constraints that prevent saturation and reset windup can be included. In fact, a complete control law can be synthesized by using a general performance index, but optimal control theory (Bryson and Ho, 1969) is beyond our scope.

EXAMPLE 7.12

We consider the adjustment of the gains of the controller of Example 7.10, but without the derivative mode. The following specifications are to be satisfied. The system must be type 1, the bandwidth $\omega_b \leq 20$ rad/s, and the characteristic equation of the closed-loop system should be that of a type 1 ITAE form.

In Example 7.10, we omit the spring-damper combination shown in Fig. 7.8. The controller is then PI and the resulting system is type 1 as required by the specifications. The closed-loop transfer function relating the output h to the command input h_r can be obtained from Fig. 7.9 after omitting the derivative mode as

$$\frac{h}{h_r} = G_c(s) = \frac{(Rc_1/\rho g)(k_p s + k_i)}{s(\tau_1 s + 1) + (Rc_1/\rho g)(k_p s + k_i)} \tag{7.73}$$

Let the plant parameters be such that $Rc_1/\rho g = 100$ and $\tau_1 = 50$ seconds. We then obtain

$$G_c(s) = \frac{2k_p s + 2k_i}{s^2 + (0.02 + 2k_p)s + 2k_i} \tag{7.74}$$

A table of the standard ITAE forms is given in the next chapter. For a type 1 second-order system, the ITAE form for the denominator of Eq. (7.74) is given by (see Table 8.1)

$$s^2 + 1.4\omega_0 s + \omega_0^2$$

where ω_0 is as yet a free parameter. We note that the damping ratio $\zeta = 0.7$. On comparing this form with the denominator of Eq. (7.74), it is seen that

$2k_i = \omega_0^2$ and $(0.02 + 2k_p) = 1.4\omega_0$. Hence, Eq. (7.74) can be expressed as

$$G_c(s) = \frac{(1.4\omega_0 - 0.02)s + \omega_0^2}{s^2 + 1.4\omega_0 + \omega_0^2} = \frac{\tau s + 1}{\dfrac{1}{\omega_0^2}s^2 + \dfrac{2(0.7)}{\omega_0}s + 1} \tag{7.75}$$

where the time constant τ is defined by

$$\tau = 1.4/\omega_0 - 0.02/\omega_0^2$$
$$\approx 1.4/\omega_0 \qquad \text{for } \omega_0 > 1$$

A Bode diagram of the closed-loop transfer function, Eq. (7.75), is shown in Fig. 7.20, which shows that the magnitude is -3 dB at $\omega/\omega_0 = 2$. Hence, the bandwidth is $\omega_b = 2\omega_0$. As ω_0 is increased, the rise time decreases and the system is fast responding, but the bandwidth increases. Hence, to meet the specifications, we let $\omega_0 = 10$ rad/s. The controller gains are then obtained as $k_i = \omega_0^2/2 = 50$ and $(0.02 + 2k_p) = 1.4\omega_0$, that is, $k_p = 6.99$.

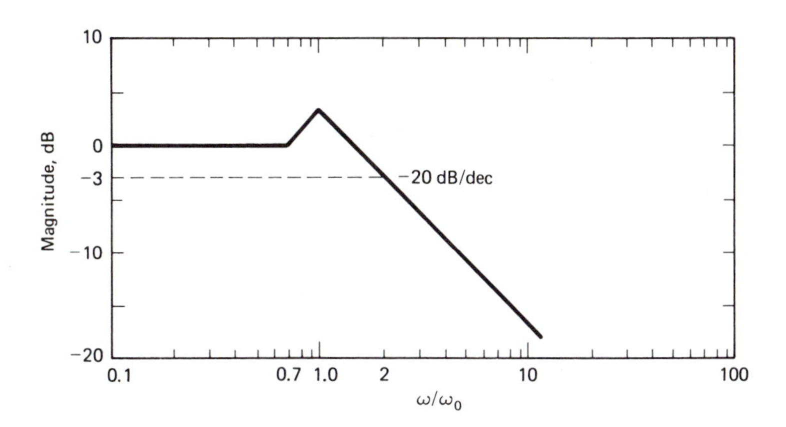

Figure 7.20 Bode diagram of a closed-loop transfer function.

The closed-loop transfer function, Eq. (7.75), then becomes

$$\frac{h}{h_r} = \frac{13.98s + 100}{s^2 + 14s + 100} \tag{7.76}$$

For h_r, a unit step input, and zero initial conditions, we obtain the response $h(t)$ by using the techniques of Chapter 4 as

$$h(t) = 1 + 1.398\, e^{-7t} \sin(7.14t - 45.66°) \tag{7.77}$$

7.6.1 The Ultimate-Cycle Method of Ziegler and Nichols

Based on experiments and analysis, Ziegler and Nichols (1943) have developed rules of thumb that are simple to employ and are quite popular in practice. Here, we discuss their ultimate-cycle method. Their analysis is based on minimizing the IAE performance index of Eq. (7.69). They observed that the optimal controller gains that minimized the IAE performance index yielded a unit step response as shown in Fig. 7.21, where the second overshoot is approximately one-quarter of the maximum overshoot. This is called the quarter-decay criterion.

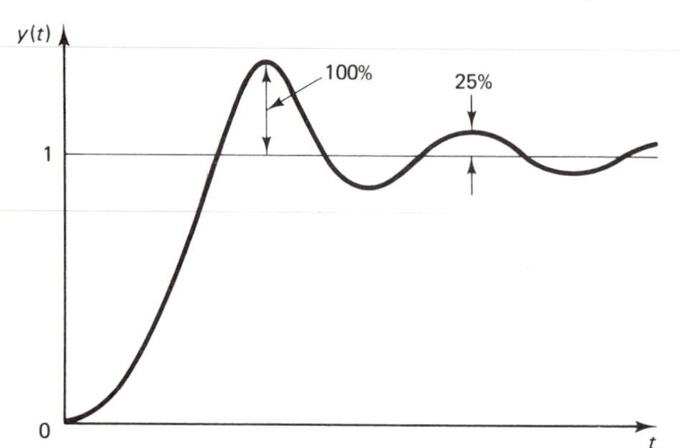

Figure 7.21 Step response.

The quarter-decay response is often a good compromise among small rise time, small settling time, and adequate margin of stability. Their ultimate-cycle method has the following two restrictions. When all the control modes except the proportional mode are turned off, in the resulting system with proportional control, there exists a value of gain k_{pu} at which the system is on the margin of stability. This value of the proportional gain is called the ultimate gain k_{pu}. This restriction is not satisfied, for example, by a second-order linear system that is stable for $0 < k_p < \infty$. The second restriction is that at the ultimate gain, a pair of roots of the characteristic equation must lie on the imaginary axis of the s-plane, whereas the remaining roots have negative real parts so that the response at the stability margin exhibits sustained oscillations.

Let T_u be the period of these oscillations. The rules of thumb for tuning the gains of a PID control law with transfer function

$$k_p + \frac{k_i}{s} + \frac{k_d s}{\tau s + 1}$$

are as follows:

P control law:

$$k_p = 0.5k_{pu}, \qquad k_i = 0, \qquad k_d = 0 \tag{7.78}$$

PI control law:

$$k_p = 0.45k_{pu}, \qquad k_i = 0.45k_{pu}/0.83T_u, \qquad k_d = 0 \tag{7.79}$$

PID control law:

$$k_p = 0.6k_{pu}, \qquad k_i = 0.6k_{pu}/0.5T_u, \qquad k_d = (0.6k_{pu})(0.125T_u) \tag{7.80}$$

EXAMPLE 7.13

An electronic PI controller has been employed in Example 7.11 for the electrical position-control system of Example 3.4. The gains k_p and k_i of this controller are to be tuned by the ultimate-cycle method. It is assumed that the mathematical model of the rest of the system has been validated and the parameter values identified by experimental frequency response, if necessary. It is interesting to note that the gains can be tuned by this method purely from experiments in the field when an analytical method is not available.

To find the ultimate gain and the period of oscillation, we set $k_i = 0$ so that the resultant system of Fig. 7.19 becomes a proportional control system of third order. The characteristic equation can be obtained directly from Fig. 7.19 with $k_i = 0$ or from the equation $|s\mathbf{I} - \mathbf{A} + \mathbf{BK}| = 0$, where $\mathbf{A} - \mathbf{BK}$ is given by Eq. (3.50) with k_a replaced by k_p. This equation is obtained as

$$s^3 + \left(\frac{1}{\tau_1} + \frac{1}{\tau_2}\right) s^2 + \left(\frac{1}{\tau_1 \tau_2} + \frac{k_b k_t}{\tau_1 \tau_2 R_a c}\right) s + k_p \left(\frac{k_t k_r c_1}{\tau_1 \tau_2 R_a c}\right) = 0 \qquad (7.81)$$

Let the plant parameters be such that the preceding equation becomes

$$s^3 + 4s^2 + 40s + 80k_p = 0 \qquad (7.82)$$

We can use the Routh criterion to determine the value of k_p for which Eq. (7.82) is on the margin of stability. The Routh array yields

$$\begin{array}{cc} 1 & 40 \\ 4 & 80k_p \\ 40 - 80k_p/4 & 0 \\ 80k_p & \end{array}$$

The ultimate gain k_p is given by

$$40 - 80k_p/4 = 0 \qquad \text{i.e., } k_{pu} = 2 \qquad (7.83)$$

A root locus of Eq. (7.82) can be drawn for $0 \le k_p < \infty$ from the equation

$$\frac{80k_p}{s(s + 4s + 40)} = -1$$

i.e.,

$$\frac{80k_p}{s(s + 2 + j6)(s + 2 - j6)} = -1 \qquad (7.84)$$

The angle of the asymptotes is given by

$$\begin{aligned} \measuredangle s &= \pm i\pi/3 \qquad i = 1, 3, 5, \ldots \\ &= \pm 60° \end{aligned}$$

and the point where the asymptotes intersect the real axis by

$$\sigma_c = \frac{-2 - 2}{3} = -\frac{4}{3}$$

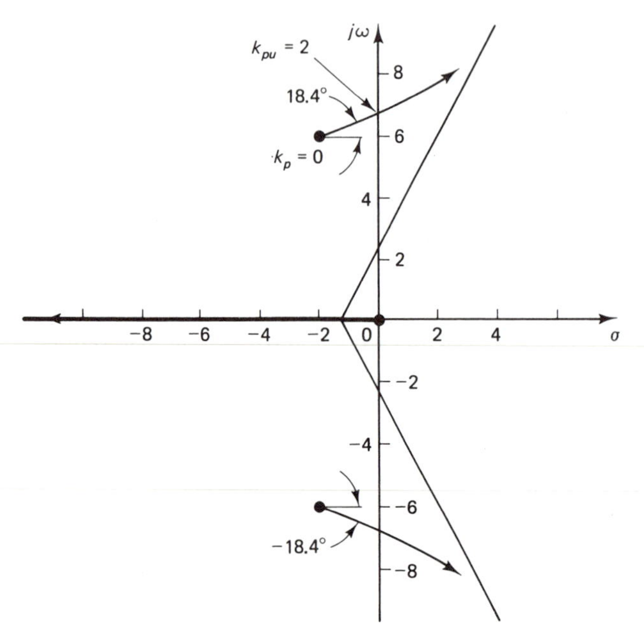

Figure 7.22 Root locus of Eq. (7.82).

The root locus is shown in Fig. 7.22. It is seen that as k_p is increased, the pair of complex conjugate roots crosses the imaginary axis at the ultimate gain k_{pu}, whereas the real root remains negative.

The pair of purely imaginary roots $\pm j\omega_u$ can be obtained from the root locus or by setting $s = j\omega_u$ in Eq. (7.82), with $k_p = k_{pu}$. We obtain

$$-j\omega_u^3 - 4\omega_u^2 + 40j\omega_u + 80k_{pu} = 0$$

Equating the real and imaginary parts individually to zero, we get

$$-4\omega_u^2 + 80k_{pu} = 0$$

$$40\omega_u - \omega_u^3 = 0$$

From the second equation, we obtain $\omega_u = (40)^{1/2}$ and substituting this value in the first equation, it can be checked that $k_{pu} = 2$. The ultimate period becomes $T_u = 2\pi/\omega_u = 0.993$ second. The gains k_p and k_i of the PI controller of Fig. 7.19 are now obtained from Eq. (7.79) as

$$k_p = 0.45(2) = 0.90$$

and

$$k_i = \frac{0.45(2)}{0.83(0.993)} = 1.09$$

The characteristic equation of the system of Fig. 7.19 with PI control law is obtained as

$$s^4 + \left(\frac{1}{\tau_1} + \frac{1}{\tau_2}\right)s^3 + \left(\frac{1}{\tau_1\tau_2} + \frac{k_b k_t}{\tau_1\tau_2 R_a c}\right)s^2 + k_p\left(\frac{k_t k_r c_1}{\tau_1\tau_2 R_a c}\right)s + k_i\left(\frac{k_t k_r c_1}{\tau_1\tau_2 R_a c}\right) = 0 \tag{7.85}$$

Using the previous plant parameter values and $k_p = 0.9$, the preceding equation becomes

$$s^4 + 4s^3 + 40s^2 + 72s + 80k_i = 0 \tag{7.86}$$

The root locus of Eq. (7.86) can be drawn for $0 \le k_i < \infty$ from the equation

$$\frac{80k_i}{s(s^3 + 4s^2 + 40s + 72)} = -1$$

i.e.,

$$\frac{80k_i}{s(s+2)(s+1+j5.92)(s+1-j5.92)} = -1 \tag{7.87}$$

The angle of asymptotes is given by

$$\measuredangle s = \pm i180/4 \qquad i = 1, 3, 5, \ldots$$
$$= \pm 45°, \pm 135°$$

The point where the asymptotes intersect the real axis is obtained as

$$\sigma_c = \frac{-2-1-1}{4} = -1$$

The root locus for $0 \le k_i < \infty$ is shown in Fig. 7.23. The locus crosses to the right half of the s-plane at $k_i = 4.95$ at $\omega = 4.24$. The roots of the characteristic equation, Eq. (7.86), for $k_i = 1.09$ may now be obtained from this figure.

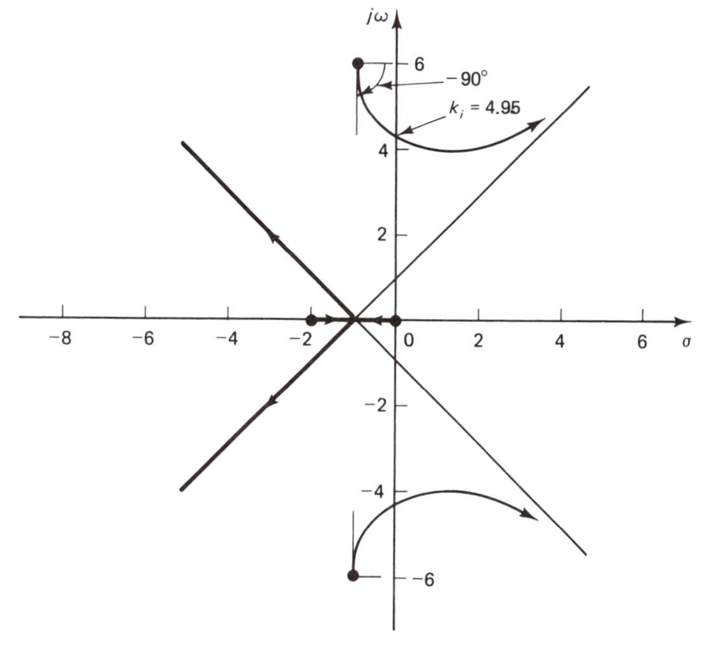

Figure 7.23 Root locus of Eq. (7.86).

The value of k_i for the margin of stability can also be obtained from the Routh criterion. The Routh array becomes

$$\begin{array}{ccc} 1 & 40 & 80k_i \\ 4 & 72 & 0 \\ 22 & 80k_i & \\ 72 - 14.55k_i & 0 & \\ 80k_i & & \end{array}$$

The value of k_i for margin of stability is again obtained as $k_i = 72/14.55 = 4.95$. The value of k_i that we have set is $k_i = 1.09$ and hence there is adequate margin of stability.

7.7 COMPENSATION

In some control systems, the performance specifications cannot be satisfied by adjusting the controller gains. In other cases, it is desirable to enhance the controller performance. For this purpose, additional components are inserted within the structure of the control system to make up for the deficiency of the controller. This alteration of the control system to improve its performance is called compensation. A compensator may consist of electrical, mechanical, hydraulic, or pneumatic components. The three commonly employed methods of compensation are cascade or series compensation, feedback compensation, and feedforward compensation.

7.7.1 Cascade Compensation

In this method, the compensator is inserted in the forward path of the control loop, usually after the controller. The problem then reduces to the selection of the poles and zeros of the compensator.

EXAMPLE 7.14

We consider a system for the attitude control of a satellite. Actually, satellites require three-axes attitude control for their proper orientation. In this example,

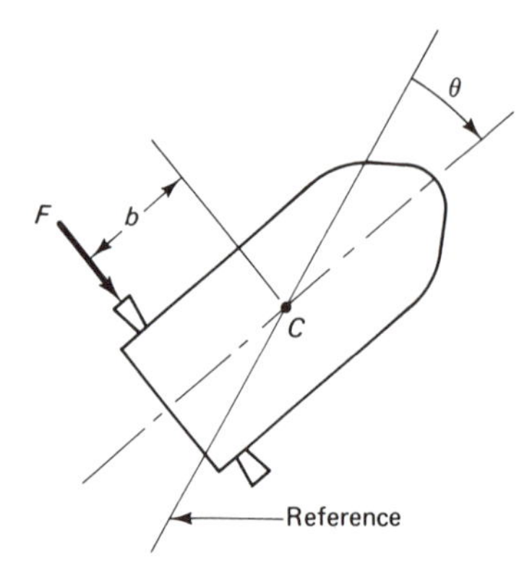

Figure 7.24 Attitude control of a satellite.

we discuss only a single-axis control, where the rotation is about an axis perpendicular to the plane of Fig. 7.24. The error between the reference attitude $\theta_r(t)$ and the actual attitude $\theta(t)$ is sensed and a PI control law is used to actuate the thrustors.

The torque produced is given by $T = Fb$, where F is the thrust force, and b is the distance shown in Fig. 7.24. Since there is no drag, the torque balance yields

$$T = I\ddot{\theta} + T_d \tag{7.88}$$

where I is the mass moment of inertia about the mass center C, and T_d is the disturbance torque caused by the pressure of solar wind and other causes. Neglecting the dynamics of the actuator, a block diagram of the attitude-control system is shown in Fig. 7.25.

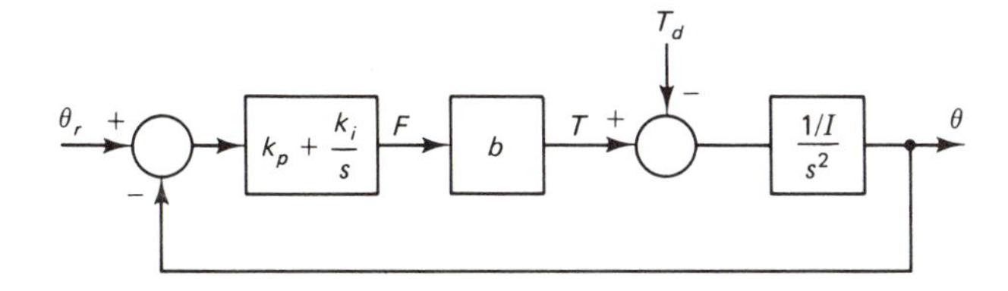

Figure 7.25 Block diagram of attitude-control system.

It is seen that the system is type 3 and we expect problems with stability. The characteristic equation is obtained as

$$s^3 + \frac{b}{I} k_p s + \frac{b}{I} k_i = 0 \tag{7.89}$$

The coefficient of the s^2 term is zero and hence the Routh criterion will indicate that the system is unstable and cannot be stabilized by adjusting the controller gains. In case the integral mode is omitted and only proportional control is employed, the characteristic equation becomes

$$s^2 + \frac{b}{I} k_p = 0 \tag{7.90}$$

which is an undamped quadratic with purely imaginary roots and adjusting k_p affects only the natural frequency.

We propose to employ proportional control only with a cascade compensator as shown in Fig. 7.26. The resulting system is type 2.

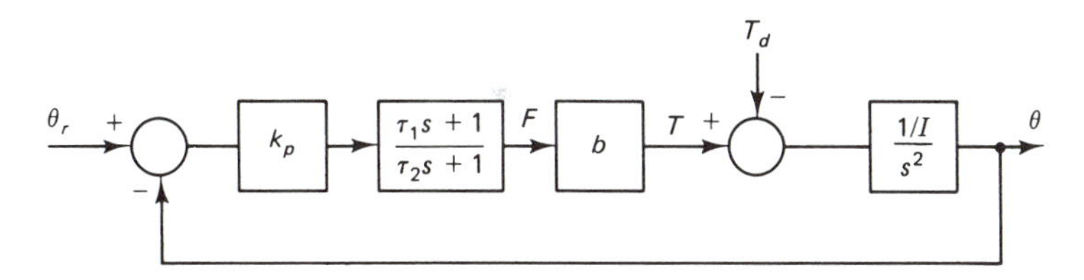

Figure 7.26 Attitude-control system with cascade compensation.

Its characteristic equation is obtained as

$$\tau_2 s^3 + s^2 + \frac{b\tau_1}{I} k_p s + \frac{b}{I} k_p = 0$$

Let $b/I = 100$. Then the preceding equation becomes

$$s^3 + \frac{1}{\tau_2} s^2 + 100 \frac{\tau_1}{\tau_2} k_p s + \frac{100}{\tau_2} k_p = 0 \tag{7.91}$$

It can be shown from the Routh criterion that for stability we need $\tau_1 > \tau_2$. This compensator is then called a phase-lead network because its phase angle is positive for finite frequencies. The three parameters τ_1, τ_2, and k_p in Eq. (7.91) can be adjusted by using the methods of the previous section. The characteristic equation of a standard ITAE, third-order, type 2 system, is given by (see Chapter 8, Table 8.2)

$$s^3 + 1.75\omega_0 s^2 + 3.25\omega_0^2 s + \omega_0^3 = 0 \tag{7.92}$$

On comparing the corresponding coefficients of Eqs. (7.91) and (7.92), we obtain $\tau_1 = 3.25/\omega_0$, $\tau_2 = 0.57/\omega_0$, and $k_p = 0.0057\omega_0^2$. The single parameter ω_0 can now be chosen as a compromise between fast step response and adequate bandwidth.

7.7.2 Feedback Compensation

In this method of compensation, the derivative of the controlled output is also measured and added to the controller output with an additional inner feedback loop to enhance the control law. This compensation can also serve as a substitute for the derivative mode with a PI controller. In a position-control system, the angular velocity is usually sensed by a tachometer for feedback compensation, and, in this special case, the rate feedback compensation is also called tachometer feedback.

EXAMPLE 7.15

It is proposed to employ rate feedback compensation instead of cascade compensation for the attitude-control system of Example 7.14 with a proportional control law. For this purpose, the attitude rate $\dot{\theta}$ is sensed by a rate sensor and added to the controller output as shown in Fig. 7.27.

With an inner feedback around one of the integrators, the resulting system is now type 1. Its characteristic equation is obtained as

$$s^2 + \frac{a}{I} k_1 s + \frac{a}{I} k_p = 0$$

i.e.,

$$s^2 + 100k_1 s + 100k_p = 0 \tag{7.93}$$

where we have let $a/I = 100$. There are two parameters k_1 and k_p to be

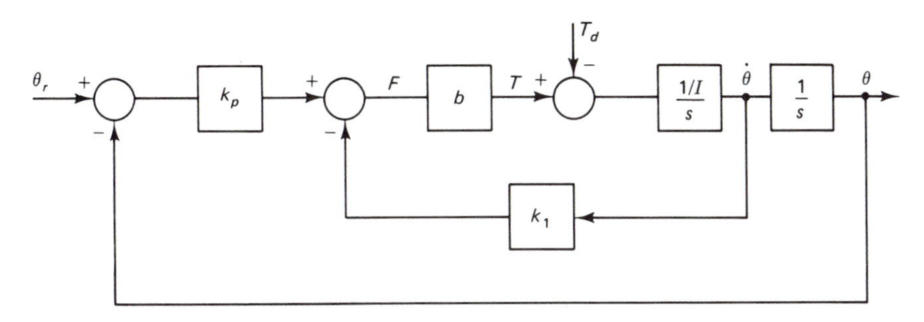

Figure 7.27 Attitude-control system with rate feedback.

adjusted. We note that $\omega_n^2 = 100k_p$ and $2\zeta\omega_n = 100k_1$. After selecting the natural frequency, we obtain the controller gain k_p. Then k_1 can be adjusted to yield a desired damping ratio. It is seen that rate feedback compensation has introduced damping to the system.

7.7.3 Feedforward Compensation

In the control systems considered thus far, a disturbance or load upset causes an error between the controlled output and the command input and the controller function is to reduce this error. But the effect of the disturbance must show up in the error before the controller can counteract it. In feedforward compensation, the disturbance is measured and a signal is added to the controller output to enhance the control law. Thus, a corrective action is initiated without waiting for the effect of the disturbance to show up in the error.

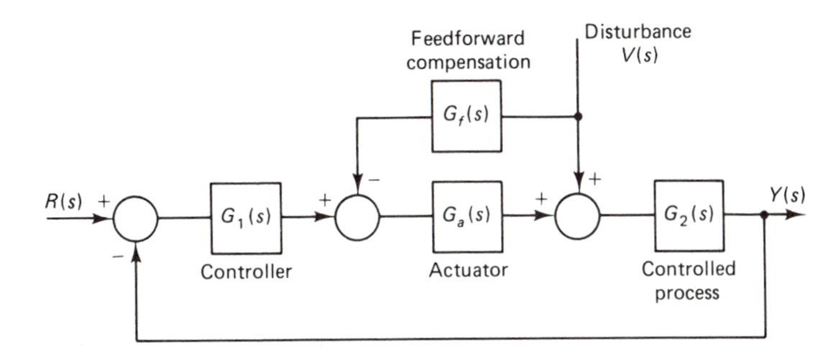

Figure 7.28 Control system with feedforward compensation.

A block diagram with feedforward compensation is shown in Fig. 7.28. It is noted that feedforward compensation has no effect on the system stability and it does not alter the transfer function relating the output to the command input.

Without feedforward compensation, the output is given by

$$Y(s) = \left(\frac{G_1 G_a G_2}{1 + G_1 G_a G_2}\right) R(s) + \left(\frac{G_2}{1 + G_1 G_a G_2}\right) V(s) \tag{7.94}$$

and with feedforward compensation, we obtain

$$Y(s) = \left(\frac{G_1 G_a G_2}{1 + G_1 G_a G_2}\right) R(s) + \left(\frac{G_2(1 - G_f G_a)}{1 + G_1 G_a G_2}\right) V(s) \tag{7.95}$$

If $G_f(s)$ is chosen such that $G_f(s) = G_a^{-1}(s)$, then it is seen from Eq. (7.95) that the effect of the disturbance is completely nullified. In case this $G_f(s)$ is noncausal, then $G_f(s)$ is chosen such that $G_f(s)G_a(s)$ approaches unity. We note that in Fig. 7.28, the feedforward signal is added to the controller output with a sign that is opposite to the sign with which the disturbance enters the process.

There are two difficulties associated with feedforward compensation. First, in some applications the disturbance may not be available for measurement. Second, feedforward compensation is an open-loop technique, and if the actuator transfer function $G_a(s)$ is not known accurately, then compensation may not be achieved at all.

EXAMPLE 7.16

It is desired to add feedforward compensation to enhance the action of the PID controller of the liquid-level control system of Example 7.10. For this purpose, the disturbance flow q_d in Fig. 7.8 is measured and an additional connection is introduced to the flow-control valve such that its opening decreases depending on q_d. From Fig. 7.9, a signal $G_f(s)q_d$ is subtracted from the controller output. Since, $G_a(s) = c_1$, we choose $G_f = 1/c_1$, so that $G_f G_a = 1$ and thus completely nullify the effect of the disturbance flow.

7.8 MULTIVARIABLE SYSTEMS

The controller design with output feedback discussed so far in this chapter is restricted to single-input, single-output systems. A simple approach for multivariable systems is to neglect the coupling terms in the mathematical model and design individual controllers for each of the variables to be controlled. The coupling terms are considered as "disturbances," but the performance cannot be improved beyond a certain level without taking the interactions into account.

One approach to the design of a controller for multivariable systems is to use state feedback, and this topic is studied in the next chapter. The design of a controller with output feedback uses the technique of decoupling or diagonalization so that each command input affects only one controlled output. Let a multivariable system be described by the state equations

$$\begin{aligned} \dot{\mathbf{x}} &= \mathbf{A}\mathbf{x} + \mathbf{B}\mathbf{u} + \mathbf{B}_1\mathbf{v} \\ \mathbf{y} &= \mathbf{C}\mathbf{x} \end{aligned} \tag{7.96}$$

where $\mathbf{y}$ and $\mathbf{u}$ are $m \times 1$ vectors. The plant matrix relating the output $\mathbf{Y}(s)$ to the control $\mathbf{U}(s)$ is given by

$$\mathbf{G}_p(s) = \mathbf{C}(s\mathbf{I} - \mathbf{A})^{-1}\mathbf{B} \tag{7.97}$$

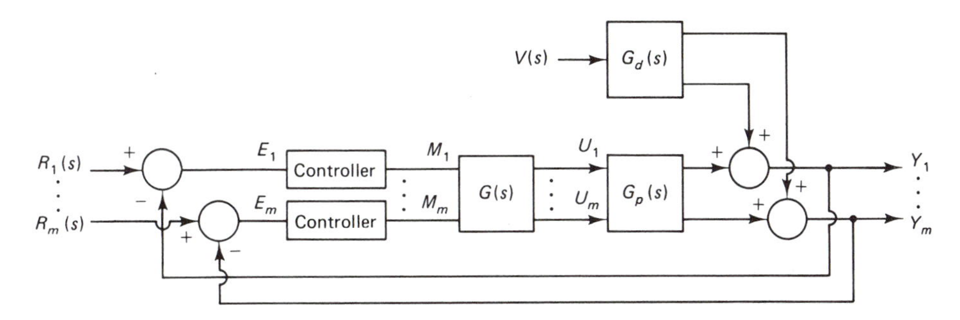

Figure 7.29 Multivariable system with output feedback control.

We design a compensator matrix $\mathbf{G}(s)$, shown in the block diagram of Fig. 7.29, so that

$$\mathbf{Y}(s) = \mathbf{G}_p(s)\mathbf{G}(s)\mathbf{M}(s) \tag{7.98}$$

The matrix $\mathbf{G}(s)$ is chosen such that $\mathbf{G}_p(s)\mathbf{G}(s)$ is a diagonal matrix. For the general case, the necessary and sufficient conditions for decoupling are given by Falb and Wolovich (1967). The problem is simplified when both $\mathbf{y}$ and $\mathbf{u}$ are $(m \times 1)$ vectors, and we assume that this is the case so that both $\mathbf{G}_p(s)$ and $\mathbf{G}(s)$ are $(m \times m)$ matrices. After decoupling is achieved in this manner, we can design the individual controllers for the m single-input, single-output uncoupled systems.

The procedure is illustrated by the following examples. We point out that several difficulties may arise. The decoupling compensation may require noncausal (i.e., physically unrealizable) networks. The transfer-function matrix is not known very accurately and hence accurate uncoupling may not be achieved. Some poles of the plant may be cancelled by the zeros of the compensator and this may render the system uncontrollable (see the following chapter).

EXAMPLE 7.17

A two-input–two-output, electrical temperature-control system was discussed in Example 3.8, where the structure of the controller was omitted. A PI controller is to be designed for each of the two outputs. The plant and disturbance matrices are given by Eq. (3.92). Let $\tau_1 = \tau_2 = 10$ and $b_1 = b_2 = 100$. With these parameter values, the matrices become

$$\mathbf{G}_p(s) = \begin{bmatrix} \dfrac{100}{s+0.1} & 0 \\ \dfrac{10}{(s+0.1)^2} & \dfrac{100}{s+0.1} \end{bmatrix} \qquad \mathbf{G}_d(s) = \begin{bmatrix} \dfrac{0.1}{s+0.1} \\ \dfrac{0.01}{(s+0.1)^2} \end{bmatrix} \tag{7.99}$$

The decoupling compensator matrix is chosen as

$$\mathbf{G}(s) = \begin{bmatrix} 1 & G_{12}(s) \\ G_{21}(s) & 1 \end{bmatrix} \tag{7.100}$$

We now obtain

$$\mathbf{G}_p\mathbf{G} = \begin{bmatrix} \dfrac{100}{s+0.1} & \dfrac{100G_{12}}{s+0.1} \\ \dfrac{10}{(s+0.1)^2} + \dfrac{100G_{21}}{s+0.1} & \dfrac{100}{s+0.1} \end{bmatrix} \tag{7.101}$$

We want this matrix to be diagonal. Hence, we set

$$G_{12}(s) = 0 \qquad \text{and} \qquad G_{21}(s) = -\frac{0.1}{s+0.1}$$

In this case, the diagonal elements of Eq. (7.101) are identical to the diagonal elements of the plant matrix in Eq. (7.99). Now, individual PI controllers can be designed for the two uncoupled systems by employing the techniques of single-input, single-output systems. The structure of the control

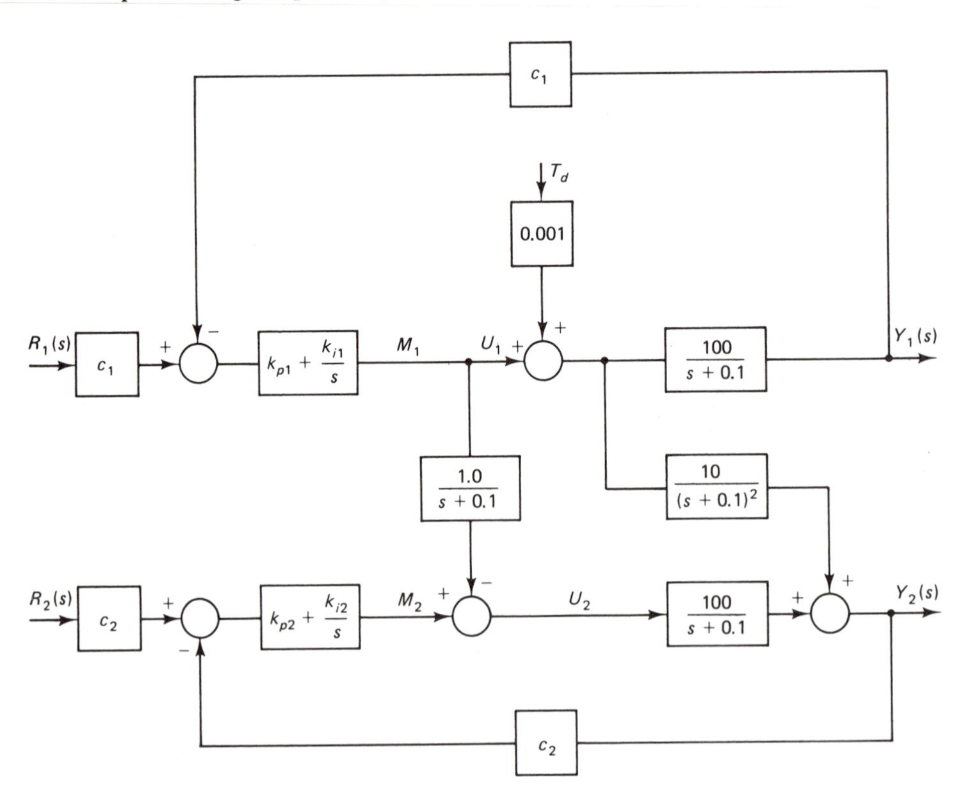

Figure 7.30 Structure of the control system.

system is shown in Fig. 7.30. Each of the two uncoupled systems is of second order, and the two controller gains can be set to yield the desired natural frequency and damping ratio.

EXAMPLE 7.18

The plant transfer function matrix of a two-input, two-output system is described by

$$\begin{bmatrix} Y_1(s) \\ Y_2(s) \end{bmatrix} = \begin{bmatrix} \dfrac{1}{s+1} & \dfrac{1}{s+2} \\ \dfrac{4}{s+3} & \dfrac{5}{s+4} \end{bmatrix} \begin{bmatrix} U_1(s) \\ U_2(s) \end{bmatrix} \tag{7.102}$$

A compensator is designed to decouple the plant. The compensator matrix is chosen as in Eq. (7.100), and we obtain

$$\mathbf{G}_p(s)\mathbf{G}(s) = \begin{bmatrix} \dfrac{1}{s+1} + \dfrac{G_{21}}{s+2} & \dfrac{G_{12}}{s+1} + \dfrac{1}{s+2} \\ \dfrac{4}{s+3} + \dfrac{5G_{21}}{s+4} & \dfrac{4G_{12}}{s+3} + \dfrac{5}{s+4} \end{bmatrix} \tag{7.103}$$

We want this matrix to be diagonal. Hence, we set

$$G_{12}(s) = -\frac{s+1}{s+2} \qquad \text{and} \qquad G_{21}(s) = -\frac{0.8(s+4)}{s+3}$$

Substituting these results in the diagonal elements of Eq. (7.103), we obtain

$$\begin{bmatrix} Y_1(s) \\ Y_2(s) \end{bmatrix} = \begin{bmatrix} \dfrac{0.2(s^2+5s+14)}{(s+1)(s+2)(s+3)} & 0 \\ 0 & \dfrac{s^2+5s+14}{(s+2)(s+3)(s+4)} \end{bmatrix} \begin{bmatrix} M_1 \\ M_2 \end{bmatrix} \tag{7.104}$$

The diagonal elements of Eq. (7.104) are different from the original diagonal elements of the plant matrix. The structure of the control system

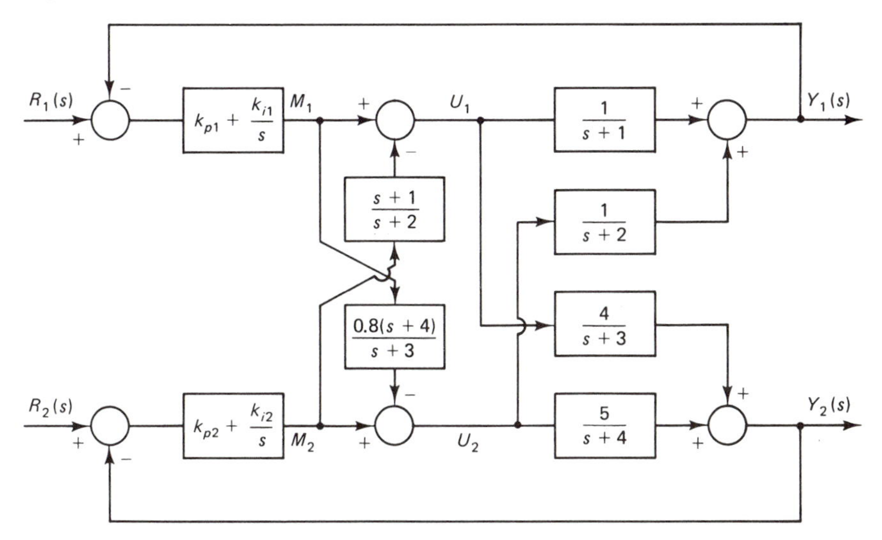

Figure 7.31 Structure of the control system.

with individual PI controllers for each of the two uncoupled systems is shown in Fig. 7.31. The uncoupled control systems are each of fourth order and the methods of the previous sections can be employed to set the gains of the controllers.

7.9 SUMMARY

In this chapter, we have studied the techniques of controller design with output feedback. Controller design with state feedback is the subject of the next chapter. The selection of an appropriate control law depends to a large extent on the system type that results. For this reason, we have classified the control systems into types and analyzed their steady-state errors caused by a change in the command input and/or disturbance.

The commonly employed control laws belong to the PID family. We have shown how the PID control laws can be implemented by analog components. The digital-computer implementation of PID control laws is covered in Chapter 9. We have discussed the techniques of tuning the controller gains to satisfy the performance specifications.

Finally, we have studied the methods of compensation where additional components are inserted within the structure of a control system to enhance the controller performance. For some additional reading in this connection, we refer to Phelan (1977). The emphasis in this chapter has been on single-input-single-output systems, but controller design for multivariable systems by decoupling is included. Another method of controller design for multivariable systems by state feedback is covered in the next chapter.

REFERENCES

BRYSON, A. E., and HO, Y. C. (1969). *Applied Optimal Control.* Waltham, MA: Blaisdell.

FALB, P. L., and WOLOVICH, W. A. (1967, December). "Decoupling in the Design and Synthesis of Multivariable Control Systems." *IEEE Trans. on Automatic Control AC*(12), 651-659.

GRAHAM, D., and LATHROP, R. C. (1953). "The Synthesis of Optimum Response: Criteria and Standard Forms." *Trans. AIEE* 72 (Part 2), 273-288.

NEWTON, G. C., GOULD, L. A., and KAISER, J. F. (1957). *Analytical Design of Linear Feedback Controls.* New York: John Wiley & Sons.

PALM, W. J. (1983). *Modeling, Analysis and Control of Dynamic Systems.* New York: John Wiley & Sons.

PHELAN, R. (1977). *Automatic Control Systems.* Ithaca, NY: Cornell University Press.

TAKAHASHI, Y., RABINS, M., and AUSLANDER, D. (1970). *Control.* Reading, MA: Addison-Wesley.

ZIEGLER, J. G., and NICHOLS, N. B. (1942). "Optimum Settings for Automatic Controllers." *Trans. ASME* 64(8), 759.

ZIEGLER, J. G., and NICHOLS, N. B. (1943). "Process Lags in Automatic Control Circuits." *Trans. ASME* 65(5), 433.

PROBLEMS

7.1. State the type of the control system of Example 3.2, whose block diagram is shown in Fig. 3.6. Find the conditions on the system parameters for asymptotic stability. Assuming that the conditions for asymptotic stability are satisfied, obtain the steady-state error in terms of system parameters when the disturbance torque T_d is a unit step function. How should the spring stiffness be changed to minimize the magnitude of the steady-state error?

7.2. State the type of the control system of Example 3.3, whose block diagram is shown in Fig. 3.9. Let $c_1 = c_2$ and assume that the system parameters are such that it is asymptotically stable. Find the total steady-state error in terms of the system parameters due to unit step-function inputs in both ω_r and the disturbance T_d.

7.3. Obtain the block diagram for the system of Problem 3.5 for controlling the thickness of rolled steel. Determine the system type. Find the steady-state error for a unit step-function change in the normal force on the roller as a load disturbance.

7.4. Obtain the block diagram for the temperature-control system of Problem 3.12, considering the ambient temperature T_a as a disturbance. Determine the system type. Find the steady-state error for a unit step-function change in the disturbance temperature T_a.

7.5. Identify the system types for the control systems shown in Fig. P7.5(a) and (b). Obtain the steady-state errors for both systems when the disturbance d is a unit step function. Give reasons for the difference in the steady-state errors for the two cases.

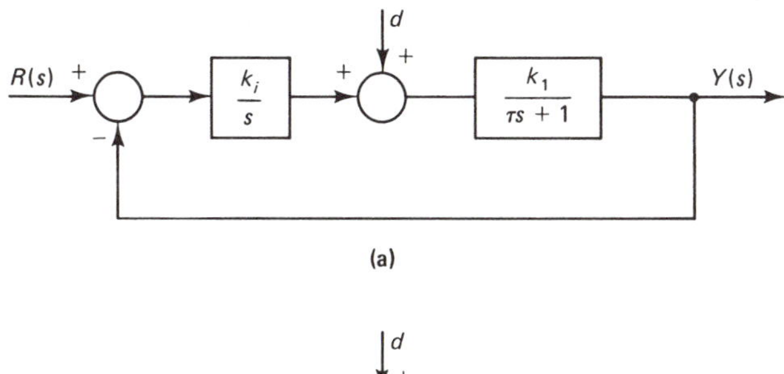

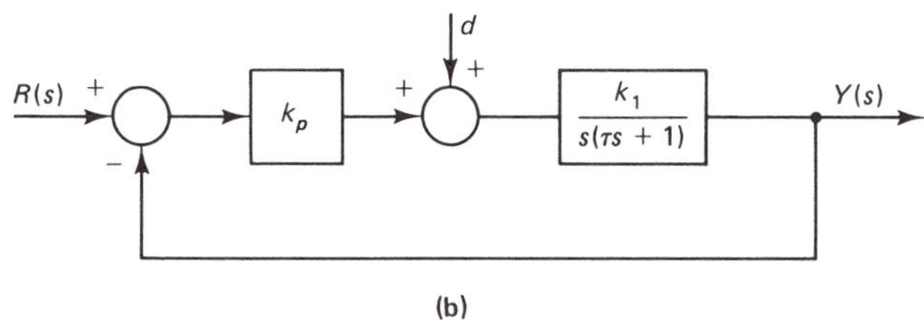

Figure P7.5 Block diagram of a control system.

7.6. The block diagram of a control system is shown in Fig. P7.6.

(a) When $k_i = 10$, $r(t)$ is a unit step function and $d(t) = 0$. Obtain the value of the steady-state error.

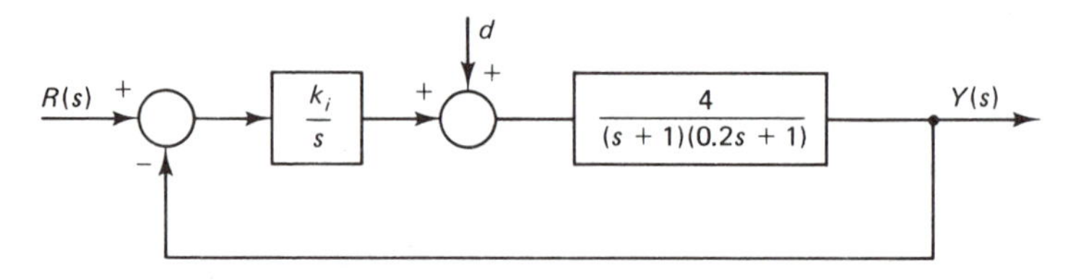

Figure P7.6 Block diagram of a control system.

(b) When the input is a ramp such that $r(t) = t$, and $d = 0$, it is desired to limit the steady-state error to a value equal to or less than 0.2. Obtain the value of k_i and determine if this requirement is consistent with the requirement of stability.

7.7. A pneumatic controller, which also acts as an error detector, is shown in Fig. P7.7. Here, p_r is the desired reference pressure, $y_r = c_1 p_r$, p_0 is controlled output pressure, $p_r - p_0$ is the error e, p is the output pressure of the controller. Obtain the transfer function of the controller and the error detector, and determine the control law that is implemented.

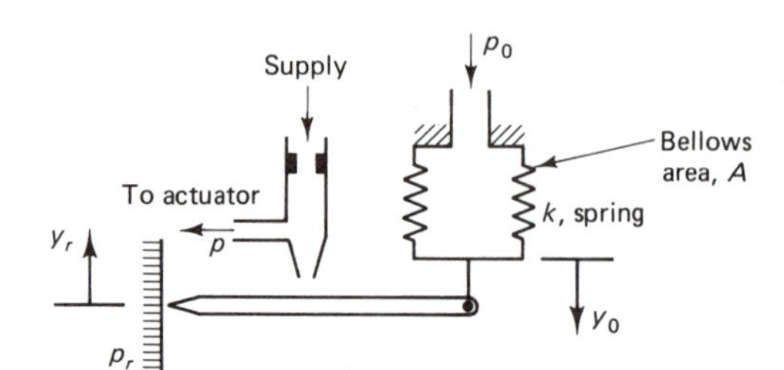

Figure P7.7 Pneumatic controller and error detector.

7.8. A control system with a PI control law is shown in Fig. P7.8. Determine the values of the gains k_p and k_i so that the characteristic equation satisfies the third-order ITAE form for a type 1 system. (See Tables 8.1 and 8.3 for ITAE forms for a type 1 system.) For these values of k_p and k_i, detemine the steady-state error when $r(t) = t$, i.e., a unit ramp function.

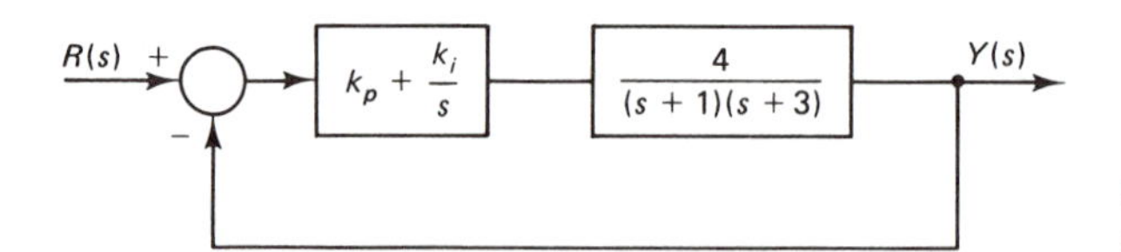

Figure P7.8 Control system with PI control law.

7.9. Consider the control system shown in Fig. P7.8 with a PI control law. Determine the values of k_p and k_i such that when $r(t) = t$, i.e., a unit ramp function, the steady-state error is 0.03 and the quadratic roots of the characteristic equation have real parts equal to -1.

7.10. An electrical position-control system with a PI control law is shown in Fig. 7.19. Now, it is desired to replace the PI control law with an ideal PID control law. For certain parameter values, when the disturbance $T_d = 0$, the block diagram is shown in Fig. P7.10. Determine the controller gains k_p, k_i, and k_d by the ultimate method of Ziegler-Nichols. Investigate the stability of the resultant system.

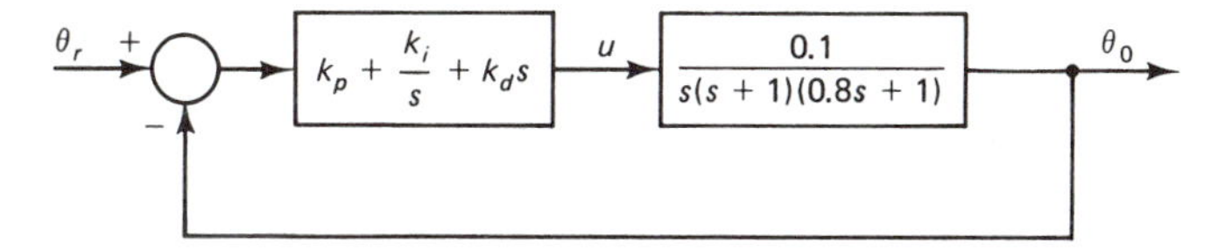

Figure P7.10 Position control system with PID control law.

7.11. Consider the hydraulic speed-control system of Example 3.5, whose block diagram is shown in Fig. 3.19. Assuming that the spool-valve natural frequency is high and much beyond the system bandwidth, the servo valve is represented by a gain constant as in Example 4.4. For certain parameter values, the block diagram is shown in Fig. P7.11.

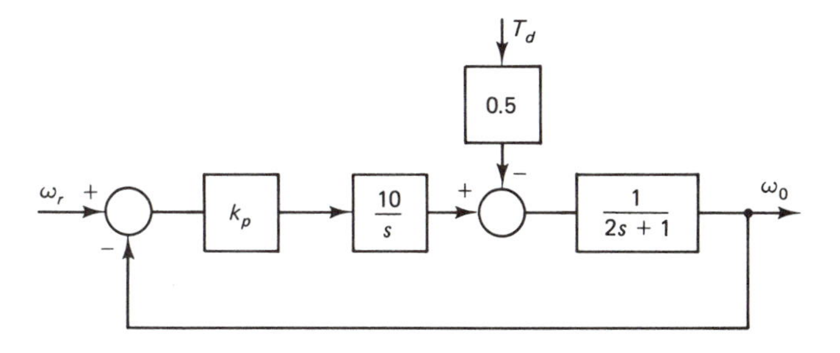

Figure P7.11 Block diagram of a hydraulic control system.

The following specifications are required to be satisfied:

(1) When the set point is not changed so that $\omega_r = 0$ and the disturbance torque $T_d = 50t$, i.e., a ramp of magnitude 50, the magnitude of the error is equal to or less than 2 rad/s.

(2) The real parts of the roots of the characteristic equation are not greater than -1.

Show that the amplifier gain k_p can be selected to satisfy the first specification, but not the second without additional compensation.

7.12. In order to satisfy both specifications of Problem 7.11, it is proposed to replace the proportional control of Fig. P7.11 by a practical PD control law as shown in Fig. P7.12. Obtain the values of the controller gains k_p and k_d to satisfy both specifications of Problem 7.11.

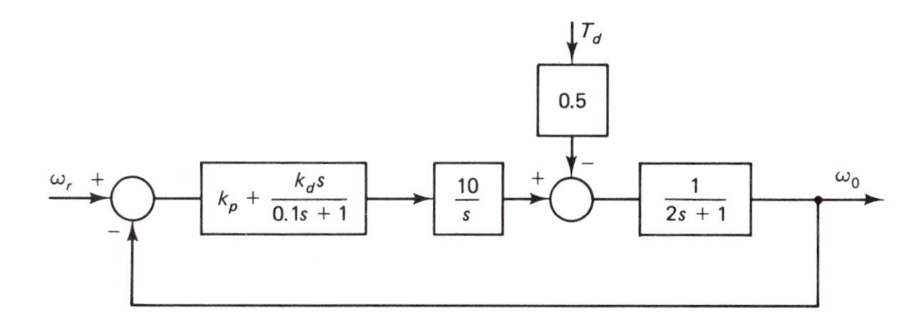

Figure P7.12 Hydraulic control system with practical PD control law.

7.13. In order to satisfy both specifications of Problem 7.11, it is proposed to employ a cascade compensator as shown in Fig. P7.13. Obtain the values of k_p and τ to satisfy both specifications of Problem 7.11. Is the resulting compensator a phase-lead network?

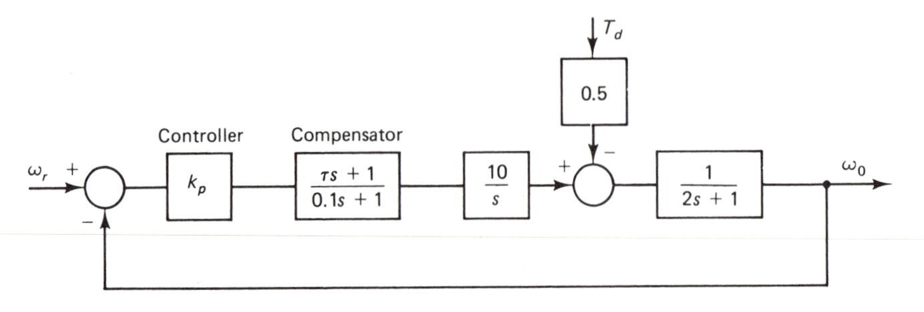

Figure P7.13 Hydraulic control system with cascade compensator.

7.14. A type 1 control system with a proportional control law is shown in Fig. P7.14(a). Its response is considered to be too oscillatory. Hence, the following two suggestions have been made. The first suggestion is to use a practical PD control law as shown in Fig. P7.14(b). The second suggestion is to employ a feedback compensation as shown in Fig. P7.14(c).

(a) For the system of Fig. P7.14(b), determine the value of k_p, k_d, and τ so that its characteristic equation satisfies the third-order ITAE form for a type 1 system of Table 8.1 with $\omega_0 = 10$.

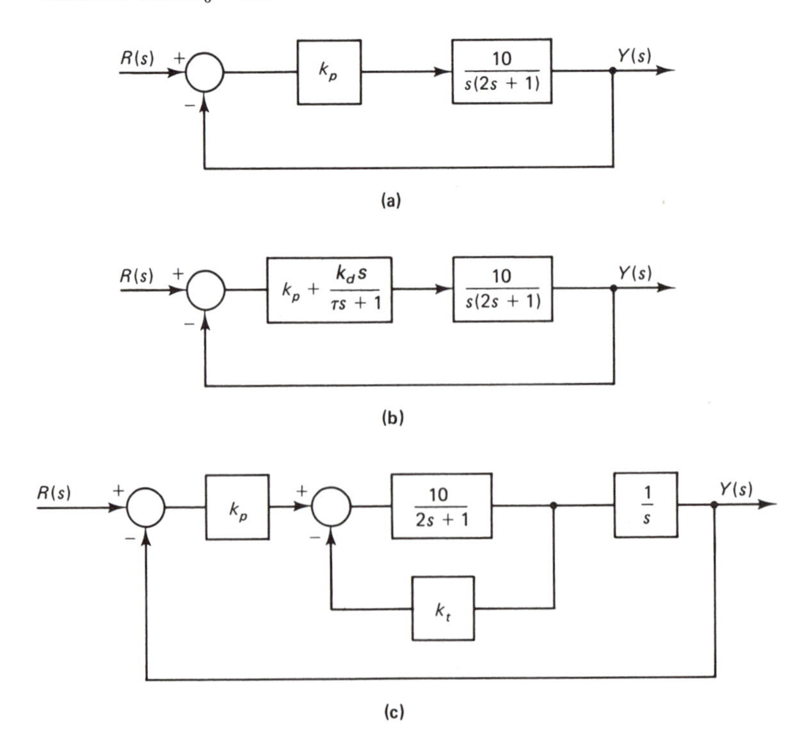

Figure P7.14 Block diagram of a type 1 control system.

(b) For the system of Fig. P7.14(c), determine the values of k_p and k_t so that its characteristic equation satisfies the second-order ITAE form for a type 1 system of Table 8.1 with $\omega_0 = 10$.

7.15. Simulate the three systems of Fig. P7.14(a), (b), and (c) on a computer and obtain their unit step responses when $r(t)$ is a unit step function and all initial conditions are zero. For the system of Fig. P7.14(a), let $k_p = 20$, and for the other two systems use the parameter values found in Problem 7.14.

7.16. A two-input-two-output, multivariable control system is shown in Fig. P7.16. Determine a compensator matrix **G** to achieve decoupling. Then determine suitable proportional controller gains k_{p1} and k_{p2} from considerations of stability.

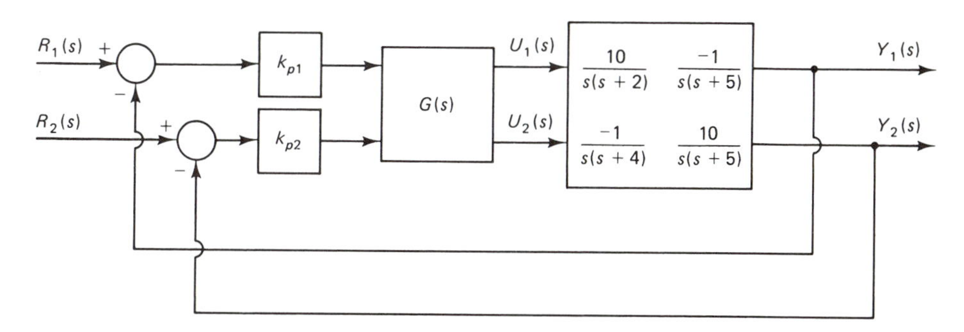

Figure P7.16 Block diagram of a multivariable system.

7.17. Simulate the system of Fig. P7.16 on a computer and obtain the unit step response when $r_1(t)$ is a unit step, $r_2(t) = 0$, and all initial conditions are zero. Use the values of the decoupling compensator matrix **G** and the controller gains determined in Problem 7.16.

8

Controller Design with State Feedback

8.1 INTRODUCTION

In this chapter, we study the design of controllers using state-variables feedback to achieve the desired s-plane location of the roots of the characteristic equation of the closed-loop system. It is an alternative approach to controller design with output feedback of the previous chapter. But a controller with state feedback is superior to a PID controller with compensation, since it permits us to place all the roots of the characteristic equation at the desired locations.

The roots of the characteristic equation are located to satisfy the performance specifications. Hence, the first task is to translate all the performance specifications into the desired root locations. The control law is then obtained by feeding back all the state variables with constant gains. The values of the gains depend on the desired root locations. Hence, the method requires the measurement of all the state variables.

As a first step, our approach is to develop the control law, assuming that the measurements of all the state variables are available for feedback purpose. But some of the state variables may not be available for measurement. Besides, the use of so many sensors in high-order systems may not be economically justifiable, especially since they are not needed for a controller with output feedback. Hence, as a second step, we study the design of an observer whose function is to estimate the values of the state variables from whatever few measurements are available.

The final controller consists of the combined control-law mechanization and observer, where the control-law implementation is based on the estimated values of the state variables rather than on their measured values. In some cases, the implementation of this controller with analog devices may not be cost effective, especially when an inexpensive controller with output feedback is adequate. But microprocessor-based digital controllers have become quite inexpensive and are replacing the traditional analog controllers in many applications. Digital computer implementation of controllers with state feedback is studied in Chapter 9.

Some of the developments in this chapter are at first restricted to single-input, single-output systems. The results are then generalized to multivariable systems. Controller design with state feedback for multivariable systems does not require a decoupling compensator that is needed by a controller with output feedback.

8.2 PERFORMANCE SPECIFICATIONS AND ROOT LOCATIONS

Before proceeding with the controller design, the problem is to transfer the performance specifications to the requirement that the roots of the characteristic equation of the closed-loop system must be placed at certain locations in the s-plane. Some of the performance specifications that we have discussed in the previous chapters are summarized as follows.

1. Stable operation with adequate stability margin (Chapter 6).
2. Step-response specifications: rise time, settling time, and maximum overshoot (Chapter 4).
3. Frequency-response specifications: bandwidth for noise filtering and peak resonance (Chapter 5).
4. Desired system type, steady-state tracking accuracy, and disturbance rejection (Chapter 7).
5. Sensitivity of roots to parameter variations (Chapter 6).

The specifications concerning steady-state errors for command inputs and disturbances dictate the system type that is selected. In many applications, a type 0 system would be unacceptable and type 1 systems are commonly employed. However in a type 1 system, if the integrator in the forward path is located after the disturbance enters the system, then it will not reject a step change in the disturbance, and type 2 system may be desirable.

Some of the specifications may require a compromise. For example, a small rise time may produce a large overshoot and increase the bandwidth so that the noise is not filtered adequately. As stated in the previous chapter, a good approach is to use one of the performance criteria given by Eqs. (7.69) to (7.72). The ITAE performance index can be employed for this purpose. The coefficients of the characteristic equation that minimize the ITAE criterion have been evaluated by Graham and Lathrop (1953).

The optimum coefficients of the denominator of the closed-loop transfer function

$$\frac{Y(s)}{R(s)} = \frac{a_0}{s^n + a_{n-1}s^{n-1} + \cdots + a_1 s + a_0} \tag{8.1}$$

of a type 1 system subjected to a step input are listed in Table 8.1 for systems up

TABLE 8.1 MINIMUM ITAE FORMS OF TYPE 1 SYSTEMS FOR STEP INPUT

$s + \omega_0$
$s^2 + 1.4\omega_0 s + \omega_0^2$
$s^3 + 1.75\omega_0 s^2 + 2.15\omega_0^2 s + \omega_0^3$
$s^4 + 2.1\omega_0 s^3 + 3.4\omega_0^2 s^2 + 2.7\omega_0^3 s + \omega_0^4$
$s^5 + 2.8\omega_0 s^4 + 5.0\omega_0^2 s^3 + 5.5\omega_0^3 s^2 + 3.4\omega_0^4 s + \omega_0^5$
$s^6 + 3.25\omega_0 s^5 + 6.6\omega_0^2 s^4 + 8.6\omega_0^3 s^3 + 7.45\omega_0^4 s^2 + 3.95\omega_0^5 s + \omega_0^6$

to the sixth order. For a type 2 system, the optimum coefficients of the denominator of the closed-loop transfer function

$$\frac{Y(s)}{R(s)} = \frac{a_1 s + a_0}{s^n + a_{n-1}s^{n-1} + \cdots + a_1 s + a_0} \tag{8.2}$$

subjected to a ramp input are given in Table 8.2. These standard forms have a free parameter ω_0 that can be selected by trial and error as a compromise among rise time, settling time, and bandwidth. Finally, the sensitivity of the roots to parameter changes must be studied separately.

TABLE 8.2 MINIMUM ITAE FORMS OF TYPE 2 SYSTEMS FOR RAMP INPUT

$s^2 + 3.2\omega_0 s + \omega_0^2$
$s^3 + 1.75\omega_0 s^2 + 3.25\omega_0^2 s + \omega_0^3$
$s^4 + 2.41\omega_0 s^3 + 4.93\omega_0^2 s^2 + 5.14\omega_0^3 s + \omega_0^4$
$s^5 + 2.19\omega_0 s^4 + 6.5\omega_0^2 s^3 + 6.3\omega_0^3 s^2 + 5.24\omega_0^4 s + \omega_0^5$
$s^6 + 6.12\omega_0 s^5 + 13.42\omega_0^2 s^4 + 17.16\omega_0^3 s^3 + 14.14\omega_0^4 s^2 + 6.76\omega_0^5 s + \omega_0^6$

EXAMPLE 8.1

Determine the roots of the characteristic equation of a second-order system to satisfy the following specifications. The system must be type 1, minimum ITAE form. The bandwidth required is $\omega_b \leq 50$ rad/s, and the rise time and settling time must be small so that the system is fast acting.

The desired closed-loop transfer function of a type 1 second-order system that minimizes the ITAE performance index is obtained from Table 8.1 as

$$\frac{Y(s)}{R(s)} = \frac{\omega_0^2}{s^2 + 1.4\omega_0 s + \omega_0^2} \tag{8.3}$$

We note that the natural frequency $\omega_n = \omega_0$ and the damping ratio $\zeta = 0.7$. The unit step response when $R(s) = 1/s$ is obtained by using the method of Chapter 4 as

$$y(t) = 1 + 1.4\, e^{-0.7\omega_0 t} \sin(0.71\omega_0 t - 134.43°) \tag{8.4}$$

The maximum overshoot can be calculated from this response and is 4.6%. The rise time for the response to go from 10% to 90% of its final value is

$T_r = 2.46/\omega_0$. The settling time for the response to settle within ±5% of its final value is $T_s = 4.76/\omega_0$. It is seen from Eq. (8.3) that the bandwidth is approximately given by $\omega_b = \omega_0$.

Hence, if we let ω_0 have a large value, the rise time and settling time will be small, but the bandwidth will be large. Consequently, we choose the maximum possible value of ω_0 to satisfy the bandwidth specification, that is, $\omega_0 = 50$ rad/s, and obtain the rise time as 0.05 second and the settling time as 0.095 second. The desired roots of the characteristic equation are now obtained as

$$s_1, s_2 = -35 \pm j35.71$$

To facilitate the selection of the value of ω_0, it is convenient to represent Table 8.1 by factoring the polynomials. Letting $d = s/\omega_0$ and denoting $(d + a + jb) \times (d + a - jb)$ by $(d + a \pm jb)$ for compactness, the standard ITAE forms for a type 1 system are given in Table 8.3 (Franklin and Powell, 1980). The unit step responses of the ITAE transfer functions, Eq. (8.1), are shown in Fig. 8.1, where normalized time $\omega_0 t$ has been used.

TABLE 8.3 NORMALIZED AND FACTORED ITAE FORMS FOR TYPE 1 SYSTEMS

$d + 1$
$d + 0.707 \pm j0.707$
$(d + 0.7081)(d + 0.521 \pm j1.068)$
$(d + 0.424 \pm j1.263)(d + 0.626 \pm j0.4141)$
$(d + 0.8955)(d + 0.376 \pm j1.292)(d + 0.5758 \pm j0.5339)$
$(d + 0.3099 \pm j1.263)(d + 0.5805 \pm j0.7828)(d + 0.7346 \pm j0.2873)$

EXAMPLE 8.2

Determine the roots of the characteristic equation of a third-order system to satisfy the following specifications. The system must be type 1, minimum ITAE form. The bandwidth required is $\omega_b \leq 36$ rad/s, and the rise time and settling time must be sufficiently small.

The desired closed-loop transfer function is obtained from Table 8.3 as

$$\frac{Y(s)}{R(s)} = \frac{\omega_0^3}{(s + 0.7081\omega_0)(s + 0.521\omega_0 + j1.068\omega_0)(s + 0.521\omega_0 - j1.068\omega_0)} \tag{8.5}$$

It is seen from Fig. 8.1 that both the rise time and settling time are inversely proportional to ω_0. The quadratic term in Eq. (8.5) has a natural frequency $\omega_n = 1.19\omega_0$. The bandwidth of Eq. (8.5) is therefore approximately equal to $0.7081\omega_0$. Hence, we choose the maximum value of ω_0 to satisfy the bandwidth specification, and let $0.7081\omega_0 = 36$, that is, $\omega_0 \approx 50$ rad/s. With

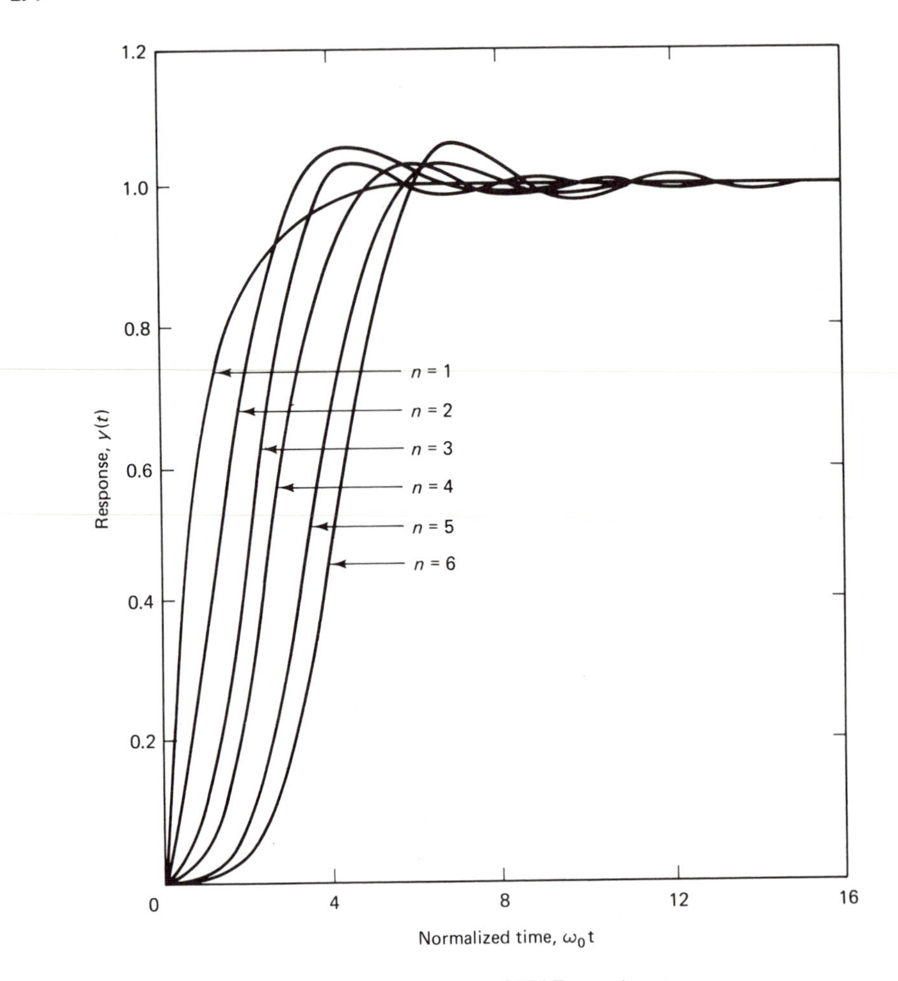

Figure 8.1 Step response of ITAE type 1 systems.

this value of ω_0, Eq. (8.5) becomes

$$\frac{Y(s)}{R(s)} = \frac{125{,}000}{(s+35.4)(s+26.05+j53.4)(s+26.05-j53.4)} \tag{8.6}$$

The quadratic term in Eq. (8.6) has a natural frequency $\omega_n = 59.39$ rad/s and a damping ratio $\zeta = 0.44$. From Fig. 8.1, the maximum overshoot is approximately 3%, and with $\omega_0 = 50$ rad/s, the rise time is approximately 0.04 second.

8.3 CONTROLLABILITY

The existence of a solution to the control law with state feedback, to be studied in the next section, depends on the property of controllability. Let an open-loop plant be described by the state equation

$$\dot{\mathbf{x}}(t) = \mathbf{A}\mathbf{x}(t) + \mathbf{B}\mathbf{u}(t) + \mathbf{B}_1\mathbf{v}(t) \tag{8.7}$$

where $\mathbf{x}(t)$ is the $n \times 1$ state vector, $\mathbf{u}(t)$ the $m \times 1$ control input, $\mathbf{v}(t)$ represents the disturbance, and $\mathbf{A}$ and $\mathbf{B}$ are constant matrices of dimensions $n \times n$ and $n \times m$, respectively. We propose to design a control law by feeding back all the state variables through an $m \times n$ constant matrix $\mathbf{K}$ as

$$\mathbf{u}(t) = -\mathbf{K}\mathbf{x}(t) + \mathbf{N}\mathbf{r}(t) \tag{8.8}$$

where $\mathbf{r}(t)$ is a $m \times 1$ command input vector, and $\mathbf{N}$ is an $m \times m$ constant scaling matrix. Using Eq. (8.8) in Eq. (8.7), we describe the closed-loop control system by

$$\dot{\mathbf{x}}(t) = (\mathbf{A} - \mathbf{B}\mathbf{K})\mathbf{x}(t) + \mathbf{B}\mathbf{N}\mathbf{r}(t) + \mathbf{B}_1\mathbf{v}(t) \tag{8.9}$$

The characteristic equation of this closed-loop system is obtained as

$$\det|s\mathbf{I} - \mathbf{A} + \mathbf{B}\mathbf{K}| = 0 \tag{8.10}$$

In the previous section, we have described how the performance specifications can be converted into the desired root locations of the characteristic equation, Eq. (8.10). Given the roots of Eq. (8.10), the question now is whether or not it is possible to solve for the matrix $\mathbf{K}$. The answer to this question depends upon the property of controllability studied in this section.

Definition. The state $\mathbf{x}(t)$ of Eq. (8.7) is said to be controllable at time t_0 if there exists a piecewise continuous input $\mathbf{u}(t)$ that will drive the state from its initial value $\mathbf{x}(t_0)$ to any final value $\mathbf{x}(t_f)$ in a finite time interval $t_f - t_0 > 0$. We consider only time invariant systems and we let $t_0 = 0$.

The property of controllability can also be defined for the outputs of a system. The preceding definition refers to the state and is sometimes called state controllability, which we shall call simply controllability. To appreciate this concept, we consider the special case where matrix $\mathbf{A}$ has distinct eigenvalues. We know from Chapter 4 that, for this case, there exists a similarity transformation $\mathbf{x} = \mathbf{P}\mathbf{x}^*$ so that Eq. (8.7) can be transformed to the Jordan normal form

$$\dot{\mathbf{x}}^* = \mathbf{\Lambda}\mathbf{x}^* + \mathbf{B}^*\mathbf{u} \tag{8.11}$$

where $\mathbf{\Lambda}$ is a diagonal matrix, $\mathbf{B}^* = \mathbf{P}^{-1}\mathbf{B}$, and we have neglected the disturbance since it does not affect controllability.

For a multivariable system, if $\mathbf{B}^*$ has a row of all zero elements, then the state variable associated with that row is not influenced by the input since the equations are uncoupled. For a single-input, single-output system, where u is a scalar, if $\mathbf{B}^*$ has a zero element, then the state variable associated with that element is uncontrollable.

A general test for controllability may be stated as follows. The system of Eq. (8.7) is controllable if and only if the $n \times nm$ matrix

$$\mathbf{Q} = [\mathbf{B} \quad \mathbf{A}\mathbf{B} \quad \mathbf{A}^2\mathbf{B} \quad \cdots \quad \mathbf{A}^{n-1}\mathbf{B}] \tag{8.12}$$

is of rank n. The proof can be found in Kreindler and Sarachik (1964). The rank of a matrix is defined as the highest-ordered nonzero determinant that can be extracted from the matrix. For multivariable systems, an easier method of determining the rank is to form an $n \times n$ matrix $\mathbf{QQ}^T$, and if its determinant is nonzero, then the rank of $\mathbf{Q}$ is n. When u is a scalar, $\mathbf{Q}$ is an $n \times n$ matrix, and if its determinant is nonzero, then its rank is n.

Controllability is a property of the pair ($\mathbf{A}$, $\mathbf{B}$). We note that in general, most systems are controllable, but there are some pathological cases that are exceptions. One of the reasons why a system is not controllable is that a pole of the system has been cancelled by a zero.

EXAMPLE 8.3

Let a single-input, single-output system have a pole at -4. Suppose that this pole location is undesirable and is cancelled by a zero of the compensator that introduces a pole at -2, which is desirable as shown in Fig. 8.2(a). The two state variables are chosen by rearranging the block diagram as shown in Fig. 8.2(b).

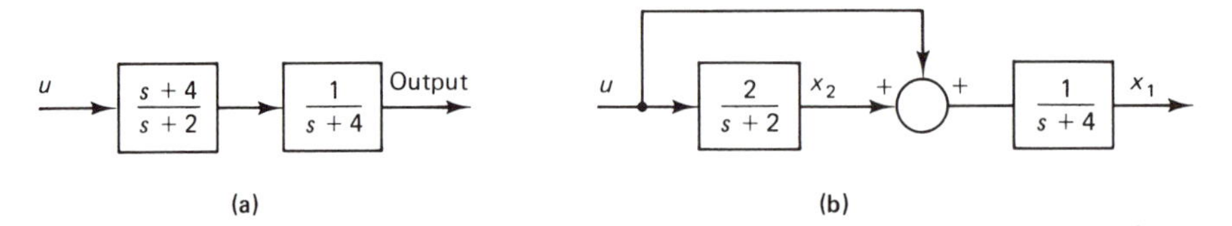

Figure 8.2 (a) Plant with compensator. (b) State variables.

The state equations are obtained from Fig. 8.2(b) as

$$\begin{aligned} \dot{x}_1 &= -4x_1 + x_2 + u \\ \dot{x}_2 &= -2x_2 + 2u \end{aligned} \tag{8.13}$$

The $\mathbf{A}$ and $\mathbf{B}$ matrices are described by

$$\mathbf{A}\begin{bmatrix} -4 & 1 \\ 0 & -2 \end{bmatrix} \qquad \mathbf{B} = \begin{bmatrix} 1 \\ 2 \end{bmatrix}$$

The $\mathbf{Q}$ matrix of Eq. (8.12) is obtained as

$$\mathbf{Q} = [\mathbf{B} \quad \mathbf{AB}] = \begin{bmatrix} 1 & -2 \\ 2 & -4 \end{bmatrix}$$

The determinant of $\mathbf{Q}$ is zero, and hence its rank is not 2 and the system is not controllable. To investigate the uncontrollable mode, we diagonalize matrix $\mathbf{A}$. Its determinant is given by

$$|s\mathbf{I} - \mathbf{A}| = (s + 2)(s + 4)$$

The eigenvectors associated with the eigenvalues -2 and -4 are given, respectively, by

$$\begin{bmatrix} 1 \\ 2 \end{bmatrix} \quad \text{and} \quad \begin{bmatrix} 1 \\ 0 \end{bmatrix}$$

Hence, the similarity transformation matrix and its inverse are

$$\mathbf{P} = \begin{bmatrix} 1 & 1 \\ 2 & 0 \end{bmatrix} \qquad \mathbf{P}^{-1} = \begin{bmatrix} 0 & 0.5 \\ 1 & -0.5 \end{bmatrix}$$

and

$$\mathbf{\Lambda} = \mathbf{P}^{-1}\mathbf{A}\mathbf{P} = \begin{bmatrix} -2 & 0 \\ 0 & -4 \end{bmatrix} \qquad \mathbf{B}^* = \mathbf{P}^{-1}\mathbf{B} = \begin{bmatrix} 1 \\ 0 \end{bmatrix}$$

The uncoupled state equations become

$$\begin{aligned} \dot{x}_1^* &= -2x_1^* + u \\ \dot{x}_2^* &= -4x_2^* \end{aligned} \tag{8.14}$$

It is seen that $\mathbf{B}^*$ has a zero element and the state variable x_2^* is not controllable. This example illustrates the folly of cancelling an undesirable pole, especially an unstable one, by a zero of the compensator since it renders the system uncontrollable.

EXAMPLE 8.4

In a single-input, single-output third-order system, the **A** and **B** matrices are described by

$$\mathbf{A} = \begin{bmatrix} 0 & 1 & 0 \\ 0 & 0 & 1 \\ 0 & -3 & -2 \end{bmatrix} \qquad \mathbf{B} = \begin{bmatrix} 0 \\ 0 \\ 1 \end{bmatrix}$$

We get

$$\mathbf{A}^2 = \begin{bmatrix} 0 & 0 & 1 \\ 0 & -3 & -2 \\ 0 & 6 & 1 \end{bmatrix}$$

and

$$\mathbf{Q} = [\mathbf{B} \quad \mathbf{AB} \quad \mathbf{A}^2\mathbf{B}] = \begin{bmatrix} 0 & 0 & 1 \\ 0 & 1 & -2 \\ 1 & -2 & 1 \end{bmatrix}$$

The determinant of **Q** is not zero but is equal to -1. Hence, its rank is 3, and the pair $(\mathbf{A}, \mathbf{B})$ is controllable.

EXAMPLE 8.5

We consider a two-input, two-output system with four state variables described by

$$\dot{\mathbf{x}} = \mathbf{A}\mathbf{x} + \mathbf{B}\mathbf{u}$$

where the **A** and **B** matrices are given by

$$\mathbf{A} = \begin{bmatrix} -1 & 1 & 2 & 0 \\ 1 & -2 & 0 & 2 \\ 0 & 0 & 0 & 0 \\ 0 & 0 & 0 & 0 \end{bmatrix} \qquad \mathbf{B} = \begin{bmatrix} 0 & 0 \\ 0 & 0 \\ 1 & 0 \\ 0 & 1 \end{bmatrix} \tag{8.15}$$

The matrix **Q** of Eq. (8.12) is obtained from

$$\mathbf{Q} = [\mathbf{B} \quad \mathbf{AB} \quad \mathbf{A}^2\mathbf{B} \quad \mathbf{A}^3\mathbf{B}] \tag{8.16}$$

Now,

$$\mathbf{A}^2 = \begin{bmatrix} 2 & -3 & -2 & 2 \\ -3 & 5 & 2 & -4 \\ 0 & 0 & 0 & 0 \\ 0 & 0 & 0 & 0 \end{bmatrix} \qquad \mathbf{A}^3 = \begin{bmatrix} -5 & 8 & 4 & -6 \\ 8 & -13 & -6 & 10 \\ 0 & 0 & 0 & 0 \\ 0 & 0 & 0 & 0 \end{bmatrix}$$

$$\mathbf{AB} = \begin{bmatrix} 2 & 0 \\ 0 & 2 \\ 0 & 0 \\ 0 & 0 \end{bmatrix} \qquad \mathbf{A}^2\mathbf{B} = \begin{bmatrix} -2 & 2 \\ 2 & -4 \\ 0 & 0 \\ 0 & 0 \end{bmatrix} \qquad \mathbf{A}^3\mathbf{B} = \begin{bmatrix} 4 & -6 \\ -6 & 10 \\ 0 & 0 \\ 0 & 0 \end{bmatrix}$$

Substituting these results in Eq. (8.16),

$$\mathbf{Q} = \begin{bmatrix} 0 & 0 & 2 & 0 & -2 & 2 & 4 & -6 \\ 0 & 0 & 0 & 2 & 2 & -4 & -6 & 10 \\ 1 & 0 & 0 & 0 & 0 & 0 & 0 & 0 \\ 0 & 1 & 0 & 0 & 0 & 0 & 0 & 0 \end{bmatrix}$$

and

$$\mathbf{QQ}^T = \begin{bmatrix} 64 & -96 & 0 & 0 \\ -96 & 160 & 0 & 0 \\ 0 & 0 & 1 & 0 \\ 0 & 0 & 0 & 1 \end{bmatrix}$$

The determinant of $\mathbf{QQ}^T$ is not zero and is equal to 1024. Hence, the rank of **Q** is 4 and the system is controllable.

8.4 CONTROL LAW DESIGN

The objective of this section is to design a control law by feeding back all the state variables through a constant gain matrix. The gains are chosen to place the roots of the characteristic equation of the closed-loop system at the desired locations in the s-plane. For this purpose, it is assumed that the measurement of all the state variables is available for implementing the control law. In a later section, we discard this assumption and design an observer that yields an estimate of the state variables for feedback from the measurement of the controlled outputs.

Since all the roots are placed at their desired locations, we can consider the state feedback control scheme as a generalization of the PD controller with output feedback studied in the previous chapter. The integral control mode increases the order of the system by one and cannot be achieved by state feedback through constant gains. The effect of the state feedback control law on the resulting system type is discussed later in section 8.4.2. In case the control law is required to improve the system type, we discuss the modification that includes integral control with state feedback.

First, we consider single-input, single-output systems and later generalize the results to multivariable systems.

8.4.1 Single-Input, Single-Output Systems

Let the plant be described by the state equations

$$\dot{\mathbf{x}}(t) = \mathbf{A}\mathbf{x}(t) + \mathbf{B}u(t) + \mathbf{B}_1 v(t) \tag{8.17}$$

$$y(t) = \mathbf{C}\mathbf{x}(t) \tag{8.18}$$

where $\mathbf{x}$ is an $n \times 1$ state vector and u and y are scalars. The feedback control law to be designed is given by

$$u(t) = -\mathbf{K}\mathbf{x}(t) + Nr(t) \tag{8.19}$$

where $\mathbf{K}$ is a $1 \times n$ row matrix, N is a scalar scaling constant, and r is the scalar command input. Substituting for u from Eq. (8.19) in Eq. (8.17), we describe the closed-loop control system by

$$\dot{\mathbf{x}}(t) = (\mathbf{A} - \mathbf{B}\mathbf{K})\mathbf{x}(t) + \mathbf{B}Nr(t) + \mathbf{B}_1 v(t) \tag{8.20}$$

The characteristic equation is obtained as

$$|s\mathbf{I} - \mathbf{A} + \mathbf{B}\mathbf{K}| = 0 \tag{8.21}$$

The desired roots $-s_1, -s_2, \ldots, -s_n$ of this characteristic equation are selected to satisfy the performance specifications as shown in section 8.2. We now solve for the elements of the feedback gain matrix $\mathbf{K}$ from the equation

$$|s\mathbf{I} - \mathbf{A} + \mathbf{B}\mathbf{K}| = (s + s_1)(s + s_2)\cdots(s + s_n) \tag{8.22}$$

The n unknown elements of $\mathbf{K}$ can be solved by matching the corresponding coefficients of this n^{th}-order polynomial. The calculation of the gains by matching the corresponding coefficients of Eq. (8.22) becomes cumbersome for high-order systems. A formula that is convenient for computer solution has been derived by Ackermann for this purpose. Expanding the right-hand side of Eq. (8.22), we obtain the desired characteristic polynomial as

$$\begin{aligned}\Delta(s) &= (s + s_1)(s + s_2)\cdots(s + s_n) \\ &= s^n + a_{n-1}s^{n-1} + \cdots + a_0\end{aligned} \tag{8.23}$$

We define $\Delta(\mathbf{A})$ as

$$\Delta\mathbf{A} = \mathbf{A}^n + a_{n-1}\mathbf{A}^{n-1} + \cdots a_0\mathbf{I} \tag{8.24}$$

The Ackermann's formula is then given by

$$\mathbf{K} = \lfloor 0 \quad 0 \quad \cdots \quad 0 \quad 1 \rfloor \mathbf{Q}^{-1}\Delta(\mathbf{A}) \tag{8.25}$$

where $\mathbf{Q}$ is the controllability matrix defined by Eq. (8.12). A proof of Eq. (8.25) is given by Franklin and Powell (1980). The inversion of matrix $\mathbf{Q}$ in Eq. (8.25) can be avoided by solving for a $1 \times n$ row matrix $\mathbf{d}^T$ from

$$\mathbf{d}^T\mathbf{Q} = \lfloor 0 \quad 0 \quad \cdots \quad 0 \quad 1 \rfloor \tag{8.26}$$

and then solving for $\mathbf{K}$ from the equation

$$\mathbf{K} = \mathbf{d}^T\Delta(\mathbf{A}) \tag{8.27}$$

A question now arises whether Eq. (8.22) has a unique solution for $\mathbf{K}$. The answer is in the affirmative when the pair $(\mathbf{A}, \mathbf{B})$ is controllable. The proof becomes obvious when we note that Eq. (8.25) requires the inverse of the controllability matrix $\mathbf{Q}$. If the determinant of $\mathbf{Q}$ is nonzero, then its inverse exists and its rank is n, which is the condition for controllability. We note again that, with the exception of a few pathological cases, most systems are controllable.

EXAMPLE 8.6

The mathematical model of an electrical position-control system has been obtained in Example 3.4. A PI controller with output feedback for this system has been discussed in Example 7.11. In this example, we consider a controller with state feedback, such that the closed-loop system satisfies the performance specifications of Example 8.2. A block diagram of the plant obtained from Fig. 3.15 is shown in Fig. 8.3(a).

The matrices $\mathbf{A}$ and $\mathbf{B}$ of the plant state equations are given by Eq. (3.49), namely,

$$\mathbf{A} = \begin{bmatrix} 0 & 1 & 0 \\ 0 & -1/\tau_2 & k_t/\tau_2 c \\ 0 & -k_b/R_a\tau_1 & -1/\tau_1 \end{bmatrix} \qquad \mathbf{B} = \begin{bmatrix} 0 \\ 0 \\ k_r/R_a\tau_1 \end{bmatrix}$$

Let the parameter values be such that these matrices become

$$\mathbf{A} = \begin{bmatrix} 0 & 1 & 0 \\ 0 & -0.5 & 10 \\ 0 & -0.1 & -10 \end{bmatrix} \qquad \mathbf{B} = \begin{bmatrix} 0 \\ 0 \\ 100 \end{bmatrix} \tag{8.28}$$

First, we check whether the system is controllable. The controllability matrix is obtained as

$$\mathbf{Q} = (\mathbf{B} \quad \mathbf{AB} \quad \mathbf{A}^2\mathbf{B}) = \begin{bmatrix} 0 & 0 & 1{,}000 \\ 0 & 1{,}000 & -10{,}500 \\ 100 & -1{,}000 & 9{,}900 \end{bmatrix} \tag{8.29}$$

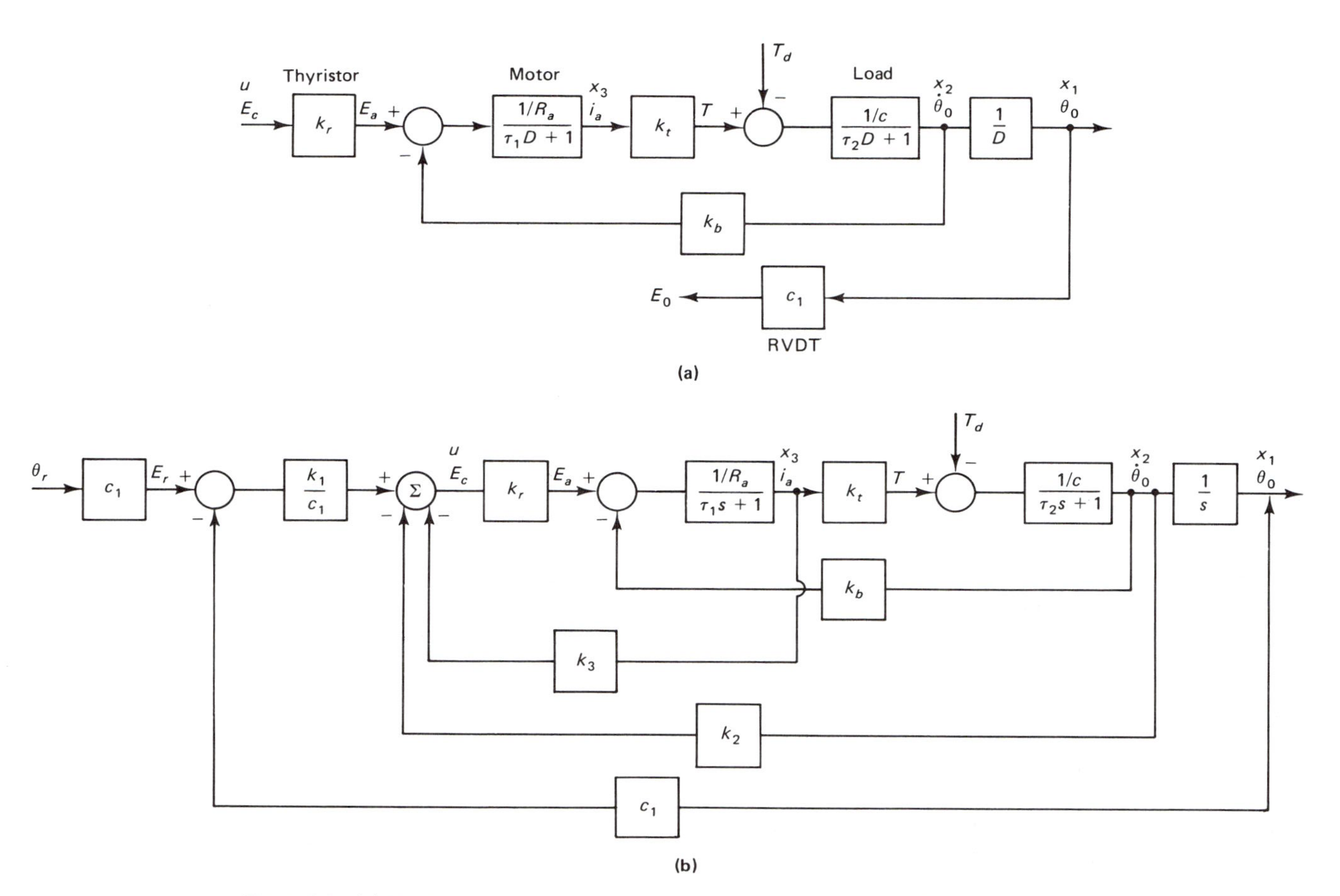

Figure 8.3 (a) Plant of a position-control system. (b) Structure of a control system with state feedback.

The determinant of this matrix is $|\mathbf{Q}| = 10^8 \neq 0$. Hence, the rank of $\mathbf{Q}$ is 3, and the pair $(\mathbf{A}, \mathbf{B})$ is controllable. It is seen from Fig. 8.3 that the system is of third order and type 1, but the integrator is located after the entrance of the disturbance. To satisfy the other specifications of Example 8.2, we obtain the characteristic equation of the closed-loop system from Eq. (8.6) as

$$(s + 35.4)(s + 26.05 + j53.4)(s + 26.05 - j53.4) = 0$$

i.e.,

$$s^3 + 87.5s^2 + 5{,}374.5s + 124{,}969 = 0 \tag{8.30}$$

The characteristic equation is also given by

$$|s\mathbf{I} - \mathbf{A} + \mathbf{BK}| = \begin{vmatrix} s & -1 & 0 \\ 0 & s + 0.5 & -10 \\ 100k_1 & 0.1 + 100k_2 & s + 10 + 100k_3 \end{vmatrix}$$

$$= s^3 + (10.5 + 100k_3)s^2 + (6 + 50k_3 + 1000k_2)s + 1000k_1 = 0 \tag{8.31}$$

Matching the corresponding coefficients of Eqs. (8.30) and (8.31), we obtain

$$k_1 = 124.97 \qquad k_2 = 5.33 \qquad k_3 = 0.77 \tag{8.32}$$

We also obtain the gains from the Ackermann formula to illustrate its use for high-order systems. From Eq. (8.26), we get

$$\lfloor d_1 \quad d_2 \quad d_3 \rfloor \begin{bmatrix} 0 & 0 & 1{,}000 \\ 0 & 1{,}000 & -10{,}500 \\ 100 & -1{,}000 & 9{,}900 \end{bmatrix} = \lfloor 0 \quad 0 \quad 1 \rfloor$$

Hence, $d_1 = 0.001$, $d_2 = 0$, and $d_3 = 0$. We solve for $\mathbf{K}$ from Eq. (8.27) and get

$$\lfloor k_1 \quad k_2 \quad k_3 \rfloor = \lfloor 0.001 \quad 0 \quad 0 \rfloor \Delta(\mathbf{A}) \tag{8.33}$$

where

$$\Delta(\mathbf{A}) = \mathbf{A}^3 + 87.5\mathbf{A}^2 + 5{,}374.5\mathbf{A} + 124{,}969\mathbf{I}$$

$$= \begin{bmatrix} 124{,}969 & 5{,}331.5 & 770 \\ 0 & 122{,}227 & 45{,}600 \\ 0 & -456 & 80{,}866 \end{bmatrix}$$

Substituting this result in Eq. (8.33), we obtain $k_1 = 124.97$, $k_2 = 5.33$, and $k_3 = 0.77$, which is the same result given by Eq. (8.32). The structure of the state feedback control system is shown in Fig. 8.3(b), assuming that all state variables have been measured. The output $y = x_1 = \theta_0$ is sensed by a RVDT, and the scale factor N in Eq. (8.19) is chosen as $N = k_1$. The structure of Fig. 8.3(b) is that of a proportional control with output feedback and additional feedback compensation from all the remaining state variables.

8.4.2 State Feedback with PI Control

The preceding control law is equivalent to proportional control with output feedback and feedback compensation from all the remaining state variables. In Example 8.6, the resulting control system is type 1, but the integrator is located after the disturbance enters the system and hence there will be a steady-state error due to a change in the disturbance torque. Therefore, it may be desirable to convert the system to type 2 by modifying the control law.

We consider a system such as the hydraulic speed-control system of Example 4.4. For certain parameter values, the block diagram of the plant is shown in Fig. 8.4(a). In case proportional control law is employed with output feedback, the

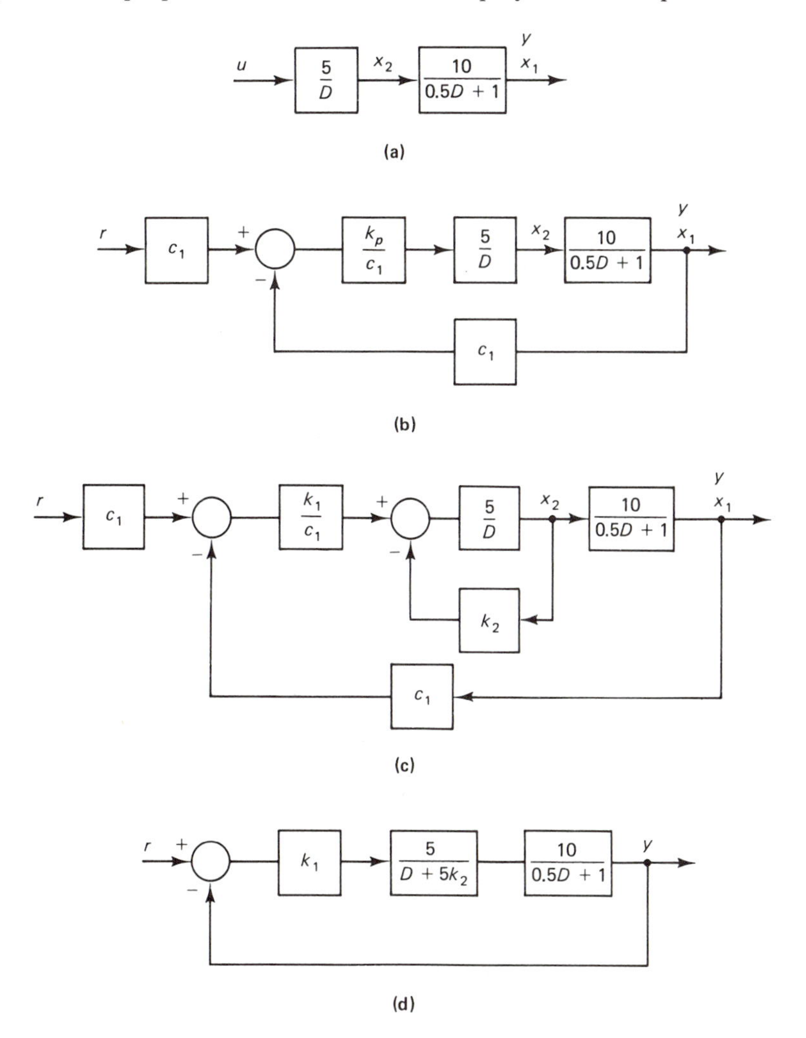

Figure 8.4 (a) Plant of a hydraulic speed control system. (b) Output feedback control system with proportional control. (c) Control system with state feedback. (d) Resultant control system with state feedback.

control system will be type 1 as shown in Fig. 8.4(b). The resulting control system with state feedback is shown in Fig. 8.4(c). Now, there is a feedback around the integrator in the forward path and it is seen from Fig. 8.4(d) that the control system is type 0.

From this discussion, it is seen that a control law with state feedback may have a detrimental effect on the system type. We now discuss a modified control law that improves the system type, but it also raises the order of the system by one. It is called state feedback with PI control. Considering a single-input, single-output system, let the plant be described by the state equations

$$\begin{aligned} \dot{\mathbf{x}} &= \mathbf{A}\mathbf{x} + \mathbf{B}u + \mathbf{B}_1 v \\ y &= \mathbf{C}\mathbf{x} \end{aligned} \tag{8.34}$$

We define an additional state variable x_{n+1} as

$$x_{n+1}(t) = \int_0^t (y - r)\,dt$$

i.e.,

$$\begin{aligned} \dot{x}_{n+1} &= y - r \\ &= \mathbf{C}\mathbf{x} - r \end{aligned} \tag{8.35}$$

where r is the command input, and $(r - y)$ is the error. The augmented state equations are expressed as

$$\begin{bmatrix} \dot{\mathbf{x}} \\ \dot{x}_{n+1} \end{bmatrix} = \begin{bmatrix} \mathbf{A} & \mathbf{0} \\ \lfloor C \rfloor & 0 \end{bmatrix} \begin{bmatrix} \mathbf{x} \\ x_{n+1} \end{bmatrix} + \begin{bmatrix} \mathbf{B} \\ 0 \end{bmatrix} u + \begin{bmatrix} \mathbf{B}_1 \\ 0 \end{bmatrix} v + \begin{bmatrix} \mathbf{0} \\ -r \end{bmatrix} \tag{8.36}$$

For compactness, we define

$$\hat{\mathbf{x}} = \begin{bmatrix} \mathbf{x} \\ x_{n+1} \end{bmatrix} \qquad \hat{\mathbf{A}} = \begin{bmatrix} \mathbf{A} & \mathbf{0} \\ \mathbf{C} & 0 \end{bmatrix} \qquad \hat{\mathbf{B}} = \begin{bmatrix} \mathbf{B} \\ 0 \end{bmatrix} \qquad \hat{\mathbf{B}}_1 = \begin{bmatrix} \mathbf{B}_1 \\ 0 \end{bmatrix} \tag{8.37}$$

and express Eq. (8.36) as

$$\dot{\hat{\mathbf{x}}} = \hat{\mathbf{A}}\hat{\mathbf{x}} + \hat{\mathbf{B}}u + \begin{bmatrix} \mathbf{0} \\ -r \end{bmatrix} + \hat{\mathbf{B}}_1 v \tag{8.38}$$

The control law is

$$u = -\mathbf{K}\hat{\mathbf{x}} + Nr \tag{8.39}$$

where $\mathbf{K} = \lfloor k_1 \cdots k_n \ \ k_{n+1} \rfloor$, and N is a scalar scaling factor. Substituting for u in Eq. (8.38) from Eq. (8.39), we obtain

$$\dot{\hat{\mathbf{x}}} = (\hat{\mathbf{A}} - \hat{\mathbf{B}}\mathbf{K})\hat{\mathbf{x}} + \hat{\mathbf{B}}Nr + \begin{bmatrix} \mathbf{0} \\ -r \end{bmatrix} + \hat{\mathbf{B}}_1 v \tag{8.40}$$

The characteristic equation of the closed-loop system is given by

$$|s\mathbf{I} - \hat{\mathbf{A}} + \hat{\mathbf{B}}\mathbf{K}| = 0 \tag{8.41}$$

After selecting its desired roots $-s_1, \ldots, -s_{n+1}$, we get

$$|s\mathbf{I} - \hat{\mathbf{A}} + \hat{\mathbf{B}}\mathbf{K}| = (s + s_1) \cdots (s + s_n)(s + s_{n+1}) \tag{8.42}$$

Assuming that the pair $(\hat{\mathbf{A}}, \hat{\mathbf{B}})$ is controllable, the $(n + 1)$ unknown elements of $\mathbf{K}$ are solved either by matching the corresponding coefficients in Eq. (8.42) or by using the Ackermann formula. We show by considering the following example that this modified control law with state feedback is equivalent to PI control with output feedback and feedback compensation from all the remaining state variables.

EXAMPLE 8.7

Let the system of Example 8.6 be required to control the speed rather than the position. For this purpose, the output speed is sensed by a tachometer. The block diagram of the plant is shown in Fig. 8.5.

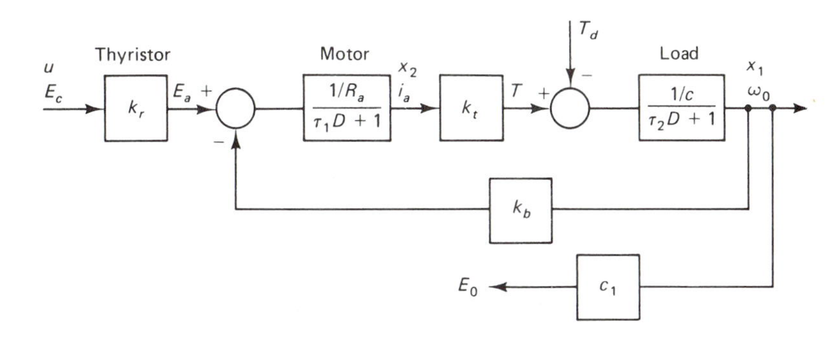

Figure 8.5 Plant of a speed-control system.

If state feedback is used as in the previous example, the control system will be type 0 with resulting steady-state errors. We assume that the performance specifications require a type 1 system that meets the conditions specified in Example 8.2. Hence, we employ state feedback with PI control. The state equations of the plant are obtained as

$$\begin{aligned} \dot{x}_1 &= -\left(\frac{1}{\tau_2}\right)x_1 + \left(\frac{k_t}{\tau_2 c}\right)x_2 - \left(\frac{1}{\tau_2 c}\right)T_d \\ \dot{x}_2 &= -\left(\frac{k_b}{R_a\tau_1}\right)x_1 - \left(\frac{1}{\tau_1}\right)x_2 + \left(\frac{k_r}{R_a\tau_1}\right)u \\ y &= x_1 \end{aligned} \tag{8.43}$$

The matrices are obtained by inspection of these equations as

$$\mathbf{A} = \begin{bmatrix} -1/\tau_2 & k_t/\tau_2 c \\ -k_b/R_a\tau_1 & -1/\tau_1 \end{bmatrix} \qquad \mathbf{B} = \begin{bmatrix} 0 \\ k_r/R_a\tau_1 \end{bmatrix} \qquad \mathbf{B}_1 = \begin{bmatrix} -1/\tau_2 c \\ 0 \end{bmatrix}$$

$$\mathbf{C} = \lfloor 1 \quad 0 \rfloor \tag{8.44}$$

With the same parameter values employed in Example 8.6, we get

$$\mathbf{A} = \begin{bmatrix} -0.5 & 10 \\ -0.1 & -10 \end{bmatrix} \qquad \mathbf{B} = \begin{bmatrix} 0 \\ 100 \end{bmatrix} \tag{8.45}$$

We define an additional state variable x_3 as

$$x_3 = \int_0^t (y - r)\, dt$$

i.e.,

$$\dot{x}_3 = y - r = x_1 - r \tag{8.46}$$

Hence, the $\hat{\mathbf{A}}$ and $\hat{\mathbf{B}}$ matrices are

$$\hat{\mathbf{A}} = \begin{bmatrix} -0.5 & 10 & 0 \\ -0.1 & -10 & 0 \\ 1 & 0 & 0 \end{bmatrix} \qquad \hat{\mathbf{B}} = \begin{bmatrix} 0 \\ 100 \\ 0 \end{bmatrix} \tag{8.47}$$

The controllability matrix is given by

$$\mathbf{Q} = (\hat{\mathbf{B}} \quad \hat{\mathbf{A}}\hat{\mathbf{B}} \quad \hat{\mathbf{A}}^2\hat{\mathbf{B}}) = \begin{bmatrix} 0 & 1{,}000 & -10{,}500 \\ 100 & -1{,}000 & 9{,}900 \\ 0 & 0 & 1{,}000 \end{bmatrix} \tag{8.48}$$

The determinant of $\mathbf{Q}$ is nonzero. Its rank therefore is 3, and the pair $(\hat{\mathbf{A}}, \hat{\mathbf{B}})$ is controllable. The characteristic equation of the closed-loop system is given by

$$|s\mathbf{I} - \hat{\mathbf{A}} + \hat{\mathbf{B}}\mathbf{K}| = \begin{vmatrix} s + 0.5 & -10 & 0 \\ 0.1 + 100k_1 & s + 10 + 100k_2 & 100k_3 \\ -1 & 0 & s \end{vmatrix}$$

$$= s^3 + (10.5 + 100k_2)s^2 + (6 + 50k_2 + 1{,}000k_1)s + 1{,}000k_3 = 0 \tag{8.49}$$

To satisfy the performance specifications of Example 8.2, the desired characteristic equation is

$$s^3 + 87.5s^2 + 5{,}374.5s + 124{,}969 = 0 \tag{8.50}$$

Matching the corresponding coefficients of Eqs. (8.49) and (8.50), we obtain

$$k_1 = 5.33 \qquad k_2 = 0.77 \qquad k_3 = 124.97 \tag{8.51}$$

The scale factor N is chosen as $N = k_1 = 5.33$. Assuming that both x_1 and x_2 are measured, the structure of the feedback control system is shown in Fig. 8.6. It is seen that the system is equivalent to PI control with output feedback

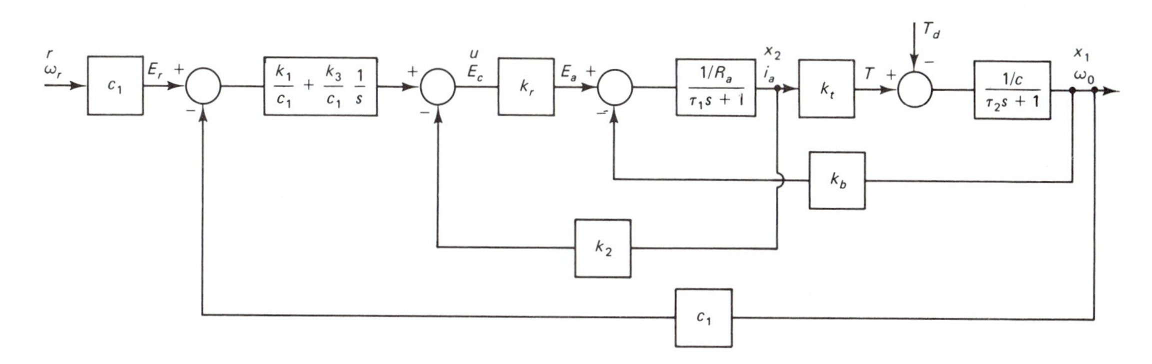

Figure 8.6 Structure with state feedback and PI control.

and feedback compensation from the motor armature current. It is a third-order, type 1 system.

8.4.3 Multivariable Systems

Let a multivariable system be described by the state equation

$$\begin{aligned} \dot{\mathbf{x}} &= \mathbf{Ax} + \mathbf{Bu} + \mathbf{B}_1\mathbf{v} \\ \mathbf{y} &= \mathbf{Cx} \end{aligned} \tag{8.52}$$

where $\mathbf{x}$ is an $n \times 1$ vector, $\mathbf{u}$ an $m \times 1$ vector, $\mathbf{y}$ an $m \times 1$ vector, and $\mathbf{A}$, $\mathbf{B}$, and $\mathbf{C}$ are $n \times n$, $n \times m$, and $m \times n$ matrices, respectively. We desire a control law of the form

$$\mathbf{u} = -\mathbf{Kx} + \mathbf{Nr} \tag{8.53}$$

where $\mathbf{r}$ is an $m \times 1$ command input vector, $\mathbf{K}$ an $m \times n$ gain matrix, and $\mathbf{N}$ is an $m \times m$ scaling matrix. Now, there are mn unknown elements of $\mathbf{K}$ and only n roots to be specified. Hence, the problem is not well posed. We discuss a method of reducing the multivariable problem to the single-input case (Owens, 1981). We set

$$\begin{aligned} \mathbf{K} &= \mathbf{pk}^T \\ &= \begin{bmatrix} p_1 \\ \cdot \\ p_m \end{bmatrix} \lfloor k_1 \cdots k_n \rfloor \end{aligned} \tag{8.54}$$

where the m elements of $\mathbf{p}$ are chosen arbitrarily, and the n elements of $\mathbf{k}^T$ are to be obtained from the n roots that are specified. Substitution for $\mathbf{u}$ from Eqs. (8.53) and (8.54) in Eq. (8.52) yields

$$\dot{\mathbf{x}} = (\mathbf{A} - \mathbf{Bpk}^T)\mathbf{x} + \mathbf{BNr} + \mathbf{B}_1\mathbf{v}$$

The resulting closed-loop characteristic equation becomes

$$|s\mathbf{I} - \mathbf{A} + \mathbf{Bpk}^T| = 0 \tag{8.55}$$

This is identical to the characteristic equation of a single-input system

$$\dot{\mathbf{x}} = \mathbf{A}\mathbf{x} + \mathbf{B}\mathbf{p}u \tag{8.56}$$

After arbitrarily choosing the $\mathbf{p}$ vector, we equate the left-hand side of Eq. (8.55) to the desired characteristic polynomial and solve for the n elements of $\mathbf{k}^T$. The method of sections 8.4.1 and 8.4.2 can be used to determine the state feedback controller. A difficulty can arise because the equivalent single-input system of Eq. (8.56) may not be controllable. For a solution to exist, the determinant of the controllability matrix

$$\mathbf{Q} = [\mathbf{B}\mathbf{p} \quad \mathbf{A}\mathbf{B}\mathbf{p} \quad \cdots \quad \mathbf{A}^{n-1}\mathbf{B}\mathbf{p}] \tag{8.57}$$

must be nonzero. This requirement depends on the choice of $\mathbf{p}$ and also on the original multivariable system.

It is reasonable to expect the multivariable system of Eq. (8.52) to be controllable, but there may be no choice of $\mathbf{p}$ for which Eq. (8.56) is controllable. In that case, the optimal control theory (Bryson and Ho, 1969; Sage, 1968) may be used to design a controller with state feedback, but this theory is beyond our scope. Here, we merely state the results of a linear, unconstrained, optimal servomechanism problem with a quadratic performance index.

For the system of Eq. (8.52), with the error vector as $\mathbf{e} = \mathbf{r} - \mathbf{y} = \mathbf{r} - \mathbf{C}\mathbf{x}$, a control law is desired that minimizes the quadratic performance index

$$I = \tfrac{1}{2}\langle \mathbf{e}(t_f), \mathbf{P}\mathbf{e}(t_f)\rangle + \tfrac{1}{2}\int_0^{t_f} [\langle \mathbf{e}(t), \mathbf{Q}\mathbf{e}(t)\rangle + \langle \mathbf{u}(t), \mathbf{R}\mathbf{u}(t)\rangle]\, dt \tag{8.58}$$

where t_f is the terminal time, $\mathbf{R}$ is a positive definite matrix, and $\mathbf{P}$ and $\mathbf{Q}$ are positive semidefinite matrices. For a controllable system, the optimal feedback control law is

$$\mathbf{u}(t) = -\mathbf{R}^{-1}\mathbf{B}^T[\mathbf{K}(t)\mathbf{x}(t) - \mathbf{f}(t)] \tag{8.59}$$

where $\mathbf{K}(t)$ and $\mathbf{f}(t)$ are obtained, respectively, from the solutions of

$$\dot{\mathbf{K}}(t) = -\mathbf{K}(t)\mathbf{A} - \mathbf{A}^T\mathbf{K}(t) + \mathbf{K}(t)\mathbf{B}\mathbf{R}^{-1}\mathbf{B}^T\mathbf{K}(t) - \mathbf{C}^T\mathbf{Q}\mathbf{C} \tag{8.60}$$

$$\dot{\mathbf{f}}(t) = (\mathbf{K}(t)\mathbf{B}\mathbf{R}^{-1}\mathbf{B}^T - \mathbf{A}^T)\mathbf{f}(t) - \mathbf{C}^T\mathbf{Q}\mathbf{r}(t) \tag{8.61}$$

with terminal conditions

$$\mathbf{K}(t_f) = \mathbf{C}^T\mathbf{P}\mathbf{C} \tag{8.62}$$

$$\mathbf{f}(t_f) = \mathbf{C}^T\mathbf{P}\mathbf{r}(t_f) \tag{8.63}$$

Equation (8.60) is known as the matrix Ricatti equation. The structure of the optimal servomechanism is shown in Fig. 8.7. We note the several difficulties of implementing the optimal controller. The knowledge of $\mathbf{r}(t)$ for $0 \le t \le t_f$ is required for precomputing the optimal control law. The gain matrix $\mathbf{K}(t)$ is time varying. The matrices $\mathbf{P}$, $\mathbf{Q}$, and $\mathbf{R}$ of the performance index must be selected suitably to satisfy the specifications. For these reasons, state feedback control law for placement of roots appears as an attractive alternative.

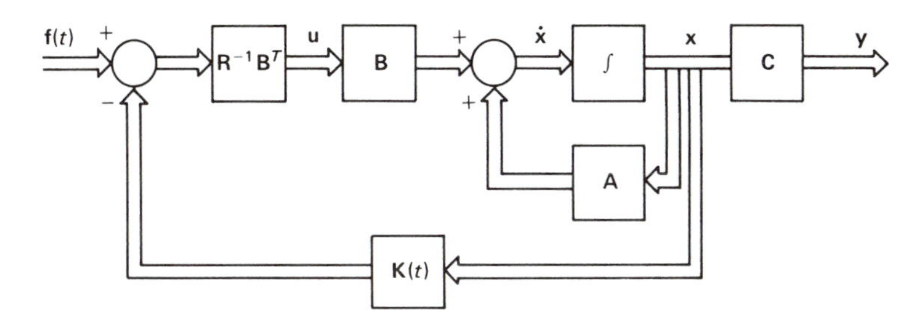

Figure 8.7 Structure of an optimal servomechanism.

When the command input vector $\mathbf{r} = \mathbf{0}$, an optimal state regulator is obtained by replacing $\mathbf{e}$ by $\mathbf{x}$ in Eq. (8.58) and setting $\mathbf{f}(t) = \mathbf{0}$. The control law is given by

$$\mathbf{u}(t) = -\mathbf{R}^{-1}\mathbf{B}^T\mathbf{K}(t)\mathbf{x}(t) \tag{8.64}$$

where $\mathbf{K}(t)$ is obtained from the solution of Eq. (8.60) with conditions of Eq. (8.62) after replacing matrix $\mathbf{C}$ by the identity matrix in both equations. For this case, when the terminal time $t_f \to \infty$ and $\mathbf{P} = \mathbf{0}$, the gain matrix $\mathbf{K}$ becomes a constant. For a detailed development of the optimal control theory and applications, the reader is referred to books on optimal control by Bryson and Ho (1969), Sage (1968), and others.

EXAMPLE 8.8

Let the $\mathbf{A}$ and $\mathbf{B}$ matrices of a two input system with two state variables be described by

$$\mathbf{A} = \begin{bmatrix} -2 & 0 \\ 0 & -2 \end{bmatrix} \qquad \mathbf{B} = \begin{bmatrix} 1 & 1 \\ 0 & 1 \end{bmatrix} \tag{8.65}$$

The controllability matrix $\mathbf{Q}$ of Eq. (8.16) becomes

$$\mathbf{Q} = [\mathbf{B} \quad \mathbf{AB}] = \begin{bmatrix} 1 & 1 & -2 & -2 \\ 0 & 1 & 0 & -2 \end{bmatrix}$$

and

$$\mathbf{QQ}^T = \begin{bmatrix} 10 & 5 \\ 5 & 5 \end{bmatrix}$$

The determinant of $\mathbf{QQ}^T$ is not zero and hence the pair $(\mathbf{A}, \mathbf{B})$ is controllable. The controllability matrix for the equivalent single-input system is obtained as

$$\mathbf{Q} = [\mathbf{Bp} \quad \mathbf{ABp}] = \begin{bmatrix} p_1 + p_2 & -2(p_1 + p_2) \\ p_2 & -2p_2 \end{bmatrix}$$

The determinant of $\mathbf{Q}$ is zero, and hence the pair $(\mathbf{A}, \mathbf{Bp})$ is not controllable. For this system, we have to design either a decoupled, output feedback controller of Chapter 7, or a controller with state feedback from optimal control theory.

EXAMPLE 8.9

A noninteracting controller with output feedback was considered in Example 7.17 for the two-input, two-output temperature control system of Example 3.8. With the parameter values used in Example 7.17, the state equations, Eqs. (3.87) and (3.88), become

$$\begin{aligned} \dot{x}_1 &= -0.1x_1 + 100u_1 + 0.1T_d \\ \dot{x}_2 &= 0.1x_1 - 0.1x_2 + 100u_2 \end{aligned} \tag{8.66}$$

Hence, the **A** and **B** matrices of Eq. (3.89) are obtained as

$$\mathbf{A} = \begin{bmatrix} -0.1 & 0 \\ 0.1 & -0.1 \end{bmatrix} \qquad \mathbf{B} = \begin{bmatrix} 100 & 0 \\ 0 & 100 \end{bmatrix} \tag{8.67}$$

With no change in the set points so that $\mathbf{r} = \mathbf{0}$, it is required to design a regulator. In the control law $\mathbf{u} = -\mathbf{Kx}$, we arbitrarily select $p_1 = p_2 = 1$ and obtain

$$\mathbf{Bp} = \begin{bmatrix} 100 \\ 100 \end{bmatrix} \qquad \mathbf{Q} = [\mathbf{Bp} \quad \mathbf{ABp}] = \begin{bmatrix} 100 & -10 \\ 100 & 0 \end{bmatrix}$$

The determinant of **Q** is not zero and the pair $(\mathbf{A}, \mathbf{Bp})$ is controllable. The characteristic equation of the closed-loop system becomes

$$\begin{aligned} |s\mathbf{I} - \mathbf{A} + \mathbf{Bpk}^T| &= \begin{vmatrix} s + 0.1 + 100k_1 & 100k_2 \\ -0.1 + 100k_1 & s + 0.1 + 100k_2 \end{vmatrix} \\ &= s^2 + (0.2 + 100k_1 + 100k_2)s \\ &\quad + 0.01 + 10k_1 + 20k_2 = 0 \end{aligned} \tag{8.68}$$

For this second-order system, let the desired roots be such that the natural frequency is 5 rad/s and the damping ratio is 0.7. Thus, the desired characteristic equation is

$$s^2 + 7s + 25 = 0 \tag{8.69}$$

Matching the corresponding coefficients of Eq. (8.68) and (8.69), we obtain $k_1 = -2.363$ and $k_2 = 2.431$. The structure of this regulator is shown in Fig. 8.8.

The transfer function relating the output vector to the disturbance is obtained from

$$\begin{aligned} \mathbf{Y}(s) = \mathbf{X}(s) &= (s\mathbf{I} - \mathbf{A} + \mathbf{Bpk}^T)^{-1} \begin{bmatrix} 0.1 \\ 0 \end{bmatrix} T_d \\ &= \frac{1}{s^2 + 7s + 25} \begin{bmatrix} s + 243.2 & -243.1 \\ 236.4 & s - 236.2 \end{bmatrix} \begin{bmatrix} 0.1 \\ 0 \end{bmatrix} T_d \end{aligned} \tag{8.70}$$

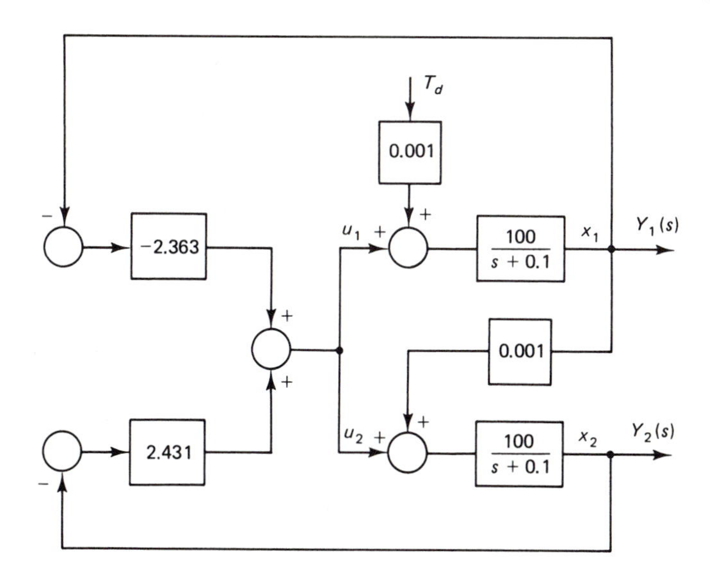

Figure 8.8 Multivariable regulator.

Hence, we obtain

$$Y_1(s) = \left(\frac{0.1(s + 243.2)}{s^2 + 7s + 25}\right) T_d$$

$$Y_2(s) = \left(\frac{23.64}{s^2 + 7s + 25}\right) T_d$$

For a step change in the disturbance T_d of magnitude H, the final-value theorem yields the limits as $t \to \infty$ of y_1 and y_2 as $24.32H/25$ and $23.64H/25$, respectively. These are the steady-state errors. Obviously, we have a type 0 system that is not able to reject a step change in the disturbance.

To eliminate this steady-state error in the output $y_1 = x_1$, we introduce an integral control mode by including an additional state variable x_3, where $\dot{x}_3 = x_1$. The augmented matrices become

$$\hat{\mathbf{A}} = \begin{bmatrix} -0.1 & 0 & 0 \\ 0.1 & -0.1 & 0 \\ 1 & 0 & 0 \end{bmatrix} \qquad \hat{\mathbf{B}} = \begin{bmatrix} 100 & 0 \\ 0 & 100 \\ 0 & 0 \end{bmatrix} \tag{8.71}$$

Arbitrarily choosing $p_1 = p_2 = 1$, we obtain

$$\mathbf{K} = \begin{bmatrix} 1 \\ 1 \end{bmatrix} \lfloor k_1 \quad k_2 \quad k_3 \rfloor \qquad \hat{\boldsymbol{B}}\boldsymbol{p} = \begin{bmatrix} 100 \\ 100 \\ 0 \end{bmatrix}$$

The controllability matrix for the equivalent single-input system becomes

$$\mathbf{Q} = [\hat{\mathbf{B}}\mathbf{p} \quad \hat{\mathbf{A}}\hat{\mathbf{B}}\mathbf{p} \quad \hat{\mathbf{A}}^2\hat{\mathbf{B}}\mathbf{p}] = \begin{bmatrix} 100 & -10 & 1 \\ 100 & 0 & -1 \\ 0 & 100 & -10 \end{bmatrix} \tag{8.72}$$

The determinant of $\mathbf{Q}$ is not zero, and hence the pair $(\hat{\mathbf{A}}, \hat{\mathbf{B}}\mathbf{p})$ is controllable. The characteristic equation is

$$|s\mathbf{I} - \hat{\mathbf{A}} + \hat{\mathbf{B}}\mathbf{p}\mathbf{k}^T| = s^3 + (0.2 + 100k_1 + 100k_2)s^2 + (10k_1 + 20k_2 + 100k_3 + 0.01)s + 10k_3 = 0 \tag{8.73}$$

For this third-order system, the characteristic equation is obtained from Table 8.1. Let the desired value of ω_0 be 5 rad/s. The desired characteristic equation becomes

$$s^3 + 8.75s^2 + 53.75s + 125 = 0 \tag{8.74}$$

Matching the corresponding coefficients of Eqs. (8.73) and (8.74), we obtain $k_1 = 119.798$, $k_2 = -119.712$, and $k_3 = 12.5$.

8.5 OBSERVABILITY

The development of the control law in the previous section has assumed that all the state variables are available for feedback. In the next section, we study the design of an observer whose function is to obtain an estimate of the state variables for feedback purpose from the measurement of the output vector $\mathbf{y}(t)$ and knowledge of the control input vector $\mathbf{u}(t)$. The existence of a solution to the observer equation depends on the property of observability studied in this section.

Let the state equations of a linear, time-invariant, multivariable system be described by

$$\dot{\mathbf{x}} = \mathbf{A}\mathbf{x} + \mathbf{B}\mathbf{u} + \mathbf{B}_1\mathbf{v} \tag{8.75}$$

$$\mathbf{y} = \mathbf{C}\mathbf{x}$$

It is assumed that the plant parameters have already been identified so that $\mathbf{A}$, $\mathbf{B}$, and $\mathbf{C}$ are known matrices.

Definition. The system described by Eq. (8.75) is said to be completely state observable, or simply observable, if for any time t_0, there exists a finite time $t_1 > t_0$, such that the knowledge of the output vector $\mathbf{y}(t)$ and the input vector $\mathbf{u}(t)$ in the time interval $t_0 \leq t \leq t_1$ is sufficient to uniquely determine the initial state $\mathbf{x}(t_0)$. For time-invariant systems, we can let $t_0 = 0$.

Let us consider a special case where matrix $\mathbf{A}$ has distinct eigenvalues. By similarity transformation $\mathbf{x} = \mathbf{P}\mathbf{x}^*$, we can uncouple the state variables so that we obtain

$$\dot{\mathbf{x}}^* = \mathbf{\Lambda x}^* + \mathbf{B}^*\mathbf{u} + \mathbf{B}_1^*\mathbf{v}$$
$$\mathbf{y} = \mathbf{C}^*\mathbf{x}^* \tag{8.76}$$

where $\mathbf{\Lambda}$ is diagonal, $\mathbf{B}^* = \mathbf{P}^{-1}\mathbf{B}$, $\mathbf{B}_1^* = \mathbf{P}^{-1}\mathbf{B}_1$, and $\mathbf{C}^* = \mathbf{CP}$. Suppose that $\mathbf{C}^*$ has a jth column of all zero elements. Then the output vector $\mathbf{y}$ is not affected by x_j^* whose behavior therefore cannot be observed from the measurement of $\mathbf{y}$. Hence, the system described by Eq. (8.76) is observable if $\mathbf{C}^*$ has no column with all zero elements. When u and y are scalars, the system of Eq. (8.76) is observable if the row matrix $\mathbf{C}^*$ has no zero element.

A general test for observability may be stated as follows. The system described by Eq. (8.75) is observable if and only if the $n \times nm$ matrix

$$\mathbf{M} = [\mathbf{C}^T \quad \mathbf{A}^T\mathbf{C}^T \quad (\mathbf{A}^T)^2\mathbf{C}^T \quad \cdots \quad (\mathbf{A}^T)^{n-1}\mathbf{C}^T] \tag{8.77}$$

has rank n. Here, $\mathbf{C}^T$ denotes the transpose of $\mathbf{C}$ and so on. The proof can be found in reference (Kreindler and Sarachik, 1964). We note that observability is a property of the pair $(\mathbf{A}, \mathbf{C})$. In general, most systems are observable, but there are some exceptions, which are illustrated by the following two examples.

EXAMPLE 8.10

A single-input, single-output system has a pole at -2 and a zero at -4. The zero at -4 is cancelled by a pole of the compensator as shown in Fig. 8.9(a). The two state variables are chosen by rearranging the block diagram as shown in Fig. 8.9(b).

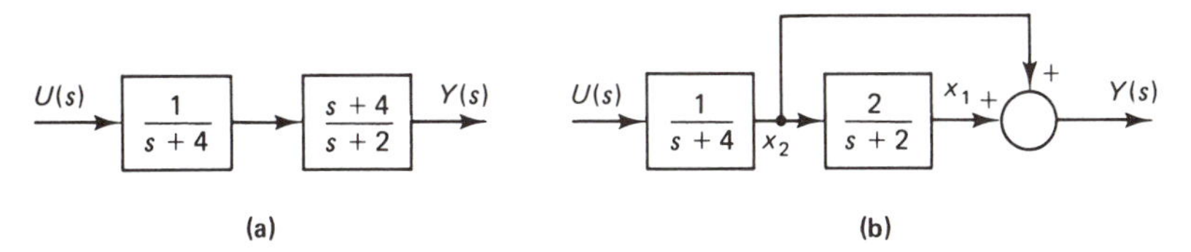

Figure 8.9 (a) Plant with compensator. (b) State variables.

The state equations are obtained from Fig. 8.9(b) as

$$\dot{x}_1 = -2x_1 + 2x_2$$
$$\dot{x}_2 = -4x_2 + u$$
$$y = x_1 + x_2 \tag{8.78}$$

The $\mathbf{A}$, $\mathbf{B}$, and $\mathbf{C}$ matrices are described by

$$\mathbf{A} = \begin{bmatrix} -2 & 2 \\ 0 & -4 \end{bmatrix} \qquad \mathbf{B} = \begin{bmatrix} 0 \\ 1 \end{bmatrix} \qquad \mathbf{C} = \lfloor 1 \quad 1 \rfloor \tag{8.79}$$

The $\mathbf{M}$ matrix is obtained as

$$\mathbf{M} = [\mathbf{C}^T \quad \mathbf{A}^T\mathbf{C}^T] = \begin{bmatrix} 1 & -2 \\ 1 & -2 \end{bmatrix} \tag{8.80}$$

The determinant of **M** is zero. Hence, its rank is not 2 and the system described by Eq. (8.78) is not observable. The reason for this behavior is that a zero has been cancelled by a pole. To investigate the unobservable mode, we diagonalize matrix **A**. Its determinant can be obtained as

$$|s\mathbf{I} - \mathbf{A}| = (s+2)(s+4)$$

The eigenvectors associated with the eigenvalues -2 and -4 are given, respectively, by

$$\begin{bmatrix} 1 \\ 0 \end{bmatrix} \qquad \begin{bmatrix} -1 \\ 1 \end{bmatrix}$$

Hence, the similarity transformation matrix **P** and its inverse are

$$\mathbf{P} = \begin{bmatrix} 1 & -1 \\ 0 & 1 \end{bmatrix} \qquad \mathbf{P}^{-1} = \begin{bmatrix} 1 & 1 \\ 0 & 1 \end{bmatrix}$$

and

$$\mathbf{\Lambda} = \mathbf{P}^{-1}\mathbf{A}\mathbf{P} = \begin{bmatrix} -2 & 0 \\ 0 & -4 \end{bmatrix} \qquad \mathbf{B}^* = \mathbf{P}^{-1}\mathbf{B} = \begin{bmatrix} 1 \\ 1 \end{bmatrix} \qquad \mathbf{C}^* = \mathbf{C}\mathbf{P} = \lfloor 1 \quad 0 \rfloor \tag{8.81}$$

It is seen that $\mathbf{C}^*$ has a zero element, and hence the state variable x_2^* is not observable. We note that $\mathbf{B}^*$ does not have a zero element, and hence the system is controllable. Also, from the controllability matrix, we obtain

$$\mathbf{Q} = (\mathbf{B} \quad \mathbf{A}\mathbf{B}) = \begin{bmatrix} 0 & 2 \\ 1 & -4 \end{bmatrix}$$

Its determinant is not zero, and again we conclude that the system is controllable.

EXAMPLE 8.11

A system for the attitude control of a satellite was studied in Example 7.14. The equation of motion was given by Eq. (7.88), where the torque $T = bF$, and F is the thrust force. Hence, we have

$$I\ddot{\theta} = bF - T_d \tag{8.82}$$

Choosing the state variables as $x_1 = \theta$ and $x_2 = \dot{\theta}$, and letting the output measurement be the attitude rate $\dot{\theta}$, we obtain

$$\begin{aligned} \dot{x}_1 &= x_2 \\ \dot{x}_2 &= \frac{b}{I}F - \frac{1}{I}T_d \\ y &= x_2 \end{aligned} \tag{8.83}$$

The associated matrices are

$$\mathbf{A} = \begin{bmatrix} 0 & 1 \\ 0 & 0 \end{bmatrix} \qquad \mathbf{B} = \begin{bmatrix} 0 \\ b/I \end{bmatrix} \qquad \mathbf{B}_1 = \begin{bmatrix} 0 \\ -1/I \end{bmatrix} \qquad \mathbf{C} = \lfloor 0 \quad 1 \rfloor$$

The observability matrix $\mathbf{M}$ becomes

$$\mathbf{M} = [\mathbf{C}^T \quad \mathbf{A}^T\mathbf{C}^T] = \begin{bmatrix} 0 & 0 \\ 1 & 0 \end{bmatrix} \tag{8.84}$$

The determinant of $\mathbf{M}$ is zero, and hence the system is not observable. The reason why this system is not observable is that we can determine the position by integrating the velocity only up to an arbitrary constant. Hence, a bad choice of the variable has been made for the output measurement.

8.6 OBSERVER DESIGN

The control law of section 8.4 requires all the state variables for feedback. It has been indicated earlier that some of the state variables may not be available for measurement and the use of many sensors may not be justified economically. The purpose of this section is to study the design of an observer to reconstruct all the states, given the measurement of the output $\mathbf{y}$ and knowing the input $\mathbf{u}$. The observer produces an estimate $\hat{\mathbf{x}}(t)$ of the state variables $\mathbf{x}(t)$ and these estimates are then used to implement the control law.

The concept of an observer was introduced by Luenberger (1966, 1971), and the ideas have been further developed since that time. We first consider single-input, single-output systems and then generalize the results to multivariable systems.

8.6.1 Single-Input Single-Output Systems

Let the plant be described by the state equations

$$\begin{aligned} \dot{\mathbf{x}}(t) &= \mathbf{A}\mathbf{x}(t) + \mathbf{B}u(t) \\ y(t) &= \mathbf{C}\mathbf{x}(t) \end{aligned} \tag{8.85}$$

where $\mathbf{x}$ is an $n \times 1$ vector, u and y are scalars, and $\mathbf{A}$, $\mathbf{B}$, and $\mathbf{C}$ are $n \times n$, $n \times 1$, and $1 \times n$ matrices, respectively. It is assumed that the plant parameters have already been identified so that the matrices are known.

A simple method of estimating the states is to construct or simulate the plant dynamics

$$\dot{\hat{\mathbf{x}}}(t) = \mathbf{A}\hat{\mathbf{x}}(t) + \mathbf{B}u(t) \tag{8.86}$$

where $\hat{\mathbf{x}}$ denotes an estimate of $\mathbf{x}$. Since $\mathbf{A}$, $\mathbf{B}$, and u are known, we can solve this equation on line and obtain $\hat{\mathbf{x}}$. Defining the estimation error as $\mathbf{e}(t) = \mathbf{x} - \hat{\mathbf{x}}$, we subtract Eq. (8.86) from the plant equation, Eq. (8.85), to obtain the error differential equation

$$\dot{\mathbf{e}}(t) = \mathbf{A}\mathbf{e}(t) \tag{8.87}$$

However, this open-loop method of estimator design has several deficiencies. We demand that the closed-loop characteristic equation $|s\mathbf{I} - \mathbf{A} + \mathbf{BK}| = 0$ be asymptotically stable. But the characteristic equation of the error obtained from Eq. (8.87) is the open-loop equation $|s\mathbf{I} - \mathbf{A}| = 0$. If **A** is unstable, the error will grow with time. The parameter values and hence the matrices **A** and **B** in Eq. (8.86) may not be known accurately, and hence the estimate will not represent the state adequately.

We make use of the measured output to correct these deficiencies. The difference between the measured output and the estimated output is fed back to correct the model. With $\mathbf{H} = \lfloor h_1 \; h_2 \cdots h_n \rfloor^T$ as an $n \times 1$ column matrix, the equation of the estimator is

$$\dot{\hat{\mathbf{x}}}(t) = \mathbf{A}\hat{\mathbf{x}}(t) + \mathbf{B}u(t) + \mathbf{H}[y(t) - \mathbf{C}\hat{\mathbf{x}}(t)] \tag{8.88}$$

The block diagram is shown in Fig. 8.10.

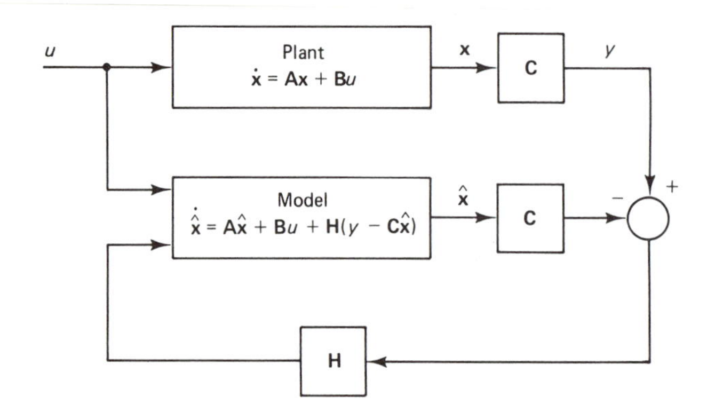

Figure 8.10 Closed-loop estimator.

The plant equation may be written as

$$\dot{\mathbf{x}} = \mathbf{A}\mathbf{x} + \mathbf{B}u + \mathbf{H}(y - \mathbf{C}\mathbf{x}) \tag{8.89}$$

Since $y = \mathbf{C}\mathbf{x}$, the plant equation, Eq. (8.89), is unchanged. Subtracting Eq. (8.88) from Eq. (8.89), we obtain the error equation as

$$\dot{\mathbf{e}}(t) = (\mathbf{A} - \mathbf{HC})\mathbf{e}(t) \tag{8.90}$$

and the characteristic equation of the error becomes

$$|s\mathbf{I} - \mathbf{A} + \mathbf{HC}| = 0 \tag{8.91}$$

We select the n roots of this equation as $-\alpha_1, -\alpha_2, \ldots, -\alpha_n$ to ensure that they have negative real parts so that Eq. (8.90) is asymptotically stable. Hence, we obtain

$$|s\mathbf{I} - \mathbf{A} + \mathbf{HC}| = (s + \alpha_1)(s + \alpha_2) \cdots (s + \alpha_n) \tag{8.92}$$

The n unknown elements of **H** can now be solved by matching the corresponding coefficients of this n^{th}-order polynomial as in the case of the control law. It may be more convenient to employ the Ackermann formula for high-order systems.

Let the desired characteristic polynomial of the observer be denoted by

$$\Delta_0(s) = (s+\alpha_1)(s+\alpha_2)\cdots(s+\alpha_n)$$
$$= s^n + b_{n-1}s^{n-1} + \cdots + b_0$$

We define $\Delta_0(\mathbf{A})$ as

$$\Delta_0(\mathbf{A}) = \mathbf{A}^n + b_{n-1}\mathbf{A}^{n-1} + \cdots + b_0\mathbf{I} \tag{8.93}$$

Note that if we take the transpose of $\mathbf{A} - \mathbf{HC}$, we obtain $\mathbf{A}^T - \mathbf{C}^T\mathbf{H}^T$, which is the same form as the control law matrix $\mathbf{A} - \mathbf{BK}$. The Ackermann formula for the solution of the observer matrix $\mathbf{H}$ is given by Franklin and Powell (1980) as

$$\begin{aligned}\mathbf{H}^T &= \lfloor h_1 \quad h_2 \quad \cdots \quad h_n \rfloor \\ &= \lfloor 0 \quad 0 \quad \cdots \quad 0 \quad 1 \rfloor \mathbf{M}^{-1}\Delta_0(\mathbf{A})\end{aligned} \tag{8.94}$$

where $\mathbf{M}$ is the observability matrix defined by Eq. (8.77). The inversion of matrix $\mathbf{M}$ in Eq. (8.94) can be avoided by following the procedure described by Eqs. (8.26) and (8.27) for the control law matrix.

The question to be answered is whether it is possible to arbitrarily select the roots $-\alpha_1, -\alpha_2, \ldots, -\alpha_n$ and then solve for the matrix $\mathbf{H}$ from Eq. (8.92). The answer is in the affirmative when the pair $(\mathbf{A}, \mathbf{C})$ is observable. The proof becomes obvious when we note that Eq. (8.94) requires the inverse of the observability matrix $\mathbf{M}$. When $\mathbf{M}$ is not singular, its rank is n, which is the condition for observability.

8.6.2 Multivariable Systems

The preceding results can be generalized to multivariable systems. Let a multivariable plant be described by the state equations

$$\begin{aligned}\dot{\mathbf{x}}(t) &= \mathbf{Ax}(t) + \mathbf{Bu}(t) \\ \mathbf{y}(t) &= \mathbf{Cx}(t)\end{aligned} \tag{8.95}$$

where $\mathbf{x}$, $\mathbf{u}$, and $\mathbf{y}$ are vectors of dimensions $n \times 1$, $m \times 1$, and $m \times 1$, respectively. Matrices $\mathbf{A}$, $\mathbf{B}$, and $\mathbf{C}$ have dimensions $n \times n$, $n \times m$, and $m \times n$, respectively. The estimator equation analogous to Eq. (8.88) becomes

$$\dot{\hat{\mathbf{x}}}(t) = \mathbf{A}\hat{\mathbf{x}}(t) + \mathbf{Bu}(t) + \mathbf{H}[\mathbf{y}(t) - \mathbf{C}\hat{\mathbf{x}}(t)] \tag{8.96}$$

where $\mathbf{H}$ is an $n \times m$ constant matrix. Now, there are nm unknown elements of $\mathbf{H}$ and only n roots to be specified. Hence, the problem is not well posed.

We reduce the multivariable problem to an equivalent single-input, single-output case by using the procedure of control law design of section 8.4.3. We set

$$\mathbf{H} = \begin{bmatrix} h_1 \\ h_2 \\ \vdots \\ h_n \end{bmatrix} \lfloor p_1 \quad p_2 \quad \cdots \quad p_m \rfloor = \mathbf{hp}^T \tag{8.97}$$

where the m elements of $\mathbf{p}^T$ are chosen arbitrarily, and the n elements of $\mathbf{h}$ are to be obtained from the n observer roots that are specified. The characteristic equation for the error vector analogous to Eq. (8.91) becomes

$$|s\mathbf{I} - \mathbf{A} + \mathbf{h}\mathbf{p}^T\mathbf{C}| = 0 \tag{8.98}$$

After arbitrarily choosing the m elements of $\mathbf{p}^T$, we equate the left-hand side of Eq. (8.98) to the desired characteristic polynomial of the observer and solve for the n elements of $\mathbf{h}$. For a solution to exist, the determinant of the $n \times n$ observability matrix

$$\mathbf{M} = [\mathbf{C}^T\mathbf{p} \quad \mathbf{A}^T\mathbf{C}^T\mathbf{p} \quad (\mathbf{A}^T)^2\mathbf{C}^T\mathbf{p} \quad \cdots \quad (\mathbf{A}^T)^{n-1}\mathbf{C}^T\mathbf{p}] \tag{8.99}$$

must be nonzero. This requirement depends on the choice of $\mathbf{p}$ and on the original multivariable system. A difficulty can arise as in the case of the multivariable control law. The multivariable pair $(\mathbf{A}, \mathbf{C})$ may be observable but there may be no choice of $\mathbf{p}^T$ for which the equivalent single-input, single-output pair $(\mathbf{A}, \mathbf{p}^T\mathbf{C})$ of Eq. (8.99) is observable.

8.6.3 Closed-Loop Dynamics and Observer Roots

The combination of the control law mechanization and the observer may be considered as the controller. The structure of the controlled system is shown in Fig. 8.11. Since the control law was designed assuming that the true state $\mathbf{x}$ was available for feedback instead of $\hat{\mathbf{x}}$, we examine what effect this has on the system dynamics.

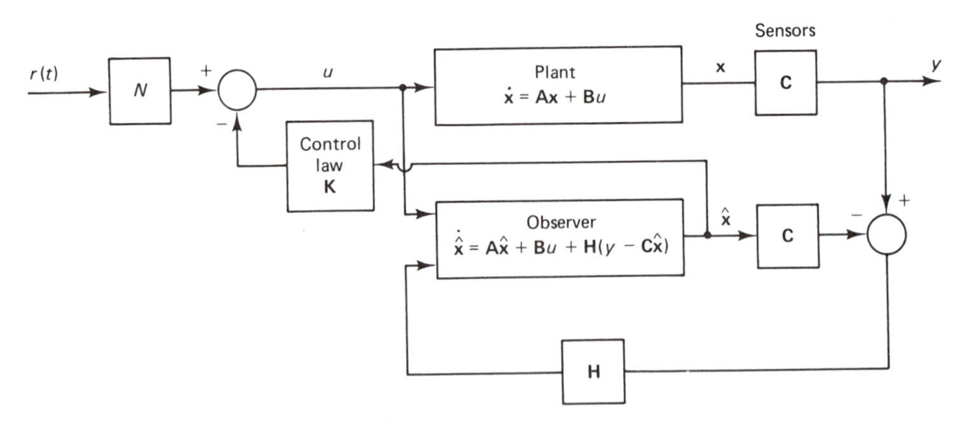

Figure 8.11 Control-law mechanization and observer.

With the control law of the form

$$u(t) = Nr(t) - \mathbf{K}\hat{\mathbf{x}}(t) \tag{8.100}$$

the controlled system becomes

$$\dot{\mathbf{x}}(t) = \mathbf{A}\mathbf{x}(t) - \mathbf{B}\mathbf{K}\hat{\mathbf{x}}(t) + \mathbf{B}Nr(t) \tag{8.101}$$

Using the observer error $\mathbf{e}(t) = \mathbf{x} - \hat{\mathbf{x}}$, the preceding equation can be expressed as

$$\dot{\mathbf{x}}(t) = (\mathbf{A} - \mathbf{B}\mathbf{K})\mathbf{x}(t) + \mathbf{B}\mathbf{K}\mathbf{e}(t) + \mathbf{B}Nr(t) \tag{8.102}$$

Combining this equation with the observer error equation, Eq. (8.90), we can describe the complete system by two coupled equations

$$\begin{bmatrix} \dot{\mathbf{x}}(t) \\ \dot{\mathbf{e}}(t) \end{bmatrix} = \left[\begin{array}{c|c} \mathbf{A} - \mathbf{BK} & \mathbf{BK} \\ \hline \mathbf{0} & \mathbf{A} - \mathbf{HC} \end{array}\right] \begin{bmatrix} \mathbf{x} \\ \mathbf{e} \end{bmatrix} + \begin{bmatrix} \mathbf{B}N \\ \mathbf{0} \end{bmatrix} r(t) \tag{8.103}$$

The characteristic equation of the closed-loop system becomes

$$\begin{vmatrix} s\mathbf{I} - \mathbf{A} + \mathbf{BK} & -\mathbf{BK} \\ \mathbf{0} & s\mathbf{I} - \mathbf{A} + \mathbf{HC} \end{vmatrix} = |s\mathbf{I} - \mathbf{A} + \mathbf{BK}| \quad |s\mathbf{I} - \mathbf{A} + \mathbf{HC}| = \Delta(s)\Delta_0(s) = 0 \tag{8.104}$$

The roots of this characteristic equation of the combined system consist of the sum of the control roots and the estimator roots. The control roots are unchanged from those obtained by assuming state feedback $\mathbf{x}(t)$. Hence, the control law and the observer can be designed separately and then used jointly.

The final matter to be settled is the specification of the desired roots of the observer characteristic equation, Eq. (8.91). The estimation error decays at a rate dependent on these roots. In Eq. (8.102), we require the observer error $\mathbf{e}(t)$ to decay at a fast rate with time constants that are much smaller than the time constants of the controlled system so that the total response is dominated by the slower control roots. Hence, the observer roots should be placed to the left of the control roots in the s-plane. But, if the observer roots are placed too far to the left of the control roots, then the observer gains represented by the elements of $\mathbf{H}$ will be high. Hence, the measurement noise will not be filtered out and may even be amplified.

Clearly, a compromise is required in selecting the roots of the observer characteristic equation. Optimal estimation theory can be employed for this purpose (Bryson and Ho, 1969). A rule of thumb is to let $-\alpha_i$ be approximately equal to $-4s_i$, where $-\alpha_i$ are the observer roots, and $-s_i$ are the control roots. The control roots are of course chosen to satisfy the performance specifications as indicated in section 8.2.

The controller, including the control law and observer, can be constructed with analog components, such as the operational amplifiers studied in the previous chapter. It is expected that a controller with state feedback would be more expensive than a controller with output feedback of Chapter 7. However, digital computer implementation of the controller with state feedback involves software and hence would be cost effective.

EXAMPLE 8.12

In Example 8.6, we have obtained a control law for the electrical position-control system of Example 3.4. We now add an observer for this system. The numerical values of the $\mathbf{A}$ and $\mathbf{B}$ matrices are given by Eq. (8.28). Only the state variable $x_1 = \theta$ is measured as shown in Fig. 8.3 so that $y = \mathbf{Cx}$, where

$$\mathbf{C} = \lfloor 1 \quad 0 \quad 0 \rfloor \tag{8.105}$$

First, we check whether the system is observable. The observability matrix is obtained as

$$\mathbf{M} = [\mathbf{C}^T \quad \mathbf{A}^T\mathbf{C}^T \quad (\mathbf{A}^T)^2\mathbf{C}^T] = \begin{bmatrix} 1 & 0 & 0 \\ 0 & 1 & -0.5 \\ 0 & 0 & 10 \end{bmatrix} \tag{8.106}$$

The determinant of this matrix is not zero. Hence, its rank is 3 and the system is observable. Now, with

$$\mathbf{HC} = \begin{bmatrix} h_1 \\ h_2 \\ h_3 \end{bmatrix} \lfloor 1 \quad 0 \quad 0 \rfloor$$

the characteristic equation of the observer becomes

$$\begin{aligned} |s\mathbf{I} - \mathbf{A} + \mathbf{HC}| &= \begin{vmatrix} s + h_1 & -1 & 0 \\ h_2 & s + 0.5 & -10 \\ h_3 & 0.1 & s + 10 \end{vmatrix} \\ &= s^3 + (10.5 + h_1)s^2 + (6 + 10.5h_1 + h_2)s \\ &\quad + 6h_1 + 10h_2 + 10h_3 = 0 \end{aligned} \tag{8.107}$$

The control roots that satisfy the performance specifications are given by Eq. (8.30). Multiplying these roots by 4, we obtain the desired observer characteristic equation as

$$\begin{aligned} \Delta_0(s) &= (s + 141.6)(s + 104.2 + j213.6)(s + 104.2 - j213.6) \\ &= s^3 + 350s^2 + 85{,}992s + 7{,}997{,}936 = 0 \end{aligned} \tag{8.108}$$

Matching the corresponding coefficients of Eqs. (8.107) and (8.108), we obtain $h_1 = 339.5$, $h_2 = 82{,}421.3$, and $h_3 = 717{,}168.6$. It is noted that these gains are quite large.

The observer that we have studied reconstructs the entire state, given measurements of some of the states. It is possible to design a reduced-order observer that reconstructs only the states that are not measured. The state vector is partioned into two parts: the state variables directly measured and the remaining state variables to be estimated. For the design of reduced-order estimators, we refer to Luenberger (1971) and Gopinath (1971). However, it is known (Franklin and Powell, 1980) that when there is significant noise on the measurements, the observer for the full state vector yields results that are superior to those of a reduced-order observer.

8.7 SUMMARY

In this chapter, we have presented a method of controller design wherein all the state variables are fed back through constant gains to form the control law. The gains are chosen to place the roots of the characteristic equation of the closed loop system at the desired locations in the s-plane. The root locations must be chosen to satisfy the performance specifications and standard ITAE forms can be used for this purpose. It is shown that when a system is controllable, a control gain matrix can be found to place the roots at arbitrary locations. We have discussed the method of including integral control with state feedback when it is desired to improve the system type.

The measurement of all the state variables can be avoided by including an observer. We have studied the design of an observer that produces an estimate of the state variables, given the measurement of the output and the knowledge of the control input u. The roots of the characteristic equation of the observer must be chosen such that the estimation error decays at a much faster rate than the system response. It is shown that when a system is observable, the observer gain matrix can be found to place the observer roots at any arbitrary locations in the s-plane.

Finally, the control law and estimator design are combined and the control law is implemented by using an estimate of the state. It is shown that the control roots are unchanged from those obtained by assuming that the actual state is available for feedback instead of its estimate. Hence, the control law and the observer can be designed separately and then used jointly. If the control law and observer are constructed with analog components, it is expected that this controller will be more expensive than one with output feedback. But, if a microprocessor-based digital computer is employed, the controller can be implemented by software. This topic is studied in the next chapter.

An important observation is that with the state feedback control law, the resulting control system in most cases will be type 0. As seen from Chapter 7, a type 0 system is undesirable regarding steady-state errors and disturbance rejection capability. Therefore, state feedback with PI control is recommended for most applications.

REFERENCES

BRYSON, A. E., and HO, Y. C. (1969). *Applied Optimal Control.* Waltham, MA: Blaisdell.

FRANKLIN, G. F., and POWELL, J. D. (1980). *Digital Control of Dynamic Systems.* Reading, MA: Addison-Wesley.

GOPINATH, B. (1971, March). "On the Control of Linear Multiple Input-Output Systems." *Bell System Technical Journal 50.*

GRAHAM, D., and LATHROP, R. C. (1953). "The Synthesis of Optimum Response: Criteria and Standard Forms." *Trans. AIEE 72* (Part 2), 273–288.

KREINDLER, E., and SARACHIK, P. E. (1964, April). "On the Concepts of Controllability and Observability of Linear Systems." *IEEE Trans. on Automatic Control,* 129–136.

LUENBERGER D. G. (1966, April). "Observers for Multivariable Systems." *IEEE Trans. on Automatic Control AC*(11) 190–197.

LUENBERGER, D. G. (1971, December). "An Introduction to Observers." *IEEE Trans. on Automatic Control AC*(16), 596–602.

OWENS, D. H. (1981). *Multivariable and Optimal Systems.* New York: Academic Press.
SAGE, A. P. (1968). *Optimum Systems Control.* Englewood Cliffs, NJ: Prentice-Hall.

PROBLEMS

8.1. Consider a closed-loop transfer function given by Eq. (8.1). Determine the roots of the characteristic equation of a second-order system, where $n = 2$ in Eq. (8.1) to satisfy the following specifications. The system must be type 1 and minimum ITAE form. The bandwidth must be 35 rad/s.

8.2. Solve Problem 8.1 for a third-order system, where $n = 3$ in Eq. (8.1).

8.3. Suppose that the open-loop system described by $\dot{\mathbf{x}} = \mathbf{Ax} + \mathbf{B}u$ is completely state controllable. Show that the system $\dot{\mathbf{x}} = \mathbf{PAP}^{-1}\mathbf{x} + \mathbf{PB}u$ is also completely state controllable, where $\mathbf{P}$ is an arbitrary $n \times n$ nonsingular constant matrix.

8.4. Investigate the state controllability of the following open-loop system. In case it is not state controllable, determine the open-loop transfer function and investigate whether there is pole-zero cancellation.

$$\dot{\mathbf{x}} = \begin{bmatrix} 0 & 1 & 0 \\ 5 & 0 & 2 \\ -2 & 0 & -2 \end{bmatrix}\mathbf{x} + \begin{bmatrix} -1 \\ 1 \\ -1 \end{bmatrix}u$$

$$y = \lfloor 0 \quad 1 \quad 1 \rfloor \mathbf{x}$$

8.5. Solve Problem 8.4 for the system described by

$$\dot{\mathbf{x}} = \begin{bmatrix} -4 & 2 & 0 \\ 0 & -3 & 0 \\ 0 & 0 & -2 \end{bmatrix}\mathbf{x} + \begin{bmatrix} 0 \\ 1 \\ 1 \end{bmatrix}u$$

$$y = \lfloor 1 \quad 0 \quad 0 \rfloor \mathbf{x}$$

8.6. Investigate the state controllability of a two-input, two-output open-loop system described by

$$\dot{\mathbf{x}} = \begin{bmatrix} 0 & 1 & 0 \\ 0 & 0 & 1 \\ -2 & -1 & -4 \end{bmatrix}\mathbf{x} + \begin{bmatrix} 0 & 0.5 \\ 0.2 & 0 \\ 2 & 1 \end{bmatrix}\begin{bmatrix} u_1 \\ u_2 \end{bmatrix}$$

$$\mathbf{y} = \begin{bmatrix} 1 & 0 & 0 \\ 0 & 1 & 1 \end{bmatrix}\mathbf{x}$$

8.7. Consider the hydraulic speed-control system of Example 4.4. Let the parameter values be such that the system matrices given by Eq. (4.64) become

$$\mathbf{A} = \begin{bmatrix} -3 & 10 \\ 0 & 0 \end{bmatrix} \qquad \mathbf{B} = \begin{bmatrix} 0 \\ 5 \end{bmatrix}$$

$$\mathbf{C} = \lfloor 1 \quad 0 \rfloor$$

Obtain the gain matrix $\mathbf{K}$ of the state feedback law such that for the closed-loop system, $\omega_n = 29.7$ rad/s and $\zeta = 0.7$. Determine the control system type and the steady-state error to unit step input. Let $N = k_1$.

8.8. It is desired to convert the type 0 system of Problem 8.7 to type 1 by using state feedback with PI control as discussed in section 8.4.2. To satisfy the specifications, a third-order, minimum ITAE type 1 form is selected from Table 8.1 with $\omega_0 = 12.5$. Obtain the gain matrix **K** of the state feedback law.

8.9. An open-loop system has the following system matrices.

$$\mathbf{A} = \begin{bmatrix} 0 & 1 & 0 \\ 0 & -3 & 8 \\ 0 & -2 & -10 \end{bmatrix} \qquad \mathbf{B} = \begin{bmatrix} 0 \\ 0 \\ 50 \end{bmatrix}$$

$$\mathbf{C} = \lfloor 1 \quad 0 \quad 0 \rfloor$$

Find the state feedback gain matrix such that the roots of the closed-loop characteristic equation are located at -4, $-8 \pm j16$.

8.10. Obtain the state feedback gain matrix for the multivariable system of Problem 8.6 such that the roots of the closed-loop characteristic equation are located at -4, $-8 \pm j16$.

8.11. Consider the attitude-control system of Example 7.14. From Example 8.11, assuming that x_1 is measured instead of x_2, the system matrices are given by

$$\mathbf{A} = \begin{bmatrix} 0 & 1 \\ 0 & 0 \end{bmatrix} \qquad \mathbf{B} = \begin{bmatrix} 0 \\ 100 \end{bmatrix} \qquad \mathbf{C} = \lfloor 1 \quad 0 \rfloor$$

The specifications require that the closed-loop characteristic equation satisfy a minimum ITAE type 2 form with $\omega_0 = 30$ rad/s. Obtain the gain matrix **K** of the state feedback control law.

8.12. Investigate the observability of the open-loop system of Problem 8.4. In case it is not observable, determine the open-loop transfer function, and investigate whether there is pole-zero cancellation.

8.13. Solve Problem 8.12 for the system of Problem 8.5.

8.14. Solve Problem 8.12 for the multivariable system of Problem 8.6.

8.15. Design an observer for the system of Problem 8.7 such that the roots of the observer characteristic equation are four times the control roots.

8.16. Solve Problem 8.15 for the system of Problem 8.8.

8.17. Solve Problem 8.15 for the system of Problem 8.9.

8.18. Solve Problem 8.15 for the system of Problem 8.10.

8.19. Solve Problem 8.15 for the system of Problem 8.11.

9

Digital Control

9.1 INTRODUCTION

With the availability of inexpensive microprocessors, the use of digital control has become attractive and economical, especially when the complexity of the control law increases. A digital controller also has other advantages over an analog controller. These include the flexibility of modifying the control law by changing the software, time-sharing of components, better accuracy since the drift associated with analog components is avoided, and the use of the control computer for data reduction, monitoring, logic, safety features, and start-up and shut-down procedures. A digital computer can control many separate processes by scanning the variables with a multiplexer and generating control laws on a time-sharing basis.

The main objective of this chapter is to discuss the digital implementation of the analog controller designs covered in Chapters 7 and 8. For the controller design with output feedback of Chapter 7, the computer is used to digitally implement a PID family of control laws and a digital compensator, if one is required, from a program stored in its memory. In the case of a controller design with state feedback of Chapter 8, the computer is used to digitally mechanize the control law with state feedback and an observer. This function is called first-level control. In addition, a supervisory computer communicating with the first-level controllers can be used to perform functions such as optimization, trajectory planning in robotics, and generation of command inputs. Here, we are concerned only with first-level control.

In our case, the plant, i.e., the process to be controlled and the actuator, is described by ordinary, linear differential equations and varies continuously with time. On the other hand, in the computer mathematical operations, the signals vary at discrete instants of time. In addition, the amplitude of the signal is rounded off to one of a finite number of levels, depending on the number of binary digits (bits) of the device, and this process is called quantization.

A signal that is both discrete-time and quantized is called a *digital signal.* A system containing only discrete-time variables is called a *discrete-time system,* whereas a system containing both discrete-time and continuous-time variables is called a *sampled-data system.* Hence, in this chapter, we deal with sampled-data systems.

We begin our study by considering the mathematical modeling of the major components of a digital control system, including the analog-to-digital and digital-to-analog converters. The state differential equations of the plant are then converted to state difference equations, relating the values of the states, inputs, and outputs at discrete instants of time so that the plant model is compatible with the digital controller. The z-transformation is introduced, and an alternate plant model employing discrete transfer functions is presented.

The design of a controller with output feedback is considered next for the purpose of digital implementation of the PID family of control laws. We then present methods of analysis of the transient response and stability of closed-loop digital control systems. The controllability and observability properties of discrete-time systems are analyzed. The final topic involves the digital implementation of a control law with state feedback and an observer.

9.2 MATHEMATICAL MODELING OF DIGITAL CONTROL SYSTEMS

The configuration of a digital control system can take on many forms. The two common configurations with output feedback are shown in Figs. 9.1 and 9.2. In Fig. 9.1, the command input $r(t)$ is a continuous-time signal, whereas in Fig. 9.2, $r(kT)$ is a digital signal, which may be supplied externally or retrieved from computer memory. We first give a brief description of the operation of the block diagram.

The analog-to-digital converter (ADC) acts on a continuous-time signal, commonly an electrical voltage, and converts it into a digital signal that is supplied to the computer. In Fig. 9.1, the ADC acts on the error signal $e(t)$ between the

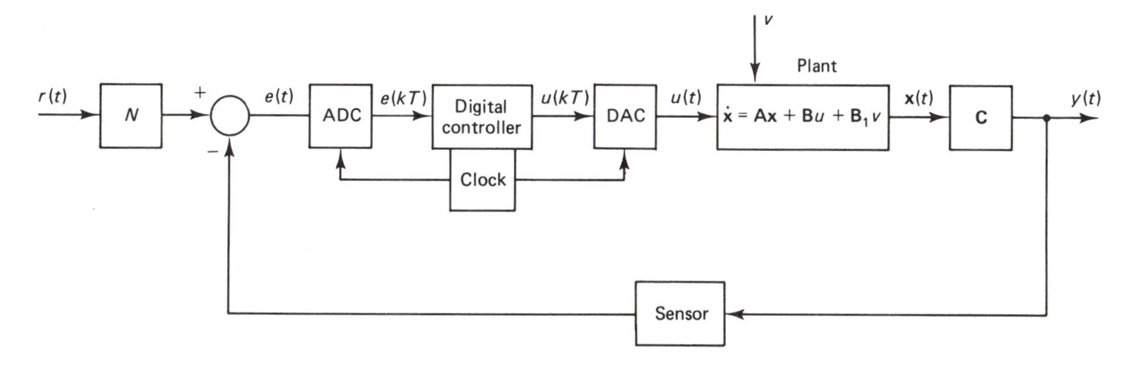

Figure 9.1 Digital-control system with error sampling.

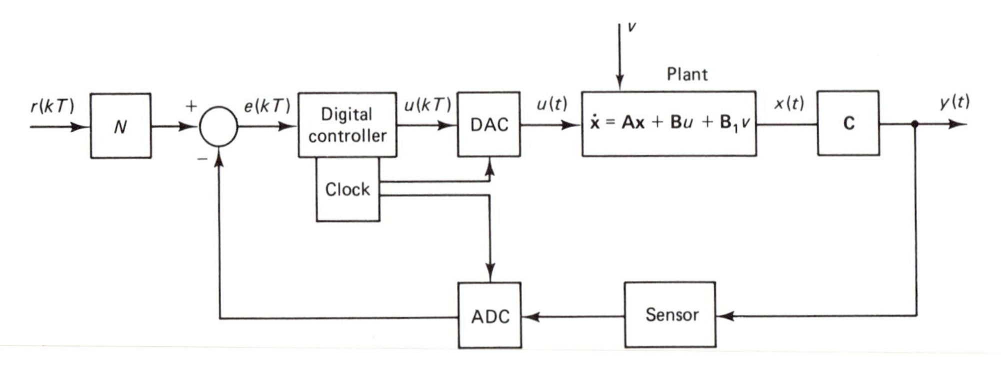

Figure 9.2 Digital-control system with output sampling.

command input $r(t)$ and the output $y(t)$. In Fig. 9.2, the ADC acts on the sensor output and the error is formed in the computer. A clock incorporated in the computer sends out a pulse every T seconds, and the ADC supplies a coded number to the computer every time it receives a pulse. The time interval T between any two successive pulses is called the sampling period. In general, the sampling period may be variable, but here we consider the case where it is fixed.

The digital controller is used to implement a control law digitally from a stored program. In this chapter, we first study the case of output feedback, where we are concerned with a PID family of control laws and compensation. Later, we cover the case of controller design with state feedback, where our concern is to digitally mechanize a control law and an observer.

The output signal of the controller is intermittent and is in coded form. It must be converted into an analog signal unless the actuator operates on discrete signals, as in the case of a pulse motor. The digital-to-analog converter (DAC) takes the controller output every T seconds, decodes the signal, and converts it into a continuous or piecewise continuous form, which is then supplied to the actuator of the process to be controlled.

In multivariable systems and in cases where a single digital computer is used to control several independent outputs, the appropriate variables can be scanned by a multiplexer and/or multiple ADC and DAC channels can be used. The control laws are then generated on a time-sharing basis by the computer. In the following, we develop mathematical models of these components of a digital control system.

9.2.1 Analog-to-Digital Converter

The function of an analog-to-digital converter is to sample a continuous-time signal, quantize its amplitude to one of a finite number of levels, and code it into an equivalent binary number that can be accepted by the digital computer. In the sampling process, a continuous-time signal $f(t)$ is used to modulate the amplitudes of a pulse train $p(t)$ as shown in Fig. 9.3. This process is called pulse-amplitude modulation. The pulse train $p(t)$ has a period T and width Δt. The original analog signal $f(t)$ and the pulse-amplitude-modulated signal $f^*(t) = f(t)p(t)$ are shown in Fig. 9.4.

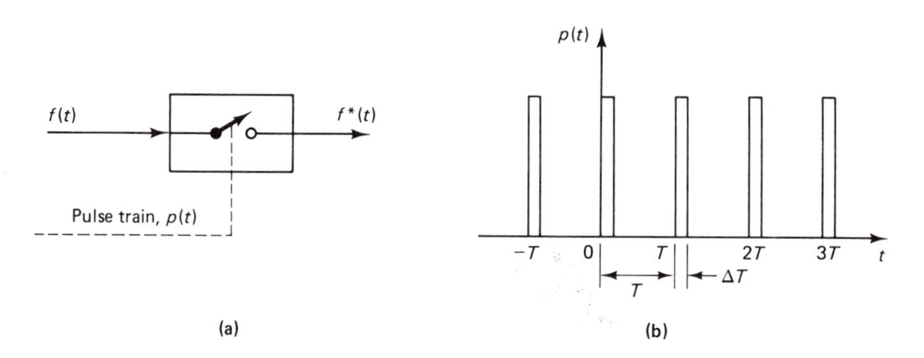

Figure 9.3 (a) Sampler. (b) Pulse train.

Usually, the pulse duration time Δt is very small when compared with the time constant of the plant. Hence, to simplify the mathematical representation, it is commonly assumed that $p(t)$ is a train of impulses and impulse-amplitude modulation is used as a mathematical model of the sampling process.

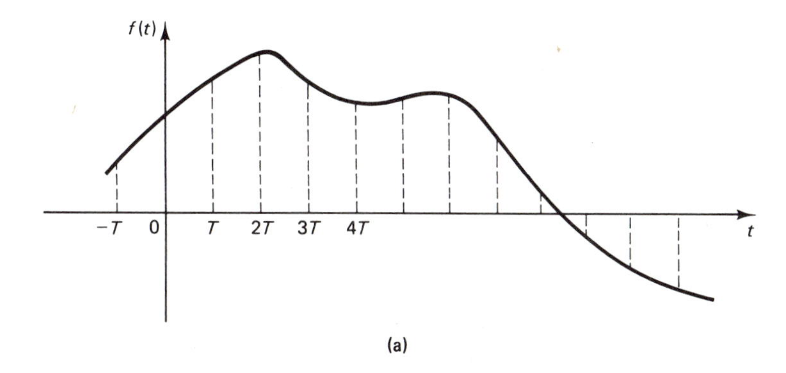

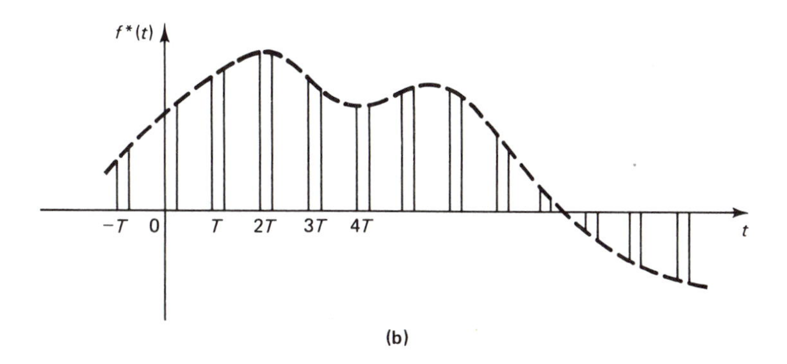

Figure 9.4 (a) Analog signal. (b) Pulse-amplitude-modulated signal.

Letting $p(t)$ be an impulse train described by

$$p(t) = \sum_{k=-\infty}^{\infty} \delta(t - kT) \tag{9.1}$$

we get

$$\begin{aligned} f^*(t) &= \sum_{k=-\infty}^{\infty} f(t)\delta(t - kT) \\ &= \cdots + f(-T)\delta(t + T) + f(0)\delta(t) + f(T)\delta(t - T) + \cdots \end{aligned} \tag{9.2}$$

where $f(kT)$ is the average value of $f(t)$ between kT and $kT + \Delta t$. The pulse-amplitude-modulated signal is shown in Fig. 9.5. Only the values of $f(t)$ at the sampling instants are retained by the digital memory.

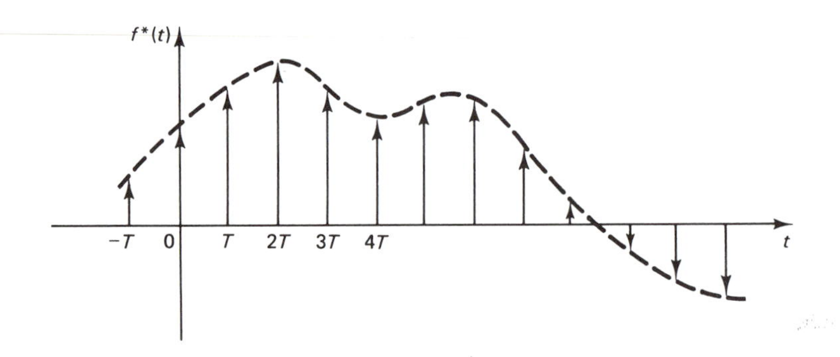

Figure 9.5 Impulse-amplitude-modulated signal.

In quantization, the amplitudes of the impulse train are rounded off to a finite number of digits. Digital computers use binary code and a binary digit is called a bit. The resolution depends on the word length in bits. A word length of 10 bits is usually sufficient for the ADC so that the error introduced by quantization can be neglected. With a word length of 10 bits, a signal can be resolved to one part in 2^{10} i.e., 1 in 1,024. Here, we assume that the ADC has a sufficient word length and, neglecting quantization as a secondary effect, represent its mathematical model as a sampler. For a study of quantization effects, the reader may consult Franklin and Powell (1980).

For the configuration shown in Fig. 9.2, the analog sensor and ADC can be replaced by a digital sensor. For example, in the computer control of robotic manipulators and of machine tools, a digital sensor called an encoder is frequently used to sense angular displacement and/or velocity. The most common type is the optical encoder which may be either incremental or absolute (deSilva, 1985).

In an incremental encoder, a disk attached to a rotating shaft has a circular track of equally spaced, alternately transparent and opaque areas. A light source such as a light emitting diode is placed on one side of the disk as shown in Fig. 9.6(a). On the other side of the track, two light detectors such as photocells are located at half-window width offset (1/4 pitch). An alternate construction employs

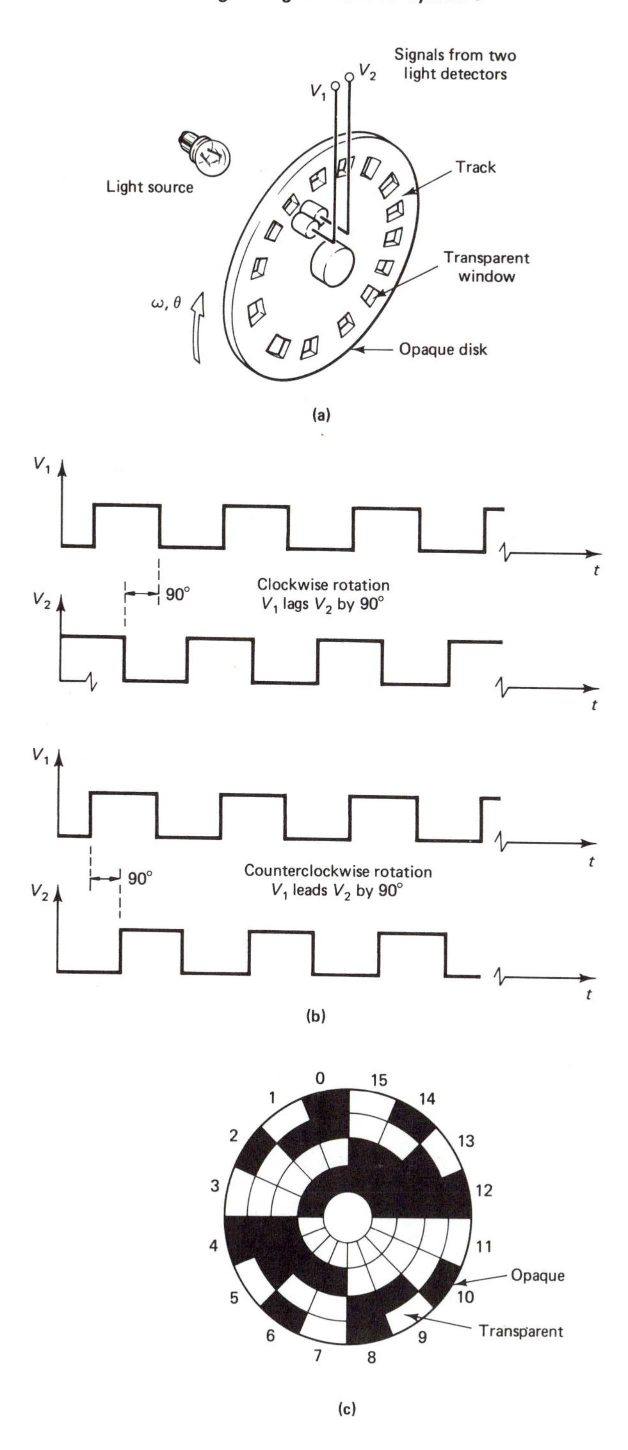

Figure 9.6 (a) Optical incremental encoder. (b) Pulse signals from incremental encoder. (c) Optical absolute encoder with binary code.

two identical tracks with a 1/4 pitch offset and two light detectors aligned radially, one for each track.

As the disk rotates, the light is interrupted by the opaque areas, thus generating two pulse trains. The direction of rotation is sensed by detecting which pulse train leads the other as shown in Fig. 9.6(b). The angular displacement is determined by pulse counting and angular velocity by either counting the pulses per unit-time or by timing the pulses.

The pulse signals are supplied to a up/down counter which determines the direction of rotation and codes the count. The coded signal is transferred to a buffer which is read by the control computer every sampling instant. A cumulative count is required for displacement measurement, and the counter is not cleared after the count is read by the computer. A resolution of 0.01 degree or better can be obtained, depending on the number of windows and word length in bits. For velocity measurement by pulse counting, the counter is cleared once the count is transferred to the buffer which is read at intervals of counting-cycle time.

An absolute encoder yields a coded signal directly without any pulse counting or pulse timing. The disk of an absolute encoder which employs a binary code is shown in Fig. 9.6(c). For simplicity, only four tracks are shown in this figure but in practice, the number of tracks n is about 14. The disk is divided into 2^n sectors. A transparent area corresponds to binary 1 and an opaque area to binary 0. A light source is placed on one side of the disk and n light detectors are located on the other side, one for each track.

The output from the light detectors provides a digitally coded signal that determines the position of the disk. In Fig. 9.6(c), the word size of data is four bits. In each sector, the outermost element is the least significant bit (LSB) and the innermost element is the most significant bit (MSB). For a 14-track absolute encoder, the resolution is 0.022 degree. The accuracy can be improved by using a gray coded disk, but an additional logic device is required to convert the gray code to the corresponding binary code.

The absolute encoder can also be used to measure the angular velocity either by dividing the angle of rotation by the sampling period or by timing the interval between two consecutive readings. An absolute encoder is more expensive than an incremental one. However, an absolute encoder does not require a counter and a buffer, and a missed reading does not affect the next reading.

9.2.2 Sampling Period and Aliasing

The choice of the sampling period is one of the important considerations in digital control systems. The equipment cost decreases when the sampling period T is large, since the speed demands for multiplexing and computation are reduced. But the sampling period must also be small enough for effective control, depending on the time constants of the plant and on how fast the command input changes with time. A large sampling period can also introduce instability in control systems. Hence, in selecting the sampling period, a compromise is required between the equipment cost and the system performance.

A criterion for the selection of a suitable sampling period is Shannon's sampling theorem. This theorem states that the sampling operation results in no loss of information provided the sampled signal $f(t)$ is band limited such that in the frequency domain $|F(j\omega)| \to 0$ for $\omega > \omega_m$, and the sampling frequency ω_s in rad/s is chosen such that $\omega_s > 2\omega_m$, that is, the sampling period $T < 2\pi/2\omega_m$. A proof of this theorem follows.

Since the impulse train of Eq. (9.1) is periodic, we can expand it in complex Fourier series as

$$\sum_{k=-\infty}^{\infty} \delta(t - kT) = \sum_{k=-\infty}^{\infty} C_k \exp(jk2\pi t/T) \tag{9.3}$$

where the constants C_k are obtained from

$$\begin{aligned} C_k &= \frac{1}{T}\int_{-T/2}^{T/2} \left[\sum_{k=-\infty}^{\infty} \delta(t - kT) \exp(-jk2\pi t/T)\right] dt \\ &= 1/T \end{aligned} \tag{9.4}$$

Substituting the result from Eqs. (9.3) and (9.4) in Eq. (9.2) and letting $\omega_s = 2\pi/T$, where ω_s is the sampling frequency in rad/s, we get

$$f^*(t) = f(t)\frac{1}{T}\sum_{k=-\infty}^{\infty} \exp(jk\omega_s t) \tag{9.5}$$

The two-sided Laplace transform of Eq. (9.5) yields

$$\begin{aligned} F^*(s) &= \frac{1}{T}\sum_{k=-\infty}^{\infty}\int_{-\infty}^{\infty} f(t) \exp[-(s - jk\omega_s)t]\, dt \\ &= \frac{1}{T}\sum_{k=-\infty}^{\infty} F(s - jk\omega_s) \end{aligned} \tag{9.6}$$

Hence, the Laplace transform $F(s)$ of the original signal $f(t)$ is related to the Laplace transform of the sampled signal by Eq. (9.6). Letting $s = j\omega$ in the frequency domain, we obtain

$$F^*(j\omega) = \frac{1}{T}\sum_{k=-\infty}^{\infty} F(j\omega - jk\omega_s) \tag{9.7}$$

This expression reveals that the spectrum of the sampled signal contains, in addition to the original signal frequency ω, an infinite number of sideband frequencies $\omega \pm k\omega_s$. Assuming that $|F(j\omega)| \to 0$ for $\omega > \omega_m$, the spectrum of $F^*(j\omega)$ for the case where the sampling frequency $\omega_s < 2\omega_m$ is shown in Fig. 9.7.

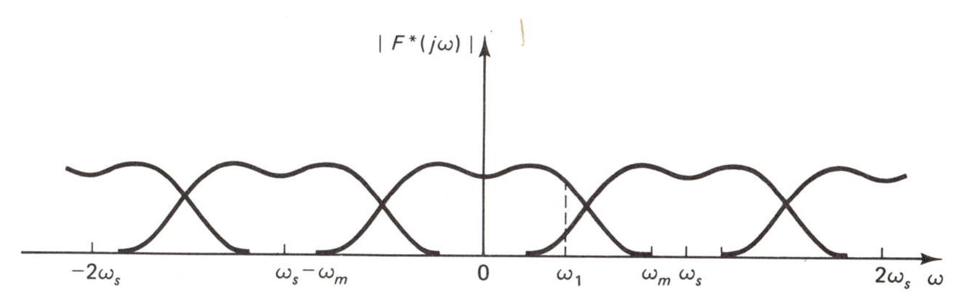

Figure 9.7 Spectrum of $F^*(j\omega)$ with aliasing.

It is seen that there is an overlap between the original signal spectrum and the sideband spectrum. At frequency ω_1 in Fig. 9.7, the larger amplitude is the value of $|F(j\omega_1)|$. The other is the amplitude of $|F(j\omega_1 - j\omega_s)|$, which is centered at ω_s. The frequency $\omega_1 - \omega_s$, which shows up at ω_1, is called an alias of ω_1, and the process is called aliasing. The spectrum of $F^*(j\omega)$ when $\omega_s > 2\omega_m$ is shown in Fig. 9.8. We note that now there is no overlap between the original spectrum and the sideband spectrum and hence there is no aliasing. This is in essense the requirement of the sampling theorem.

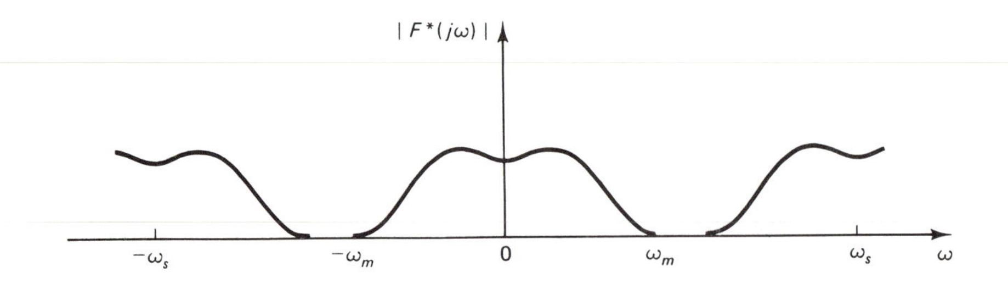

Figure 9.8 Spectrum of $F^*(j\omega)$ without aliasing.

A low-pass filter, called a guard filter, is usually used with the analog-to-digital converter as an antialiasing device. A simple first-order filter of the form

$$G_f(s) = \frac{1}{\tau_f s + 1} \tag{9.8}$$

is satisfactory. The filter time constant is chosen such that the break frequency $1/\tau_f$ is higher than the system bandwidth. Thus, the guard filter can be ignored in the basic controller design.

A safety factor is often used in practice so that the sampling frequency $\omega_s > 2\omega_m$. Most signals have no sharp cutoff frequency ω_m, and we may find that $|F(j\omega)| \to 0$ as $\omega \to \infty$. In such cases, we have to estimate ω_m such that $|F(j\omega)|$ is negligibly small for $\omega > \omega_m$. The bandwidth ω_b of the closed-loop control system can be used as a guide in selecting the sampling period. In practice, the sampling frequency is chosen between ten and twenty times the closed-loop bandwidth frequency as a safety factor. The safety factor is also necessary since a pulse train is approximated by an impulse train in the analysis.

The sampling period is therefore selected in the range

$$2\pi/20\omega_b \leq T \leq 2\pi/10\omega_b$$

For further discussion of the sampling period, we refer to Franklin and Powell (1980) and to Peatman (1972). As stated in Takahashi, et al. (1970), a sampling period of 1 second for liquid-flow control, 5 seconds for pressure and level control, and about 20 seconds for temperature control are representative of process-control applications, where the plant time constants are usually large.

EXAMPLE 9.1

Let us determine the period required for sampling the exponential function $f(t) = e^{-t}$. Laplace transforming this function, we obtain

$$F(s) = \frac{1}{s+1} \tag{9.9}$$

The amplitude in the frequency domain is given by

$$|F(j\omega)| = \frac{1}{(\omega^2 + 1)^{1/2}} \tag{9.10}$$

The bandwidth of this function is $\omega_b = 1$. Choosing the sampling frequency as $20\omega_b$, we get $\omega_s = 20\,\text{rad/s}$, and the sampling period as $T = 2\pi/\omega_s = 0.314$ second. Hence, we can select 0.3 second as a suitable sampling period.

9.2.3. Digital-to-Analog Converter

The output of a digital controller is a control law that is coded in the binary form. This digital signal must be first converted to a sequence of voltage pulses with pulse-amplitude modulation. Unless a pulse motor is used for an actuator, it is not possible to drive an analog actuator with the pulse sequence. Hence, the pulse sequence is converted into an analog signal that is then supplied to an actuator. Thus, the function of a digital-to-analog converter (DAC) is twofold. It decodes a digital signal from the binary form to a sequence of pulses and then converts the pulse sequence into an analog signal.

A DAC consists of a resistor network, an operational amplifier, transistor switches, and reference voltage sources. Two types of networks are commonly used: weighted-resistor and resistor-ladder networks. The schematic diagram of a 5-bit DAC with a weighted-resistor network is shown in Fig. 9.9. An n-bit DAC employs n input resistors, and the value of each resistor is inversely proportional to the weight of the particular bit in the binary word.

In the weighted-resistor network, the values of the resistors are that of the binary system, with the least significant bit (LSB) resistance being 2^{n-1} times that of

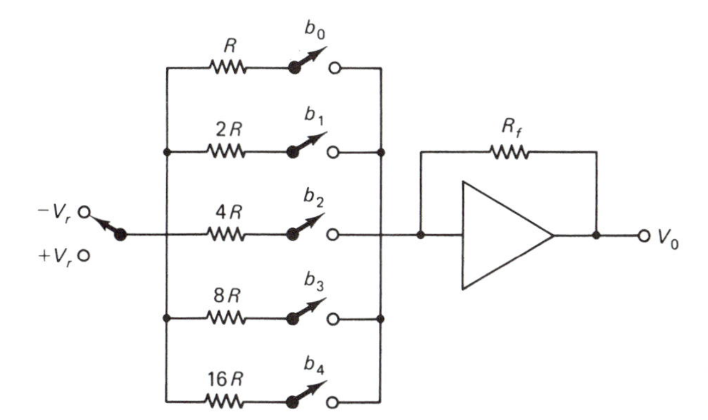

Figure 9.9 A weighted-resistor digital-to-analog converter.

the most significant bit (MSB). The switches are moved electronically with transistors, depending on logical 1 or logical 0 appearing at the corresponding digital input. From our study of operational amplifiers in Chapter 7, it is seen that the output of an n-bit converter is given by

$$V_0 = \pm V_r R_f / R[b_0 + b_1/2 + \cdots + b_{n-1}/2^{n-1}] \tag{9.11}$$

where b_i is the logic state of the ith bit and can be either 1 or 0. The change in output voltage per bit is called the gain of the DAC and is given by

$$k_g = \left| \frac{R_f V_r}{R 2^{n-1}} \right| \tag{9.12}$$

EXAMPLE 9.2

The input voltage V_r of a 5-bit weighted-resistor DAC is 2 volts. The input network resistors are 10, 20, 40, 80, and 160 kohms. The feedback resistor is 40 kohms. Calculate the maximum absolute output voltage and the DAC gain.

From Eqs. (9.11) and (9.12), we obtain

$$\max|V_0| = \frac{2(40)}{10}\left(1 + \frac{1}{2} + \frac{1}{4} + \frac{1}{8} + \frac{1}{16}\right)$$

$$= 15.5 \text{ volts}$$

$$k_g = \frac{40(2)}{10(2^4)} = 0.5 \text{ volt per bit}$$

An analog signal is constructed for the time interval $kT < t \leq (k+1)T$ from the known values of the signal at the past sampling instants kT, $(k-1)T$, $(k-2)T, \ldots$. The Newton–Gregory extrapolation for $kT < \Delta t \leq (k+1)T$ yields

$$u(kT + \Delta t) = u(kT) + \dot{u}(kT)\Delta t + \ddot{u}(kT)\left(\frac{T + \Delta t}{2}\right)\Delta t + \cdots \tag{9.13}$$

where the time derivatives $\dot{u}$ and $\ddot{u}$ at $t = kT$ are evaluated from backward differences as

$$\dot{u}(kT) = \frac{u(kT) - u[(k-1)T]}{T} \tag{9.14}$$

$$\begin{aligned} \ddot{u}(kT) &= \frac{\dot{u}(kT) - \dot{u}[(k-1)T]}{T} \\ &= \frac{u(kT) - 2u[(k-1)T] + u[(k-2)T]}{T^2} \end{aligned} \tag{9.15}$$

Zero-Order Hold. In practice, it becomes necessary to truncate the series. If only the first term on the right-hand side of Eq. (9.13) is retained, the extrapolator is called a zero-order hold (ZOH) or clamp. Its advantage is that it is easy to construct physically. The impulse sequence is converted into an analog signal by

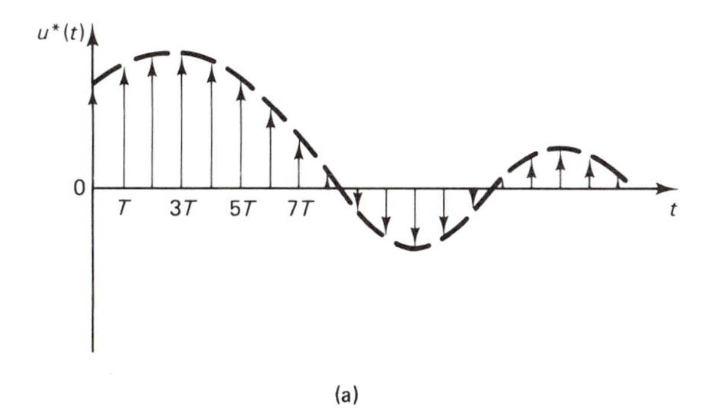

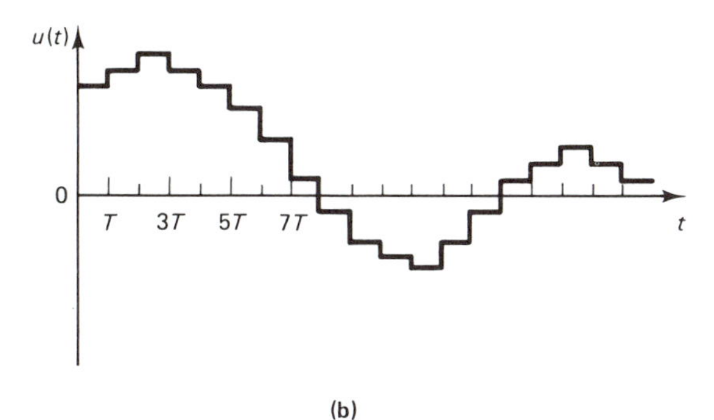

Figure 9.10 (a) Amplitude-modulated impulse sequence. (b) Analog output of a zero-order hold.

holding the value of the impulse until the next one arrives in a staircase fashion, as shown in Fig. 9.10, so that

$$u(kT + \Delta t) = u(kT) \qquad kT \leq \Delta t < (k+1)T \tag{9.16}$$

If the input to a zero-order hold is a unit impulse, its response is a pulse of unit amplitude and width T as shown in Fig. 9.11. The transfer function can be

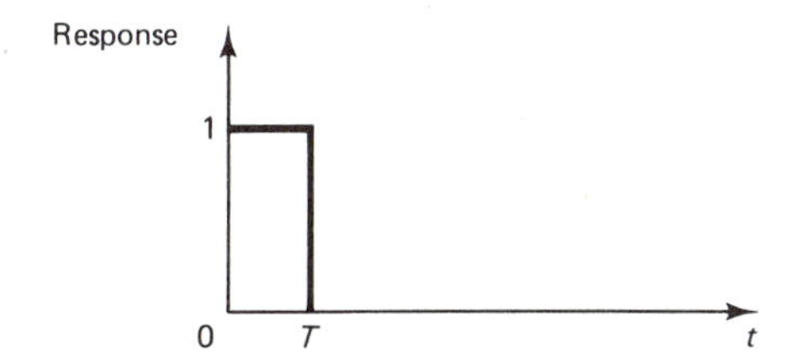

Figure 9.11 Unit-impulse response of a zero-order hold.

obtained from this impulse response as

$$G(s) = \frac{1}{s} - \frac{e^{-Ts}}{s} = \frac{1 - e^{-Ts}}{s} \tag{9.17}$$

In the frequency domain, it is a low-pass filter and introduces a phase lag, which is a linear function of the frequency.

First- and Higher-Order Holds. When the first two terms are retained in the series of Eq. (9.13), the extrapolator is called a first-order hold. In this case, we have

$$u(kT + \Delta t) \approx u(kT) + \left(\frac{u(kT) - u[(k-1)T]}{T} \right) \Delta t$$

$$kT < \Delta t < (k+1)T \qquad (9.18)$$

Its transfer function can be obained as

$$G(s) = \left(\frac{Ts + 1}{T} \right) \left(\frac{1 - e^{-Ts}}{s} \right)^2 \qquad (9.19)$$

A second-order hold, exponential, and other sophisticated data extrapolators can be designed. But it is found that their added complexity is not justified in terms of improvement in control system performance. For this reason, a zero-order hold is the most widely used data extrapolator and we use it here to represent a mathematical model of the digital-to-analog converter.

The block diagram of a digital control system, with output feedback and error sampling, shown in Fig. 9.1, may now be reexpressed as shown in Fig. 9.12. The analog-to-digital converter is modeled as a sampler and the quantization effects and coding operation are neglected. The control law is implemented by a computer program that acts on the error samples $e^*(t)$ and produces the control signal $u^*(t)$.

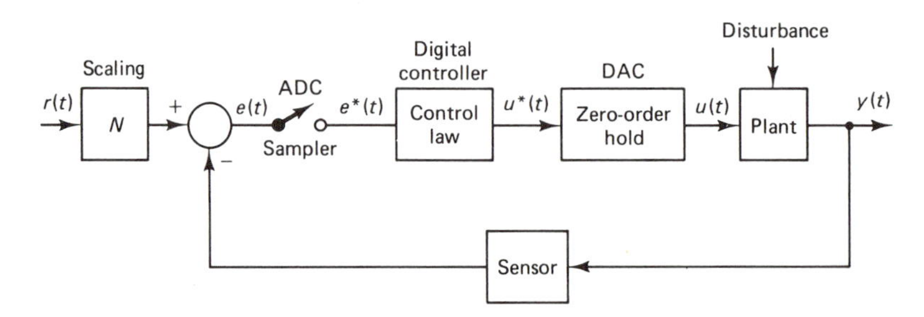

Figure 9.12 Digital control system with output feedback.

The digital-to-analog converter is modeled as a zero-order hold that converts the impulse train $u^*(t)$ into a piecewise constant analog signal $u(t)$. The coding operation of the ADC and the decoding operation of the DAC are not shown so that the computer operations are represented as if they are performed on impulses. We develop discrete-time models of the plant in the following section.

9.3 DISCRETE-TIME MODELS OF THE PLANT

The mathematical modeling of the plant, which includes the actuator and the process to be controlled, has been studied in the previous chapters. The models have been developed in the form of ordinary, linear differential equations with constant coefficients and can be represented as state differential equations or transfer functions. These continuous-time models are now converted into discrete-time models so that they can be combined with the control law to obtain the overall model of the control system. We represent the models in two alternative forms: state difference equations and discrete transfer functions.

9.3.1 State-Difference Equations of the Plant

Let the plant model be described by the state differential equations

$$\begin{aligned} \dot{\mathbf{x}}(t) &= \mathbf{A}\mathbf{x}(t) + \mathbf{B}\mathbf{u}(t) + \mathbf{B}_1\mathbf{v}(t) \\ \mathbf{y}(t) &= \mathbf{C}\mathbf{x}(t) \end{aligned} \tag{9.20}$$

where $\mathbf{u}(t)$ is the control input, and $\mathbf{v}(t)$ is the disturbance. This model is now converted to state difference equations relating the values of the state, inputs, and outputs at discrete instants of time. From Chapter 4, the solution of Eq. (9.20) can be written as

$$\begin{aligned} \mathbf{x}(t) = \mathbf{\Phi}(t-t_0)\mathbf{x}(t_0) &+ \int_{t_0}^{t} \mathbf{\Phi}(t-t')\mathbf{B}\mathbf{u}(t')\,dt' \\ &+ \int_{t_0}^{t} \mathbf{\Phi}(t-t')\mathbf{B}_1\mathbf{v}(t')\,dt' \end{aligned} \tag{9.21}$$

We now let $t_0 = kT$ and $t = (k+1)T$. With a zero-order hold, $\mathbf{u}(t')$ is a constant for $kT < t' < (k+1)T$. Defining $\mathbf{\Psi}(T)$ and $\mathbf{w}(kT)$ as

$$\mathbf{\Psi}(T) = \int_{kT}^{(k+1)T} \mathbf{\Phi}[(k+1)T - t']\mathbf{B}\,dt' \tag{9.22}$$

and

$$\mathbf{w}(kT) = \int_{kT}^{(k+1)T} \mathbf{\Phi}[(k+1)T - t']\mathbf{B}_1\mathbf{v}(t')\,dt' \tag{9.23}$$

we express Eq. (9.21) as

$$\begin{aligned} \mathbf{x}[(k+1)T] &= \mathbf{\Phi}(T)\mathbf{x}(kT) + \mathbf{\Psi}(T)\mathbf{u}(kT) + \mathbf{w}(kT) \\ \mathbf{y}(kT) &= \mathbf{C}\mathbf{x}(kT) \end{aligned} \tag{9.24}$$

If an extrapolator other than a zero-order hold is employed, then we have to redefine $\mathbf{\Psi}(T)$. After choosing the sampling period T, $\mathbf{\Phi}$, and $\mathbf{\Psi}$ become constant matrices. To simplify the notation, we represent $\mathbf{x}(kT)$ by $\mathbf{x}(k)$ and $\mathbf{y}(kT)$ by $\mathbf{y}(k)$. In this notation, it is not assumed that the sampling period is unity. Addtional state variables may have to be introduced to account for the control law. Hence, in Eq. (9.24) we represent $\mathbf{x}(kT)$ by $\mathbf{x}_p(k)$ to indicate explicitly that it describes the state of the plant only. The plant state difference equations, Eq. (9.24), are therefore represented as

$$\begin{aligned} \mathbf{x}_p(k+1) &= \mathbf{\Phi}\mathbf{x}_p(k) + \mathbf{\Psi}\mathbf{u}(k) + \mathbf{w}(k) \\ \mathbf{y}(k) &= \mathbf{C}\mathbf{x}_p(k) \end{aligned} \tag{9.25}$$

EXAMPLE 9.3

The plant of a single-input, single-output digital control system is shown in the block diagram of Fig. 9.13, where $u(t)$ is the control input, and $v(t)$ is a unit step load disturbance. Obtain the state difference equations of the plant.

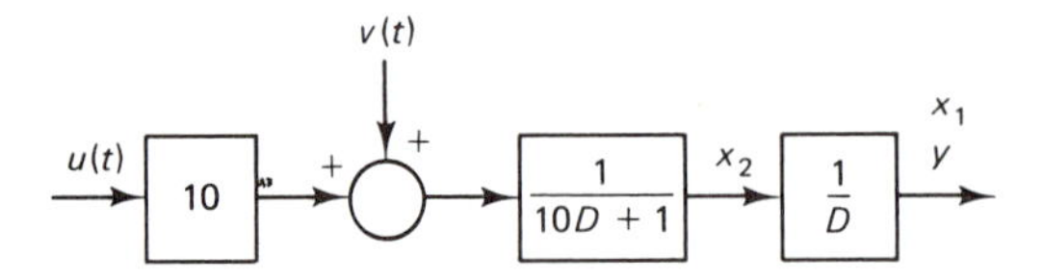

Figure 9.13 Block diagram of a plant.

Choosing x_1 and x_2 as the state variables as shown in Fig. 9.13, the state differential equations are

$$\dot{x}_1 = x_2$$

$$\dot{x}_2 = -0.1x_2 + u(t) + 0.1v(t) \tag{9.26}$$

$$y = x_1$$

Hence, we get

$$\mathbf{A} = \begin{bmatrix} 0 & 1 \\ 0 & -0.1 \end{bmatrix} \qquad \mathbf{B} = \begin{bmatrix} 0 \\ 1 \end{bmatrix} \qquad \mathbf{B}_1 = \begin{bmatrix} 0 \\ 0.1 \end{bmatrix} \qquad \mathbf{C} = \lfloor 1 \quad 0 \rfloor$$

Now,

$$(s\mathbf{I} - \mathbf{A})^{-1} = \frac{1}{s(s+0.1)} \begin{bmatrix} s+0.1 & 1 \\ 0 & s \end{bmatrix} = \begin{bmatrix} \frac{1}{s} & \frac{1}{s(s+0.1)} \\ 0 & \frac{1}{s+0.1} \end{bmatrix}$$

The inverse Laplace transform of the preceding equation yields

$$\mathbf{\Phi}(t) = \begin{bmatrix} 1 & 10(1 - e^{-0.1t}) \\ 0 & e^{-0.1t} \end{bmatrix}$$

and hence

$$\mathbf{\Phi}(T) = \begin{bmatrix} 1 & 10(1 - e^{-0.1T}) \\ 0 & e^{-0.1T} \end{bmatrix} \tag{9.27}$$

Also

$$\mathbf{\Psi}(t) = \int_0^t \mathbf{\Phi}(t - t')\mathbf{B}\, dt'$$

$$= \int_0^t \begin{bmatrix} 10(1 - e^{-0.1(t-t')}) \\ e^{-0.1(t-t')} \end{bmatrix} dt'$$

$$= \begin{bmatrix} 10t - 100 + 100e^{-0.1t} \\ 10 - 10\, e^{-0.1t} \end{bmatrix}$$

and

$$\mathbf{\Psi}(T) = \begin{bmatrix} 10T - 100 + 100\, e^{-0.1T} \\ 10 - 10\, e^{-0.1T} \end{bmatrix} \tag{9.28}$$

Since $v(t) = 1$ for $t \geq 0$, we get

$$\begin{aligned} \mathbf{w}(t) &= \int_0^t \mathbf{\Phi}(t - t')\mathbf{B}_1\, dt' \\ &= \begin{bmatrix} t - 10 + 10\, e^{-0.1t} \\ 1 - e^{-0.1t} \end{bmatrix} \end{aligned} \tag{9.29}$$

and $\mathbf{w}(T)$ is obtained by letting $t = T$ in Eq. (9.29). Assuming that a suitable sampling period is $T = 0.1$ second, we obtain

$$\mathbf{\Phi} = \begin{bmatrix} 1 & 0.10 \\ 0 & 0.99 \end{bmatrix} \qquad \mathbf{\Psi} = \begin{bmatrix} 0.005 \\ 0.100 \end{bmatrix} \qquad \mathbf{w} = \begin{bmatrix} 0 \\ 0.01 \end{bmatrix}$$

Hence, the plant state difference equations are obtained as

$$\mathbf{x}_p(k+1) = \begin{bmatrix} 1 & 0.10 \\ 0 & 0.99 \end{bmatrix} \mathbf{x}_p(k) + \begin{bmatrix} 0.005 \\ 0.100 \end{bmatrix} u(k) + \begin{bmatrix} 0 \\ 0.01 \end{bmatrix} \tag{9.30}$$

EXAMPLE 9.4

A mathematical model of the plant of a two-input, two-output, temperature-control system is given in Example 3.8. The state differential equations are described by Eqs. (3.87) and (3.88) and the matrix $(s\mathbf{I} - \mathbf{A})^{-1}$ is given by Eq. (3.91). Choosing the parameter values as $\tau_1 = \tau_2 = 10$ seconds as in Examples 7.17 and 8.9, the numerical values of matrices **A** and **B** are given by Eq. (8.67). For computer control of this system, we obtain the plant state difference equations neglecting the disturbance. We have

$$(s\mathbf{I} - \mathbf{A})^{-1} = \begin{bmatrix} \dfrac{1}{s + 0.1} & 0 \\ \dfrac{0.1}{(s + 0.1)^2} & \dfrac{1}{s + 0.1} \end{bmatrix} \tag{9.31}$$

Taking the inverse Laplace transformation of the preceding equation, we obtain

$$\mathbf{\Phi}(t) = \begin{bmatrix} e^{-0.1t} & 0 \\ 0.1t\, e^{-0.1t} & e^{-0.1t} \end{bmatrix} \tag{9.32}$$

Now,

$$\begin{aligned} \mathbf{\Psi}(t) &= \int_0^t \mathbf{\Phi}(t - t')\mathbf{B}\, dt' \\ &= \int_0^t \begin{bmatrix} e^{-0.1t'} & 0 \\ 0.1t'\, e^{-0.1t'} & e^{-0.1t'} \end{bmatrix} \begin{bmatrix} 100 & 0 \\ 0 & 100 \end{bmatrix} dt' \\ &= \begin{bmatrix} 1{,}000(1 - e^{-0.1t}) & 0 \\ 1{,}000(1 - e^{-0.1t}) - 100t\, e^{-0.1t} & 1{,}000(1 - e^{-0.1t}) \end{bmatrix} \end{aligned} \tag{9.33}$$

Let the bandwidth of the closed-loop system be specified as 0.1 rad/s. Choosing the sampling frequency as $\omega_s = 20(0.1) = 2$ rad/s, the sampling period becomes $2\pi/\omega_s = \pi$ seconds. We choose $T = 3$ seconds. Hence, from Eqs. (9.32) and (9.33), we obtain

$$\mathbf{\Phi} = \begin{bmatrix} 0.741 & 0 \\ 0.222 & 0.741 \end{bmatrix} \tag{9.34}$$

$$\mathbf{\Psi} = \begin{bmatrix} 259.182 & 0 \\ 36.936 & 259.182 \end{bmatrix} \tag{9.35}$$

The state difference equations of the plant are now obtained as

$$\begin{aligned} \mathbf{x}_p(k+1) &= \mathbf{\Phi}\mathbf{x}_p(k) + \mathbf{\Psi}\mathbf{u}(k) \\ \mathbf{y}(k) &= \begin{bmatrix} 1 & 0 \\ 0 & 1 \end{bmatrix} \mathbf{x}_p(k) \end{aligned} \tag{9.36}$$

Hence,

$$\begin{aligned} x_1(k+1) &= 0.741x_1(k) + 259.182u_1(k) \\ x_2(k+1) &= 0.222x_1(k) + 0.741x_2(k) + 36.936u_1(k) + 259.182u_2(k) \end{aligned} \tag{9.37}$$

9.3.2 *Discrete Transfer Functions of the Plant*

An alternative model of linear, time-invariant discrete-time systems consists of discrete transfer functions obtained by using the z-transform. Laplace transforming both sides of Eq. (9.2), we get

$$F^*(s) = \sum_{k=0}^{\infty} f(kT)\, e^{-kTs} \tag{9.38}$$

where we have assumed that $f(t) = 0$ for $t < 0$. This expression represents an alternative expression of $F^*(s)$ to that of Eq. (9.6). We note that Eq. (9.38) is a transcendental function of s. The reason for using the z-transform is to convert transcendental functions of s into algebraic functions of z. The complex variable z is defined by

$$z = e^{Ts}$$

i.e.,

$$s = \frac{1}{T} \ln z \tag{9.39}$$

From this definition, it is seen that the operator z is an advance operator representing an advance of one sampling period. Since $z^{-1} = e^{-Ts}$, we note that z^{-1} is a delay operator representing a delay of one sampling period. The transformation, Eq. (9.39), is not one to one, that is, there are many values of s for each value of z. If $z = e^{s_1 T}$ and $s_2 = s_1 + jn2\pi/T$, where $n = 1, 2, 3, \ldots$, then also $z = e^{s_2 T}$.

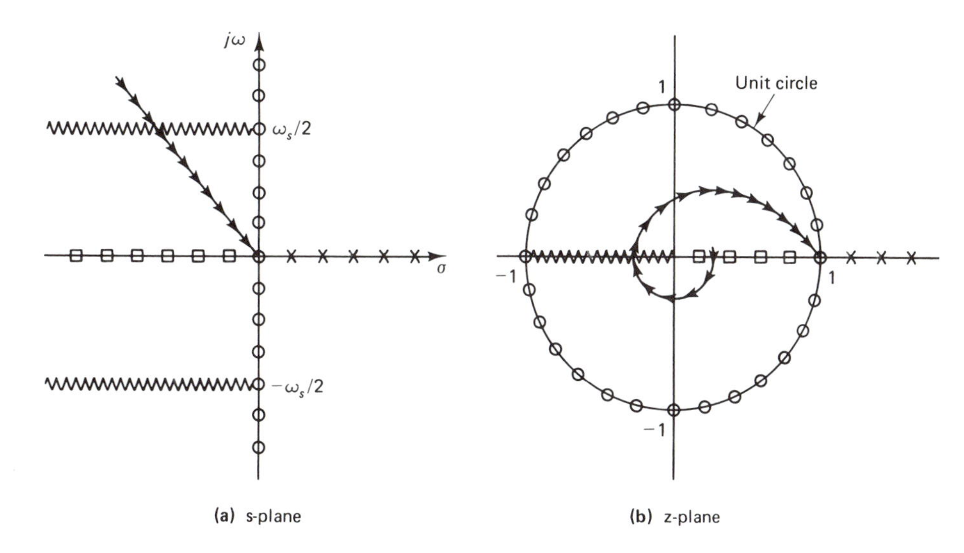

Figure 9.14 Mapping from (a) the s-plane into (b) the z-plane.

It is useful to examine how the s-plane is mapped into the z-plane by the transform of Eq. (9.39). This is illustrated in Fig. 9.14. The s-plane is divided into an infinite number of periodic strips determined by the sampling frequency $\omega_s = 2\pi/T$. The strip, where $-\omega_s/2 \leq \omega \leq \omega_s/2$, is called the primary strip. If s is a point inside the primary strip, then $s \pm jn\omega_s$, where $n = 1, 2, 3, \ldots$, is also mapped into the same point in the z-plane. For $s = j\omega$, $z = e^{j\omega T}$, i.e., $|z| = 1$ and $\measuredangle z = \omega T$. Hence, the entire imaginary axis of the s-plane is mapped into the unit circle in the z-plane centered at its origin. The origin $s = 0$ corresponds to $z = 1$.

On the real axis of the s-plane, $s = \pm\sigma$, and we get $z = e^{\sigma T} > 1$ for the positive real axis of the s-plane and $z = e^{-\sigma T} < 1$ for the negative real axis. When $s = \sigma + j\omega$, $z = e^{\sigma T}e^{j\omega T} = e^{\sigma T}\underline{/\omega T}$. A straight line in the left half of the s-plane with a constant damping ratio is mapped into a logarithmic spiral inside the unit circle of the z-plane. The whole left half of the s-plane is mapped inside the unit circle of the z-plane, and the whole right half of the s-plane is mapped outside the unit circle. This is an important point for the stability investigation of discrete-time systems.

The z-transform of a discrete time function $f^*(t)$ is denoted by $F(z)$, and from Eqs. (9.38) and (9.39), we get

$$F(z) = \sum_{k=0}^{\infty} f(kT)z^{-k} \tag{9.40}$$

It is seen from the definition that the z-transform is a linear operator. Any function that has a Laplace transform also has a z-transform. For many functions, the infinite series of Eq. (9.40) can be expressed in closed form.

EXAMPLE 9.5

Consider the discrete-time exponential function

$$f(kT) = e^{-akT} \qquad k = 0, 1, 2, \ldots \tag{9.41}$$

Its Laplace transform is obtained from Eq. (9.38) as

$$F^*(s) = \sum_{k=0}^{\infty} e^{-akT} e^{-kTs} \tag{9.42}$$

From Eq. (9.40), the z-transform is

$$\begin{aligned} F(z) &= \sum_{k=0}^{\infty} e^{-akT} z^{-k} \\ &= 1 + (e^{-aT}z^{-1}) + (e^{-aT}z^{-1})^2 + \cdots \\ &= \frac{1}{1 - e^{-aT}z^{-1}} \qquad |e^{-aT}z^{-1}| < 1 \\ &= \frac{z}{z - e^{-aT}} \qquad |z| > e^{-aT} \end{aligned} \tag{9.43}$$

EXAMPLE 9.6

Consider the discrete-time ramp function

$$f(kT) = akT$$

Its z-transform is obtained from Eq. (9.40) as

$$\begin{aligned} F(z) &= \sum_{k=0}^{\infty} akTz^{-k} \\ &= aTz^{-1}(1 + 2z^{-1} + 3z^{-2} + \cdots) \\ &= \frac{aTz^{-1}}{(1 - z^{-1})^2} \qquad |z^{-1}| < 1 \\ &= \frac{aTz}{(z-1)^2} \end{aligned} \tag{9.44}$$

The z-transforms of some elementary functions are given in Appendix A. A more extensive table is given by Kuo (1980). In addition to the linearity properties of the z-transform, a property that is very useful is the real translation discussed in the following.

$$\begin{aligned} z\text{-transform of } f(kT - T) &= \sum_{k=0}^{\infty} f(kT - T)z^{-k} \\ &= f(0)z^{-1} + f(T)z^{-2} + f(2T)z^{-3} + \cdots \\ &= z^{-1}F(z) \end{aligned} \tag{9.45}$$

Similarly, for a positive integer n, we get

$$z\text{-transform of } f(kT - nT) = z^{-n}F(z) \tag{9.46}$$

$$
\begin{aligned}
z\text{-transform of } f(kT+T) \\
&= \sum_{k=0}^{\infty} f(kT+T)z^{-k} \\
&= z[f(0)+f(T)z^{-1}+f(2T)z^{-2}+\cdots]-zf(0) \qquad (9.47)\\
&= zF(z)-zf(0)
\end{aligned}
$$

Similarly, for a positive integer n, we get

$$
z\text{-transform of } f(kT+nT) = z^n F(z) - z^n \sum_{k=0}^{n-1} f(kT)z^{-k} \tag{9.48}
$$

The second term on the right-hand side of Eq. (9.48) appears because we employ a one-sided z-transform defined for $k \geq 0$. If a two-sided z-transform were employed, then the right-hand side of Eq. (9.48) would be $z^n F(z)$.

We now employ the preceding development of the z-transform to obtain a discrete transfer function model of the plant. We take the z-transform of both sides of Eq. (9.25). Letting $\mathbf{X}_p(z)$ denote the z-transform of $\mathbf{x}_p(k)$, from Eq. (9.47), the z-transform of $\mathbf{x}_p(k+1)$ is given by $z\mathbf{X}_p(z) - z\mathbf{x}_p(0)$. Letting the disturbance $\mathbf{w}(k) = 0$ and the initial conditions $\mathbf{x}_p(0) = \mathbf{0}$, the z-transform of both sides of Eq. (9.25) yields

$$
\begin{aligned}
z\mathbf{X}_p(z) &= \mathbf{\Phi}\mathbf{X}_p(z) + \mathbf{\Psi}\mathbf{U}(z) \\
\mathbf{Y}(z) &= \mathbf{C}\mathbf{X}_p(z)
\end{aligned} \tag{9.49}
$$

Solving for $\mathbf{X}_p(z)$ from the first equation, we get

$$
\mathbf{X}_p(z) = (z\mathbf{I} - \mathbf{\Phi})^{-1}\mathbf{\Psi}\mathbf{U}(z)
$$

and

$$
\mathbf{Y}(z) = \mathbf{C}(z\mathbf{I} - \mathbf{\Phi})^{-1}\mathbf{\Psi}\mathbf{U}(z) \tag{9.50}
$$

Hence, the discrete transfer function matrix relating the input to the output of the plant is obtained as

$$
\mathbf{G}_p(z) = \mathbf{C}(z\mathbf{I} - \mathbf{\Phi})^{-1}\mathbf{\Psi} \tag{9.51}
$$

and is shown in the block diagram of Fig. 9.15.

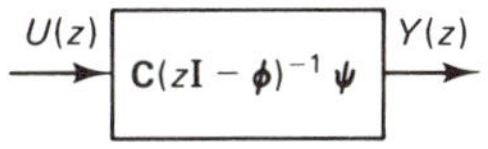

$U(z)$ → $\mathbf{C}(z\mathbf{I} - \phi)^{-1}\psi$ → $Y(z)$

Figure 9.15 Discrete transfer function matrix of the plant.

EXAMPLE 9.7

We obtain the discrete transfer-function matrix for the two-input, two-output plant of Example 9.4. For this system, matrices $\mathbf{\Phi}$ and $\mathbf{\Psi}$ are given by Eq. (9.34) and (9.35), respectively. We know from Eq. (3.89) that $\mathbf{C}$ is a (2×2)

identity matrix. Substituting these results in Eq. (9.51), we obtain

$$\mathbf{G}_p(z) = \begin{bmatrix} 1 & 0 \\ 0 & 1 \end{bmatrix} \begin{bmatrix} z - 0.741 & 0 \\ -0.222 & z - 0.741 \end{bmatrix}^{-1} \begin{bmatrix} 259.182 & 0 \\ 36.936 & 259.182 \end{bmatrix}$$

$$= \frac{1}{(z - 0.741)^2} \begin{bmatrix} z - 0.741 & 0 \\ 0.222 & z - 0.741 \end{bmatrix} \begin{bmatrix} 259.182 & 0 \\ 36.936 & 259.182 \end{bmatrix} \tag{9.52}$$

$$= \begin{bmatrix} \dfrac{259.182}{z - 0.741} & 0 \\ \dfrac{36.936z + 30.168}{(z - 0.741)^2} & \dfrac{259.182}{z - 0.741} \end{bmatrix}$$

Hence,

$$\begin{aligned} Y_1(z) &= \left(\frac{259.182}{z - 0.741}\right) U_1(z) \\ Y_2(z) &= \left(\frac{36.936z + 30.168}{(z - 0.741)^2}\right) U_1(z) + \left(\frac{259.182}{z - 0.741}\right) U_2(z) \end{aligned} \tag{9.53}$$

The difference equations relating the inputs and outputs can be obtained from the discrete transfer functions by making use of the real translation property. For example, from the first equation in Eq. (9.53), we obtain

$$(z - 0.741)\,Y_1(z) = 259.182\,U_1(z)$$

or

$$y_1(k + 1) = 0.741y_1(k) + 259.182u_1(k)$$

Since $y_1(k) = x_1(k)$, it is seen that this equation is identical to the first equation in Eq. (9.37).

9.4 DIGITAL CONTROLLER DESIGN WITH OUTPUT FEEDBACK

In this section, we discuss the digital computer implementation of a PID family of control laws for systems with output feedback. The digital computer implementation of control laws with state feedback is covered in a later section. The performance specifications for digital control systems can be stated in terms of the discrete-time response. These specifications, if required, can be converted to the desired locations of the roots of the closed-loop characteristic equation in the z-plane, which is analogous to the s-plane of continuous-time systems.

However, our applications involve analog plants and the techniques of linear, continuous-time controller design are well established. Therefore, there is a strong motivation in practice to state the performance specifications for digital control systems in terms of desired continuous-time response, which has already been covered in the previous chapters. Assuming that this is the case, we employ the following two methods of using performance specifications stated in terms of continuous-time response for the design of digital controllers.

Method 1. The control law and compensator, if one is required, are designed for continuous-time systems as in Chapter 7, simply ignoring the fact that a digital controller will eventually be used. The control law and compensator, if any, are then converted to discrete-time form by approximation techniques.

Method 2. The performance specifications, which are stated in terms of continuous-time response, are first converted to the desired locations of the roots of the characteristic equation as in Chapter 8. From these, we obtain the corresponding root locations of the discrete-time system in the z-plane by the mapping $z = e^{Ts}$, and then design a discrete-time control law.

Here, we use Method 1 for the digital implementation of a PID family of control laws and compensator, if any, for the case of controller design with output feedback, and Method 2 for the case of control law design with state feedback.

9.4.1 PID Control Law

For the output feedback case studied in this section, the design of the control law, including the setting of controller gains and compensator for continuous-time systems, is covered in Chapter 7. The two approximation techniques used in conjunction with Method 1 are the difference-equations approximation of differential equations and pole-zero mapping of transfer functions. Since the method involves approximation techniques, the controller gains obtained for the continuous case may require some final adjustment for the discrete case.

A PID control law as shown in Chapter 7 has the form

$$U(s) = (k_p + k_i/s + k_d s)E(s) \tag{9.54}$$

where the error is defined as $e(t) = r(t) - y(t)$. We disregard the fact that the derivative term in Eq. (9.54) is noncausal since our aim is to implement the control law digitally.

The integral term in Eq. (9.54) is approximated by using numerical integration based on the Newton–Coates quadrature formula. A simple algorithm is the rectangular rule where

$$k_i \int_0^{kT} e(t)\,dt = k_i T \sum_{j=1}^{k} e[(j-1)T] \tag{9.55}$$

An elementary area from $(j-1)T$ to jT is approximated by a rectangle whose height is the value of the integrand at $(j-1)T$ and width is T as shown in Fig. 9.16(a). This rectangular rule, also known as the Euler rule, is not accurate and a better approximation is obtained by the trapezoidal rule, where the height is chosen as the average value of the integrand at $(j-1)T$ and jT as shown in Fig. 9.16(b).

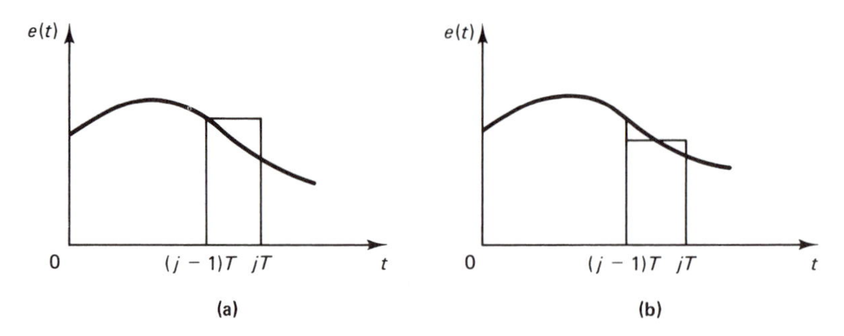

Figure 9.16 (a) Rectangular rule integration, and (b) trapezoidal rule integration.

Hence,

$$k_i \int_0^{kT} e(t)\,dt = k_i T/2 \sum_{j=1}^{k} \{e[(j-1)T] + e(jT)\} \tag{9.56}$$

Higher-order approximations, such as parabolic rule (Simpson rule), are available, but the accuracy of Eq. (9.56) is adequate for control purposes and is used here. The derivative term in Eq. (9.54) is approximated by

$$k_d \frac{de}{dt} = \frac{k_d}{T}\{e(kT) - e[(k-1)T]\} \tag{9.57}$$

Hence, a difference-equation approximation of Eq. (9.54) is obtained as

$$\begin{aligned} u(kT) = k_p e(kT) + k_i T/2 \sum_{j=1}^{k} \{e[(j-1)T] + e(jT)\} \\ + k_d/T\{e(kT) - e[(k-1)T]\} \end{aligned} \tag{9.58}$$

Changing index k to $k-1$ and subtracting the resulting equation from both sides of Eq. (9.58), we get

$$\begin{aligned} u(k) - u(k-1) = k_p[e(k) - e(k-1)] + k_i T/2[e(k) + e(k-1)] \\ + k_d/T[e(k) - 2e(k-1) + e(k-2)] \end{aligned} \tag{9.59}$$

Defining new constants as

$$\begin{aligned} a_0 &= k_p + k_i T/2 + k_d/T \qquad a_1 = -k_p + k_i T/2 - 2k_d/T \\ a_2 &= k_d/T \end{aligned} \tag{9.60}$$

we express Eq. (9.59) in the form

$$u(k) - u(k-1) = a_0 e(k) + a_1 e(k-1) + a_2 e(k-2) \tag{9.61}$$

This is the difference equation relating the input e and the output u of the controller for a digital PID control law. The discrete transfer function of the controller is obtained as follows. Taking the z-transform of both sides of Eq. (9.59) and using the real translation property, Eq. (9.46), we get

$$(1 - z^{-1})U(z) = k_p(1 - z^{-1})E(z) + (k_iT/2)(1 + z^{-1})E(z) + (k_d/T)(1 - z^{-1})^2E(z)$$

$$U(z) = k_pE(z) + k_i\frac{T}{2}\left(\frac{1 + z^{-1}}{1 - z^{-1}}\right)E(z) + \frac{k_d}{T}(1 - z^{-1})E(z) \quad (9.62)$$

$$\frac{U(z)}{E(z)} = k_p + k_i\frac{T}{2}\left(\frac{z + 1}{z - 1}\right) + \frac{k_d}{T}\left(\frac{z - 1}{z}\right)$$

$$= \frac{a_0z^2 + a_1z + a_2}{z(z - 1)}$$

where the coefficients a_0, a_1, and a_2 are defined by Eq. (9.60). A block diagram of the controller is shown in Fig. 9.17. We note that the coefficients of the controller

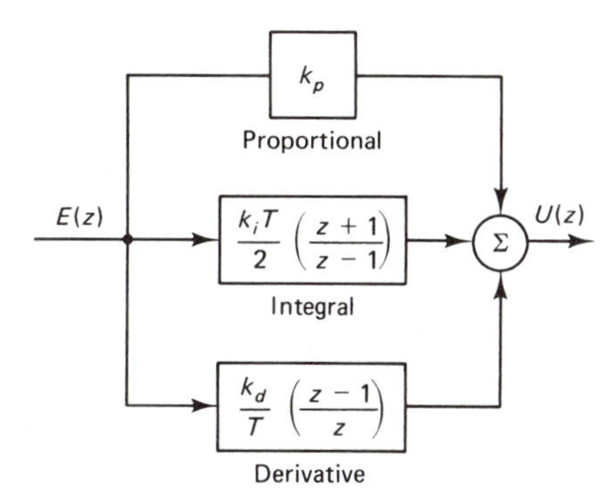

Figure 9.17 Block diagram of a digital PID controller.

depend not only on the three gains of the analog PID control law, but also on the sampling period T.

EXAMPLE 9.8

In Example 7.11, we have studied the design of a PI controller for the electrical position-control system of Example 3.4. The controller gains selected in Example 7.13 are $k_p = 0.90$ and $k_i = 1.09$. Hence, the analog PI control law is given by

$$U(s) = (0.90 + 1.09/s)E(s)$$

We now consider the implementation of this control law by a digital computer. Setting $k_d = 0$ in Eq. (9.60), we obtain $a_0 = (0.90 + 0.545T)$, $a_1 = (-0.90 + 0.545T)$, and $a_2 = 0$. Hence, the difference equation to be implemented by the digital computer is obtained from Eq. (9.61) as

$$u(k) - u(k - 1) = (0.90 + 0.545T)e(k) + (-0.90 + 0.545T)e(k - 1) \quad (9.63)$$

The discrete transfer function of the controller is obtained from Eq. (9.62) and becomes

$$G_1 = \frac{(0.90 + 0.545T)z + (-0.90 + 0.545T)}{z - 1} \tag{9.64}$$

The gains of the digital controller depend on the sampling period T. First, we have to determine the bandwidth ω_b of the closed-loop system of Fig. 7.19. Choosing the sampling frequency as $\omega_s = 20\omega_b$, the sampling period is obtained as $T = 2\pi/20\omega_b$.

9.4.2 Compensator Design

We have seen in Chapter 7 that a compensator is required in those cases where the performance of a control system with a PID family of control laws is inadequate. According to Method 1 that we use in this section, the compensator is first designed for a continuous-time system by using the techniques discussed in Chapter 7. In case the compensator is implemented by analog components, then its mathematical model is included in the plant state-difference equations, Eq. (9.25), or its transfer function is included in the plant discrete transfer function, Eq. (9.51).

In many cases, it is economical to implement the compensator digitally. The method that we have used to implement a PID control law digitally is based on the difference-equations approximation of differential equations. We now present another approximate method of obtaining a discrete transfer function from a given transfer function in the Laplace domain. This approximation is based on the trapezoidal rule of integration given by Eq. (9.56). An integral relationship in the Laplace domain is

$$U(s) = \frac{1}{s} E(s) \tag{9.65}$$

Employing the trapezoidal rule of integration, we obtain

$$u(k) = u(k-1) + T/2[e(k) + e(k-1)]$$

Taking the z-transform of both sides of this equation and using the translation property, Eq. (9.45), it follows that

$$(1 - z^{-1})U(z) = T/2(1 + z^{-1})E(z)$$

Hence,

$$U(z) = \frac{1}{\dfrac{2}{T}\left(\dfrac{z-1}{z+1}\right)} E(z) \tag{9.66}$$

On comparing Eq. (9.65) with Eq. (9.66), it can be seen that a discrete transfer-function approximation can be obtained from a given transfer function in the Laplace domain by the substitution

$$s = \frac{2}{T}\left(\frac{z-1}{z+1}\right) \tag{9.67}$$

This substitution based on the trapezoidal rule of integration is also known as Tustin's method. The difference equation of the compensator, if required, can then be obtained from the discrete transfer function. The transformation, Eq. (9.67), may be viewed as a mapping from the s-plane to the z-plane. Solving for z in terms of s, we get

$$z = \frac{1 + Ts/2}{1 - Ts/2} \tag{9.68}$$

On the imaginary axis of the s-plane, where $s = j\omega$, Eq. (9.68) becomes

$$z = \frac{1 + jT\omega/2}{1 - jT\omega/2} \tag{9.69}$$

It can be seen that in Eq. (9.69), $|z| = 1$. Hence, the substitution of Eq. (9.67) maps the imaginary axis of the s-plane into the unit circle of the z-plane, just like the original transformation, Eq. (9.39). Other approximations can be developed, but the Tustin approximation is commonly used and is also known as the bilinear transformation.

EXAMPLE 9.9

We have designed a cascade compensator for the continuous-time attitude-control system of Example 7.14. The transfer function of the compensator is given by

$$G(s) = \frac{\tau_1 s + 1}{\tau_2 s + 1} = \frac{0.325s + 1}{0.057s + 1} \tag{9.70}$$

where we have let $\omega_0 = 10$ rad/s in Example 7.14. In this example, we consider its digital implementation. Substituting for s in Eq. (9.70) from Eq. (9.67), the discrete transfer function is obtained as

$$G(z) = \frac{(0.325)\dfrac{2}{T}\left(\dfrac{z-1}{z+1}\right) + 1}{(0.057)\dfrac{2}{T}\left(\dfrac{z-1}{z+1}\right) + 1} = \frac{(0.65 + T)z + (T - 0.65)}{(0.114 + T)z + (T - 0.114)} \tag{9.71}$$

The difference equation relating the input and output of the compensator may now be obtained from Eq. (9.71).

9.5 CLOSED-LOOP RESPONSE

The response of a closed-loop digital control system can be investigated by using either the closed-loop discrete transfer function and z-transform analysis or the closed-loop state difference equations and analysis in the discrete-time domain. In the following, we use both methods, but first we discuss the methods of obtaining discrete transfer functions and state difference equations for closed-loop systems.

9.5.1 Closed-Loop Transfer Function

The block diagram of a control system with output feedback can be obtained by combining the block diagrams of Figs. 9.15 and 9.17 as shown in Fig. 9.18, where

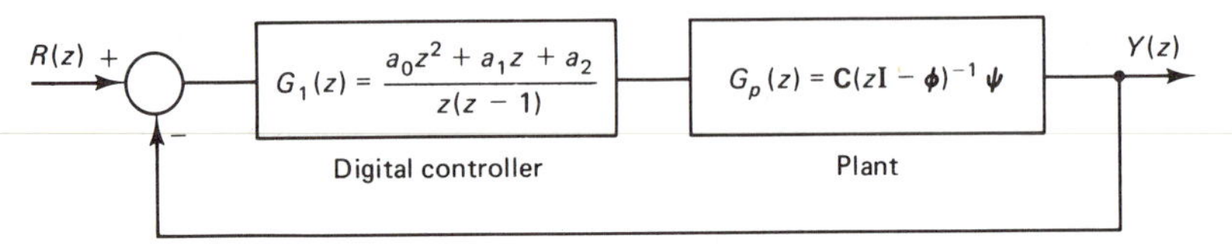

Figure 9.18 Block diagram of a closed-loop system.

the control law of Fig. 9.17 has been expressed by Eq. (9.62). Let the plant transfer-function matrix be denoted by $\mathbf{G}_p(z)$ and that of the controller by $\mathbf{G}_1(z)$. In case a digital compensator is required, we include its transfer function with that of the control law in $G_1(z)$. From Fig. 9.18, we obtain

$$\mathbf{Y}(z) = [\mathbf{I} + \mathbf{G}_p(z)\mathbf{G}_1(z)]^{-1}\mathbf{G}_p(z)\mathbf{G}_1(z)\mathbf{R}(z) \tag{9.72}$$

The closed-loop, discrete transfer-function matrix is then defined by Eq. (9.72). For a single-input, single-output system, we obtain

$$\frac{Y(z)}{R(z)} = G_c(z) = \frac{G_p(z)G_1(z)}{1 + G_p(z)G_1(z)} \tag{9.73}$$

EXAMPLE 9.10

Let a PD control law be desired for the plant of Example 9.3. As shown in Fig. 9.19, the control system is type 1, and its closed-loop transfer function is

$$\frac{Y(s)}{R(s)} = \frac{k_d s + k_p}{s^2 + (k_d + 0.1)s + k_p} \tag{9.74}$$

Let the performance specifications require a natural frequency $\omega_n = 3$ rad/s and a damping ratio $\zeta = 0.7$. To satisfy these specifications, we obtain $k_p = 9$ and $k_d = 4.1$. Hence, Eq. (9.74) becomes

$$\frac{Y(s)}{R(s)} = \frac{0.456s + 1}{s^2/9 + (2/3)(0.7)s + 1} \tag{9.75}$$

It can be verified, by sketching a Bode diagram, that the bandwidth of the closed-loop system is approximately 6 rad/s. Choosing the sampling

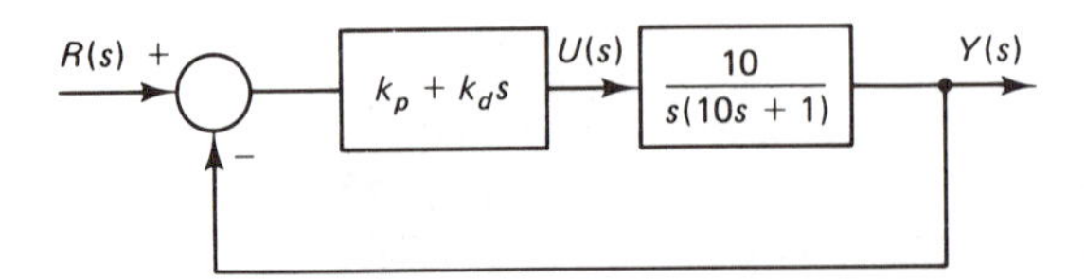

Figure 9.19 Closed-loop analog control system.

frequency as 60 rad/s, the sampling period becomes $T = 2\pi/60 = 0.1$ second. Setting $k_i = 0$ in Eq. (9.62), we obtain the controller discrete transfer function as

$$G_1(z) = \frac{U(z)}{E(z)} = \frac{(k_p + k_d/T)z - k_d/T}{z} = \frac{50z - 41}{z} \tag{9.76}$$

The plant transfer function is obtained by substituting from Eq. (9.30) in Eq. (9.51). It follows that

$$\begin{aligned} G_p(z) &= \lfloor 1 \quad 0 \rfloor \begin{bmatrix} z-1 & -0.10 \\ 0 & z-0.99 \end{bmatrix}^{-1} \begin{bmatrix} 0.005 \\ 0.100 \end{bmatrix} \\ &= \frac{0.005z + 0.00505}{(z-1)(z-0.99)} \end{aligned} \tag{9.77}$$

A block diagram of the digital control system is shown in Fig. 9.20. The

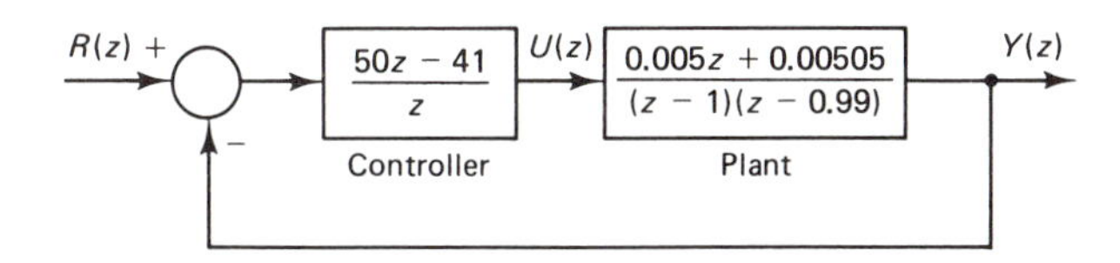

Figure 9.20 Closed-loop digital control system.

closed-loop discrete transfer function is obained by substituting for $G_p(z)$ from Eq. (9.77) and for $G_1(z)$ from Eq. (9.76) in Eq. (9.73). After simplification, we obtain

$$\frac{Y(z)}{R(z)} = G_c(z) = \frac{0.25z^2 + 0.0475z - 0.207}{z^3 - 1.74z^2 + 1.0375z - 0.207} \tag{9.78}$$

9.5.2 Closed-Loop State Difference Equations

We have obtained the state-difference equations of the plant in section 9.3.1 and expressed them in Eq. (9.25) as

$$\mathbf{x}_p(k+1) = \mathbf{\Phi}\mathbf{x}_p(k) + \mathbf{\Psi}\mathbf{u}(k) + \mathbf{w}(k) \tag{9.79}$$

Now, we choose additional state variables $\mathbf{x}_c(k)$ to describe the controller, including the control law and digital compensator if any, and express the controller-difference equations in the form

$$\mathbf{x}_c(k+1) = \mathbf{F}\mathbf{x}_c(k) + \mathbf{H}\mathbf{e}(k) \tag{9.80}$$

$$\mathbf{u}(k) = \mathbf{M}\mathbf{x}_c(k) + \mathbf{N}\mathbf{e}(k) \tag{9.81}$$

where the error is defined by

$$\mathbf{e}(k) = \mathbf{r}(k) - \mathbf{y}(k) = \mathbf{r}(k) - \mathbf{C}\mathbf{x}_p(k) \tag{9.82}$$

Substituting for $\mathbf{u}(k)$ and $\mathbf{e}(k)$ from Eqs. (9.81) and (9.82) in Eqs. (9.79) and (9.80), we obtain

$$\mathbf{x}_p(k+1) = (\mathbf{\Phi} - \mathbf{\Psi NC})\mathbf{x}_p(k) + \mathbf{\Psi M x}_c(k) + \mathbf{\Psi N r}(k) + \mathbf{w}(k) \tag{9.83}$$

$$\mathbf{x}_c(k+1) = -\mathbf{HC x}_p(k) + \mathbf{F x}_c(k) + \mathbf{H r}(k) \tag{9.84}$$

We combine Eqs. (9.83) and (9.84) in the form

$$\mathbf{x}(k+1) = \mathbf{P x}(k) + \mathbf{Q r}(k) + \begin{bmatrix} \mathbf{w}(k) \\ \hline \mathbf{0} \end{bmatrix} \tag{9.85}$$

where we define

$$\mathbf{x}(k) = \begin{bmatrix} \mathbf{x}_p(k) \\ \hline \mathbf{x}_c(k) \end{bmatrix} \qquad \mathbf{P} = \left[\begin{array}{c|c} \mathbf{\Phi} - \mathbf{\Psi NC} & \mathbf{\Psi M} \\ \hline -\mathbf{HC} & \mathbf{F} \end{array}\right] \qquad \mathbf{Q} = \begin{bmatrix} \mathbf{\Psi N} \\ \hline \mathbf{H} \end{bmatrix} \tag{9.86}$$

As a special case, when only a proportional control law is used and a digital compensator is not required, the additional controller state variables are not defined. We simply get

$$\mathbf{u}(k) = k_p\mathbf{e}(k) = k_p[\mathbf{r}(k) - \mathbf{C x}_p(k)] \tag{9.87}$$

Substituting for $u(k)$ from Eq. (9.87) in Eq. (9.79) and letting $\mathbf{x}(k) = \mathbf{x}_p(k)$, we obtain

$$\mathbf{x}(k+1) = (\mathbf{\Phi} - k_p\mathbf{\Psi C})\mathbf{x}(k) + k_p\mathbf{\Psi r}(k) + \mathbf{w}(k) \tag{9.88}$$

This equation is in the form of Eq. (9.85) with $\mathbf{P} = \mathbf{\Phi} - k_p\mathbf{\Psi C}$ and $\mathbf{Q} = k_p\mathbf{\Psi}$.

EXAMPLE 9.11

We now obtain the state-difference equations for the closed-loop digital control system of Example 9.10. From Eq. (9.76), the equation of the controller is given by

$$U(z) = (50 - 41z^{-1})E(z)$$

and from property of Eq. (9.45), the difference equation becomes

$$u(k) = -41e(k-1) + 50e(k) \tag{9.89}$$

In this case, we choose one controller state variable $x_c(k) = e(k-1)$ and express Eq. (9.89) as

$$x_c(k+1) = e(k) \tag{9.90}$$

$$u(k) = -41x_c(k) + 50e(k) \tag{9.91}$$

On comparing Eq. (9.90) with Eq. (9.80) and Eq. (9.91) with Eq. (9.81), we see that for this example F, H, M, and N are scalars whose values are given by $F = 0$, $H = 1$, $M = -41$, $N = 50$, and $\mathbf{C} = \lfloor 1 \quad 0 \rfloor$, where $y(k) = \mathbf{C x}_p(k)$.

The plant-difference equations are given by Eq. (9.30). Hence,

$$\mathbf{\Phi} - \mathbf{\Psi} NC = \begin{bmatrix} 1 & 0.10 \\ 0 & 0.99 \end{bmatrix} - \begin{bmatrix} 0.005 \\ 0.100 \end{bmatrix} 50 \lfloor 1 \quad 0 \rfloor$$

$$= \begin{bmatrix} 0.75 & 0.10 \\ -5 & 0.99 \end{bmatrix}$$

$$\mathbf{\Psi} M = \begin{bmatrix} -0.205 \\ -0.41 \end{bmatrix} \qquad -HC = \lfloor -1 \quad 0 \rfloor \qquad \mathbf{x}(k) = \begin{bmatrix} x_1(k) \\ x_2(k) \\ x_c(k) \end{bmatrix}$$

Substituting these results in Eq. (9.86), we obtain

$$\mathbf{x}(k+1) = \begin{bmatrix} 0.75 & 0.10 & -0.205 \\ -5 & 0.99 & -4.1 \\ -1 & 0 & 0 \end{bmatrix} \mathbf{x}(k) + \begin{bmatrix} 0.25 \\ 5 \\ 1 \end{bmatrix} r(k) + \begin{bmatrix} \mathbf{w}(k) \\ \hline 0 \end{bmatrix} \tag{9.92}$$

We can also obtain the closed-loop discrete transfer function of Eq. (9.73) from the closed-loop state-difference equations of Eq. (9.85). Neglecting the disturbance and assuming zero initial conditions, the z-transform of both sides of Eq. (9.85) yields

$$\begin{aligned} z\mathbf{X}(z) &= \mathbf{PX}(z) + \mathbf{QR}(z) \\ \mathbf{X}(z) &= (z\mathbf{I} - \mathbf{P})^{-1}\mathbf{QR}(z) \\ \mathbf{Y}(z) &= \mathbf{CX}(z) \\ &= \mathbf{C}(z\mathbf{I} - \mathbf{P})^{-1}\mathbf{QR}(z) \end{aligned} \tag{9.93}$$

Hence, it is seen that the closed-loop transfer matrix of Eq. (9.72) is indeed given by

$$\mathbf{G}_c(z) = \mathbf{C}(z\mathbf{I} - \mathbf{P})^{-1}\mathbf{Q} \tag{9.94}$$

It can be easily verified that we can obtain the closed-loop transfer function, Eq. (9.78), of Example 9.10 from the state-difference equations, Eq. (9.92), of Example 9.11 by using Eq. (9.94).

9.5.3 Response from State-Difference Equations

We now obtain the response of a closed-loop digital control system by using its state-difference equations model. In the preceding, we have expressed the state-difference equations of a closed-loop system in the form

$$\mathbf{x}[(k+1)T] = \mathbf{Px}(kT) + \mathbf{Qr}(kT) \tag{9.95}$$

$$\mathbf{y}(kT) = \mathbf{Cx}(kT) \tag{9.96}$$

where we have neglected the disturbance. The response is obtained by a recursion procedure. Setting $k = 0, 1, 2, \ldots$ in Eq. (9.95), we obtain the following result.

$$k = 0: \quad \mathbf{x}(T) = \mathbf{P}\mathbf{x}(0) + \mathbf{Q}\mathbf{r}(0) \tag{9.97}$$

$$k = 1: \quad \mathbf{x}(2T) = \mathbf{P}\mathbf{x}(T) + \mathbf{Q}\mathbf{r}(T) = \mathbf{P}^2\mathbf{x}(0) + \mathbf{P}\mathbf{Q}\mathbf{r}(0) + \mathbf{Q}\mathbf{r}(T) \tag{9.98}$$

$$k = 3: \quad \mathbf{x}(3T) = \mathbf{P}\mathbf{x}(2T) + \mathbf{Q}\mathbf{r}(2T) = \mathbf{P}^3\mathbf{x}(0) + \mathbf{P}^2\mathbf{Q}\mathbf{r}(0) + \mathbf{P}\mathbf{Q}\mathbf{r}(T) + \mathbf{Q}\mathbf{r}(2T) \tag{9.99}$$

where we have made use of Eq. (9.97) in Eq. (9.98) and of Eq. (9.98) in Eq. (9.99). Hence, in general for any $k > 0$, the response can be expressed in the form

$$\mathbf{x}(kT) = \mathbf{P}^k\mathbf{x}(0) + \sum_{i=0}^{k-1} \mathbf{P}^i\mathbf{Q}\mathbf{r}[(k-i-1)T] = \mathbf{P}^k\mathbf{x}(0) + \sum_{i=0}^{k-1} \mathbf{P}^{k-i-1}\mathbf{Q}\mathbf{r}(iT) \tag{9.100}$$

$$\mathbf{y}(kT) = \mathbf{C}\mathbf{x}(kT) \tag{9.101}$$

The first term on the right-hand side of Eq. (9.100) is contributed by the initial conditions and the second by the inputs over the interval from 0 to $(k-1)T$. This second term represents the convolution summation and is analogous to the convolution integral for continuous-time systems. From Eq. (9.100), we note that the state-transition matrix for the system of Eq. (9.95) is given by $\mathbf{P}^k$. In case a disturbance term $\mathbf{Q}_1\mathbf{w}(kT)$ is included on the right-hand side of Eq. (9.95), it is a simple matter to add a corresponding term

$$\sum_{i=0}^{k-1} \mathbf{P}^{k-i-1}\mathbf{Q}_1\mathbf{w}(iT)$$

to the right-hand side of Eq. (9.100).

EXAMPLE 9.12

For a closed-loop digital control system with single input and single output, let **P**, **Q**, and **C** matrices be described by

$$\mathbf{P} = \begin{bmatrix} 0.704 & 0.533 \\ -0.582 & 0.081 \end{bmatrix} \qquad \mathbf{Q} = \begin{bmatrix} 0.296 \\ 0.582 \end{bmatrix} \qquad \mathbf{C} = \lfloor 1 \quad 0 \rfloor \tag{9.102}$$

We obtain the system response when the initial condition $\mathbf{x}(0) = \mathbf{0}$ and $r(kT) = 1$ for $k \geq 0$. Using the data of Eq. (9.102) in Eq. (9.100), we get the response of the system as follows:

$kT =$	0	1	2	3	4	5
$x_1(kT) =$	0	0.296	0.814	1.113	1.157	1.081
$x_2(kT) =$	0	0.582	0.458	0.145	−0.054	−0.096

$kT =$	6	7	8	9	10
$x_1(kT) =$	1.006	0.975	0.978	0.992	1.001
$x_2(kT) =$	−0.056	−0.009	0.014	0.014	0.005

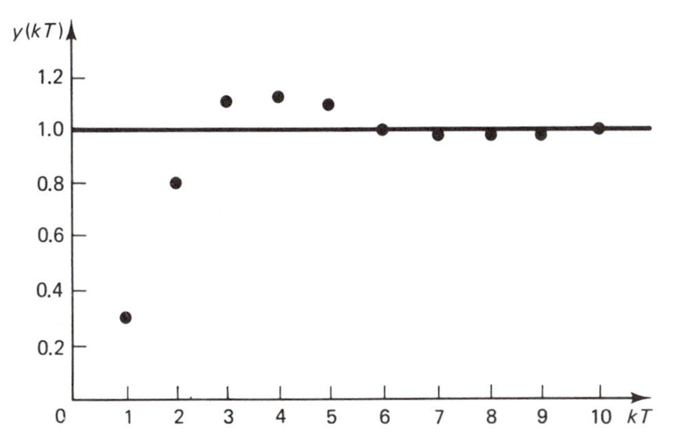

Figure 9.21 Response of the digital control system.

Since $y(kT) = x_1(kT)$, the output is shown plotted in Fig. 9.21.

9.5.4 Response from Discrete Transfer Function

The response of a digital control system can also be obtained by using its closed-loop discrete transfer function and z-transform analysis. The discrete transfer function matrix is shown in Eq. (9.72). For a single-input, single-output system, Eq. (9.73) can be represented in the form

$$\frac{Y(z)}{R(z)} = G_c(z) = \frac{b_m z^m + \cdots + b_0}{a_n z^n + a_{n-1} z^{n-1} + \cdots + a_0} \tag{9.103}$$

where $n \geq m$ for causal systems. The characteristic equation is given by

$$a_n z^n + a_{n-1} z^{n-1} + \cdots + a_0 = 0 \tag{9.104}$$

which is identical to $|z\mathbf{I} - \mathbf{P}| = 0$. Letting the command input $R(z)$ be a known testing signal, the response $y(kT)$ of Eq. (9.103) can be determined from the inverse z-transform.

There are several methods of obtaining the inverse z-transformation, but here we consider only the method of partial fraction expansion. The discrete transfer function is decomposed by a partial fraction expansion, and the inverse z-transform of each elementary term is obtained by consulting a table. The following development is useful for this purpose. Let a^{k-1} be defined for $k \geq 1$, where a is a real or complex constant. We obtain

$$\begin{aligned} z\text{-transform } a^{k-1} &= \sum_{k=1}^{\infty} a^{k-1} z^{-k} \\ &= z^{-1} \sum_{k=0}^{\infty} (az^{-1})^k \\ &= \frac{z^{-1}}{1 - az^{-1}} \qquad |az^{-1}| < 1 \\ &= \frac{1}{z - a} \qquad |z| > a \end{aligned} \tag{9.105}$$

Hence, the inverse z-transform of $1/(z-a)$ is a^{k-1} for $k \geq 1$. Since,

$$\frac{1}{z-a} = \sum_{k=1}^{\infty} a^{k-1} z^{-k} \tag{9.106}$$

and the series converges uniformly, we differentiate both sides of Eq. (9.106) with respect to z and obtain

$$\frac{-1}{(z-a)^2} = -z^{-1} \sum_{k=1}^{\infty} k a^{k-1} z^{-k}$$

i.e.,

$$\frac{z}{(z-a)^2} = z\text{-transform } [k a^{k-1}] \tag{9.107}$$

Proceeding in this manner, we can construct Table 9.1, which is useful for obtaining the inverse z-transform of elementary terms when the characteristic equation has a repeated root.

TABLE 9.1

Elementary term	Time sequence	
$\frac{1}{z-a}$	a^{k-1}	$k \geq 1$
$\frac{z}{(z-a)^2}$	ka^{k-1}	
$\frac{z(z+a)}{(z-a)^3}$	$k^2 a^{k-1}$	
$\frac{z(z^2+4az+a^2)}{(z-a)^4}$	$k^3 a^{k-1}$	

EXAMPLE 9.13

Let the closed-loop transfer function of a digital control system be given by

$$\frac{Y(z)}{R(z)} = G_c(z) = \frac{2z^2+3z-17}{(z-2)(z-3)^2} \qquad |z| > 3 \tag{9.108}$$

Let $r(kT) = 1$ for $k \geq 0$. Then, $R(z) = z/(z-1)$, and we obtain

$$\begin{aligned} Y(z) &= \frac{2z^3+3z^2-17z}{(z-1)(z-2)(z-3)^2} \\ &= \frac{c_1}{z-1} + \frac{c_2}{z-2} + \frac{c_3}{z-3} + \frac{c_4 z}{(z-3)^2} \end{aligned} \tag{9.109}$$

The double root at $z = 3$ is expanded as shown in the preceding equation so that we can use Table 9.1 for the inverse z-transform. We get $c_1 = 3$, $c_2 = -6$,

$c_3 = 0$, and $c_4 = 5$. Using Table 9.1, we obtain the response as

$$y(kT) = 3(1)^{k-1} - 6(2)^{k-1} + 5k(3)^{k-1} \qquad k \geq 1 \tag{9.110}$$

9.6 STABILITY

We now investigate the stability of closed-loop digital control systems. Neglecting the disturbance, we express the state-difference equations of a closed-loop system in the form

$$\mathbf{x}(k+1) = \mathbf{P}\mathbf{x}(k) + \mathbf{Q}\mathbf{r}(k) \tag{9.111}$$

$$\mathbf{y}(k) = \mathbf{C}\mathbf{x}(k) \tag{9.112}$$

As in the case of continuous-time systems, there are several definitions of stability for discrete-time systems, but for systems described by linear, constant-coefficient difference equations as in Eq. (9.111), the distinctions among the different concepts of stability are not important. In the following, we consider stability in the sense of Lyapunov and in the sense of bounded input-bounded output (BIBO).

Stability in the Sense of Lyapunov. The concept of stability in the sense of Lyapunov does not admit any inputs, and the system is disturbed only by initial conditions. Hence in Eq. (9.111), we consider only

$$\mathbf{x}(k+1) = \mathbf{P}\mathbf{x}(k) \qquad \text{Initial conditions: } \mathbf{x}(0) \tag{9.113}$$

The equilibrium state of Eq. (9.113) is obtained from $\mathbf{x}(k) = \mathbf{P}\mathbf{x}(k)$, and since $\mathbf{P}$ is not an identity matrix, the only equilibrium state is $\mathbf{x}_e = \mathbf{0}$. The equilibrium state of Eq. (9.113) is said to be asymptotically stable if for every $\epsilon > 0$, there exists a $\delta > 0$, where δ may depend on ϵ, such that $\|\mathbf{x}(0)\| \leq \delta$ implies that $\|\mathbf{x}(k)\| < \epsilon$ for all $k \geq 0$ and $\lim_{k\to\infty} \|\mathbf{x}(k)\| = 0$.

If $\mathbf{P}$ has distinct eigenvalues, we can transform the state variables to the normal form as shown in Chapter 4. The state-transition matrix of the transformed state variables is the diagonal matrix $\mathbf{\Lambda}^k$, where

$$\mathbf{\Lambda}^k = \begin{bmatrix} z_1^k & 0 & \cdots & 0 \\ 0 & z_2^k & \cdots & 0 \\ \vdots & \vdots & \vdots\vdots\vdots & \vdots \\ 0 & 0 & & z_n^k \end{bmatrix} \tag{9.114}$$

and z_i are the eigenvalues of $\mathbf{P}$. Hence, the equilibrium state of Eq. (9.113) is asymptotically stable if and only if all eigenvalues of $\mathbf{P}$, which are the roots of the closed-loop characteristic equation $|z\mathbf{I} - \mathbf{P}| = 0$, satisfy the condition that $|z_i| < 1$. If $\mathbf{P}$ has repeated eigenvalues, we can reduce it to a Jordan normal form and show that $|z_i| < 1$ is the condition for stability for any general $\mathbf{P}$ matrix. If one or more $|z_i| > 1$, it can be seen from Eq. (9.114) that $\lim_{k\to\infty} \|\mathbf{x}(k)\| \to \infty$.

Bounded Input, Bounded Output (BIBO) Stability. A discrete-time system is called BIBO stable if for an input sequence such that $\|\mathbf{r}(k)\| \leq \delta < \infty$ for all $k \geq 0$, the output sequence satisfies the condition $\|\mathbf{y}(k)\| \leq \epsilon < \infty$ for all $k \geq 0$.

We have seen that the solution of Eq. (9.111) is given by Eq. (9.100). Applying the Hölder inequality to Eq. (9.100) and noting that the norm of the input sequence is bounded by δ, we obtain

$$\|\mathbf{x}(k)\| \leq \|\mathbf{P}^k\| \, \|\mathbf{x}(0)\| + \delta \sum_{i=0}^{k-1} \|\mathbf{P}^{k-i-1}\mathbf{Q}\| \tag{9.115}$$

Hence, $\|\mathbf{x}(k)\|$ will be bounded when $\|\mathbf{P}^k\|$ remains bounded for all $k \geq 0$.[1] If z_i is an eigenvalue of $\mathbf{P}$ and $\mathbf{v}_i$ is a corresponding eigenvector, we get

$$\mathbf{P}\mathbf{v}_i = z_i \mathbf{v}_i$$

$$\|\mathbf{P}\mathbf{v}_i\| = |z_i| \, \|\mathbf{v}_i\|$$

$$\|\mathbf{P}\| \leq |z_{\max}|$$

where $z_{\max}$ is the maximum eigenvalue of $\mathbf{P}$. Since $\|\mathbf{P}^k\| \leq \|\mathbf{P}\|^k \leq |z_{\max}|^k$ and $\|\mathbf{y}(k)\| = \|\mathbf{C}\mathbf{x}(k)\|$, the condition for BIBO stability is that all eigenvalues of $\mathbf{P}$ must satisfy the condition $|z_i| < 1$.

Hence, we have shown that for both asymptotic stability of the equilibrium state and for BIBO stability, all eigenvalues of $\mathbf{P}$ must lie inside the unit circle in the complex z-plane. We note that the coefficients of the characteristic equation are affected by the sampling period T. Hence, the stability of a digital control system may depend also on the choice of the sampling period.

EXAMPLE 9.14

In Example 9.12, the closed-loop $\mathbf{P}$ matrix is given by Eq. (9.102). The characteristic equation becomes

$$|z\mathbf{I} - \mathbf{P}| = z^2 - 0.785z + 0.367 = 0$$

The two roots of this equation are $z_1 = 0.393 + j0.462$ and $z_2 = 0.393 - j0.462$. Hence, $|z_1| = |z_2| = 0.607 < 1$ and the closed-loop system of Example 9.12 is stable.

EXAMPLE 9.15

We consider the system of Example 9.13 whose closed-loop discrete transfer function is given by Eq. (9.108). The characteristic equation is obtained by setting its denominator equal to zero and is given by

$$|z\mathbf{I} - \mathbf{P}| = (z-2)(z-3)^2 = 0 \tag{9.116}$$

Now, $|z_1| = 2 > 1$ and $|z_2| = 3 > 1$. Hence, the closed-loop system of Example 9.13 is unstable. It can be seen from the response given by Eq. (9.110) that when the input is a unit step sequence, $y(kT) \to \infty$ as $k \to \infty$.

It is advantageous to use a method that yields the stability information without the trouble of factoring a high-order polynomial. We note that Routh's criterion yields information regarding the locations of the roots of the characteristic equation relative to the imaginary axis of the complex s-plane. To employ the Routh criterion to the discrete-time case, we have first to use a transformation that maps the region outside the unit circle in the z-plane to the right half of the s-plane. This mapping is provided by the bilinear or Tustin transformation given by Eqs. (9.67) and (9.68), where we now omit the factor $2/T$. Hence, the transformation we use for this purpose is

$$s = \frac{z-1}{z+1} \qquad \text{i.e., } z = \frac{1+s}{1-s} \tag{9.117}$$

We have seen earlier that this transformation maps the unit circle in the z-plane into the imaginary axis of the s-plane. Substituting $s = \sigma + j\omega$ in Eq. (9.117), we obtain

$$|z| = \left[\frac{(1+\sigma)^2 + \omega^2}{(1-\sigma)^2 + \omega^2}\right]^{1/2} \tag{9.118}$$

When $|z| > 1$, it is seen that $\sigma > 0$ and when $|z| < 1$, we get $\sigma < 0$. Hence, the transformation of Eq. (9.117) maps the region inside the unit circle of the z-plane into the left half of the s-plane. The region outside the unit circle is mapped into the right half of the s-plane. There is one-to-one correspondence between finite poles and zeros. The procedure is to first transform the characteristic equation from the z-plane to the s-plane by using Eq. (9.117) and then to employ the Routh criterion.

EXAMPLE 9.16

Let the closed-loop discrete transfer function of a system be given by

$$G_c(z) = \frac{z^2 + 4z + 3}{z^3 + 2z^2 - 0.5z - 1} \tag{9.119}$$

The characteristic equation is

$$z^3 + 2z^2 - 0.5z - 1 = 0 \tag{9.120}$$

Using the transformation of Eq. (9.117), we obtain

$$\left(\frac{1+s}{1-s}\right)^3 + 2\left(\frac{1+s}{1-s}\right)^2 - 0.5\left(\frac{1+s}{1-s}\right) - 1 = 0$$

i.e.,

$$s^3 + 3s^2 - 17s - 3 = 0 \tag{9.121}$$

The coefficients of this characteristic equation do not have the same sign. The Routh array will therefore show that Eq. (9.121) has at least one root with a positive real part. This implies that the characteristic equation, Eq. (9.120), has at least one root outside the unit circle in the z-plane and therefore the system is unstable.

9.7 CONTROLLABILITY AND OBSERVABILITY

As discussed in Chapter 8, the existence of a solution to the design of a controller with state feedback depends on the property of controllability. The existence of a solution to the design of an observer depends on observability. In this section, we investigate these properties for the case of linear discrete-time systems. For this purpose, we consider the open-loop plant to be controlled, before the controller and observer have been designed. The continuous-time model of a plant to be controlled digitally has been converted to state-difference equations in section 9.3. Neglecting the disturbance and ignoring the subscript p, which denotes plant state variables in Eq. (9.25), we obtain the plant state-difference equations as

$$\mathbf{x}[(k+1)T] = \mathbf{\Phi x}(kT) + \mathbf{\Psi u}(kT) \tag{9.122}$$

$$\mathbf{y}(kT) = \mathbf{Cx}(kT) \tag{9.123}$$

Definition of Controllability. The system described by Eq. (9.122) is said to be controllable if for every initial state $\mathbf{x}(0)$, there is a finite N and a sequence of controls $\mathbf{u}(0), \mathbf{u}(T), \ldots, \mathbf{u}[(N-1)T]$ such that the initial state $\mathbf{x}(0)$ at $k = 0$ can be transferred to some arbitrary state $\mathbf{x}(NT)$ at $k = N$.

Consider the case where a similarity transformation matrix $\mathbf{P}$ exists that uncouples Eq. (9.122) and $\mathbf{\Phi}$ is transformed to the diagonal matrix $\mathbf{\Lambda}$. If $\mathbf{P}^{-1}\mathbf{\Psi}$ has a jth row of all zero elements, then the jth normal state variable is not affected by the input and is not controllable. A general test for controllability of Eq. (9.122) can be developed as follows. In our definition, we let $N = n$, where n is the order of the system. In a manner analogous to Eq. (9.100), the solution of Eq. (9.122) can be obtained as

$$\mathbf{x}(nT) = \mathbf{\Phi}^n\mathbf{x}(0) + \sum_{i=0}^{n-1} \mathbf{\Phi}^{n-1-i}\mathbf{\Psi u}(iT) \tag{9.124}$$

We then obtain

$$\begin{aligned}\mathbf{x}(nT) - \mathbf{\Phi}^n\mathbf{x}(0) &= \mathbf{\Psi u}[(n-1)T] + \cdots + \mathbf{\Phi}^{n-1}\mathbf{\Psi u}(0) \\ &= [\mathbf{\Psi} \quad \mathbf{\Phi\Psi} \quad \cdots \quad \mathbf{\Phi}^{n-1}\mathbf{\Psi}] \begin{bmatrix} \mathbf{u}[(n-1)T] \\ \mathbf{u}[(n-2)T] \\ \vdots \\ \mathbf{u}(0) \end{bmatrix}\end{aligned} \tag{9.125}$$

When the number of state variables is n and the number of inputs is m, we define an $(n \times nm)$ controllability matrix $\mathbf{Q}$ as

$$\mathbf{Q} = [\mathbf{\Psi} \quad \mathbf{\Phi\Psi} \quad \cdots \quad \mathbf{\Phi}^{n-1}\mathbf{\Psi}] \tag{9.126}$$

Since the left-hand side of Eq. (9.125) is known, we can solve for $\mathbf{u}(kT)$ when the rank of $\mathbf{Q}$ is n. Hence, the system of Eq. (9.122) is controllable if and only if the rank of the $(n \times nm)$ matrix $\mathbf{Q}$ is n. For the case of a single input, we have $m = 1$, and if the determinant of the $(n \times n)$ matrix $\mathbf{Q}$ is nonzero, then its rank is n.

Definition of Observability. The system described by Eqs. (9.122) and (9.123) is said to be observable if for any initial state $\mathbf{x}(0)$, there is a finite N such that $\mathbf{x}(0)$

can be computed from the observations of $\mathbf{y}(0), \mathbf{y}(T), \ldots, \mathbf{y}[(N-1)T]$ and knowledge of the control sequence $\mathbf{u}(0), \mathbf{u}(T), \ldots, \mathbf{u}[(N-1)T]$.

In case there is a similarity transformation matrix $\mathbf{P}$ that uncouples the state variables and $\mathbf{CP}$ has a jth column of all zero elements, then the jth normal state variable does not affect the output and is not observable. A test for observability can be developed as follows. In our definition of observability, we let $N = n$, where n is the order of the system. Using Eqs. (9.123) and (9.124),

$$\mathbf{y}(nT) = \mathbf{C\Phi}^n\mathbf{x}(0) + \mathbf{C}\sum_{i=0}^{n-1} \mathbf{\Phi}^{n-1-i}\mathbf{\Psi u}(iT)$$

Since the second term on the right-hand side is known, let us represent it compactly by $\mathbf{f}(iT)$ for $i = 0, \ldots, n-1$. We then obtain the following sequence.

$$\begin{bmatrix} \mathbf{y}(0) - \mathbf{f}(0) \\ \mathbf{y}(T) - \mathbf{f}(T) \\ \vdots \\ \mathbf{y}[(n-1)T] - \mathbf{f}[(n-1)T] \end{bmatrix} = \begin{bmatrix} \mathbf{C} \\ \mathbf{C\Phi} \\ \vdots \\ \mathbf{C\Phi}^{n-1} \end{bmatrix} \mathbf{x}(0) \tag{9.127}$$

When the number of state variables is n and the number of outputs is m, we define an $(nm \times n)$ observability matrix $\mathbf{M}$ as

$$\mathbf{M} = \begin{bmatrix} \mathbf{C} \\ \mathbf{C\Phi} \\ \vdots \\ \mathbf{C\Phi}^{n-1} \end{bmatrix} \tag{9.128}$$

Since the left-hand side of Eq. (9.127) is known, we can solve for $\mathbf{x}(0)$ when the rank of $\mathbf{M}$ is n. Hence, the system of Eqs. (9.122) and (9.123) is observable if and only if the rank of the $(nm \times n)$ observability matrix $\mathbf{M}$ is n. In the case of a single output, we have $m = 1$, and if the determinant of the $(n \times n)$ matrix $\mathbf{M}$ is nonzero then its rank is n. We may also define the observability matrix as

$$\mathbf{M}^T = [\mathbf{C}^T \quad \mathbf{\Phi}^T\mathbf{C}^T \quad \cdots \quad (\mathbf{\Phi}^{n-1})^T\mathbf{C}^T]$$

If rank of $\mathbf{M}$ is n, so is that of $\mathbf{M}^T$.

As in the case of continuous-time systems, most discrete-time systems are controllable and observable. Pathological cases however do arise mostly due to the concellation of a pole by a zero and in the case of observability, by the wrong choice of the state variables for measurement.

EXAMPLE 9.17

The state-difference equations for the two-input, two-output plant of a temperature-control system are obtained in Example 9.4. The $\mathbf{\Phi}$ and $\mathbf{\Psi}$ matrices are given by Eqs. (9.34) and (9.35), respectively. The controllability matrix becomes

$$\mathbf{Q} = [\boldsymbol{\Psi} \quad \boldsymbol{\Phi\Psi}]$$

$$= \begin{bmatrix} 259.18 & 0 & 192.05 & 0 \\ 36.94 & 259.18 & 84.91 & 192.05 \end{bmatrix} \tag{9.129}$$

$$\mathbf{QQ}^T = \begin{bmatrix} 104{,}057.47 & 25{,}881.08 \\ 25{,}881.08 & 112{,}631.75 \end{bmatrix}$$

The determinant of $\mathbf{QQ}^T$ is not zero. Hence, the rank of $\mathbf{Q}$ is 2 and the system is controllable. The $\mathbf{C}$ matrix of the output as given by Eq. (9.36) is a (2×2) identity matrix. Hence, the observability matrix becomes

$$\mathbf{M} = \begin{bmatrix} \mathbf{C} \\ \mathbf{C\Phi} \end{bmatrix}$$

$$= \begin{bmatrix} 1 & 0 \\ 0 & 1 \\ 0.741 & 0 \\ 0.222 & 0.741 \end{bmatrix} \tag{9.130}$$

$$\mathbf{M}^T\mathbf{M} = \begin{bmatrix} 1.598 & 0.165 \\ 0.165 & 1.549 \end{bmatrix}$$

The determinant of $\mathbf{M}^T\mathbf{M}$ is not zero. Hence, the rank of $\mathbf{M}$ is 2 and the system is observable.

EXAMPLE 9.18

For a single-input, single-output plant with three state variables, let

$$\boldsymbol{\Phi} = \begin{bmatrix} 0 & 0 & 0 \\ 0 & 0 & 1 \\ 1/24 & -9/24 & 26/24 \end{bmatrix} \qquad \boldsymbol{\Psi} = \begin{bmatrix} 7 \\ 3 \\ 4/3 \end{bmatrix} \qquad \mathbf{C} = \lfloor 3 \quad -17 \quad 24 \rfloor$$

For this system, it can be shown that both the $\mathbf{Q}$ and $\mathbf{M}$ matrices of Eqs. (9.126) and (9.128), respectively, are singular. Hence, this system is neither controllable nor observable. The eigenvalues of $\boldsymbol{\Phi}$ are 1/2, 1/3, and 1/4. Using the similarity transformation matrix,

$$\mathbf{P} = \begin{bmatrix} 4 & 9 & 16 \\ 2 & 3 & 4 \\ 1 & 1 & 1 \end{bmatrix}$$

the transformed normal state variables are described by

$$\mathbf{x}^*(k+1) = \begin{bmatrix} 1/2 & 0 & 0 \\ 0 & 1/3 & 0 \\ 0 & 0 & 1/4 \end{bmatrix} \mathbf{x}^*(k) + \begin{bmatrix} 1 \\ 1/3 \\ 0 \end{bmatrix} u(k)$$

$$y(k) = \lfloor 2 \quad 0 \quad 4 \rfloor \mathbf{x}^*(k) \tag{9.131}$$

It is seen that the normal state variable $x_3^*(k)$ cannot be controlled and $x_2^*(k)$ cannot be observed. The discrete transfer function of the plant can be obtained from Eq. (9.51) as

$$\begin{aligned}\frac{Y(z)}{U(z)} &= G_p(s) = \mathbf{C}(z\mathbf{I} - \mathbf{\Phi})^{-1}\mathbf{\Psi} \\ &= \frac{2(z-1/3)(z-1/4)}{(z-1/2)(z-1/3)(z-1/4)}\end{aligned} \tag{9.132}$$

We note that two poles of the transfer function are cancelled by two zeros and this accounts for the plant being neither controllable nor observable.

9.8 CONTROLLER DESIGN WITH STATE FEEDBACK

This section covers the design of a digital controller with state feedback. For this purpose, it is assumed that all state variables are available for feedback, and, in the next section, we show how an observer can be designed to yield an estimate of the state variables from the measured output. The procedure we use here is very similar to that used in Chapter 8 for continuous-time systems.

9.8.1 Single-Input, Single-Output Systems

We first restrict the development to single-input, single-output systems and later generalize the results to multivariable systems. The state-difference equations for a single-input, single-output plant are described by

$$\begin{aligned}\mathbf{x}(k+1) &= \mathbf{\Phi}\mathbf{x}(k) + \mathbf{\Psi}u(k) + \mathbf{w}(k) \\ y(k) &= \mathbf{C}\mathbf{x}(k)\end{aligned} \tag{9.133}$$

where $u(k)$ is the control, and $\mathbf{w}(k)$ is the disturbance. The desired feedback control law has the form

$$u(k) = -\mathbf{K}\mathbf{x}(k) + Nr(k) \tag{9.134}$$

where $r(k)$ is the command input, $\mathbf{K}$ a $1 \times n$ row matrix, and N is a scalar scaling constant. The state-difference equations of the closed-loop system are obtained by substituting for $u(k)$ from Eq. (9.134) in Eq. (9.133) as

$$\mathbf{x}(k+1) = (\mathbf{\Phi} - \mathbf{\Psi}\mathbf{K})\mathbf{x}(k) + \mathbf{\Psi}Nr(k) + \mathbf{w}(k) \tag{9.135}$$

This is in the form of Eq. (9.85), with $\mathbf{P} = \mathbf{\Phi} - \mathbf{\Psi}\mathbf{K}$ and $\mathbf{Q} = \mathbf{\Psi}N$. The characteristic equation of the closed-loop system is

$$|z\mathbf{I} - \mathbf{\Phi} + \mathbf{\Psi}\mathbf{K}| = 0 \tag{9.136}$$

We now use Method 2 described in section 9.4. The performance specifications that are stated in terms of continuous-time response are first converted to the desired locations of the roots of the characteristic equation in the complex s-plane as shown in Chapter 8. Let these roots be $-s_1, -s_2, \ldots, -s_n$. The desired locations of the roots of Eq. (9.136) in the complex z-plane are obtained from the mapping $z_i = e^{-Ts_i}$ for $i = 1, \ldots, n$. Equating the desired characteristic equation to Eq. (9.136), we obtain

$$|z\mathbf{I} - \mathbf{\Phi} + \mathbf{\Psi}\mathbf{K}| = (z - z_1)(z - z_2) \cdots (z - z_n) \tag{9.137}$$

The n unknown elements of $\mathbf{K}$ can now be obtained by matching the corresponding coefficients in Eq. (9.137). It may be convenient to use the Ackermann formula for high-order systems. In a manner analogous to Eq. (8.25), we obtain

$$\mathbf{K} = \lfloor 0 \quad 0 \quad \cdots \quad 0 \quad 1 \rfloor \mathbf{Q}^{-1}\Delta(\mathbf{\Phi}) \tag{9.138}$$

where $\mathbf{Q}$ is the controllability matrix defined by Eq. (9.126) and $\Delta(\mathbf{\Phi})$ is obtained by substituting $\mathbf{\Phi}$ for z on the right-hand side of Eq. (9.137). We note that Eq. (9.138) requires the inverse of $\mathbf{Q}$, which must therefore be nonsingular. Hence, to obtain a unique solution for $\mathbf{K}$, the pair $(\mathbf{\Phi}, \mathbf{\Psi})$ must be controllable.

EXAMPLE 9.19

A digital controller with state feedback is to be designed for the plant of Example 9.3. As stated in Example 9.10, the continuous-time response specifications for the closed-loop system are as follows. The natural frequency $\omega_n = 3$ rad/s, the damping ratio $\zeta = 0.7$, and the closed-loop bandwidth ω_b to be approximately 6 rad/s. Hence, the desired roots in the s-plane are

$$-s_1, -s_2 = -\zeta\omega_n \pm j\omega_n\sqrt{1-\zeta^2} = -2.1 \pm j2.14 \tag{9.139}$$

It is shown in Example 9.10 that a suitable sampling period for this system is $T = 0.1$ second. The roots in the z-plane are obtained from the mapping $z_i = e^{-Ts_i}$ and we get

$$\begin{aligned} z_1 &= \exp[0.1(-2.1 + j2.14)] \\ &= 0.792 + j0.172 \\ z_2 &= 0.792 - j0.172 \end{aligned}$$

We have $|z_1| = |z_2| = 0.81 < 1$, and hence the closed-loop system is stable. The desired characteristic equation becomes

$$(z - z_1)(z - z_2) = z^2 - 1.584z + 0.657 = 0 \tag{9.140}$$

For $T = 0.1$ second, the $\mathbf{\Phi}$ and $\mathbf{\Psi}$ matrices are given by Eq. (9.30). It can easily be verified that the pair $(\mathbf{\Phi}, \mathbf{\Psi})$ is controllable. Substituting these values in Eq. (9.136), we obtain the characteristic equation as

$$\begin{aligned} |z\mathbf{I} - \mathbf{\Phi} + \mathbf{\Psi}\mathbf{K}| &= z^2 + (-1.99 + 0.005k_1 + 0.1k_2)z \\ &\quad + 0.99 + 0.005k_1 - 0.1k_2 = 0 \end{aligned} \tag{9.141}$$

Equating the corresponding coefficients in Eqs. (9.140) and (9.141), we obtain $k_1 = 7.3$ and $k_2 = 3.7$. Assuming that both state variables are available for feedback, the structure of the control system is shown in Fig. 9.22, where the

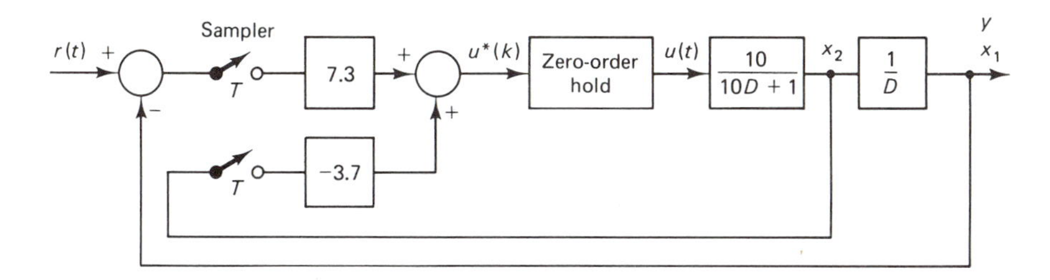

Figure 9.22 Structure of the digital control system.

scale factor $N = 1$ since it is assumed that the sensor that measures x_1 has unity gain.

9.8.2 State Feedback with PI Control

If the plant has no pure integrator, then its discrete transfer function does not have a pole at $z = e^{T_0} = 1$. With the control law of Eq. (9.134), the control system will be type 0. If a type 1 system is required, we have to modify the control law as was done for continuous-time systems in section 8.4.2. Using the rectangular rule of integration, we define an additional controller state variable $x_c(k)$ as

$$x_c(k+1) = x_c(k) - Te(k) \tag{9.142}$$

and since $e(k) = r(k) - y(k) = r(k) - \mathbf{C}\mathbf{x}_p(k)$, we get

$$x_c(k+1) = T\mathbf{C}\mathbf{x}_p(k) + x_c(k) - Tr(k) \tag{9.143}$$

The n plant state variables are described by

$$\mathbf{x}_p(k+1) = \mathbf{\Phi}\mathbf{x}_p(k) + \mathbf{\Psi}u(k) \tag{9.144}$$

Defining

$$\mathbf{x}(k) = \begin{bmatrix} \mathbf{x}_p(k) \\ \hline x_c(k) \end{bmatrix} \qquad \hat{\mathbf{\Phi}} = \left[\begin{array}{c|c} \mathbf{\Phi} & 0 \\ \hline T\mathbf{C} & 1 \end{array}\right] \qquad \hat{\mathbf{\Psi}} = \begin{bmatrix} \mathbf{\Psi} \\ \hline 0 \end{bmatrix} \tag{9.145}$$

we combine Eqs. (9.144) and (9.143) in the form

$$\mathbf{x}(k+1) = \hat{\mathbf{\Phi}}\mathbf{x}(k) + \hat{\mathbf{\Psi}}u(k) + \begin{bmatrix} \mathbf{0} \\ -T \end{bmatrix} r(k) \tag{9.146}$$

Assuming that the pair $(\hat{\mathbf{\Phi}}, \hat{\mathbf{\Psi}})$ is controllable, the desired control law is

$$u(k) = -\mathbf{K}\mathbf{x}(k) + Nr(k) \tag{9.147}$$

where $\mathbf{K} = \lfloor k_1 \cdots k_n \;\; k_{n+1} \rfloor$. Substituting for $u(k)$ from Eq. (9.147) in Eq. (9.146), the closed-loop system is described by

$$\mathbf{x}(k+1) = (\hat{\mathbf{\Phi}} - \hat{\mathbf{\Psi}}\mathbf{K})\mathbf{x}(k) + \hat{\mathbf{\Psi}}Nr(k) + \begin{bmatrix} \mathbf{0} \\ -T \end{bmatrix} r(k) \tag{9.148}$$

Equating the characteristic equation to the desired characteristic equation, we obtain

$$|z\mathbf{I} - \hat{\mathbf{\Phi}} + \hat{\mathbf{\Psi}}\mathbf{K}| = (z - z_1)(z - z_2)\cdots(z - z_n)(z - z_{n+1}) \tag{9.149}$$

The $(n + 1)$ unknown gains of $\mathbf{K}$ are then obtained from Eq. (9.149).

9.8.3 Control of Multivariable Systems

For the extension of the preceding method to multivariable systems, we follow a procedure similar to that of section 8.4.3. In Eq. (9.133), we let $\mathbf{u}$ be an $m \times 1$ vector, and $\mathbf{y}$ an $m \times 1$ vector. In the control law of Eq. (9.134), $\mathbf{r}$ is an $m \times 1$ command input vector, $\mathbf{K}$ an $m \times n$ gain matrix, and $\mathbf{N}$ is an $m \times m$ scaling matrix. Now, there are mn unknown elements of $\mathbf{K}$ and only n roots to be specified. As in Eq. (8.54), we convert the problem to a single-input, single-output equivalent system by letting

$$\mathbf{K} = \mathbf{p}\mathbf{k}^T = \begin{bmatrix} p_1 \\ \vdots \\ p_m \end{bmatrix} \lfloor k_1 \quad \cdots \quad k_n \rfloor \tag{9.150}$$

where the m elements of $\mathbf{p}$ are chosen arbitrarily, and the n elements of $\mathbf{k}^T$ are obtained from the n roots that are specified. The characteristic equation of the closed-loop system becomes

$$|z\mathbf{I} - \mathbf{\Phi} + \mathbf{\Psi}\mathbf{p}\mathbf{k}^T| = 0 \tag{9.151}$$

which is equivalent to that of a single-input–single-output equivalent system

$$\mathbf{x}(k + 1) = \mathbf{\Phi}\mathbf{x}(k) + \mathbf{\Psi}\mathbf{p}u(k) \tag{9.152}$$

In order to solve for $\mathbf{k}^T$, the determinant of the controllability matrix

$$\mathbf{Q} = [\mathbf{\Psi}\mathbf{p} \quad \mathbf{\Phi}\mathbf{\Psi}\mathbf{p} \quad \cdots \quad \mathbf{\Phi}^{n-1}\mathbf{\Psi}\mathbf{p}] \tag{9.153}$$

must be nonzero. This procedure will fail when the original multivariable system is controllable, but there is no choice of $\mathbf{p}$ for which Eq. (9.152) is controllable.

EXAMPLE 9.20

A regulator with state feedback was studied in Example 8.9 for the two-input–two-output temperature-control system of Example 3.8. In this example, we consider the digital implementation of the regulator. For this system, the plant matrices $\mathbf{\Phi}$ and $\mathbf{\Psi}$ are given by Eqs. (9.34) and (9.35), respectively, in Example 9.4 for $T = 3$ seconds. In Example 9.17, we have shown that the multivariable system is controllable.

We arbitrarily select $p_1 = p_2 = 1$ in Eq. (9.150) and obtain

$$\mathbf{\Psi}\mathbf{p} = \begin{bmatrix} 259.182 \\ 296.118 \end{bmatrix} \qquad \mathbf{Q} = [\mathbf{\Psi}\mathbf{p} \quad \mathbf{\Phi}\mathbf{\Psi}\mathbf{p}] = \begin{bmatrix} 259.182 & 192.054 \\ 296.118 & 276.962 \end{bmatrix}$$

The determinant of $\mathbf{Q}$ is nonzero, and hence the pair $(\mathbf{\Phi}, \mathbf{\Psi}\mathbf{p})$ is controllable. Let the continuous-time response specifications require a closed-loop natural frequency of 0.1 rad/s and a damping ratio $\zeta = 0.7$. Hence, the desired roots in the s-plane are

$$
\begin{aligned}
-s_1, -s_2 &= -\zeta\omega_n \pm j\omega_n\sqrt{1-\zeta^2} \\
&= -0.07 \pm j0.071
\end{aligned}
\tag{9.154}
$$

With a sampling period of $T = 3$ seconds, the corresponding roots in the z-plane are

$$
\begin{aligned}
z_1 &= \exp[3(-0.07 + j0.071)] \\
&= 0.792 + j0.172 \\
z_2 &= 0.792 - j0.172
\end{aligned}
$$

We have $|z_1| = |z_2| = 0.81 < 1$, and hence the closed-loop system is stable. The desired characteristic equation becomes

$$z^2 - 1.584z + 0.657 = 0 \tag{9.155}$$

The characteristic equation is also given by

$$
\begin{aligned}
|z\mathbf{I} - \mathbf{\Phi} + \mathbf{\Psi}\mathbf{p}\mathbf{k}^T| = z^2 &+ (-1.482 + 296.118k_2 + 259.182k_1)z \\
&- 192.054k_1 - 161.885k_2 + 0.549 = 0
\end{aligned}
\tag{9.156}
$$

Equating the corresponding coefficients of Eqs. (9.155) and (9.156), we obtain $k_1 = -1.038 \times 10^{-3}$ and $k_2 = 0.563 \times 10^{-3}$.

9.9 OBSERVER DESIGN

The control law of the preceding section requires all the state variables for feedback. We now study the digital implementation of an observer that produces an estimate $\hat{\mathbf{x}}(k)$ of the state variables, given the measurement of the output $\mathbf{y}(k)$ and knowing the input $\mathbf{u}(k)$. This estimate is then used to obtain the control law. For reasons mentioned in connection with the observer design for continuous-time systems in section 8.6, we make use of the measured output to improve the estimation.

First we consider only single-input, single-output systems. Letting $\mathbf{H}$ be an $(n \times 1)$ column matrix, the plant difference equations are expressed by

$$\mathbf{x}(k+1) = \mathbf{\Phi}\mathbf{x}(k) + \mathbf{\Psi}u(k) + \mathbf{H}[y(k) - \mathbf{C}\mathbf{x}(k)] \tag{9.157}$$

Since $y(k) = \mathbf{C}\mathbf{x}(k)$, the plant equation is unchanged. Since $u(k)$ and $y(k)$ are known quantities, the estimator equation is expressed as

$$\hat{\mathbf{x}}(k+1) = \mathbf{\Phi}\hat{\mathbf{x}}(k) + \mathbf{\Psi}u(k) + \mathbf{H}[y(k) - \mathbf{C}\hat{\mathbf{x}}(k)] \tag{9.158}$$

Defining the estimator error as $\mathbf{e}(k) = \mathbf{x}(k) - \hat{\mathbf{x}}(k)$, the error-difference equation is obtained by subtracting Eq. (9.158) from Eq. (9.157). It follows that

$$\mathbf{e}(k+1) = (\mathbf{\Phi} - \mathbf{HC})\mathbf{e}(k) \tag{9.159}$$

The characteristic equation of the error becomes

$$|z\mathbf{I} - \mathbf{\Phi} + \mathbf{HC}| = 0 \tag{9.160}$$

We specify the n roots of this equation as $\beta_1, \ldots, \beta_n$ and solve for the n elements of $\mathbf{H}$ from the equation

$$|z\mathbf{I} - \mathbf{\Phi} + \mathbf{HC}| = (z - \beta_1) \cdots (z - \beta_n) \tag{9.161}$$

The observer matrix $\mathbf{H}$ can also be obtained in the case of high-order systems from the Ackermann formula as

$$\begin{aligned} \mathbf{H}^T &= \lfloor h_1 \quad h_2 \quad \cdots \quad h_n \rfloor \\ &= \lfloor 0 \quad 0 \quad \cdots \quad 0 \quad 1 \rfloor \mathbf{M}^{-1} \Delta_0(\mathbf{\Phi}) \end{aligned} \tag{9.162}$$

where $\mathbf{M}$ is the observabilty matrix defined by Eq. (9.128) and $\Delta_0(\mathbf{\Phi})$ is obtained from the right-hand side of Eq. (9.161) by replacing z by the matrix $\mathbf{\Phi}$ in the desired characteristic polynomial. We can obtain a unique solution for the estimator matrix $\mathbf{H}$ either from Eq. (9.161) or from Eq. (9.162) when the pair $(\mathbf{\Phi}, \mathbf{C})$ is observable.

The observer obtained in this manner is called the prediction estimator because the estimate $\hat{\mathbf{x}}(k+1)$ in Eq. (9.158) is one cycle ahead of the measurement $y(k)$. It has advantages in those cases where the computation time required to evaluate Eq. (9.158) is large compared to the sampling period. For the design of current estimators and reduced order estimators, we refer to Franklin and Powell (1980).

For continuous-time systems, as was already noted in section 8.6, a good rule is to choose the observer roots to be approximately equal to $-4s_i$, where $-s_i$ are the control roots. In the previous section, we have shown that in the case of digital control, the control roots are chosen by the mapping $z_i = e^{-Ts_i}$. Following this rule, we select the observer roots to be approximately given by the mapping $\beta_i = e^{-4Ts_i}$.

EXAMPLE 9.21

An observer is to be designed for the second-order digital control system of Example 9.19. For this system, the plant matrices have been obtained in Example 9.3 for a sampling $T = 0.1$ second. These are given by

$$\mathbf{\Phi} = \begin{bmatrix} 1 & 0.10 \\ 0 & 0.99 \end{bmatrix} \qquad \mathbf{\Psi} = \begin{bmatrix} 0.005 \\ 0.100 \end{bmatrix} \qquad \mathbf{C} = \lfloor 1 \quad 0 \rfloor \tag{9.163}$$

The observability matrix becomes

$$\mathbf{M} = \begin{bmatrix} \mathbf{C} \\ \mathbf{C\Phi} \end{bmatrix} = \begin{bmatrix} 1 & 0 \\ 1 & 0.1 \end{bmatrix}$$

The determinant of $\mathbf{M}$ is nonzero, and hence the pair $(\mathbf{\Phi}, \mathbf{C})$ is observable. The desired roots in the s-plane as given by Eq. (9.139) are $-s_1, -s_2 = -2.1 \pm j2.14$. The observer roots are selected as

$$\begin{aligned} \beta_1 &= \exp[(0.1)(4)(-2.1 + j2.14)] \\ &= 0.283 + j0.326 \\ \beta_2 &= 0.283 - j0.326 \end{aligned}$$

The desired characteristic equation becomes

$$(z - \beta_1)(z - \beta_2) = z^2 - 0.566z + 0.186 = 0 \tag{9.164}$$

We have

$$\mathbf{HC} = \begin{bmatrix} h_1 \\ h_2 \end{bmatrix} \lfloor 1 \quad 0 \rfloor = \begin{bmatrix} h_1 & 0 \\ h_2 & 0 \end{bmatrix} \tag{9.165}$$

Using the results from Eqs. (9.163) and (9.165), the characteristic equation is also given by

$$|z\mathbf{I} - \mathbf{\Phi} + \mathbf{HC}| = z^2 + (-1.99 + h_1)z + 0.99 - 0.99h_1 + 0.1h_2 = 0 \tag{9.166}$$

Matching the corresponding coefficients in Eqs. (9.164) and (9.166), we obtain $h_1 = 1.424$ and $h_2 = 6.058$.

The preceding observer design can be extended to multivariable systems by following a procedure similar to that of section 8.6.2. Let the observer matrix be

$$\mathbf{H} = \begin{bmatrix} h_1 \\ \vdots \\ h_n \end{bmatrix} \lfloor p_1 \cdots p_m \rfloor = \mathbf{hp}^T \tag{9.167}$$

where the m elements of $\mathbf{p}$ are chosen arbitrarily, and the n elements of $\mathbf{h}$ are obtained from the n observer roots that are specified. The observer characteristic equation becomes

$$|z\mathbf{I} - \mathbf{\Phi} + \mathbf{hp}^T\mathbf{C}| = 0 \tag{9.168}$$

For the existence of a solution to $\mathbf{h}$, the requirement is that the pair $(\mathbf{\Phi}, \mathbf{p}^T\mathbf{C})$ must be observable. We note that difficulties may arise in this connection.

The controller is considered as the combination of the control law and observer. Now, with the control law described by

$$u(k) = Nr(k) - \mathbf{K}\hat{\mathbf{x}}(k) \tag{9.169}$$

the closed-loop control system becomes

$$\mathbf{x}(k+1) = \mathbf{\Phi x}(k) - \mathbf{\Psi K}\hat{\mathbf{x}}(k) + \mathbf{\Psi} Nr(k)$$

Using the observer error $\mathbf{e}(k) = \mathbf{x}(k) - \hat{\mathbf{x}}(k)$, the preceding equation becomes

$$\mathbf{x}(k+1) = (\mathbf{\Phi} - \mathbf{\Psi K})\mathbf{x}(k) + \mathbf{\Psi K e}(k) + \mathbf{\Psi} Nr(k)$$

This equation can be combined with the observer error equation, Eq. (9.159), to yield

$$\begin{bmatrix} \mathbf{x}(k+1) \\ \mathbf{e}(k+1) \end{bmatrix} = \begin{bmatrix} \mathbf{\Phi} - \mathbf{\Psi K} & \mathbf{\Psi K} \\ \mathbf{0} & \mathbf{\Phi} - \mathbf{HC} \end{bmatrix} \begin{bmatrix} \mathbf{x}(k) \\ \mathbf{e}(k) \end{bmatrix} + \begin{bmatrix} \mathbf{\Psi} N \\ \mathbf{0} \end{bmatrix} r(k)$$

Its characteristic equation becomes

$$|z\mathbf{I} - \mathbf{\Phi} + \mathbf{\Psi K}|\,|z\mathbf{I} - \mathbf{\Phi} + \mathbf{HC}| = 0$$

The roots of this characteristic equation of the combined system consist of the sum of the control roots and the estimator roots. The control roots are unchanged from those obtained by assuming exact state feedback. Hence, just as in the continuous-time case, the control law and observer can be designed separately and then used jointly. A block diagram of the controlled system is shown in Fig. 9.23.

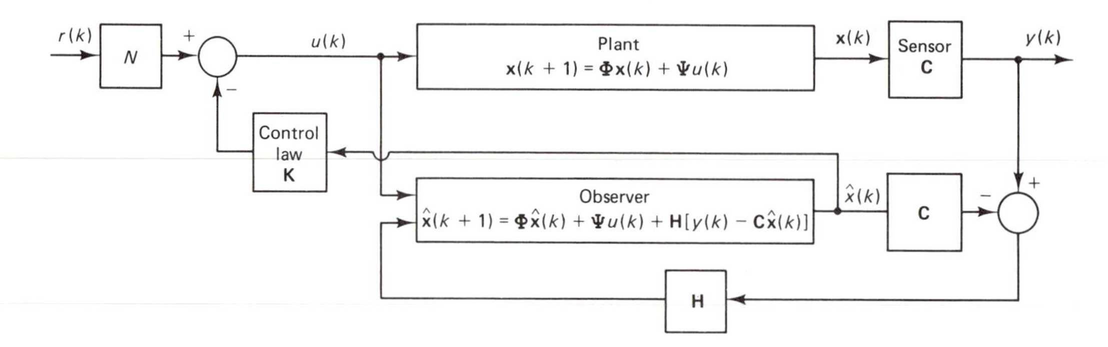

Figure 9.23 Block diagram of a controlled system.

9.10 SUMMARY

This chapter provides an analytical foundation for the use of digital computers in control applications. A digital control system consists of a plant modeled by differential equations, digital control elements modeled by difference equations, and the analog-to-digital and digital-to-analog converters that form the interface. We have modeled the interface as sampler and zero-order hold. It is shown that the sampling period must be carefully selected to satisfy the sampling theorem.

The plant differential equations have been converted to difference equations so that they are compatible with the controller model. We have also introduced the z-transform and used an alternate representation in the form of discrete transfer functions. We have studied the digital implementation of a PID family of control laws for systems with output feedback. Our approach has been to first design the control law and compensator, if one is required, for continuous-time systems. The control law and compensator are then converted to discrete-time models for digital implementation.

Two methods of investigating the response of closed-loop digital control systems are covered based on the use of state-difference equations and discrete transfer functions with inverse z-transform. We have discussed the stability of discrete-time systems and shown how the Routh criterion can be extended for their stability investigation. After a discussion of controllability and observability of discrete-time systems, the final topic covered is the digital implementation of a controller with state feedback and an observer that provides an estimate of the state variables for feedback.

We note that the application of these techniques of designing a controller with state feedback and an observer for multivariable systems may not always be possible.

This will happen when the equivalent single-input, single-output system is either not controllable or observable. In that case, discrete optimal control theory can be used for the design, but this theory is beyond our scope.

REFERENCES

DESILVA, C. W. (1985, June). "Motion Sensors in Industrial Robots." *Mechanical Engineering*, ASME.

FRANKLIN, G. F., and POWELL, J. D. (1980). *Digital Control of Dynamic Systems.* Reading, MA: Addison-Wesley.

KUO, B. C. (1980). *Digital Control Systems.* New York: Holt, Rinehart and Winston.

OWENS, D. H. (1981). *Multivariable and Optimal Systems.* New York: Academic Press.

PEATMAN, J. B. (1972). *The Design of Digital Systems.* New York: McGraw-Hill.

TAKAHASHI, Y., RABINS, M. J., and AUSLANDER, D. M. (1970). *Control and Dynamic Systems.* Reading, MA: Addison-Wesley.

PROBLEMS

9.1. Determine appropriate sampling periods for the following time functions:
(a) $f(t) = 12\,e^{-t} - 4\,e^{-2t}$
(b) $f(t) = 10(1 - t)\,e^{-t}$

9.2. Two sinusoidal signals $f_1(t) = \sin 10t$ and $f_2(t) = \sin 30t$ are both sampled at a frequency of $\omega = 40$ rad/s. List the frequencies present in $F_1^*(j\omega)$ and $F_2^*(j\omega)$, and explain why these frequencies are identical.

9.3. Obtain the state-difference equations for the plant of the hydraulic-control system whose system matrices are given in Problem 8.7. Use the procedure of section 9.3 with zero-order hold. Let the sampling period $T = 0.05$ second.

9.4. The plant of the electrical position-control system of Example 3.4 is shown in Fig. 8.3 of Example 8.6. The system matrices are described by Eq. (8.28). Obtain the state-difference equations of this plant with the sampling period $T = 0.05$ second.

9.5. Solve Problem 9.3 for the plant of the system whose matrices are given in Problem 8.6.

9.6. Solve Problem 9.3 for the plant of the system whose matrices are given in Problem 8.9.

9.7. Solve Problem 9.3 using a digital computer program.

9.8. Solve Problem 9.4 using a digital computer program.

9.9. Obtain the discrete transfer function of the plant of Problem 9.3 by using Eq. (9.51).

9.10. Obtain the discrete transfer function of the plant of Problem 9.4 by using Eq. (9.51).

9.11. Solve Problem 9.9 using a digital computer program.

9.12. Solve Problem 9.10 using a digital computer program.

9.13. Consider the plant of the hydraulic-control system of Problem 9.3. Its discrete transfer function is obtained from the solution of Problem 9.9. A digital PI controller is to be designed for this system. For the analog controller, the gains are chosen such that $k_p = 0.5k_i$ for stability and $k_i = 1.2$ for bandwidth.

(a) For these values of the controller gains, obtain the discrete transfer function of the digital controller represented by Eq. (9.62).

(b) Obtain the closed-loop discrete transfer function of Eq. (9.73).

(c) Investigate the stability of this closed-loop digital control system.

(d) Obtain the response of the closed-loop system by the inverse z-transform when all initial conditions are zero and the command input $r(t)$ is a unit step function.

9.14. Consider the plant of the electrical position-control system of Problem 9.4. Its discrete transfer function is obtained from the solution of Problem 9.10. A digital PI controller is to be designed for this system.

(a) Obtain the values of the controller gains k_p and k_i, assuming analog controller and using the method of Ziegler and Nichols discussed in section 7.6.

(b) For these values of the controller gains, obtain the discrete transfer function of the digital controller represented by Eq. (9.62).

(c) Obtain the closed-loop discrete transfer function of Eq. (9.73).

(d) Investigate the stability of this closed-loop digital control system.

(e) Obtain the response of the closed-loop system by the inverse z-transform when all initial conditions are zero and the command input $r(t)$ is a unit step function.

9.15. In Problem 7.13, a proportional controller and a compensator are to be designed for the analog control system of Fig. P7.13. To meet the specifications of the analog control system, let $k_p = 1.25$ and $\tau = 0.31$. With these values, the proportional controller and compensator are to be implemented digitally as shown in Fig. P9.15. Let the sampling period $T = 0.1$ second.

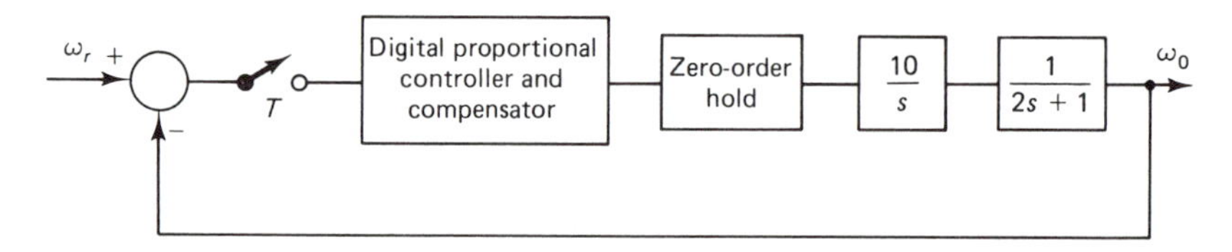

Figure P9.15 Digital control system with proportional control and compensator.

(a) Obtain the discrete transfer function of the controller and compensator.

(b) Obtain the discrete transfer function of the actuator and plant and then the discrete transfer function of the overall closed-loop digital control system.

(c) Investigate the stability of this closed-loop digital control system.

(d) Obtain the response of the closed-loop system by the inverse z-transform when all initial conditions are zero and the input $\omega_r(t)$ is a unit step function.

9.16. Investigate the controllability and observability of the following open-loop system. In case it is not controllable and/or observable, determine its discrete transfer function and determine whether there is pole-zero cancellation.

$$\mathbf{x}(k+1) = \begin{bmatrix} -1 & 2 \\ 0 & -5 \end{bmatrix} \mathbf{x}(k) + \begin{bmatrix} 1 \\ 0 \end{bmatrix} u(k)$$

$$y(k) = \lfloor 1 \quad 1 \rfloor \mathbf{x}(k)$$

9.17. Investigate the controllability and observability of the open-loop plant of Problem 9.3.

9.18. Solve Problem 9.17 for the plant of Problem 9.4.

9.19. Solve Problem 9.17 for the plant of Problem 9.5.

9.20. Solve Problem 9.17 for the plant of Problem 9.6.

9.21. Obtain the state feedback gain matrix **K** for the discrete state equations of the plant of the hydraulic system of Problem 9.3. The desired root locations of the closed-loop analog control system in the s-plane are at $-20.79 \pm j21.21$. First obtain the corresponding desired root locations in the z-plane.

9.22. It is desired to convert the type 0 system of Problem 9.21 to type 1 by using state feedback with PI control as discussed in section 9.8.2. The desired root locations of the corresponding analog system are given in Problem 8.8. Obtain the gain matrix of the state feedback control law.

9.23. Obtain the state feedback gain matrix for the discrete state equations of the plant of the electrical positioning system of Problem 9.4. The desired root locations for the corresponding analog system are discussed in Example 8.6 and given by Eq. (8.30).

9.24. Obtain the state feedback gain matrix for the discrete state equations of the multivariable plant of Problem 9.5. The desired root locations for the corresponding analog system, as discussed in Problem 8.10, are located at -4, $-8 \pm j16$.

9.25. Express the state-difference equations of the closed-loop system obtained from the solution of Problem 9.21 in the form of Eqs. (9.95) and (9.96). Obtain the response of this closed-loop system when all initial conditions are zero and the command input $r(t)$ is a unit step function. A digital computer program may be used to solve this problem.

9.26. Solve Problem 9.25 for the closed-loop system obtained from the solution of Problem 9.22.

9.27. Solve Problem 9.25 for the closed-loop system obtained from the solution of Problem 9.23.

9.28. Design an observer for the system of Problem 9.21. Let the roots of the observer characteristic equation be located at e^{-4Ts_i} in the z-plane, where $-s_i$ are the control roots of the corresponding analog control system in the s-plane.

9.29. Solve Problem 9.28 for the system of Problem 9.22.

9.30. Solve Problem 9.28 for the system of Problem 9.23.

9.31. Solve Problem 9.28 for the system of Problem 9.24.

Footnote

[1]It can be shown that this condition is also necessary by contradiction.

Appendix A

Laplace Transformation

The Laplace transformation is one of the transform methods that can be used advantageously for the solution of linear differential equations. We are concerned here with the transformation of functions of time and their time derivatives into functions of a complex variable s. It is then possible to manipulate the functions of s algebraically to facilitate a solution. The solution as a function of time is then obtained by taking the inverse Laplace transformation.

A.1 DEFINITION OF LAPLACE TRANSFORM

Given a function of time $f(t)$ specified for $t \geq 0$, its Laplace transform is defined by

$$\mathscr{L}[f(t)] = F(s) = \int_0^\infty f(t)\, e^{-st}\, dt \tag{A.1}$$

where s is a complex variable, which is also called the Laplace variable, that is, $s = \sigma + j\omega$. After performing the integration and taking the limits, the result is a function of s that is denoted by $F(s)$. For the existence of the Laplace transform of $f(t)$, it has to satisfy the following two conditions:

1. $f(t)$ must be integrable in the interval

$$0 \leq t < \infty$$

2. For some finite real $\sigma > -\infty$, the integral

$$\int_0^\infty |f(t)|\, e^{-\sigma t}\, dt < \infty$$

We note that the second condition requires that $f(t)$ be of exponential order, that is, $|f(t)| < e^{\sigma_1 t}$, such that as $t \to \infty$, $\lim f(t)\, e^{-\sigma t} \to 0$ for $\sigma > \sigma_1$. We can include functions such that as $t \to \infty$, $\lim f(t) \to \infty$, provided $f(t)$ does not grow faster than $e^{\sigma_1 t}$ for some finite real σ_1. For example, the order of $f(t) = e^{t^2}$ is higher than that of an exponential function. In a Fourier transformation, $s = j\omega$ and $\sigma = 0$. For the existence of a Fourier transform of $f(t)$, it must therefore satisfy a more stringent condition that as $t \to \infty$, $\lim f(t) \to 0$. This condition also clarifies the reason why the stability must be assured before substituting $j\omega$ for s for the frequency-response analysis in Chapter 5.

The transformation defined by Eq. (A.1) is called a one-sided Laplace transformation because time extends from 0 to ∞ and $f(t)$ for $t < 0$ is considered to be zero. When the lower limit of integration in Eq. (A.1) is replaced by $-\infty$, the transformation is called a two-sided Laplace transform. In causal systems, when the input is applied for $t \geq 0$, the system does not respond in anticipation for $t < 0$ and the past history results in the initial conditions. The two-sided Laplace transform is necessary for noncausal systems. Since we are concerned here only with causal systems, the one-sided Laplace transform is adequate for our purpose.

A.2 LAPLACE TRANSFORM OF ELEMENTARY FUNCTIONS

In the following, we derive the Laplace transform of some elementary functions by using Eq. (A.1). Table A.1 gives the Laplace transforms. For a more extensive table, the reader may consult Churchill (1958), Spiegel (1965), and Kaplan (1962).

TABLE A.1 LAPLACE AND z-TRANSFORMS*

Laplace transform $F(s)$	Time function $f(t)$	z-transform $F(z)$
$\dfrac{1}{s}$	$u(t)$ (unit step function)	$\dfrac{z}{z-1}$
$\dfrac{1}{s^2}$	t	$\dfrac{Tz}{(z-1)^2}$
$\dfrac{1}{s^3}$	$\dfrac{t^2}{2}$	$\dfrac{T^2 z(z+1)}{2(z-1)^3}$
$\dfrac{1}{s^4}$	$\dfrac{1}{3!}t^3$	$\dfrac{T^3}{6}\dfrac{z(z^2+4z+1)}{(z-1)^4}$
$\dfrac{1}{s^m}$	$\dfrac{1}{(m-1)!}t^{m-1}$	$\lim\limits_{a\to 0}\dfrac{(-1)^{m-1}}{(m-1)!}\dfrac{\partial^{m-1}}{\partial a^{m-1}}\dfrac{z}{z-e^{-aT}}$
$\dfrac{1}{s+a}$	e^{-at}	$\dfrac{z}{z-e^{-aT}}$

*$F(s)$ is the Laplace transform of $f(t)$ and $F(z)$ is the z-transform of $f(nt)$. It is assumed that $f(t) = 0$ for $t < 0$.

TABLE A.1 (CONTINUED)

Laplace transform $F(s)$	Time function $f(t)$	z-transform $F(z)$
$\dfrac{1}{(s+a)^2}$	$t\,e^{-at}$	$\dfrac{Tz\,e^{-aT}}{(z-e^{-aT})^2}$
$\dfrac{1}{(s+a)^3}$	$\dfrac{1}{2}t^2\,e^{-at}$	$\dfrac{T^2}{2}e^{-aT}\dfrac{z(z+e^{-aT})}{(z-e^{-aT})^3}$
$\dfrac{1}{(s+a)^m}$	$\dfrac{1}{(m-1)!}t^{m-1}\,e^{-at}$	$\dfrac{(-1)^{m-1}}{(m-1)!}\dfrac{\partial^{m-1}}{\partial a^{m-1}}\dfrac{z}{z-e^{-aT}}$
$\dfrac{a}{s(s+a)}$	$1-e^{-at}$	$\dfrac{z(1-e^{-aT})}{(z-1)(z-e^{-aT})}$
$\dfrac{a}{s^2(s+a)}$	$t-\dfrac{1-e^{-at}}{a}$	$\dfrac{z[(aT-1+e^{-aT})z+(1-e^{-aT}-aT\,e^{-aT})]}{a(z-1)^2(z-e^{-aT})}$
$\dfrac{b-a}{(s+a)(s+b)}$	$e^{-at}-e^{-bt}$	$\dfrac{(e^{-aT}-e^{-bT})z}{(z-e^{-aT})(z-e^{-bT})}$
$\dfrac{s}{(s+a)^2}$	$(1-at)\,e^{-at}$	$\dfrac{z[z-e^{-aT}(1+aT)]}{(z-e^{-aT})^2}$
$\dfrac{a^2}{s(s+a)^2}$	$1-(1+at)\,e^{-at}$	$\dfrac{z}{z-1}-\dfrac{z}{z-e^{-aT}}-\dfrac{aT\,e^{-aT}z}{(z-e^{-aT})^2}$
$\dfrac{(b-a)s}{(s+a)(s+b)}$	$b\,e^{-bt}-a\,e^{-at}$	$\dfrac{z[z(b-a)-(be^{-aT}-ae^{-bT})]}{(z-e^{-aT})(z-e^{-bT})}$
$\dfrac{a}{s^2+a^2}$	$\sin at$	$\dfrac{z\sin aT}{z^2-(2\cos aT)z+1}$
$\dfrac{s}{s^2+a^2}$	$\cos at$	$\dfrac{z(z-\cos aT)}{z^2-(2\cos aT)z+1}$
$\dfrac{b}{(s+a)^2+b^2}$	$e^{-at}\sin bt$	$\dfrac{z\,e^{-aT}\sin bT}{z^2-2\,e^{-aT}(\cos bT)z+e^{-2aT}}$
$\dfrac{s+a}{(s+a)^2+b^2}$	$e^{-at}\cos bt$	$\dfrac{z(z-e^{-aT}\cos bT)}{z^2-2\,e^{-aT}(\cos bT)z+e^{-2aT}}$
$\dfrac{a^2+b^2}{s[(s+a)^2+b^2]}$	$1-e^{-at}\left(\cos bt+\dfrac{a}{b}\sin bt\right)$	$\dfrac{z(Az+B)}{(z-1)(z^2-2\,e^{-aT}(\cos bT)z+e^{-2aT})}$ $A=1-e^{-aT}\cos bT-\dfrac{a}{b}e^{-aT}\sin bT$ $B=e^{-2aT}+\dfrac{a}{b}e^{-aT}\sin bT-e^{-aT}\cos bT$
$\dfrac{1}{s(s+a)(s+b)}$	$\dfrac{1}{ab}+\dfrac{e^{-at}}{a(a-b)}+\dfrac{e^{-bt}}{b(b-a)}$	$\dfrac{(Az+B)z}{(z-e^{-aT})(z-e^{-bT})(z-1)}$ $A=\dfrac{b(1-e^{-aT})-a(1-e^{-bT})}{ab(b-a)}$ $B=\dfrac{a\,e^{-aT}(1-e^{-bT})-b\,e^{-bT}(1-e^{-aT})}{ab(b-a)}$

1. Heavyside Unit Step Function. Consider the function

$$f(t) = 1 \qquad t \geq 0$$
$$= 0 \qquad t < 0$$

From Eq. (A.1), we obtain

$$F(s) = \int_0^\infty 1\, e^{-st}\, dt = -\frac{1}{s} e^{-st}\Big|_0^\infty = -\frac{0}{s} + \frac{1}{s} = \frac{1}{s} \tag{A.2}$$

provided s is restricted such that $\lim e^{-st} \to 0$ as $t \to \infty$.

2. Dirac Unit Delta Function. Let $f(t) = \delta(t)$, where $\delta(t) = 0$ for $t \neq 0$, and $\int_0^\infty \delta(t)\, dt = 1$. We get

$$F(s) = \int_0^\infty \delta(t)\, e^{-st}\, dt = \int_0^{0^+} \delta(t)\, e^0\, dt = 1 \tag{A.3}$$

3. Ramp Function. Consider a function $f(t)$, where

$$f(t) = t \qquad t \geq 0$$
$$= 0 \qquad t < 0$$

The Laplace transform is obtained as

$$F(s) = \int_0^\infty t\, e^{-st}\, dt = \left(-\frac{t\, e^{-st}}{s} - \frac{e^{-st}}{s^2} \right)\Big|_{t=0}^\infty = \frac{1}{s^2} \tag{A.4}$$

Similarly, we can show that if $f(t) = t^n$ for $t \geq 0$, where n is a positive integer, and $f(t) = 0$ for $t < 0$, we get

$$F(s) = n!/s^{n+1} \tag{A.5}$$

4. Exponential Function. Let

$$f(t) = e^{-at} \qquad t \geq 0$$
$$= 0 \qquad t < 0$$

We get,

$$F(s) = \int_0^\infty e^{-at}\, e^{-st}\, dt = \frac{1}{s+a} \tag{A.6}$$

5. Sine Function. Consider a sinusoidal function

$$f(t) = \sin \omega t \qquad t \geq 0$$
$$= 0 \qquad t < 0$$

Its Laplace transform is obtained as

$$\begin{aligned} F(s) &= \int_0^\infty (\sin \omega t)\, e^{-st}\, dt \\ &= \int_0^\infty \left(\frac{e^{j\omega t} - e^{-j\omega t}}{2j} \right) e^{-st}\, dt \qquad \text{(A.7)} \\ &= \frac{\omega}{s^2 + \omega^2} \end{aligned}$$

Similarly,

$$\mathscr{L}[\cos \omega t] = \frac{s}{s^2 + \omega^2} \tag{A.8}$$

6. Laplace Transform of Time Derivatives of a Function. The Laplace transform of df/dt is defined by

$$\mathscr{L}\left(\frac{df}{dt}\right) = \int_0^\infty \left(\frac{df}{dt}\right) e^{-st}\, dt \tag{A.9}$$

Using integration by parts and noting that $\int u\, dv = uv - \int v\, du$, we let $u = e^{-st}$ and $v = f(t)$. Hence, we obtain

$$\begin{aligned}\mathscr{L}\left(\frac{df}{dt}\right) &= \int_0^\infty \left(\frac{df}{dt}\right) e^{-st}\, dt \\ &= f(t)\, e^{-st}\Big|_0^\infty + s\int_0^\infty f(t)\, e^{-st}\, dt \\ &= sF(s) - f(0)\end{aligned} \tag{A.10}$$

Similarly, we obtain

$$\mathscr{L}\left(\frac{d^2f}{dt^2}\right) = s^2F(s) - sf(0) - \frac{df}{dt}(0) \tag{A.11}$$

$$\mathscr{L}\left(\frac{d^nf}{dt^n}\right) = s^nF(s) - s^{n-1}f(0) - \cdots - \frac{d^{n-1}f}{dt^{n-1}}(0) \tag{A.12}$$

EXAMPLE A.1

Consider $f(t) = \sin \omega t$ and $df/dt = \omega \cos \omega t$. From Eq. (A.10), we obtain

$$\mathscr{L}(\omega \cos \omega t) = s\mathscr{L}(\sin \omega t) - 0 = \frac{s\omega}{s^2 + \omega^2}$$

where we have used Eq. (A.7). This result can also be verified from Eq. (A.8).

A.3 SOME THEOREMS OF LAPLACE TRANSFORMATION

The application of Laplace transformation is often simplified by the use of certain of its properties. Some of these important properties are expressed by the following theorems.

1. Linearity of Laplace Transformation. We recall from Chapter 1 that for any transformation, operator, or mapping to be classified as linear, it has to satisfy both the additive property and the homogeneous property. It can be easily verified from the defining equation, Eq. (A.1), that

$$\mathscr{L}[f_1(t) + f_2(t)] = F_1(s) + F_2(s) \tag{A.13}$$

and

$$\mathscr{L}[kf(t)] = kF(s) \tag{A.14}$$

where k is a constant. Hence, it is verified that the Laplace transformation is a linear transformation.

2. Shifting Theorem. Consider a function $f(t)$ that is delayed by time T to obtain $f(t - T)$, where $f(t - T) = 0$ for $t < T$. The delay time T is also called dead time or transportation lag. Let $t - T = t'$ and hence $dt = dt'$. The Laplace transformation of this function is obtained as

$$\begin{aligned} \mathscr{L}[f(t-T)] &= \int_0^\infty f(t-T)\, e^{-st}\, dt \\ &= e^{-Ts} \int_0^\infty f(t')\, e^{-st'}\, dt' \\ &= e^{-Ts} F(s) \end{aligned} \tag{A.15}$$

The lower limit of integration in the preceding equation, which is $-T$, has been replaced by 0 since $f(t') = 0$ for $t' < 0$.

EXAMPLE A.2

Consider the function $f(t) = t^2$ whose Laplace transform is $2/s^3$. Let this function be delayed by 2 seconds. Then,

$$\mathscr{L}[(t-2)^2] = e^{-2s}\left(\frac{2}{s^3}\right)$$

---

3. Final-Value Theorem. This theorem is useful for the evaluation of $\lim f(t)$ as $t \to \infty$ directly from $F(s)$. The Laplace transform of a derivative as shown by Eq. (A.10) is given by

$$\int_0^\infty \frac{df}{dt} e^{-st}\, dt = sF(s) - f(0)$$

In this expression, we let $s = 0$ and obtain

$$\begin{aligned} \int_0^\infty \frac{df}{dt}\, dt &= \lim_{s\to 0} sF(s) - f(0) \\ \lim_{t\to\infty} f(t) - f(0) &= \lim_{s\to 0} sF(s) - f(0) \end{aligned} \tag{A.16}$$

Hence,

$$\lim_{t\to\infty} f(t) = \lim_{s\to 0} sF(s) \tag{A.17}$$

This result is known as the final-value theorem. It is important to note that this result is valid only when all poles of $sF(s)$ have negative real parts; otherwise, the integral on the left-hand side of Eq. (A.16) does not converge.

EXAMPLE A.3

Consider the function $f(t) = e^{3t}$, where $\lim f(t) \to \infty$ as $t \to \infty$. To use the final-value theorem, we have

$$F(s) = \frac{1}{s-3} \qquad sF(s) = \frac{s}{s-3}$$

Hence, $\lim sF(s) = 0$ as $s \to 0$. This result is not correct since $sF(s)$ has a pole at $s = 3$ and the final value theorem is not applicable.

EXAMPLE A.4

Let $f(t) = \sin \omega t$, which does not have a unique final value as $t \to \infty$. To use the final value theorem, we obtain

$$F(s) = \frac{\omega}{s^2 + \omega^2} \qquad sF(s) = \frac{s\omega}{s^2 + \omega^2}$$

Then, $\lim sF(s) = 0$ as $s \to 0$. This result is also not correct since $sF(s)$ has poles on the imaginary axis at $\pm j\omega$ and does not satisfy the requirement that all poles have negative real parts.

---

4. Convolution Theorem. Let $F(s)$ and $G(s)$ be the Laplace transforms of $f(t)$ and $g(t)$, respectively. Then, it can be shown that

$$\mathscr{L}^{-1}[F(s)G(s)] = \int_0^t f(t-t')g(t')\, dt' \qquad \text{(A.18)}$$

The left-hand side of Eq. (A.18) denotes the inverse Laplace transform and the right-hand side is called the convolution integral, which is also denoted by $f * g$. To prove this theorem, we take the Laplace transform of the right-hand side of Eq. (A.18) and get

$$\int_0^\infty e^{-st}\, dt \int_0^t f(t-t')g(t')\, dt'$$

Since $f(t-t') = 0$ for $t - t' < 0$, that is, for $t' > t$, the upper limit of integration t in Eq. (A.18) may be replaced by ∞. Then, interchanging the order of integration, we obtain

$$\mathscr{L}\left[\int_0^t f(t-t')g(t')\, dt'\right] = \int_0^\infty g(t')\, dt' \int_0^\infty f(t-t')\, e^{-st}\, dt$$

Now,

$$\int_0^\infty f(t-t')\, e^{-st}\, dt = e^{-st'}F(s)$$

Hence, we obtain

$$\mathscr{L}\left[\int_0^t f(t-t')g(t')\, dt'\right] = \left[\int_0^\infty g(t')\, e^{-st'}\, dt'\right] F(s)$$

$$= G(s)F(s)$$

This result verifies the convolution theorem given by Eq. (A.18). It can be shown that $f * g = g * f$.

EXAMPLE A.5

It is known that

$$\mathscr{L}(e^{-2t}) = \frac{1}{s+2} \qquad \mathscr{L}(e^{-3t}) = \frac{1}{s+3}$$

Using the convolution integral, we obtain

$$\begin{aligned}\mathscr{L}^{-1}\left(\frac{1}{s+2}\right)\left(\frac{1}{s+3}\right) &= \int_0^t e^{-2t'} e^{-3(t-t')}\, dt' \\ &= e^{-3t}\int_0^t e^{t'}\, dt' \\ &= e^{-3t}(e^t - 1) = e^{-2t} - e^{-3t}\end{aligned}$$

A.4 INVERSE LAPLACE TRANSFORMATION

The operation of obtaining the function $f(t)$ from its Laplace transform $F(s)$ is called the inverse Laplace transformation. It is obtained by using the complex inversion integral

$$f(t) = \mathscr{L}^{-1}[F(s)] = \frac{1}{2\pi j}\int_{c-j\infty}^{c+j\infty} F(s)\, e^{st}\, ds \qquad t \geq 0 \qquad \text{(A.19)}$$

and $f(t)$ is considered to be zero for $t < 0$. The integration is to be performed along the line $s = c$ in the complex s-plane, where c is a real constant that is greater than the real part of all singularities of $F(s)$. In practice, the integral in Eq. (A.19) is evaluated by considering an integral along a closed contour and requires the use of the residue theorem from the theory of complex variables.

In our applications, $F(s)$ has only a finite number of poles and it is of the form $F(s) = P(s)/Q(s)$, where $P(s)$ and $Q(s)$ are polynomials in s with the degree of $P(s)$ not exceeding that of $Q(s)$. Hence, we can expand $F(s)$ as a sum of elementary functions by a partial fraction expansion and obtain the inverse of each elementary function by inspection or use of tables. In the more general case where $F(s)$ consists of transcendental functions and has an infinite number of singularities, the use of Eq. (A.19) cannot be avoided.

EXAMPLE A.6

It is desired to obtain the inverse Laplace transform of

$$Y(s) = \frac{8s + 28}{(s+3)^2(s+5)}$$

Expanding in a partial fraction expansion, we obtain

$$Y(s) = \frac{k_1}{(s+3)^2} + \frac{k_2}{s+3} + \frac{k_3}{s+5}$$

The constants of the partial fraction expansion are evaluated as follows:

$$k_1 = \lim_{s \to -3} \left(\frac{8s+28}{s+5} \right) = 2$$

$$k_2 = \lim_{s \to -3} \frac{d}{ds} \left(\frac{8s+28}{s+5} \right) = \lim_{s \to -3} \frac{8(s+5) - (8s+28)}{(s+5)^2} = 3$$

$$k_3 = \lim_{s \to -5} \left(\frac{8s+28}{(s+3)^2} \right) = -3$$

Thus,

$$Y(s) = \frac{2}{(s+3)^2} + \frac{3}{s+3} + \frac{-3}{s+5}$$

Using Table A.1, we obtain the inverse Laplace transform as

$$y(t) = 2t\,e^{-3t} + 3\,e^{-3t} - 3\,e^{-5t}$$

EXAMPLE A.7

It is desired to obtain the inverse Laplace transform of

$$Y(s) = \frac{20}{s(s^2 + 2s + 5)}$$

Expanding in a partial fraction expansion, we obtain

$$Y(s) = \frac{k_1}{s} + \frac{k_2}{s+1-2j} + \frac{k_3}{s+1+2j}$$

where

$$k_1 = \lim_{s \to 0} \frac{20}{s^2 + 2s + 5} = 4$$

$$k_2 = \lim_{s \to -1+2j} \left(\frac{20}{s(s+1+2j)} \right) = \frac{10}{2j(-1+2j)} = \left(\frac{2\sqrt{5}}{2j} \right) e^{-j\phi}$$

where

$$\phi = \tan^{-1} 2/-1 = 116.56° = 2.03 \text{ rad}$$

Constant $k_3 = \bar{k}_2$, that is, k_3 is the complex conjugate of k_2. Hence, the inverse Laplace transform is given by

$$y(t) = 4 + 2\sqrt{5}\,e^{-t} \sin(2t - 2.03)$$

REFERENCES

CHURCHILL, R. V. (1958). *Operational Mathematics.* New York: McGraw-Hill.

KAPLAN, W. (1962). *Operational Methods for Linear Systems.* Reading, MA: Addison-Wesley.

SPIEGEL, M. R. (1965). *Laplace Transforms.* Schaum's Outline Series. New York: McGraw-Hill.

Appendix B

Matrix Algebra

B.1 DEFINITIONS

A matrix is defined as a rectangular array of numbers or functions arranged in rows and columns in the form

$$\mathbf{A} = \begin{bmatrix} a_{11} & a_{12} & \cdots & a_{1m} \\ a_{21} & a_{22} & \cdots & a_{2m} \\ \vdots & \vdots & \vdots & \vdots \\ a_{n1} & a_{n2} & \cdots & a_{nm} \end{bmatrix} \tag{B.1}$$

The numbers or functions a_{ij} are called elements of the matrix. The first subscript i denotes the row position of the element and the second subscript j the column position. A matrix with n rows and m columns is called an $n \times m$ matrix or a matrix of order $n \times m$. When the number of rows is equal to the number of columns, i.e., $n = m$, the matrix is called a square matrix. A square matrix having all its elements zero except those on the principal diagonal is called a diagonal matrix. For example, a 3×3 diagonal matrix $\mathbf{A}$ is described by

$$\mathbf{A} = \begin{bmatrix} a_{11} & 0 & 0 \\ 0 & a_{22} & 0 \\ 0 & 0 & a_{33} \end{bmatrix} \tag{B.2}$$

When all the elements along the principal diagonal of a diagonal matrix have the value of unity, the matrix is called a unit matrix or identity matrix $\mathbf{I}$. Given a matrix $\mathbf{A}$, its transpose is a matrix denoted by $\mathbf{A}^T$ or $\mathbf{A}'$, which is obtained by interchanging the rows and columns of the original matrix $\mathbf{A}$. For example, the transpose of the matrix defined by Eq. (B.1) is given by

$$\mathbf{A}' = \begin{bmatrix} a_{11} & a_{21} & \cdots & a_{n1} \\ a_{12} & a_{22} & \cdots & a_{n2} \\ \vdots & \vdots & \vdots\vdots\vdots & \vdots \\ a_{1m} & a_{2m} & \cdots & a_{nm} \end{bmatrix} \tag{B.3}$$

A square matrix whose elements are symmetrical with respect to the main diagonal such that $a_{ij} = a_{ji}$ is called a symmetric matrix. For example, a matrix given by

$$\mathbf{A} = \begin{bmatrix} 2 & -3 & 8 \\ -3 & 4 & -1 \\ 8 & -1 & 7 \end{bmatrix} \tag{B.4}$$

is a 3×3 symmetric matrix. It is seen that a symmetric matrix is equal to its transpose. An antisymmetric or skew symmetric matrix is a square matrix whose elements are symmetric with respect to the main diagonal but with opposite signs, that is, $a_{ij} = -a_{ji}$ for $i \neq j$. A triangular matrix is a square matrix that has all zero elements either above or below the main diagonal.

An $n \times 1$ matrix that is comprised of only one column is called a column matrix or a column vector and written as

$$\mathbf{b} = \begin{bmatrix} b_1 \\ b_2 \\ \vdots \\ b_n \end{bmatrix} \tag{B.5}$$

A $1 \times n$ matrix having a single row is called a row matrix or a row vector and is expressed as

$$\mathbf{c}' = \lfloor c_1 \quad c_2 \quad \cdots \quad c_n \rfloor \tag{B.6}$$

Thus, a row vector is denoted as the transpose of a column vector.

B.2 ADDITION AND MULTIPLICATION OF MATRICES

A matrix can be added to another or substracted from another only when the two matrices have the same order, that is, when they have equal number of rows and equal number of columns. If

$$\mathbf{C} = \mathbf{A} + \mathbf{B} \qquad \text{then} \qquad c_{ij} = a_{ij} + b_{ij} \tag{B.7}$$

$$\mathbf{D} = \mathbf{A} - \mathbf{B} \qquad \text{then} \qquad d_{ij} = a_{ij} - b_{ij} \tag{B.8}$$

For example, consider $\mathbf{D} = s\mathbf{I} - \mathbf{A}$, where s is a scalar, $\mathbf{I}$ is a 2×2 identity matrix, and $\mathbf{A}$ is a 2×2 matrix. Then

$$\mathbf{D} = s\mathbf{I} - \mathbf{A} = \begin{bmatrix} s - a_{11} & -a_{12} \\ -a_{21} & s - a_{22} \end{bmatrix} \tag{B.9}$$

The following rules govern matrix additions:

1. Commutative law: $\mathbf{A} + \mathbf{B} = \mathbf{B} + \mathbf{A}$
2. Associative law: $\mathbf{A} + (\mathbf{B} + \mathbf{C}) = (\mathbf{A} + \mathbf{B}) + \mathbf{C}$

When a matrix is multiplied by a scalar s, each element of the matrix is multiplied by the scalar so that

$$\mathbf{A}s = s\mathbf{A} = \mathbf{B} \qquad \text{where } b_{ij} = sa_{ij} \tag{B.10}$$

The multiplication **AB** of two matrices is possible only when the number of columns of **A** is equal to the number of rows of **B**. Then, matrices **A** and **B** are said to be conformable. In the multiplication **AB**, **A** is said to be postmultiplied by **B** or **B** is said to be premultiplied by **A**. If **A** is an $n \times m$ matrix and **B** is $m \times p$, the product matrix **C** is of order $n \times p$, that is,

$$\underset{(n \times p)}{\mathbf{C}} = \underset{(n \times m)}{\mathbf{A}} \times \underset{(m \times p)}{\mathbf{B}} \tag{B.11}$$

An element c_{ij} of **C** is found by multiplying the elements of the ith row of **A** with the corresponding elements of the jth column of **B** and summing the products. Thus,

$$c_{ij} = a_{i1}b_{1j} + a_{i2}b_{2j} + \cdots + a_{im}b_{mj} \tag{B.12}$$

For example,

$$\begin{aligned} \mathbf{y} = \mathbf{C}\mathbf{x} &= \begin{bmatrix} c_{11} & c_{12} & c_{13} \\ c_{21} & c_{22} & c_{23} \end{bmatrix} \begin{bmatrix} x_1 \\ x_2 \\ x_3 \end{bmatrix} \\ &= \begin{bmatrix} c_{11}x_1 + c_{12}x_2 + c_{13}x_3 \\ c_{21}x_1 + c_{22}x_2 + c_{23}x_3 \end{bmatrix} \end{aligned} \tag{B.13}$$

Here, the orders of **C**, **x**, and **y** are 2×3, 3×1, and 2×1, respectively. We note that $\mathbf{y} = \mathbf{C}\mathbf{x}$ is a compact matrix equation representation of the simultaneous linear algebraic equations

$$\begin{aligned} c_{11}x_1 + c_{12}x_2 + c_{13}x_3 &= y_1 \\ c_{21}x_1 + c_{22}x_2 + c_{23}x_3 &= y_2 \end{aligned} \tag{B.14}$$

Consider, for example, two coupled, first-order, ordinary differential equations given by

$$\dot{x}_1 = a_{11}x_1 + a_{12}x_2 + b_1u$$

$$\dot{x}_2 = a_{21}x_1 + a_{22}x_2 + b_2u$$

In matrix representation, we obtain

$$\begin{bmatrix} \dot{x}_1 \\ \dot{x}_2 \end{bmatrix} = \begin{bmatrix} a_{11} & a_{12} \\ a_{21} & a_{22} \end{bmatrix} \begin{bmatrix} x_1 \\ x_2 \end{bmatrix} + \begin{bmatrix} b_1 \\ b_2 \end{bmatrix} u \tag{B.15}$$

or

$$\dot{\mathbf{x}} = \mathbf{A}\mathbf{x} + \mathbf{b}u$$

The following rules govern multiplication of matrices:

1. Matrix multiplication is not commutative. Hence, postmultiplication of **A** by **B** does not equal premultiplication of **A** by **B**, that is, $\mathbf{AB} \neq \mathbf{BA}$. We note that when **AB** exists, **BA** may not even be defined because **B** and **A** are not conformable. However, there are some exceptions. For example, $\mathbf{AI} = \mathbf{IA} = \mathbf{A}$, where both **A** and **I** have the same order.
2. Matrix multiplication obeys the distributive law, so that
$$\mathbf{A}(\mathbf{B} + \mathbf{C}) = \mathbf{AB} + \mathbf{AC} \tag{B.16}$$
3. Matrix multiplication obeys the associative law, so that
$$\mathbf{A}(\mathbf{BC}) = \mathbf{AB}(\mathbf{C}) \tag{B.17}$$
4. The transpose of the product of two matrices is equal to the product in reverse order of their transposes, so that
$$(\mathbf{AB})' = \mathbf{B}'\mathbf{A}' \tag{B.18}$$

B.3 MATRIX INVERSION

There is no direct division of matrices and the operation of division is performed by matrix inversion. The inverse of a square matrix **A** is denoted by $\mathbf{A}^{-1}$ and is defined by
$$\mathbf{AA}^{-1} = \mathbf{A}^{-1}\mathbf{A} = \mathbf{I} \tag{B.19}$$
The requirements for the existence of a unique inverse of a matrix are as follows: (1) The matrix is square and (2) the determinant of the matrix is not zero (the matrix is not singular). A nonsquare matrix does not have an inverse, in which case it is possible to obtain nonunique right and left inverses, but this topic is omitted here.

The determinant of a square matrix is denoted by $|\mathbf{A}|$ or det **A**. The expansion can be performed by expanding along a row or column using cofactors. For example, consider det **A**, where
$$\det \mathbf{A} = \begin{vmatrix} 1 & 2 & -3 \\ 4 & 5 & -6 \\ 7 & -8 & 9 \end{vmatrix} \tag{B.20}$$
Expanding along the third column, we obtain
$$\det \mathbf{A} = (-3)\begin{vmatrix} 4 & 5 \\ 7 & -8 \end{vmatrix} - (-6)\begin{vmatrix} 1 & 2 \\ 7 & -8 \end{vmatrix} + 9\begin{vmatrix} 1 & 2 \\ 4 & 5 \end{vmatrix}$$
$$= -3(-32 - 35) + 6(-8 - 14) + 9(5 - 8) = 42$$

The cofactor of an element a_{ij} is the determinant formed by omitting the ith row and j^{th} column. Thus, the cofactor of -3 in the preceding example is the determinant formed by eliminating the first row and third column. The sign is determined from $(-1)^{i+j}$, where i and j are the row and column, respectively, of a_{ij}. A matrix whose determinant is zero is called a singular matrix.

The adjoint of a matrix is obtained by taking the transpose of the matrix formed by replacing each element of the original matrix by its cofactor. Consider

$$\mathbf{A} = \begin{bmatrix} 1 & 2 & 3 \\ 0 & -1 & 4 \\ -2 & 5 & -3 \end{bmatrix} \tag{B.21}$$

The cofactor of the element 0 is obtained as

$$(-1)^{2+1} \begin{vmatrix} 2 & 3 \\ 5 & -3 \end{vmatrix} = -(-6 - 15) = 21$$

Similarly, after determining the cofactors of each element of **A** in Eq. (B.21), we obtain the matrix formed from the cofactors as

$$\begin{bmatrix} -17 & -8 & -2 \\ 21 & 3 & -9 \\ 11 & -4 & -1 \end{bmatrix}$$

Taking the transpose of the preceding matrix, we get

$$\text{Adjoint of } \mathbf{A} = \begin{bmatrix} -17 & 21 & 11 \\ -8 & 3 & -4 \\ -2 & -9 & -1 \end{bmatrix} \tag{B.22}$$

The inverse of a square nonsingular matrix is given by

$$\mathbf{A}^{-1} = \frac{\text{Adjoint of } \mathbf{A}}{\det \mathbf{A}} \tag{B.23}$$

For example, consider the inverse of the matrix given by Eq. (B.21). Its determinant can be evaluated as -39. Substituting from Eq. (B.22) in Eq. (B.23), we obtain

$$\mathbf{A}^{-1} = \frac{1}{39} \begin{bmatrix} 17 & -21 & -11 \\ 8 & -3 & 4 \\ 2 & 9 & 1 \end{bmatrix} \tag{B.24}$$

The following are the properties of the inverse of a matrix:

1. The inverse of a matrix is unique.
2. The inverse of a symmetric matrix is itself a symmetric matrix.
3. The inverse of a triangular matrix is itself a triangular matrix of the same type.
4. The inverse of the product of two matrices is equal to the product of the inverse of the two matrices in reverse order, i.e., $(\mathbf{AB})^{-1} = \mathbf{B}^{-1}\mathbf{A}^{-1}$.
5. The inverse of the transpose of **A** is equal to the transpose of the inverse of **A**, that is, $(\mathbf{A}')^{-1} = (\mathbf{A}^{-1})'$.

B.4 EIGENVALUES AND EIGENVECTORS OF A MATRIX

We consider a set of linear simultaneous equations of the form

$$\mathbf{A}\mathbf{x} = s\mathbf{x} \tag{B.25}$$

where $\mathbf{A}$ is an $n \times n$ matrix, $\mathbf{x}$ is an $n \times 1$ column vector, and s is a scalar. Alternatively, this equation may be written as

$$s\mathbf{x} - \mathbf{A}\mathbf{x} = (s\mathbf{I} - \mathbf{A})\mathbf{x} = 0 \tag{B.26}$$

where $\mathbf{I}$ is an $n \times n$ identity matrix. A nontrivial solution of $\mathbf{x}$ exists if and only if

$$\det(s\mathbf{I} - \mathbf{A}) = 0 \tag{B.27}$$

This equation is called the characteristic equation of matrix $\mathbf{A}$. It is an nth-order polynomial in s. The roots $s_1, s_2, \ldots, s_n$ are called the eigenvalues of $\mathbf{A}$. For an eigenvalue s_i, we have

$$\mathbf{A}\mathbf{x}_i = s_i\mathbf{x}_i \tag{B.28}$$

The vector $\mathbf{x}_i$ obained from the solution of this equation as an identity is called the eigenvector of $\mathbf{A}$ corresponding to the eigenvalue s_i. We note that for a nonzero scalar α, $\alpha\mathbf{x}_i$ is also a solution of Eq. (B.28). Hence, only the direction of the eigenvector is obtained from the solution of Eq. (B.28), whereas its length is arbitrary and may be normalized to unity. Consider the example of a matrix $\mathbf{A}$, where

$$\mathbf{A} = \begin{bmatrix} 0 & 1 \\ -4 & -5 \end{bmatrix} \tag{B.29}$$

Then,

$$(s\mathbf{I} - \mathbf{A}) = \begin{bmatrix} s & -1 \\ 4 & s+5 \end{bmatrix}$$

and the characteristic equation is given by

$$\det(s\mathbf{I} - \mathbf{A}) = s(s+5) + 4 = s^2 + 5s + 4 = 0 \tag{B.30}$$

The two eigenvalues of $\mathbf{A}$ are $s_1 = -1$ and $s_2 = -4$. To find the eigenvector corresponding to the eigenvalue -1, we solve

$$\begin{bmatrix} 0 & 1 \\ -4 & -5 \end{bmatrix} \begin{bmatrix} x_1 \\ x_2 \end{bmatrix} = -1 \begin{bmatrix} x_1 \\ x_2 \end{bmatrix} \tag{B.31}$$

that is,

$$x_2 = -x_1$$

$$-4x_1 - 5x_2 = -x_2$$

The two equations are not linearly independent and only one equation is available in two unknowns. Arbitrarily choosing $x_1 = 1$, we get $x_2 = -1$. Hence,

$$\mathbf{x}_1 = \begin{bmatrix} 1 \\ -1 \end{bmatrix} = \frac{1}{\sqrt{2}} \begin{bmatrix} 1 \\ -1 \end{bmatrix} \tag{B.32}$$

where we have normalized the length of the eigenvector to unity. For the eigenvalue $s_2 = -4$, we have

$$\begin{bmatrix} 0 & 1 \\ -4 & -5 \end{bmatrix} \begin{bmatrix} x_1 \\ x_2 \end{bmatrix} = -4 \begin{bmatrix} x_1 \\ x_2 \end{bmatrix} \tag{B.33}$$

that is,

$$x_2 = -4x_1$$

$$-4x_1 - 5x_2 = -4x_2$$

Arbitrarily choosing $x_1 = 1$, we obtain $x_2 = -4$, and the normalized eigenvector corresponding to the eigenvalue $s_2 = -4$ becomes

$$\mathbf{x}_2 = \frac{1}{\sqrt{17}} \begin{bmatrix} 1 \\ -4 \end{bmatrix} \tag{B.34}$$

REFERENCES

BELLMAN, R. (1960). *Introduction to Matrix Analysis.* New York: McGraw-Hill.

WYLIE, C. R. JR. (1975). *Advanced Engineering Mathematics.* New York: McGraw-Hill.

Appendix C

Answers to Selected Problems

CHAPTER 2

2.3. **(c)** Natural frequency $= \left[\dfrac{4k}{4m_1 + m_2}\right]^{1/2}$

Damping ratio $= \dfrac{c}{[k(4m_1 + m_2)]^{1/2}}$

2.4. Natural frequency $= 1.67$ rad/s

2.6. Natural frequency $= \left[\dfrac{(k_1 + k_2)L^2}{I}\right]^{1/2}$

Damping ratio $= \dfrac{cL}{2[I(k_1 + k_2)]^{1/2}}$

2.11. $\dfrac{E_0}{E_1} = \dfrac{R_1R_2C_1C_2D^2 + R_2(C_1 + C_2)D + 1}{R_1R_2C_1C_2D^2 + (R_1C_2 + R_2C_1 + R_2C_2)D + 1}$

2.13. $\dfrac{h_2}{q_1} = \dfrac{R_2/\rho g}{R_1R_2C_1C_2D^2 + (R_1C_1 + R_2C_1 + R_2C_2)D + 1}$

2.14. $\dfrac{y}{P_1} = \dfrac{A/k}{\dfrac{1}{\omega_n^2}D^2 + \dfrac{2\zeta}{\omega_n}D + 1}$

where $\omega_n^2 = k/m$ and $\zeta = \dfrac{A^2R + c}{2(km)^{1/2}}$

2.17. $\dfrac{P_0}{P_i} = \dfrac{k_gD}{(\tau_1D + 1)(\tau_2D + 1)}$

where $k_g = \dfrac{Ac(R_2C_2 - R_1C_1)}{k_1 + k_2}$ $\quad \tau_1 = R_1C_1 \quad \tau_2 = R_2C_2$

2.21. $\left(\frac{1}{\omega_n^2} D^2 + 1\right) y = 0$

where $\omega_n^2 = \left(\frac{b}{a}\right)\left(\frac{c_1 c_2}{A_1 A_2}\right)$

CHAPTER 3

3.1. The closed-loop state equations are

$$\dot{x}_1 = -\left(\frac{1}{\tau_2}\right) x_1 + \left(\frac{n_1 n_2 k_3 k_t}{\tau_2}\right) x_2 - \left(\frac{n_1 n_2 k_3}{\tau_2}\right) T_d$$

$$\dot{x}_2 = -\left(\frac{k_b}{R_a n_1 n_2 \tau_1} + \frac{k_1 k_2 c}{R_a \tau_1}\right) x_1 - \left(\frac{1}{\tau_1}\right) x_2 + \left(\frac{k_1 k_2 c}{R_a \tau_1}\right) \omega_r$$

where $x_1 = \omega_0$, $x_2 = i_a$, $\tau_1 = L_a / R_a$

$E_r = c\omega_r$, $E_0 = c\omega_0$

$$\tau_2 = \frac{I_1 + n_1^2 I_2 + n_1^2 n_2^2 I_3}{c_1 + n_1^2 c_2 + n_1^2 n_2^2 c_3}$$

$$k_3 = \frac{1}{c_1 + n_1^2 c_2 + n_1^2 n_2^2 c_3}$$

k_2 = power amplifier gain

k_1 = differential amplifier gain

k_t = torque constant

k_b = back emf constant

3.8. For the servovalve from Fig. 2.46,

$$x_s = \left(\frac{k_1}{\frac{1}{\omega_n^2} D^2 + \frac{2\zeta}{\omega_n} D + 1}\right) i$$

For the hydraulic motor from Fig. 2.49, but including a load torque,

$$\theta_0 = \left(\frac{k_g}{D(\tau D + 1)}\right)(x_s - k_2 T_L)$$

The state variables are $x_1 = \theta_0$, $x_2 = \dot{\theta}_0$, $x_3 = x_s$, and $x_4 = \dot{x}_s$. The closed-loop state equations are

$$\dot{x}_1 = x_2$$

$$\dot{x}_2 = -\left(\frac{1}{\tau}\right) x_2 + \left(\frac{k_g}{\tau}\right) x_3 - \left(\frac{k_g k_2}{\tau}\right) T_L$$

$$\dot{x}_3 = x_4$$

$$\dot{x}_4 = -\omega_n^2 k_1 k_a c_2 x_1 - \omega_n^2 x_3 - 2\zeta\omega_n x_4 + \omega_n^2 k_1 k_a c_1 \theta_r$$

CHAPTER 4

4.1.
$$\Phi(t) = \begin{bmatrix} 2te^{-2t} + e^{-2t} & te^{-2t} \\ -4te^{-2t} & e^{-2t} - 2te^{-2t} \end{bmatrix}$$

4.5. The closed-loop transfer function is

$$G_c(s) = \frac{15}{s^3 + 9s^2 + 23s + 15}$$

4.6. The closed-loop transfer function is

$$G_c(s) = \frac{30}{s^3 + 5.5s^2 + 9.5s + 36}$$

4.10. $k = 9.356$ and rise time = 0.18 second.

4.11. Natural frequency = 16.215 rad/s and damping ratio = 0.248.

4.16. Eigenvalues are −1, −3, and −5.

$$\mathbf{P} = \begin{bmatrix} 1 & 1 & -1 \\ 2 & 0 & 2 \\ 1 & -1 & -1 \end{bmatrix}$$

CHAPTER 5

5.1. $a = 5.2°\text{C}$.

5.2. $y_0 = 0.132$ and phase angle = 7.6 degrees.

5.3. $a = 2.857$ newtons.

5.7. $M_p = 1.67$ and bandwidth = 4.56 rad.

5.8. $M_p = 5$ and bandwidth = 3.7 rad.

CHAPTER 6

6.4. $0 < k_1 < 0.925$.

6.5. $0 < k_1 < \infty$.

6.8. **(a)** Not asymptotically stable and BIBO unstable.
(b) Not asymptotically stable and BIBO unstable.
(c) Unstable.

6.17. The real root sensitivity to positive change in k_1 is $0.9306 \angle 180°$. For +20 percent change in k_1, the change in real root is $0.186 \angle 180°$.

CHAPTER 7

7.6. **(a)** Steady-state error = $-\infty$.
(b) $k_i \geq 1.25$; for stability, $k_i < 1.5$.

7.7. Control law implemented is proportional.

7.8. $k_p = 2.06$, $k_i = 2.987$, and steady-state error = 0.251.

7.12. One solution is obtained as $k_p = 1.25$, and $k_d = 0.262$.

7.13. One solution is obtained as $k_p = 1.25$, and $\tau = 0.31$ second.

CHAPTER 8

8.1. $\omega_0 = 34.65$ rad/s; roots are $-24.255 \pm j24.745$.

8.2. $\omega_0 = 34.31$ rad/s; roots are -24.295, $-17.876 \pm j36.643$.

8.4. Not state controllable.

8.7. $k_1 = 15.327$; $k_2 = 7.716$; type 0. Steady-state error $= 0.131$.

8.13. Not observable.

8.15. $h_1 = 163.32$, and $h_2 = 1411.34$.

CHAPTER 9

9.1. **(a)** $0.028 \leq T \leq 0.055$ second.
(b) $0.022 \leq T \leq 0.045$ second.

9.3.
$$\begin{aligned} x_1[(k+1)T] &= 0.8607x_1(kT) + 0.4643x_2(kT) + 0.0595u(kT) \\ x_2[(k+1)T] &= x_2(kT) + 0.25u(kT) \\ y(kT) &= x_1(kT) \end{aligned}$$

9.6.
$$\begin{aligned} x_1[(k+1)T] &= x_1(kT) + 0.0467x_2(kT) + 0.0081x_3(kT) + 0.0071u(kT) \\ x_2[(k+1)T] &= 0.8761x_2(kT) + 0.2924x_3(kT) + 0.4061u(kT) \\ x_3[(k+1)T] &= 0.0731x_2(kT) + 0.6203x_3(kT) + 2.0096u(kT) \\ y(kT) &= x_1(kT) \end{aligned}$$

9.9. $G_p(z) = \dfrac{0.0595z + 0.05658}{(z - 0.8607)(z - 1)}$

9.13. **(a)** $G_1(z) = \dfrac{0.63z - 0.57}{z - 1}$

(b) $G_c(z) = \dfrac{0.03749z^2 + 0.00173z - 0.03225}{z^3 - 2.82322z^2 + 2.72313z - 0.89295}$

(c) stable.

(d) Roots of characteristic equation are 0.89889, $0.96215 \pm j0.26009$.

9.16. Not state controllable but observable.

9.21. Control roots are $0.1727 \pm j0.3086$. $k_1 = 5.275$ and $k_2 = 4.806$.

9.28. $h_1 = 1.875$, and $h_2 = 2.185$.

Appendix D

Computer Programs

The examples we have considered in this book are generally low-order systems. The advantage of using examples of low-order systems is that the fundamental ideas can be conveyed and understood clearly without the distraction of large amounts of computations inherent in high-order systems. However, in practical applications the systems are high-order. In addition, it is desirable to vary several parameters during the design stage to investigate their effect on the system performance. Hence, computer-aided design tools are very advantageous for this purpose.

Several software packages with computer graphics are available commercially for computer-aided design of control systems with personal computers. The three programs mentioned in the references to this appendix are DACS, Program CC, and Matrix/PC. In addition to or instead of utilizing one of these software packages, it is conducive to the learning process to write one's own computer programs. Phillips and Nagle (1984) have listed several computer programs useful for the design of digital control systems.

This appendix gives the listing of some sample programs for personal computers. The programs are written in True BASIC (Kemeny and Kurtz, 1985). Many versions of BASIC are available commercially for use with personal computers. Unfortunately, they are heavily dependent upon the particular hardware being used. Besides, most versions of BASIC for personal computers do not include matrix functions. True BASIC is designed to be portable or hardware independent. Its implementation uses a compiler and it includes matrix functions which permit one to read, input, and print matrices, and to perform matrix calculations.

PROGRAM 1

```
! Resolvent
! This program calculates the resolvent,INVERSE(sI-A).
! For closed-loop systems, use (A-BK) instead of A.
! The algorithm uses the following equations.
! Inverse(sI-A)=[D(1)s^n-1 +D(2)s^n-2 +...+D(n)s^0]/p(s)
! p(s)=[p(1)s^n +p(2)s^n-1 +p(3)s^n-2 +...+p(n+1)s^0]
! where p(s) is the characteristic polynomial and D(1),...,D(n)
! are coefficient matrices.
! p(1)=1; Matrix D(1)=I
! p(k)=-Trace[D(k-1)A]/(k-1); Matrix D(k)=[D(k-1)A+p(k)I] ;k=2,...,n+1
! The program gives the matrices D(1),...,D(n) and coefficients of
! the characteristic polynomial p(1),p(2),...,p(n+1).
DIM A(10,10),H(10,10),P(1,11),D(10,10),Q(10,10),R(10,10)
PRINT "Input system order"
INPUT n
PRINT "Input A matrix by rows"
MAT INPUT A(n,n)
PRINT "System A matrix is"
MAT PRINT A
LET n1=n+1
MAT P=zer(1,n1)
MAT H=idn(n,n)
LET P(1,1)=1.0
MAT D=H
FOR k=2 to n1
    PRINT "Coef. matrix of s^";n-k+1;"is"
    MAT PRINT D
    MAT Q=D*A
    LET V=0.0
    FOR i=1 to n
```

```
        LET V=V+Q(i,i)
    NEXT i
    LET E=-V/(k-1)
    LET P(1,k)=E
    MAT R=E*H
    MAT D=Q+R
NEXT k
PRINT" The char. poly coef. in descending powers of s are"
MAT PRINT P
END
```

PROGRAM 2

```
! Transfer Function Matrix
! This program calculates the transfer function matrix
!        G(s)=C[Inverse(sI-A)]B
! The continuous system is described by
!        x(t)=Ax(t) +Bu(t)
!        y(t)=Cx(t)
! It can also be used to calculate the tranfer function matrix G(z) of a
! discrete-time system by replacing A by PHI and B by PSI.
! G(s)=C[D(1)s^n-1+D(2)s^n-2+...+D(n)s^0]B/p(s)
! where matrices D(k),characteristic poly p(s),and the algorithm used
! are described in the previous Program 1.
! The program gives the matrices CD(1)B,...,CD(n)B and coefficients of
! the characteristic polynomial p(1),p(2),...,p(n+1).
DIM A(10,10),F(10,10),H(10,10),P(1,11),D(10,10),Q(10,10),R(10,10)
DIM B(10,10),C(10,10),T(10,10)
PRINT "Input system order,no. of inputs, and no. of outputs"
INPUT n,m,w
PRINT "Input A matrix by rows"
MAT INPUT A(n,n)
```

```
PRINT " Input B matrix by rows"
MAT INPUT B(n,m)
PRINT "Input C matrix by rows"
MAT INPUT C(w,n)
PRINT "System order=";n,"No. of inputs=";m,"No. of outputs=";w
PRINT "System A matrix is"
MAT PRINT A
PRINT "System B matrix is "
MAT PRINT B
PRINT "System C matrix is"
MAT PRINT C
LET n1=n+1
MAT P=zer(1,n1)
MAT H=idn(n,n)
MAT T=zer(w,n)
MAT F=zer(w,m)
LET P(1,1)=1.0
MAT D=H
MAT T=C*D
MAT F=T*B
FOR k=2 to n1
    PRINT "Coef. matrix of s^";n-k+1;"is"
    MAT PRINT F
    MAT Q=D*A
    LET V=0.0
    FOR i=1 to n
        LET V=V+Q(i,i)
    NEXT i
    LET E=-V/(k-1)
    LET P(1,k)=E
    MAT R=E*H
```

```
    MAT D=Q+R
    MAT T=C*D
    MAT F=T*B
NEXT k
PRINT" The char. poly coef. in descending powers of s are"
MAT PRINT P
END
```

PROGRAM 3

```
! Discrete System Matrices
! This program calculates the discrete system matrices,PHI and PSI of
! Eq.(9.25) from the continuous plant matrices,A and B of Eq.(9.20).
! The program uses the method of series expansion of matrix exponential
! discussed in section 4.3.1.Hence
! PHI=I+AT+(1/2!)AATT+(1/3!)AAATTT+...
! PSI={IT+(1/2!)ATT+(1/3!)AATTT+...}B
DIM  A(12,12),B(12,12),E(12,12),F(12,12)
DIM G(12,12),PHI(12,12),K(12,12),PSI(12,12)
PRINT "Input T,system order,and number of inputs"
INPUT T,n,m
PRINT " Number of terms to be included in series"
INPUT r
PRINT "Input A matrix by rows"
MAT INPUT A(n,n)
PRINT "Input B matrix by rows"
MAT INPUT B(n,m)
PRINT
PRINT "T=";T,"System order=";n,"Number of inputs=";m
PRINT "Continuous system A matrix is"
MAT PRINT A
PRINT " Continuous system B matrix is"
```

```
MAT PRINT B
! Initialize and set dimensions of matrices
MAT E=idn(n,n)
MAT F=idn(n,n)
MAT G=zer(n,n)
MAT PHI=idn(n,n)
MAT K=zer(n,n)
MAT PSI=zer(n,m)
FOR j=1 to r
    MAT E=(T/(j+1))*E
    MAT E=E*A
    MAT F=F+E
    MAT G=(j+1)*E
    MAT PHI=PHI+G
NEXT j
MAT K=(T)*F
MAT PSI=K*B
PRINT "Discrete system PHI matrix is"
MAT PRINT PHI
PRINT "Discrete system PSI matrix is"
MAT PRINT PSI
END
```

PROGRAM 4

```
! Response
! This program calculates the response of the closed-loop difference
! Eqs. (9.95) and (9.96), namely,
!        x[(k+1)T] = Px(kT) + Qr(kT)
!           y(kT)  = Cx(kT)
! It can also be used for the solution of the open-loop plant
! difference Eq. (9.25) by replacing P by PHI , Q by PSI.and r(kT)
```

```
! by u(kT). The solution method is discussed in section 9.5.3.
DIM P(12,12),Q(12,12),C(12,12),x(12),x1(12),y(12),r(12),v(12)
! Here, x=x(kT),x1=x[(k+1)T],and v is a working matrix.Other
! matrices are identified in state equations.
PRINT "This program has a single input which is a unit step and"
PRINT "requires modification for other cases.For example,let no."
PRINT "of inputs m=3.The first input is a step of magnitude 3 so that"
PRINT "r(1)=3.The second is a ramp of slope 2 so that r(2)=2kT.The"
PRINT "third is sinusoidal so that r(3)=4sin5kT.The sampling period"
PRINT "T=o.ol second.Then in the program ,replace the statement"
PRINT "LET r(1)=1 by the following statements"
PRINT "LET r(1)=3"
PRINT "LET r(2)=.02*(k-1)"
PRINT "LET r(3)=4*sin(.05*(k-1))"
PRINT "Then run the program"
PRINT "Input system order,no. of inputs, and no. of outputs"
INPUT n,m,s
PRINT "Input desired final value of k"
INPUT J
PRINT "Input P matrix by rows"
MAT INPUT P(n,n)
PRINT "Input Q matrix by rows"
MAT INPUT Q(n,m)
PRINT "Input C matrix by rows"
MAT INPUT C(s,n)
PRINT "Input initial values of states"
MAT INPUT x(n)
PRINT "System order=";n,"No. of inputs=";m,"No. of outputs=";s
PRINT "P matrix is"
MAT PRINT P
PRINT"Q matrix is"
```

```
MAT PRINT Q
PRINT "C matrix is"
MAT PRINT C
PRINT "Initial values of state variables are"
MAT PRINT x
! Initialize and set dimensions of matrices
MAT y=zer(s)
MAT v=zer(n)
MAT x1=zer(n)
MAT r=zer(m)
FOR k=1 to J
    LET r(1)=1
    MAT x1=P*x
    MAT v=Q*r
    MAT x1=x1+v
    MAT y=C*x1
    PRINT "Output y for k=";k
    MAT PRINT y
    MAT x=x1
NEXT k
END
```

PROGRAM 5

```
! Ackermann
! This program calculates the feedback controller gain matrix K from
! the Ackermann's formula,Eq.(8.25),namely,
!     K=[0  0  ... 0  1]Inverse(Q)Delta(A)
! where Q is the controllability matrix defined by Eq.(8.12), and
! Delta(A) is the desired characteristic polynomial of Eq.(8.24).
! The program can also be used to calculate the observer gain matrix
! H of Eq.(8.94),by replacing the controllability matrix Q by the
```

```
! observability matrix M and Delta(A) by the desired observer characteristic
! polynomial defined by Eq.(8.93).
! In addition,the program can be used for computer control systems.
! To calculate the controller gain matrix of Eq.(9.138),use the
! controllability matrix Q defined by (9.126),and replace Delta(A)
! by Delta(PHI).To calculate the observer gain matrix of Eq.(9.162),
! replace the controllability matrix Q by the observability matrix M of
! Eq.(9.128),and replace Delta(A) by the characteristic polynomial
! Delta(PHI) of Eq.(9.162).
DIM A(12,12),B(12,1),C(1,13),D(12),E(1,12),F(12,1),G(12,12)
DIM H(12,12),K(1,12),P(12,12),Q(12,12),R(12,12),S(12,12)
! A and B are system matrices.All others are working matrices.
PRINT "Input system order"
INPUT n
LET n1=n+1
PRINT "Input A matrix by rows"
MAT INPUT A(n,n)
PRINT "Input B matrix"
MAT INPUT B(n,1)
PRINT "Input characteristic polynomial coefficients in descending "
PRINT "order,starting with a(n)=1"
MAT INPUT C(1,n1)
PRINT "System A matrix is"
MAT PRINT A
PRINT "System B matrix is"
MAT PRINT B
PRINT "Coefficients of desired characteristic equation are"
MAT PRINT C
! Initialize and set dimensions of matrices
MAT D=zer(n)
MAT E=zer(1,n)
```

```
MAT F=zer(n,1)
MAT G=zer(n,n)
MAT H=zer(n,n)
MAT K=zer(1,n)
MAT P=zer(n,n)
MAT Q=zer(n,n)
MAT R=idn(n,n)
MAT S=zer(n,n)
FOR i=1 to n
    LET j=i+1
    LET D(i)=C(1,j)
NEXT i
FOR i=1 to n
    LET m=n-(i-1)
    LET w=D(m)
    MAT F=R*B
    MAT G=(w)*R
    MAT P=P+G
    FOR j=1 to n
        LET Q(j,i)=F(j,1)
    NEXT j
    MAT G=A*R
    MAT R=G
NEXT i
MAT G=R
MAT P=G+P
MAT G=Inv(Q)
LET E(1,n)=1
MAT S=G*P
MAT K=E*S
MAT G=B*K
```

```
MAT H=A-G

PRINT " The feedback gain matrix K is "

MAT PRINT K

PRINT " the system closed-loop matrix A-BK is"

MAT PRINT H

END
```

REFERENCES

KEMENY, J. G., and KURTZ, T. E. (1985). *True BASIC, User's Guide.* Reading, MA: Addison-Wesley.

LEWIS, W. I. (1985). *DACS, Design and Analysis of Control Systems.* Santa Cruz, CA: Kingman Block Publishing.

Matrix/PC (1985). *Modeling, Simulation and Control Design on the PC.* Palo Alto, CA: Integrated Systems.

PHILLIPS, C. L., and NAGLE, H. T., JR. (1984). *Digital Control System Analysis and Design.* Englewood Cliffs, NJ: Prentice-Hall.

Program CC (1985). *Computer-Aided Control System Design.* Hawthorne, CA: Systems Technology.

Index

C

D

E

F

G

H

I

O

P

Q

R

S

T

U

V

W

Z